# Communications in Computer and Information Science 1977

**Rationale**
The CCIS series is devoted to the publication of proceedings of computer science conferences. Its aim is to efficiently disseminate original research results in informatics in printed and electronic form. While the focus is on publication of peer-reviewed full papers presenting mature work, inclusion of reviewed short papers reporting on work in progress is welcome, too. Besides globally relevant meetings with internationally representative program committees guaranteeing a strict peer-reviewing and paper selection process, conferences run by societies or of high regional or national relevance are also considered for publication.

**Topics**
The topical scope of CCIS spans the entire spectrum of informatics ranging from foundational topics in the theory of computing to information and communications science and technology and a broad variety of interdisciplinary application fields.

**Information for Volume Editors and Authors**
Publication in CCIS is free of charge. No royalties are paid, however, we offer registered conference participants temporary free access to the online version of the conference proceedings on SpringerLink (http://link.springer.com) by means of an http referrer from the conference website and/or a number of complimentary printed copies, as specified in the official acceptance email of the event.

CCIS proceedings can be published in time for distribution at conferences or as post-proceedings, and delivered in the form of printed books and/or electronically as USBs and/or e-content licenses for accessing proceedings at SpringerLink. Furthermore, CCIS proceedings are included in the CCIS electronic book series hosted in the SpringerLink digital library at http://link.springer.com/bookseries/7899. Conferences publishing in CCIS are allowed to use Online Conference Service (OCS) for managing the whole proceedings lifecycle (from submission and reviewing to preparing for publication) free of charge.

**Publication process**
The language of publication is exclusively English. Authors publishing in CCIS have to sign the Springer CCIS copyright transfer form, however, they are free to use their material published in CCIS for substantially changed, more elaborate subsequent publications elsewhere. For the preparation of the camera-ready papers/files, authors have to strictly adhere to the Springer CCIS Authors' Instructions and are strongly encouraged to use the CCIS LaTeX style files or templates.

**Abstracting/Indexing**
CCIS is abstracted/indexed in DBLP, Google Scholar, EI-Compendex, Mathematical Reviews, SCImago, Scopus. CCIS volumes are also submitted for the inclusion in ISI Proceedings.

**How to start**
To start the evaluation of your proposal for inclusion in the CCIS series, please send an e-mail to ccis@springer.com.

Marco Villani · Stefano Cagnoni · Roberto Serra
Editors

# Artificial Life and Evolutionary Computation

## 17th Italian Workshop, WIVACE 2023
## Venice, Italy, September 6–8, 2023
## Revised Selected Papers

*Editors*
Marco Villani
University of Modena and Reggio Emilia
Modena, Italy

Stefano Cagnoni
University of Parma
Parma, Italy

Roberto Serra
University of Modena and Reggio Emilia
Modena, Italy

ISSN 1865-0929 ISSN 1865-0937 (electronic)
Communications in Computer and Information Science
ISBN 978-3-031-57429-0 ISBN 978-3-031-57430-6 (eBook)
https://doi.org/10.1007/978-3-031-57430-6



This Springer imprint is published by the registered company Springer Nature Switzerland AG
The registered company address is: Gewerbestrasse 11, 6330 Cham, Switzerland

Paper in this product is recyclable.

# Preface

The WIVACE conference series has now reached its 17th edition, and it continues to show a vitality which may look surprising, given that it is not supported by any formal society or association. WIVACE was born out of the coalescence of two different initiatives: two workshops on Artificial Life were held in 2003 in Cosenza and in 2005 in Rome, and a workshop on Evolutionary Computation took place in Milan in 2005 (as a part of the Conference of the Italian Association for Artificial Intelligence). The organizers of the two initiatives decided to co-locate their 2006 editions in Siena, organizing a common session where it became clear that the two scientific communities had strong mutual interests, leading to the decision to merge the two workshops.

The first workshop bearing the name WIVACE took place in Samperi (Sicily) in 2007, followed by editions in Venice (2008) and in Naples (2009). At that time, some doubts were raised about the opportunity to continue this series of meetings which were so bravely cutting across disciplines. The wise decision was taken to go on, so later workshops took place in Parma (2012), Milan (2013), Vietri sul Mare (2014), Bari (2015), Salerno (2016), Venice (2017), Parma (2018) and Rende (2019).

The 2020 edition was planned to take place for the first time outside Italy, but the Covid pandemic spread all over the world and it was decided to postpone it, since WIVACE is not well-suited for remote attendance. Indeed, a large part of the interest and value of the conference lies in the many discussions that take place in the lecture room as well as in informal gatherings, often with scientists with a different background. Therefore, the WIVACE workshop in Winterthur (Switzerland) had to wait until 2021, followed by the 2022 edition in Gaeta. And in 2023 we came for the third time back to Venice, home of the European Centre for Living Technology (ECLT), which we thank for its long-lasting support, under the guidance of three different directors who followed one another in the last 15 years (Irene Poli, Marcello Pelillo and Achille Giacometti). Future editions are already planned to take place in Namur (Belgium) and Siena.

The organizers of WIVACE editions are chosen in informal meetings held during the workshops, a method that provides a continuous testing of the value of the workshop for its participants. While they all share an interest in interdisciplinary approaches to complex systems, they may come from different backgrounds (physics, computer science, mathematics, chemistry, biology etc.) and that is why the flavors of various editions may differ. However, the interest of the communities in the others' work is always high, and no parallel session has ever taken place.

This year, we have been lucky to host some invited speakers who combined their outstanding scientific merits with the capability to effectively communicate their thoughts, and thus we are deeply indebted to Wolfgang Banzhaf, Michele Vendruscolo and Joana C. Xavier for their contributions. We also thank all the contributors and all the participants for their role in making WIVACE 2023 a successful event, with several interesting and vibrant discussions.

The review process involved two phases and at least three reviewers per paper (single-blind review): a total of 55 papers were processed. The contributors were free to send an extended abstract or a full paper, and the acceptance for oral presentation was based on the quality of the submitted document. After the conference, the authors of accepted contributions were asked to send a full paper for further review for publication in the Proceedings. The 30 papers in this volume are the outcome of this selection procedure.

Special thanks are due to the members of the Program Committee and to the reviewers for their precious work. Two young researchers, Gianfranco Lombardo at the Università di Parma and Gianluca D'Addese at the Università di Modena e Reggio Emilia, not only provided their reviews, but they also were of great help in managing the review process, and in setting up the Easychair site, while Jacopo Moi and Tatiana Skrbic, both from the Università Cà Foscari, supported the local organization.

This year's organization profited from the strong and highly qualified support of ECLT, an international research center run by the Università Cà Foscari, which is associated with several universities and research institutions in Europe and in the USA. We wish to thank in particular the current director, Achille Giacometti, and the members of its staff, who managed the complicated organizational and financial aspects of the conference, and who also found a wonderful location for our lecture room. We particularly thank Roberta D'Argenio, Beatriz Barbado, Alessandra Bonesso and Giulia Brolese, who also managed the workshop website.

Thanks are also due to the host institutions of the organizers, namely, the Università di Modena e Reggio Emilia and the Università di Parma, and to the Università Cà Foscari di Venezia.

Let us finally acknowledge the precious advice of the staff at Springer, who provided their professional support through all the phases that led to this volume.

February 2024

Marco Villani
Stefano Cagnoni
Roberto Serra

# Organization

## General Chairs

| | |
|---|---|
| Marco Villani | University of Modena and Reggio Emilia, Italy |
| Stefano Cagnoni | University of Parma, Italy |
| Roberto Serra | University of Modena and Reggio Emilia, Italy |

## Local Chair

| | |
|---|---|
| Achille Giacometti | Ca' Foscari University of Venice, Italy |

## Local Organizing Committee

| | |
|---|---|
| Giulia Brolese | CESA/ECLT - Ca' Foscari University of Venice, Italy |
| Barbara del Mercato | Ca' Foscari University of Venice, Italy |
| Beatriz Barbado Gutierrez | CESA/ECLT - Ca' Foscari University of Venice, Italy |
| Roberta d'Argenio | CESA/ECLT - Ca' Foscari University of Venice, Italy |
| Gianluca d'Addese | University of Modena and Reggio Emilia, Italy |
| Gianfranco Lombardo | University of Parma, Italy |
| Jacopo Moi | Ca' Foscari University of Venice, Italy |
| Tatjana Skrbic | Ca' Foscari University of Venice, Italy |

## Program Committee

| | |
|---|---|
| Marco Baioletti | University of Perugia, Italy |
| Giulia Caravagna | University of Trieste, Italy |
| Timoteo Carletti | University of Namur, Belgium |
| Antonio Chella | University of Palermo, Italy |
| Franco Cicirelli | ICAR-CNR, Italy |
| Nicole Dalia Cilia | Kore University of Enna, Italy |
| Claudio De Stefano | University of Cassino and Southern Lazio, Italy |
| Luca Di Gaspero | University of Udine, Italy |

| | |
|---|---|
| Marco Dorigo | Université Libre de Bruxelles, Belgium |
| Alessia Faggian | University of Trento, Italy |
| Harold Fellermann | Newcastle University, UK |
| Alessandro Filisetti | Officinae Bio S.r.l., Italy |
| Francesco Fontanella | Università di Cassino, Italy |
| Mario Giacobini | University of Turin, Italy |
| Alex Graudenzi | IBFM-CNR, Italy |
| Antonio Guerrieri | ICAR-CNR, Italy |
| Giovanni Iacca | University of Trento, Italy |
| Ignazio Infantino | ICAR-CNR, Italy |
| Antonio Liotta | University of Bolzano, Italy |
| Gianfranco Lombardo | University of Parma, Italy |
| Luca Manzoni | University of Trieste, Italy |
| Roberto Marangoni | University of Parma, Italy |
| Giancarlo Mauri | University of Milan-Bicocca, Italy |
| Eric Medvet | University of Trieste, Italy |
| Giorgia Nadizar | University of Trieste, Italy |
| Mario Pavone | University of Catania, Italy |
| Riccardo Pecori | eCampus University, Italy |
| Clara Pizzuti | ICAR-CNR, Italy |
| Riccardo Righi | European Commission, Joint Research Center, Italy |
| Simone Righi | Ca' Foscari University of Venice, Italy |
| Andrea Roli | University of Bologna, Italy |
| Federico Rossi | University of Siena, Italy |
| Alessandra Scotto di Freca | University of Cassino and Southern Lazio, Italy |
| Yaroslav Sergeyev | University of Calabria, Italy |
| Roberto Serra | University of Modena and Reggio Emilia, Italy |
| Debora Slanzi | ECLT, University Ca' Foscari of Venice, Italy |
| Annalisa Socievole | ICAR-CNR, Italy |
| Giandomenico Spezzano | ICAR-CNR, Italy |
| Pasquale Stano | University of Salento, Italy |
| Thomas Stuetzle | Université Libre de Bruxelles, Belgium |
| Pietro Terna | University of Turin, Italy |
| Marco Tommasini | University of Lausanne, Switzerland |
| Elio Tuci | University of Namur, Belgium |
| Marco Villani | University of Modena and Reggio Emilia, Italy |
| Andrea Vinci | ICAR-CNR, Italy |
| Stephan Scheidegger | ZHAW School of Engineering, Switzerland |
| Rudolf Marcel Füchslin | ZHAW School of Engineering, Switzerland |
| Olli Yli-Harja | Tampere University, Finland |

# Contents

**Algorithms for Complex Systems**



**Biologically Inspired Models**

## Complex Chemical Systems



## Adaptation and Swarms

**Learning**



**Medicine**



**Social Systems**

# Algorithms for Complex Systems

# Energy Consumption of Evolutionary Algorithms in JavaScript

Juan J. Merelo-Guervós[1](✉), Mario García-Valdez[2], and Pedro A. Castillo[1]

[1] Department of Computer Engineering, Automatics and Robotics, University of Granada, Granada, Spain
{jmerelo,pacv}@ugr.es

[2] Department of Graduate Studies, National Technological Institute of Mexico, Tijuana, Mexico
mario@tectijuana.edu.mx

**Abstract.** Green computing is a methodology for saving energy when implementing algorithms. In environments where the runtime is an integral part of the application, it is essential to measure their energy efficiency so that researchers and practitioners have enough choice. In this paper, we will focus on JavaScript runtime environments for evolutionary algorithms; although not the most popular language for scientific computing, it is the most popular language for developers, and it has been used repeatedly to implement all kinds of evolutionary algorithms almost since its inception. In this paper, we will focus on the importance of measuring different versions of the same runtimes, as well as extending the EA operators that will be measured. We also like to remark on the importance of testing the operators in different architectures to have a more precise picture that tips the balance towards one runtime or another.

**Keywords:** Green computing · metaheuristics · JavaScript · energy-aware computing · evolutionary algorithms

## 1 Introduction

From the first papers using it for implementing metaheuristics [8], JavaScript is nowadays a great alternative for evolutionary algorithms and machine learning, making it a target for energy efficiency studies in those areas. However, unlike other languages, there are different interpreters with different applications: besides the well-known and established node designed for browser and server-side execution and already in its version 20, a pair of powerful runtime environments have been produced recently. There is deno [6] (written in Rust) focused on security and ease of use, and bun [16] (programmed in the relatively unknown language Zig) designed for speed and server-side applications. We already compared them in a previous paper [14]. In that paper we established


M. Villani et al. (Eds.): WIVACE 2023, CCIS 1977, pp. 3–15, 2024.
https://doi.org/10.1007/978-3-031-57430-6_1

a general methodology for measuring energy consumption: using the energy profiler pinpoint [11], that reads from the RAPL registers [9,10], and using experiments run from the command line, without any additional instrumentation, to measure power extracted by evolutionary algorithm operators from the source.

We validated this approach, and applied it to a single fitness function (OneMax) and a common operator (crossover). However, this study left a few issues open. In this paper, we will widen the results, looking at different factors

- Test new versions of the virtual machines considering energy consumption, since in general these evolve towards more efficient operation; at any rate, the balance of results might change in these new versions in unexpected ways.
- Test also in different power consumption environments, including a native Intel machine, so that the interplay between the interpreter and the power management can be observed with more precision, with a full implementation of the API used for reading the energy consumption sensors.
- A more complete evolutionary algorithm will be tested: adding the ubiquitous mutation operator to the testbed, to see what kind of power consumption profile it adds.

The rest of the paper follows this plan: next we will present the state of the art; next the results together with the experimental setup will be presented in Sect. 3, and we will end with a discussion of results, conclusions and future lines of work in Sect. 4.

## 2 State of the Art

In the last few years, evaluating the energy efficiency of algorithms as well as modifying these algorithms or their implementations so they consume less energy has increasingly become a research topic [3]; in many cases algorithms are compared with respect to their power efficiency. Since 2020, Machine learning/AI applications [19] have been extensively studied from the green perspective, primarily through experimental studies. Green computing principles have been established as a best practice, which we will also follow in this paper.

In the area of evolutionary algorithms, initial energy efficiency estimations included their behavior in different platforms [17], its interaction with cloud services [12] and how it affected genetic programming (GP) [5]; but since then, and due to the fact that metaheuristics are so extensively used in machine learning applications, studies in this area have grown. Many papers focus on analyzing how certain metaheuristics parameters have an impact on energy consumption. Díaz-Álvarez et al. [4] studies how the population size of EAs influences power consumption. In an earlier work, centered on genetic algorithms (GAs) [18], power-consumption of battery-powered devices was measured for various parameter configurations including chromosome and population sizes. The experiments used the OneMax and Trap function benchmark problems, and they concluded that execution time and energy consumption do not linearly correlate and there is a connection between the GA parameters and power consumption. In GAs, the mutation operator appears to be a power-hungry component according to Abdelhafez et al. [1], in their paper they also report that in a distributed evaluation setting, the communication scheme has a grater impact. Fernández de Vega et al. [17] experimented with different parameters for a GP algorithm and concluded that hand-held devices and single-board computers (SBCs) required an order of magnitude less energy to run the same algorithm.

These results point to several best practices in the area: first, evaluate separately different operators and fitness functions, and second, perform experiments on different devices. After the initial exploration and establishment of technology in [14], this paper will also test the mutation operator, as well as carry out tests in computers with different architectures.

## 3 Experimental Results

From the different ways to measure energy consumption [2], in [14] we chose `pinpoint` [11], a tool that uses the RAPL (Running Average Power Limit [7]) interface, to report the power consumed. This tool was on one side accurate enough to take the measurements we needed and on the other it was measuring what we needed, so it will be the one used in this paper. This tool sometimes returned 0 in energy measures; when this happened, the run was discarded and repeated until the desired number was reached. Depending on the actual processor and chipset architecture, the tool will report different quantities; however, it always reports the *package* (`PKG`) energy consumption, which includes the CPU cores and the RAM. In architectures like Intel native where the RAM consumption is available, it will report `PKG` as two separate quantities, `cores` and `ram`.

We are going to focus on command line JavaScript interpreters in these versions:

- `bun` version 0.6.4
- `deno` version 1.34, which includes the v8 library version 11.5.150.0 and typescript 5.0.4
- `node.js` version 18.16.0

These were running in an Ubuntu version 20.04.1 with kernel version 5.15.0-69. The processor is an AMD Ryzen 9 3950X 16-Core; only some registers are available in this case, as it is a processor with a different architecture than Intel's.

A Perl script was created to run the experiments; it executed the scripts and collected the results by parsing the standard output and putting it into a file with CSV format that would allow examination of the experiments. According to [14], consumption for a no-op task running in the same environment was taken out of the reading before computing the average.

As was done in [13], which was focused on wallclock performance, the experiments will be focused on the key operations performed by an EA: evaluation of fitness and "genetic" operators like mutation and crossover. What we will do in this paper is: repeat the setup in the initial exploration, check the energy consumption for the processing of 40000 chromosomes, a number chosen to take a sizable amount of memory, but also on the ballpark of the usual number of operations in an EA benchmark; it is also small enough to not create garbage collection problems with the memory, something that was detected after the initial exploration. Experiments were repeated for the same chromosome size as before, 1024, 2048, and 4096, and for the three JS runtimes used. Although the business logic is exactly the same for the experiments, the script comes in two versions, one for deno and the other for bun/node, due to the way they read command-line arguments. This does not affect the overhead in any way. Code, as well as the data resulted from the experiments and analyzed in this paper are released with a free license (along with this paper) from the repository https://github.com/JJ/energy-ga-icsoft-2023.

The charts presented in Fig. 1, which concern the simple computation related to the One Max problem, show that for all three runtimes there is indeed a variation with respect to the measurements previously made: moreover, the way they vary is different.

In the case of node.js, newer JavaScript runtimes take more or less the same time, but the consumption of energy is much less; since [14] showed that this is, indeed, one of the choke points of this runtime, it makes it more interesting in a low-consumption environment. The case of bun is more complicated: performance is the same, but energy consumption seems to decrease only for smaller chromosome sizes, getting an increase in size 4096. The case of deno is even more complicated: there is a change towards higher performance (experiments taking less time), which more or less corresponds to the decrease in consumption. However, the strange behavior of this JavaScript runtime regarding chromosome size (already observed in [14]) will have to be checked further.

The scenario that is shown in Fig. 2, represents the performance for crossover, it has some similarities with the one shown in Fig. 1, showing again the strange performance correlations for the deno JS runtime. In the other two cases, there is no great difference, it can even be slightly worse, although the difference does not seem to be significant.

Since what we are seeking is the JavaScript VM that has the best energy profile, we will first compare the energy consumption of the ones available in March 2023, and published in [14], and the ones available in May 2023, when this paper is being written. It does not seem like the slight disadvantage in energy consumption for this new version, in this specific operator, is enough to offset the savings obtained with the fitness function; let us not forget that fitness evaluation usually takes the bulk of the energy consumption and time. The fact that unexpected variations in energy consumption may occur when versions change, and sometimes dramatic ones, is probably enough to warrant re-profiling of all workloads (evolutionary algorithms or otherwise) and another round of comparison of the performance and energy consumption for all three. This is what we will do next, using the newest published versions (Fig. 3).

The figure shows how energy consumption grows less than linearly for bun; this is a change with respect to [14], when it was flat. How consumption changes for deno is weird, since it decreases as the size of the chromosome grows; this is

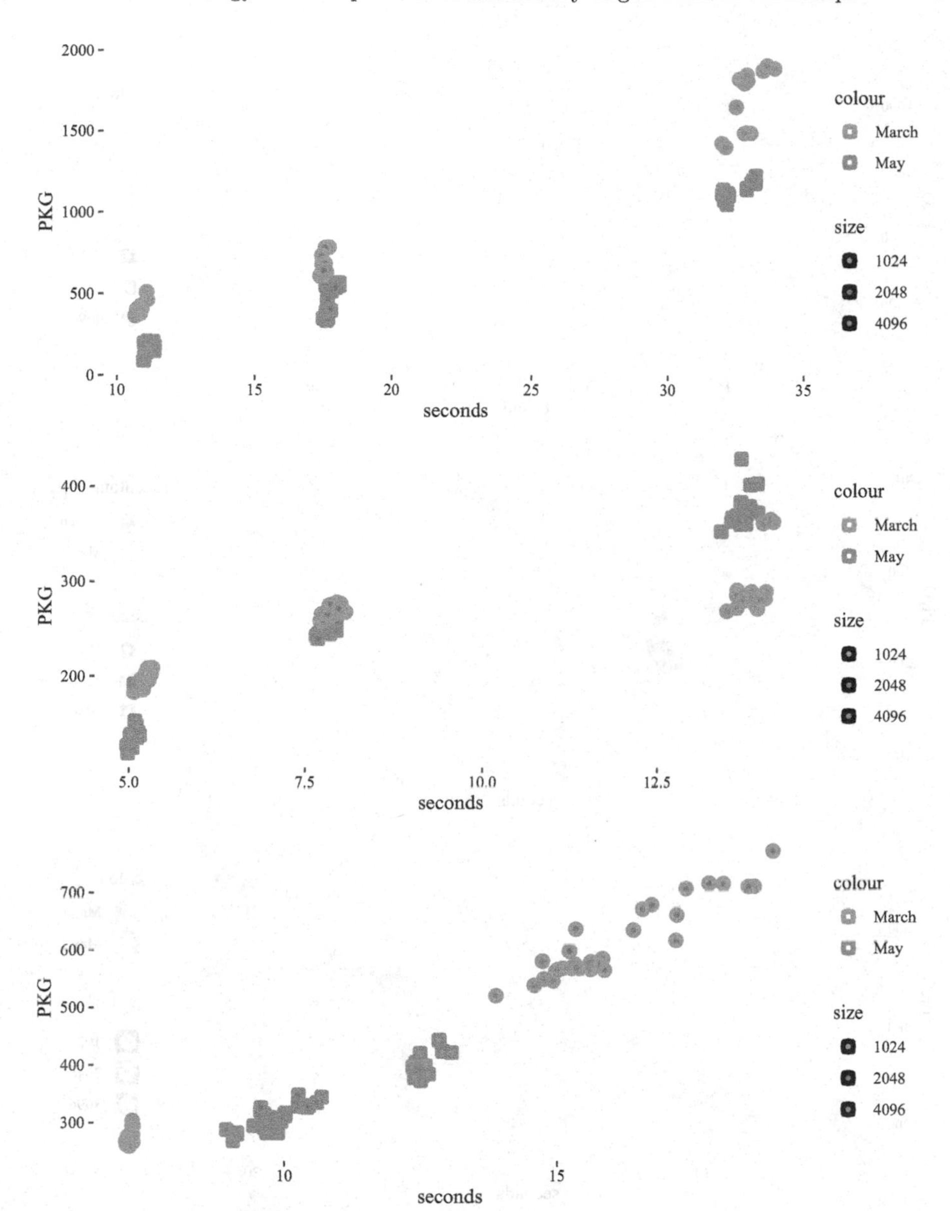


**Fig. 1.** Comparison between different versions of the VMs on the OneMax problem; from top to bottom: node, bun, deno

a slight change with respect to [14], when it decreased only for the biggest size. node is, however, the bigger energy guzzler, consuming up to 3 times more than deno on average, and more than 5 times as much as bun for the biggest size;

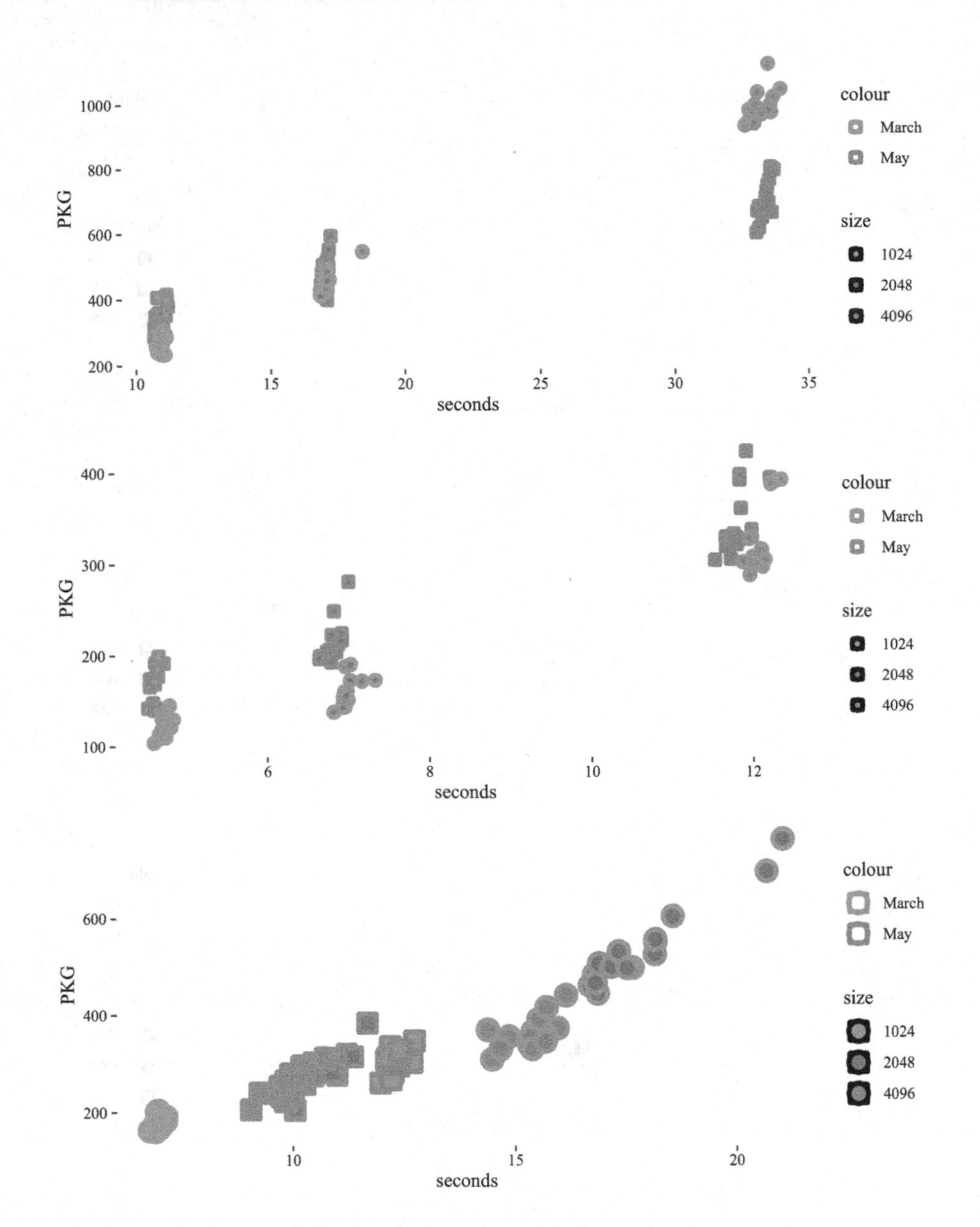


**Fig. 2.** Consumption and time for the crossover operator in the three different VMs, from top to bottom; node, bun, deno.

however, this new version of the interpreter `18.15.0` has a consumption similar to `bun` for the smallest size. Taking into account that, in most cases, we are going to deal with chromosomes with sizes that are around this order of magnitude,

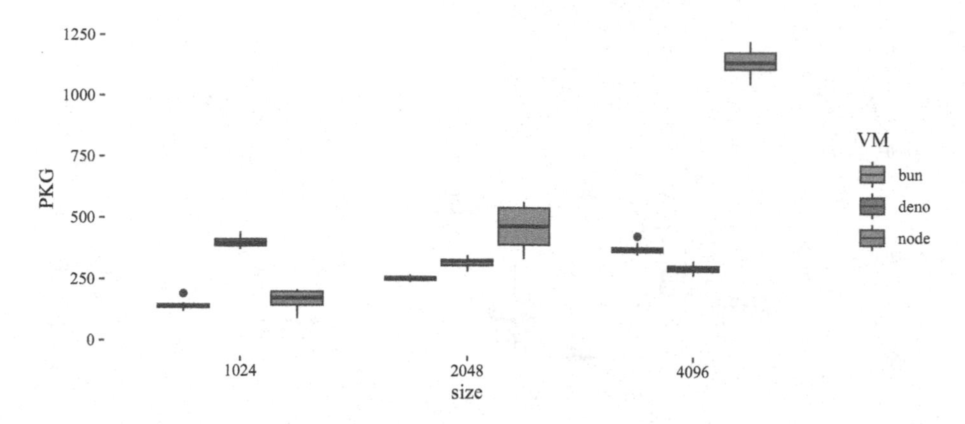


**Fig. 3.** Boxplot of PKG measurements for the OneMax fitness function and the three different virtual machines. Please note this is the same data that was labelled May in Fig. 1.

node might be a good alternative if considerations other than raw speed, such as maturity of the product, enter into consideration.

Another function we have tested, crossover, involves copy operations between strings that create new strings, since strings in JavaScript are immutable. We will again generate 40K chromosomes and group them in pairs; the strings in the pair will be crossed by interchanging a random fragment from one to the other and back in what is usually called two-point crossover. The resulting pairs will be stored in an array, which is eventually printed. The result of every experiment has been already shown in an energy vs. wallclock time chart in Fig. 2, comparing how it goes for different virtual machines. In Fig. 4 we render a boxplot for different sizes and the different virtual machines, in order to compare their energy consumption and how it grows with chromosome size.

The scenario is remarkably similar to the one shown in Fig. 1, and also similar to what we found with the previous versions of all command-line interpreters in [14]. Energy consumption for bun grows very slowly with chromosome size, less so than figures for node, whose energy consumption duplicates from chromosome size 1024 to 2048, and more than duplicates again for the bigger size, 4096, thus growing approximately in a linear way with the chromosome size; consumption is always better for bun, and the difference increases with size. But, again, the surprising energy profile for deno, which decreases with size, makes it the most energy-thrifty of the three for the biggest size. In any case, bun continues to consistently yield very low-energy consumption values across all sizes.

In this paper, we will also be testing a third operator, mutation. Mutation takes many different forms, but in its simplest form it changes a single bit in a bit string. Again, due to the fact that strings in this language are immutable,

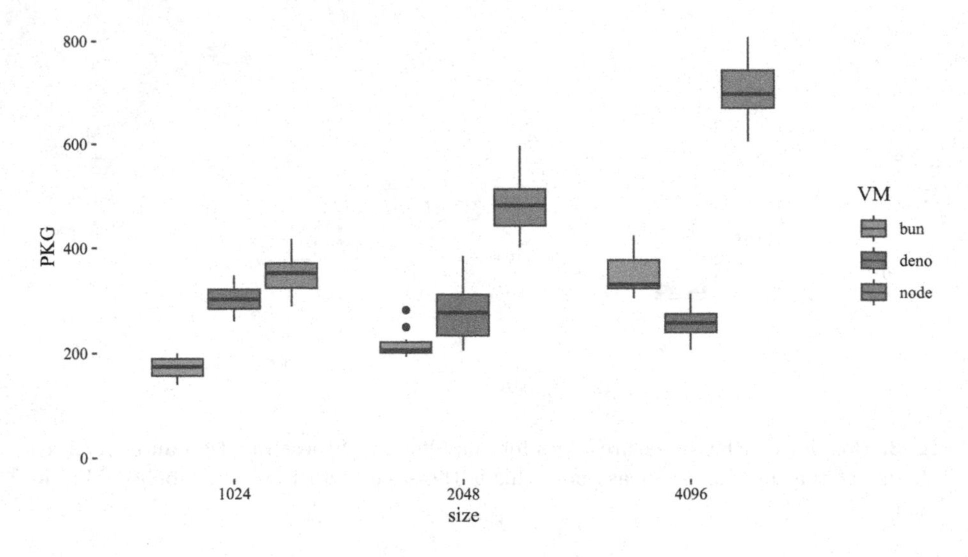


**Fig. 4.** PKG consumption for the crossover and the three different virtual machines, shown as a boxplot. Please note this is the same data that was labelled May in Figure 2.

the mutated string must be built from pieces of the original string, which will have an impact on the performance (Fig. 5).

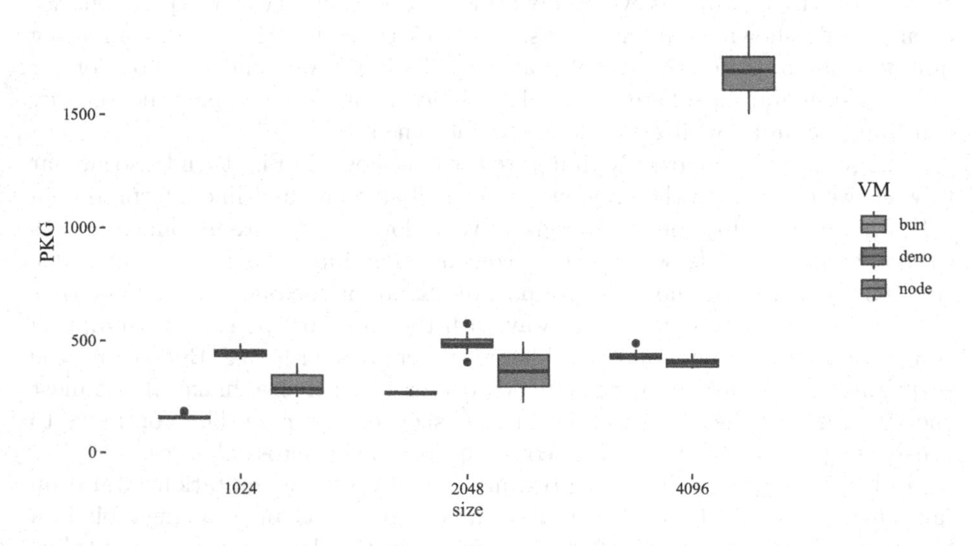


**Fig. 5.** PKG consumption for the mutation operator and the three different virtual machines, shown as a boxplot.

What these measurements show is that, since they behave essentially as a set of copy operations, its behavior is quite similar to that shown in Fig. 4. deno decreases with size, bun increases very slowly, node faster to the point that it spends three times more energy than bun for the biggest size.

It is interesting, however, to test the algorithms in different architectures, even more using a native Intel architecture with all its registers. This is why we have repeated the experiments in another computer, a Lenovo Carbon X1 with Ubuntu 22.04.1, kernel version `5.19.0-43-generic` and an Intel processor and an Intel Core i7-10610U CPU @ 1.80 GHz, with 8 cores.

One of the advantages of using the native Intel architecture is that it gives you more accurate estimations of consumption for specific parts of the system; namely, it breaks the `PKG` reading into two, `cores` and `ram`. In this case, we will be changing slightly the script so that we get separate readings for these two sensors.

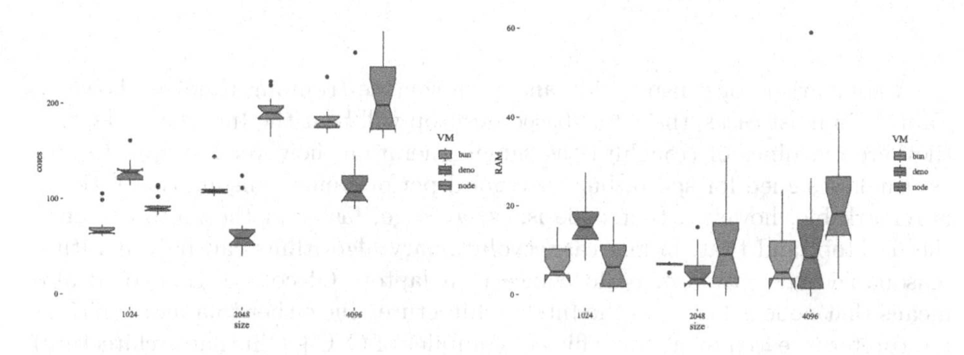


**Fig. 6.** Consumption for the OneMax fitness function in the Intel architecture and the three different virtual machines, shown as a notched boxplot; cores component (left) and RAM component (right). Please observe that the scales in the y axes are different.

Results of these experiments for the OneMax fitness function, shown in Fig. 6, allow us to check the influence on the overall energy expenses of the memory operations, which are shown in the right-hand side panel. They are first barely above the baseline, with averages rarely exceeding 20 J; but the most important thing is that there are no significant differences among the interpreters. The consumption by node seems to increase slightly with size, but it is not enough to outspend the other two interpreters. The left-hand side panel in Fig. 6 does show significant differences in every size. We can affirm that bun is, in general, better than node, although the difference is not significant at the bigger size. But the main issue here is that deno seems to be the best for any size above 1024 bits.

At any rate, comparing this graph with Fig. 1, we can see that the average consumption for the smaller size, around 100 J, is less than half what is consumed in the desktop system we have initially tested here and in [14]. The average time needed to find the solution, however, is higher, although not by an order of magnitude.

**Table 1.** Comparing times (in seconds, s) and cost (in Joules, J) for an AMD desktop and Intel-based laptop (see text for specs). These are average times and energy consumption, in seconds and Joules, respectively.

| VM | Size | AMD - s | AMD - J | Intel - s | Intel - J |
|---|---|---|---|---|---|
| bun | 1024 | 5.07 | 141.19 | 7.95 | 78.31 |
| bun | 2048 | 7.84 | 253.48 | 12.52 | 117.63 |
| bun | 4096 | 13.74 | 377.82 | 24.41 | 193.08 |
| deno | 1024 | 12.61 | 400.63 | 19.33 | 143.78 |
| deno | 2048 | 9.97 | 319.43 | 19.35 | 70.28 |
| deno | 4096 | 9.70 | 297.88 | 20.08 | 132.41 |
| node | 1024 | 11.13 | 167.40 | 15.82 | 100.29 |
| node | 2048 | 17.75 | 460.25 | 32.42 | 203.13 |
| node | 4096 | 32.63 | 1145.89 | 29.29 | 228.25 |

A comparison of consumption and performance (running time) is shown in Table 1. In most cases, the AMD-based desktop will beat the Intel-based laptop; they are machines of (roughly) the same generation, however laptops are not, as usual, designed for speed, but for a good performance/consumption ratio. It is remarkable, however, that node is, on average, faster in the laptop than in the desktop, and that, in any case, evolutionary algorithms can be run with a reasonable expectation of performance in a laptop. Of course, this could also means that node is faster on the Intel architecture, due either to a more efficient interpreter (created by a more efficient compiler of C/C++ in that architecture) or to the fact that the interpreter operations work better in that architecture. Ascertaining this, however, falls outside the scope of this paper. On the other hand, there is no single combination of interpreter and size that offers better power consumption, to the extent that, in the case mentioned above, node at a chromosome length = 4096, consumption is almost 6 times smaller in the case of the laptop. In general, it will always be less than half.

## 4 Conclusions

In this paper we set out to study the influence on energy consumption of evolutionary algorithms in three different directions: first, testing different versions of the interpreters; second, including the mutation operator, since it seems to be the one that consumes the most; and third, test different types of computers.

While in our previous paper, [14], the measurements showed clearly that bun was the less energy-consuming interpreter across all evolutionary operators and fitness functions, the experiments performed in this paper show a more nuanced scenario. The first interesting conclusion is that there is nothing inherently energy-saving in the architecture of that interpreter, and that the supremacy can change when current versions of the interpreters are compared with each other; subsequent release might increase or decrease the energy consumption, and do so differentially across different problem sizes, so this leads again to experimentation as the only possible way to really ascertain which interpreter is best. The counter-intuitive behavior of deno, which consumes less as the size increases, also leads to this conclusion. We cannot even discard node as the most energy-consuming interpreter, since it beats deno in the mutation operator, which is the one that consumes the most, at the smallest sizes, as well as the OneMax operator, also quite energy-consuming.

These new experiments with the mutation operator, which basically involves copying of large strings, do not actually show big differences between the three interpreters for the smallest size, which is probably closer to the one actually used in most EA applications. Larger sizes imply a disadvantage for node, so your mileage may vary. At any rate, using bun or node is largely, in this case, a matter of choice; a choice that should, nonetheless, be informed by actual measurements, since experiments do not give you a general answer. This is also true independently of the machine we choose: there is a slight advantage of bun over node at the smallest size, deno seems to be better when size is increased. Measuring in an Intel-powered laptop has several advantages, however: first, it gives you real register measurements, as opposed to other brands of CPUs that merely emulate them; second, it really allows you to *pinpoint*, as in the tool we use, where is the actual energy consumption, by allowing to make core consumption and memory apart; this has allowed us to find out that in the case of evolutionary algorithms, it is actually the cores that are consuming energy from the power source.

Finally, our experiment with a (powered) laptop shows that, as should be expected, opting for an energy-saving computer architecture will give you energy savings that can go from 50% to over 80%, depending on the size; these savings do not imply a decrease in performance in the same scale; even in some cases, it can be faster, in the case of deno. This leads us to encourage performing evolutionary algorithms, wherever possible, in laptops, even more so if they have an Intel processor and chipset, Apple Silicon, or any computer or processor architecture designed for energy saving.

Even if energy saving is not the main concern of the evolutionary algorithm practitioner, we encourage researchers to always follow a strategy of energy and performance profiling to be able to extract the most from the existing hardware architecture; this is always a software engineering best practice that we really need to encourage in our area.

The fact that mutation is so power-hungry leads us to designing algorithms that try and save energy in this area; this will probably imply changes in the data structures used in the evolutionary algorithm. Since in this case, we have used immutable strings, that might be the reason why it consumes so much energy. Implementation matters, [15], so exploring and measuring will always help you take the best decisions in the direction of making computing greener.

**Acknowledgements.** This work is supported by the Ministerio español de Economía y Competitividad (Spanish Ministry of Competitivity and Economy) under project PID2020-115570GB-C22 (DemocratAI::UGR).

## References

1. Abdelhafez, A., Alba, E., Luque, G.: A component-based study of energy consumption for sequential and parallel genetic algorithms. J. Supercomput. **75**, 6194–6219 (2019)
2. Cruz, L.: Tools to measure software energy consumption from your computer (2021). https://luiscruz.github.io/2021/07/20/measuring-energy.html
3. Demaine, E.D., Lynch, J., Mirano, G.J., Tyagi, N.: Energy-efficient algorithms. In: Proceedings of the 2016 ACM Conference on Innovations in Theoretical Computer Science, pp. 321–332 (2016)
4. Díaz-Álvarez, J., Castillo, P.A., Fernandez de Vega, F., Chávez, F., Alvarado, J.: Population size influence on the energy consumption of genetic programming. Measur. Control **55**(1–2), 102–115 (2022)
5. Diaz Alvarez, J., Castillo Martínez, P.A., Rodríguez Díaz, F.J., Fernández de Vega, F., et al.: A fuzzy rule-based system to predict energy consumption of genetic programming algorithms (2018)
6. Doglio, F.: Introducing Deno
7. Garcia, J.A.: Exploration of energy consumption using the intel running average power limit interface. In: 2019 IEEE Space Computing Conference (SCC), pp. 1–10 (2019). https://doi.org/10.1109/SpaceComp.2019.00005
8. González, J., Merelo-Guervós, J.J., Castillo, P.A., Rivas, V., Romero, G., Prieto, A.: Optimized web newspaper layout using simulated annealing. In: Mira, J., Sánchez-Andrés, J.V. (eds.) IWANN 1999. LNCS, vol. 1607, pp. 759–768. Springer, Heidelberg (1999). https://doi.org/10.1007/BFb0100543
9. Hähnel, M., Döbel, B., Völp, M., Härtig, H.: Measuring energy consumption for short code paths using RAPL. SIGMETRICS Perform. Eval. Rev. **40**(3), 13–17 (2012). https://doi.org/10.1145/2425248.2425252
10. Khan, K.N., Hirki, M., Niemi, T., Nurminen, J.K., Ou, Z.: RAPL in action: experiences in using RAPL for power measurements. ACM Trans. Model. Perform. Eval. Comput. Syst. (TOMPECS) **3**(2), 1–26 (2018)
11. Köhler, S., et al.: Pinpoint the Joules: unifying runtime-support for energy measurements on heterogeneous systems. In: 2020 IEEE/ACM International Workshop on Runtime and Operating Systems for Supercomputers (ROSS), pp. 31–40 (2020). https://doi.org/10.1109/ROSS51935.2020.00009
12. Maryam, K., Sardaraz, M., Tahir, M.: Evolutionary algorithms in cloud computing from the perspective of energy consumption: a review. In: 2018 14th International Conference on Emerging Technologies (ICET), pp. 1–6. IEEE (2018)
13. Merelo, J.J., et al.: Benchmarking languages for evolutionary algorithms. In: Squillero, G., Burelli, P. (eds.) EvoApplications 2016 Part II. LNCS, vol. 9598, pp. 27–41. Springer, Cham (2016). https://doi.org/10.1007/978-3-319-31153-1_3
14. Merelo-Guervós, J.J., García-Valdez, M., Castillo, P.A.: An analysis of energy consumption of JavaScript interpreters with evolutionary algorithm workloads. In: Fill, H., Mayo, F.J.D., van Sinderen, M., Maciaszek, L.A. (eds.) Proceedings of the 18th International Conference on Software Technologies, ICSOFT 2023, Rome, Italy, July 10-12, 2023, pp. 175–184. SCITEPRESS (2023). https://doi.org/10.5220/0012128100003538

15. Merelo-Guervós, J.J., Romero, G., García-Arenas, M., Castillo, P.A., Mora, A.M., Jiménez-Laredo, J.L.: Implementation matters: programming best practices for evolutionary algorithms. In: Cabestany, J., Rojas, I., Caparrós, G.J. (eds.) IWANN 2011. LNCS, vol. 6692, pp. 333–340. Springer, Heidelberg (2011). https://doi.org/10.1007/978-3-642-21498-1_42
16. Tomar, D.: Bun JS : a brand-new, lightning-quick JavaScript runtime. Medium (2022). https://devangtomar.medium.com/bun-a-brand-new-lightning-quick-javascript-runtime-e42119a306ca
17. de Vega, F.F., et al.: A cross-platform assessment of energy consumption in evolutionary algorithms. In: Handl, J., Hart, E., Lewis, P.R., López-Ibáñez, M., Ochoa, G., Paechter, B. (eds.) PPSN 2016. LNCS, pp. 548–557. Springer, Cham (2016)
18. Fernández de Vega, F., Díaz, J., García, J.Á., Chávez, F., Alvarado, J.: Looking for energy efficient genetic algorithms. In: Idoumghar, L., Legrand, P., Liefooghe, A., Lutton, E., Monmarché, N., Schoenauer, M. (eds.) Artificial Evolution. EA2019. LNCS, vol. 12052, pp. 96–109. Springer, Cham (2020). https://doi.org/10.1007/978-3-030-45715-0_8
19. Verdecchia, R., Sallou, J., Cruz, L.: A systematic review of green AI. arXiv preprint arXiv:2301.11047 (2023)

# A Tabu Search Algorithm for the Map Labeling Problem

Claudia Cavallaro, Vincenzo Cutello, Mario Pavone(✉), and Francesco Zito

Department of Mathematics and Computer Science, University of Catania, v.le Andrea Doria 6, 95125 Catania, Italy
{claudia.cavallaro,cutello}@unict.it, mpavone@dmi.unict.it, francesco.zito@phd.unict.it

**Abstract.** In this paper we propose an algorithm based on the Tabu Search metaheuristic for the Map Labeling problem, i.e. the relevant problem in cartography of assigning labels to specific points of interests in a clear and readable way. It is a combinatorial problem known to be $\mathcal{NP}$-complete and therefore it needs to be tackled by means of good and efficient heuristics. In our experiments, we used real maps of Italian cities, Rome and Venice in particular.

**Keywords:** Metaheuristics · Tabu Search · Map Labeling · Combinatorial Optimization · $\mathcal{NP}$-complete · Approximation algorithm

## 1 Introduction

The rapid growth of tourism and the mobility of people from one place to another for business, vacation, culture and other reasons, has led to an increasing use of online services to get all the information a traveler needs such as restaurants, attractions, events, open parks and much more [8,16]. The increased computing power and storage capacity of computers and portable devices, along with geolocation and connectivity capabilities, has led to a proportionally significant increase in the availability of geographic data and the overall number of users, encouraging service providers to strengthen them and make them more efficient.

A map service allows customers to visualize the main attractions of a region, such as an entire city or part of it. There, the most important buildings, streets or squares are highlighted and marked with a label placed on them. However, given the huge amount of information that can be found on a map, it could be very difficult to place these labels properly. As a consequence, new methods must be developed in order to efficiently solve this problem which despite its apparent simplicity, it is actually a computationally hard problem and, at the same time, very important to provide a high-quality service to customers.

Our goal, in this research paper, is to find the best placement of the labels of a map in order to optimize the visual aspect of such a map and provide a better service to the customers. For this purpose, we used the Tabu Search algorithm and


M. Villani et al. (Eds.): WIVACE 2023, CCIS 1977, pp. 16–28, 2024.
https://doi.org/10.1007/978-3-031-57430-6_2

we conducted several experiments to prove the efficiency of such a Metaheuristic technique, which can be used in a wide range of applications to enhance traditional methods, such as machine learning [17,18]. The remainder of this paper is organized as follows. The map labelling problem is described in Sect. 2 and related work is discussed. Section 3 briefly introduces the Tabu Search algorithm. In Sect. 4 we describe our approach. Section 5 shows the results. Finally, Sect. 6 draws some conclusions and future directions.

## 2 The Map Labeling Problem

In general a map consists of a graphic part, that is the representation of the elements of a space by symbols or conventional signs, and a textual part, that is the naming of the elements. A fundamental part of the map-making process is the careful placement of toponymy to make the map "speakable": symbols and corresponding names must be placed naturally according to rules or priorities that facilitate the use and understanding of the map. In theory, each label should be placed near its object, and overlap should be avoided. Considering that each label must be matched with all the others, the complexity is exponential.

Various methods have been proposed in literature to solve the task of automatic labeling of different types of maps or figures. The final arrangement of labels on sites and descriptive characteristics of points must comply with a series of cartographic rules, graphic requirements and constraints. To be usable, maps complete with text must be legible, clear and maintain an aesthetic balance, respecting boundaries and avoiding overlapping. Although it may seem like a problem based only on formatting style, for large areas it becomes problematic to process and test a huge number of positions relative to the labels centered on all possible pixels which are present on the map. Thus, map labeling can also be seen as a geometric combinatorial optimization problem [13,15] and has been shown to be NP-complete by reductions to 3−SAT (Boolean satisfiability problem) [7], as well as the more general Edge Label Placement (ELP) problem [9]. Thus, a good approximate solution may be found by metaheuristic methods (see for instance [14] for another example of hard geometrical problem).

Lu et al. [11] use a hybrid algorithm called DDEGA that combines Discrete Differential Evolution (DDE) and Genetic Algorithm (GA). The initial population is randomly generated, and after calculating the fitness value through a quality metric for each configuration, the best 50% individuals (solutions) participate in a new generation. More in details, 80% of the new individuals are generated using DDE, the remaining 20% using GA. After selection, individual chromosomes are combined through crossover, then go through a mutation process, following GA and DDE standard procedures. Starting from the set of possible solutions, the differential mutation produces the best positioning, that is the one with the least possible number of overlaps, following lines which are parallel to the present elements. Although such an algorithm manages to handle conflicts between labels and elements with good results, it has limitations in scalability and convergence on very large instances.

The method proposed by Lhuillier et al. [10] supports constrained labeling following the descending gradient in a density map. Dense areas of information make it more difficult to insert non-overlapping labels, because there may not be enough space. The barycentre of the label is initially positioned at the point entity to which it refers to and is directed towards a local minimum of the spatial density, where it is possible to position it in a suitable way. It may happen that the algorithm stops because the maximum fixed number of iterations has been reached without finding a suitable space or that the gradient has vanished, and in these cases the label is not positioned and will be discarded. Alternatively, curved lines called *leaders*, which connect the labels with the entities to be described, are drawn. Following the smooth gradient descent trajectories, they will not intersect, as the reference ordinary differential equations have unique solutions, but for a large number of labels they could follow a similar trajectory. So the drawback of this algorithm is that, even though it chooses not to obscure information, the associations between labels and items may not be apparent due to the convergence of leaders pointing to the areas in question. Especially for a large number of sites to be associated this method would make it difficult to recognize which label belongs to which point. The authors assume that the point labels are rectangles of uniform size, setting a default size. An adaptation of the algorithm for labeling polygonal areas is missing. Another approach concerning the cartographic labeling of points was presented in [2]. The authors use a combination of the Genetic Algorithm with a convex onion structure (COPGA), which leads to a better initialization of the labeling and evolutionary process. Given a set of points to label, the Convex Onion Peeling is formed by concentric polygons interpolating the points. The sites to be described are then divided among the different onion layers, that is, convex polygons extending from inside out. The labels of the elements are positioned outside each layer, following the standard steps of the evolutionary process. The algorithm manages to reduce the conflicts between the labels, however it does not favor the positions at the top right. Most studies have addressed the problem of labeling points, but only few have focused on the case of polygonal areas or combined the two aspects for the production of maps. In support of Geographic Information Systems, [1] presents an algorithm for the internal labeling of areas, cities, lakes or areas delimited by polygons. For narrow areas, label overlaps with edges or with the descriptive text of the adjacent area may become unavoidable, and it is also not recommended to place labels outside the areas. The proposed method finds the minimum bounding rectangle of each area and, using the distances between the intersections of auxiliary lines and boundary lines, chooses the most suitable and centered positions for the labels. The dimensions of the rectangles are set by fixing a single standard measure.

## 3 The Tabu Search Algorithm

Tabu Search is a heuristic search algorithm that is commonly used to solve optimization problems. It is particularly useful for problems where the search space is large and the search time is long. One of the advantages of Tabu Search is its ability to maintain a long memory of previously visited solutions, which can be used to avoid getting stuck in local optima and to improve the efficiency of the search (Algorithm 1). Our goal is to prove that Tabu Search can be effectively used to optimize the placement of labels on a map. By maintaining a long memory of previously visited solutions, the algorithm can avoid placing labels too close to each other, which could lead to inaccurate labeling. Additionally, Tabu Search can be used to improve the search efficiency by avoiding redundant computations and by pruning the search space based on constraints.

**Algorithm 1:** Tabu Search algorithm

**Input:** Initial solution $x$, objective function $f$, neighborhood function $nh$, tabu list size $L$, maximum number of iterations $max\_iter$

**Output:** Best found solution $x^*$

```
1  x* = x, tabu_list = [];
2  for i = 1 to max_iter do
3      best_candidate = None; best_candidate_f = float('inf');
4      for y in nh(x) do
5          if y not in tabu_list and f(y) < best_candidate_f then
6              best_candidate = y; best_candidate_f = f(y);
7          end
8      end
9      if best_candidate = None then
10         break;
11     end
12     x = best_candidate;
13     if f(x) < f(x*) then
14         x* = x;
15     end
16     Add move (x, best_candidate) to tabu list;
17     if tabu list size exceeds L then
18         Remove oldest move from tabu list;
19     end
20 end
```

We will now formally describe our approach and the methodology used to tackle the map labeling problem.

# 4 Methodology

## 4.1 Solution Format

A solution format is a description of the structure or format of a solution to a problem. In the context of map labeling, the solution format includes information such as the position of labels on the map, the orientation of labels, and how to

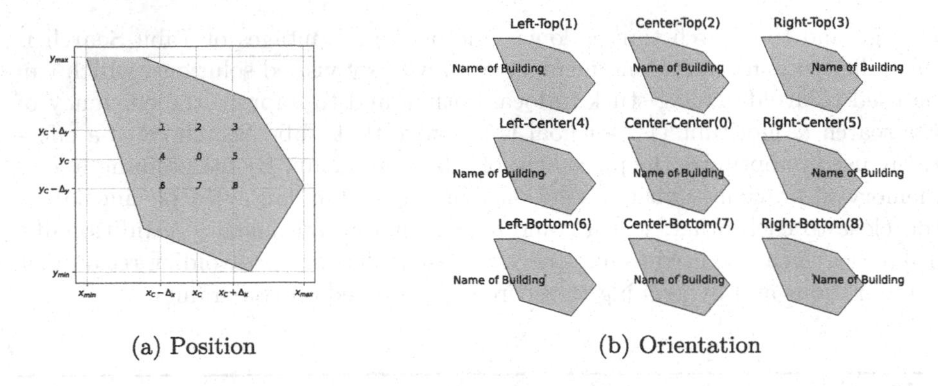


(a) Position (b) Orientation

**Fig. 1.** Different types of position and orientation of a label.

distribute the text on multiple lines. This information is encoded in a vector of integer numbers that contains $3 \cdot n$ parameters, where $n$ is the total number of labels to be placed on a map. Thus, the vector represents a solution to the problem of label placement. Each value in the solution vector ranges from 0 to 8, and each number encodes a position, orientation, and text distribution of the labels. Therefore, a solution is encoded as $(p_1, o_1, d_1, p_2, o_2, d_2, \ldots, p_n, o_n, d_n)$ where $p_i$,$o_j$ and $d_j$ represent respectively the position, orientation, and text distribution of the $i$-th label.

**Position of Labels.** A map can be used to represent a diverse range of geographic information, including buildings, monuments, squares, and points of interest. The polygons that make up a map represent the boundaries of these elements. The positions available for placing labels on a polygon are determined by a fixed set of label positions. Figure 1a provides an illustration of the available label positions for a given map element. The gray polygon represents a building that we want to label. The point $C = (x_C, y_C)$ is the center of the polygon. The values $x_{max}, x_{min}, y_{max}, y_{min}$ represent the maximum and minimum $x$ and $y$ coordinates of the polygon's bounding box, which is a rectangle that completely encloses the polygon and is used to determine the position and size of the polygon on a map. The values $\Delta_x, \Delta_y$ are defined as follows:

$$\Delta_x = (x_C - x_{min})/2, \Delta_y = (y_C - y_{min})/2 \tag{1}$$

**Text Distribution.** Another parameter that affects how text is distributed across multiple lines is the maximum number of rows allowed. This parameter is crucial in avoiding overlapping between adjacent labels and ensuring that text is properly formatted.

If the maximum number of rows is greater than the number of words, the text will be split at the word level. This ensures that each word is displayed on a separate line, making it easier to read and understand. If, instead, the maximum number of rows is smaller than the number of words, two or more words will be displayed on the same line. Choosing how to divide a text is dependent upon the goal of equal distribution of the text in all rows. More in details, the maximum number of rows, according to the solution format, is an integer value between 0 and 8. If such a value is set to $i$, the text will be distributed across $i + 1$ lines.

**Label Orientation.** Once a position for the labels in the polygon has been chosen, this parameter determines the orientation of the label. It can be either vertical (Top, Center, Bottom) or horizontal (Left, Center, Right). For each possible combination, we associate an integer value as described in Fig. 1b.

### 4.2 Label Placements Score

The quality of label placements (solution) is defined using two main concepts that will be discussed in this section: the overlapping matrix, which contains the cost of a solution, and the penalty vector, whose values represent a penalty. By multiplying the rows of the overlapping matrix with the respective values of the penalty vector we obtain a new $n \times n$ matrix, and the sum of all items contained in it represents the *label placements score*, a value which must be minimized.

**Cost Function and Overlapping Matrix.** A cost function is a mathematical function that assigns a numerical value to each possible solution to a problem. It measures the "cost" of each solution, and the algorithm aims to find the solution with the lowest cost. In the context of map labeling, the cost function may assign a higher cost to placing labels too close to each other, or to placing labels on unsuitable areas of the map. We define an *overlapping matrix* as a square matrix $Q \in \mathbb{R}^{n \times n}$ containing the percentage of overlap of the $i$-th label with the $j$-th label, normalized between 0 and 1, given a possible label placements. The cost is obtained by summing up all the values of the overlapping matrix.

**Penalty Function.** A penalty function assigns a numerical value to each infeasible solution to a problem. It measures the "penalty" of each infeasible solution, and the algorithm aims to avoid finding them. In the context of map labeling, the penalty function assigns a higher penalty in case the text of the label is split across multiple rows. Ideally, the algorithm should try to position the text on the map without any splits, to make it more readable. We define a *penalty* *vector* as a vector of $n$ elements $P \in \mathbb{R}^n$ such that the $i$-th element is calculated as follows, with $r_i$ being the number of rows of the $i$-th label:

$$p_i = 1 + \frac{r_i - 1}{8} \tag{2}$$

### 4.3 Neighbor Function

A neighbor function is a mathematical function that generates all possible solutions that are neighbors of a given solution to a problem. It is used to generate new solutions to the problem by modifying the current solution.

In the context of map labeling, the neighbor function generate new solutions by moving existing labels. The neighborhood of a possible label placement is a set of candidate solutions that are similar to the given solution $S$. The goal of the labeling algorithm is to find the optimal solution within the neighborhood.

To generate the neighborhood, we start with the given solution $S$ and randomly vary the parameters of the solution in correspondence with the most overlapped labels between them. We use the overlapping matrix to determine which labels have a high overlap. To do this, we first compute the overlapping matrix among the labels in the solution. We then find the labels that overlap the most with the other labels by taking the maximum value in each row of the overlapping matrix (which corresponds to the index of the label that overlaps with most of the other labels). In details, the steps to compute the neighborhood are the following:

1. Compute the overlapping matrix among the labels in the solution.
2. For each label $i$ in the solution, find the label $j$ that overlaps the most with $i$. Formally, $j = max_{h=1,\dots,n} Q[i,h]$ Subsequently, we determine the label $k$ that overlaps the most with label $j$, i.e. $k = max_{h=1,\dots,n} Q[j,h]$. If label $k$ is equal to $i$, then $i$ and $j$ have a strong overlapping and therefore it is necessary to alter the placement of $i$ in order to reduce the overlapping between these two labels. All labels that satisfy these two conditions are selected to be mutated in the next step.
3. Generate the neighborhood by randomly perturbing the parameters of only a subset of labels in the solution.
4. Return the resulting neighborhood.

The specific parameters that are varied are chosen randomly from a set of possible values. For example, if we are varying the position of a label, we may choose a random position within a certain range of the current position of the label. If we are varying the orientation of a label, we may choose a random orientation within a certain range of the current orientation of the label. Once the neighborhood solutions are generated, the labeling algorithm can evaluate each solution to determine its cost and penalty. The algorithm then selects the best solution from the neighborhood as the next solution to be evaluated. This process is repeated until the optimal solution is found within the neighborhood.

## 5 Results

In our experiments, we chose to evaluate the effectiveness of the proposed algorithm on real geographic data. The considered maps come from OpenStreetMap [12], an open source service that allows exporting a zone in OSM format. It contains a set of points, geographic coordinates and other additional data such as the hierarchical relationships between them and various information. To evaluate our methodology, we compared it with a default label placement strategy, which places each label as a single line in the center of a building. We scored the results of four different map areas to determine the optimal label placement. Note that the label placement score measures the overlap among the labels, so the objective is to minimize it.

We ran the algorithm for a total of 500 iterations for any instance and compared the results, denoted as Optimized Placements, with those denoted as Default Placement in terms of the computed score as seen in Sect. 4.2. This allowed us to determine the effectiveness of our methodology and identify where our approach outperforms the default label placement strategy. We also measured the number of iterations in which the algorithm found the best solution while in the remaining iterations it was not able to further improve such a solution.

The experiments on the map labeling problem show that the default strategy is suboptimal in most cases as shown in Table 1, where we can see that the label placement score achieved by the Tabu Search algorithm is lower than the score achieved by the default strategy. We can, therefore claim, that the Tabu Search algorithm can find better solutions by searching a larger and more diverse space. Due to space limitations, we show only Figs. 3 and 4 which respectively display the default label placement and the optimized label placement for Rome and Venice. For more images and a better resolution of the images presented here, we refer to the dedicated site[1].

**Table 1.** Comparison of Default and Optimized Label Placement Scores on four maps. The table includes the names of the maps, the number of labels, the score obtained by using a default placement, the score obtained by using the placement computed by our algorithm, the number of interactions required to obtain the best solution and the time taken to obtain the best solution.

| Map Name | Labels | Default Placement | Optimized Placement | Iterations | Time (s) |
|---|---|---|---|---|---|
| Rome | 40 | 2.0790E+01 | 1.0893E+00 | 60 | 5.0198E+02 |
| Syracuse | 24 | 7.6505E+00 | 3.9940E-01 | 36 | 4.6431E+01 |
| Taormina | 60 | 2.7589E+01 | 3.2716E+00 | 79 | 7.3909E+03 |
| Venice | 40 | 7.2168E+00 | 1.6048E-02 | 20 | 3.7179E+02 |

[1] https://github.com/Complex-Intelligent-Systems/Map-Labeling.

The number of labels in a map has a significant impact on the number of iterations required for Tabu Search to achieve a feasible label placement. In Fig. 2, we show the number of iterations needed by the algorithm to find an optimal solution as the number of labels increased when the search space becomes larger, and the number of iterations required to find a feasible solution increases. The data are obtained by considering four maps and for each map an optimal label placement is found by considering a different number of labels.

Summing up, the obtained results suggest that the Tabu Search algorithm can be a valuable tool for solving the label placement problem in an efficient and effective way.

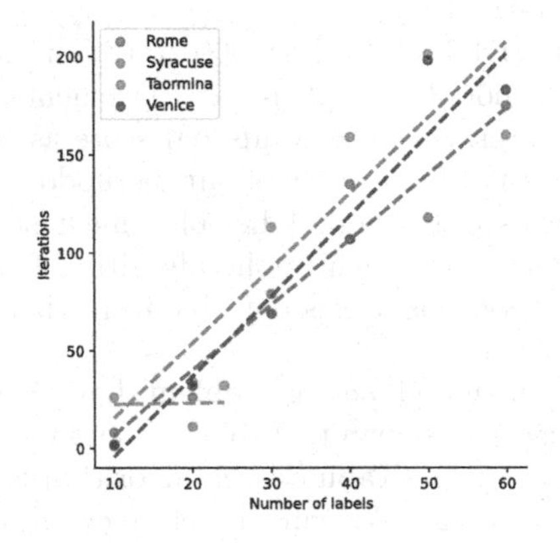


**Fig. 2.** The x-axis represents the number of labels, while the y-axis represents the number of iterations required to converge to the optimal placement.

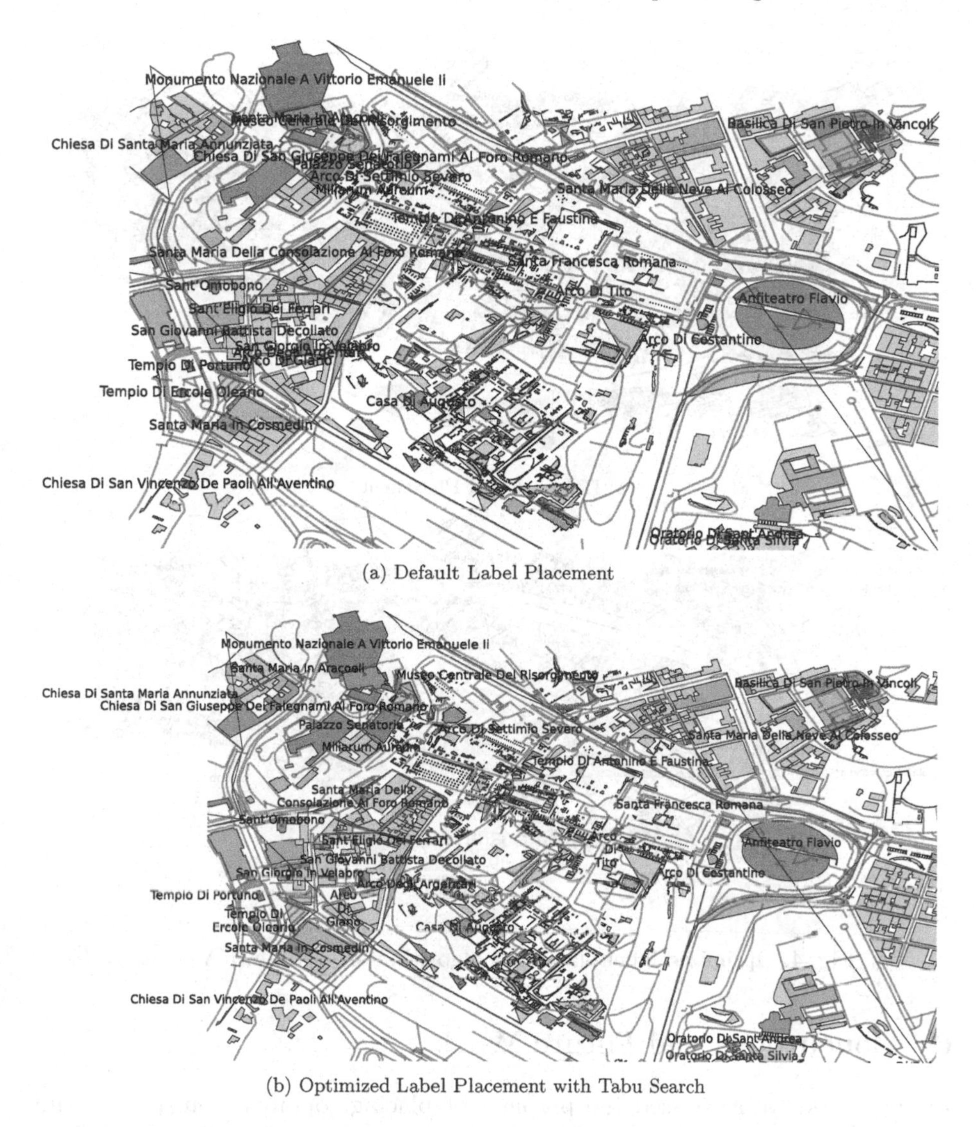


(a) Default Label Placement

(b) Optimized Label Placement with Tabu Search

**Fig. 3.** Optimized label placement algorithm for downtown Rome.

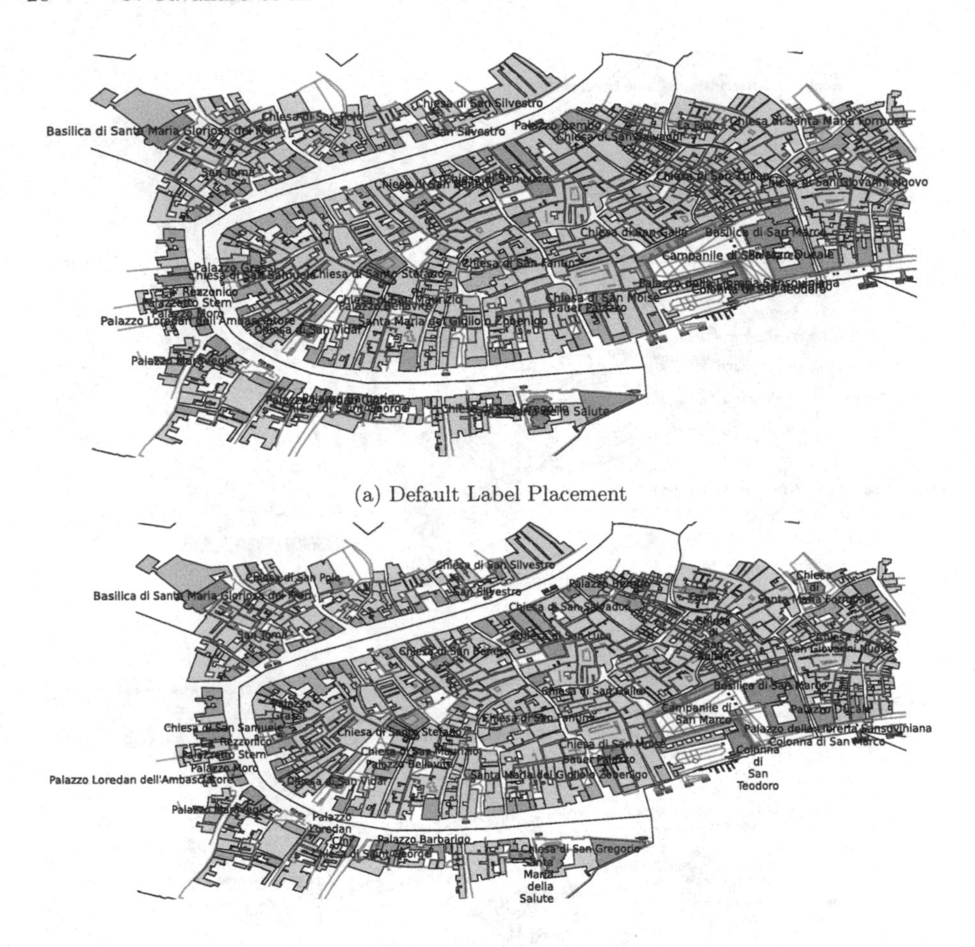


(a) Default Label Placement

**Fig. 4.** Optimized label placement algorithm for downtown Venice.

## 6 Conclusions and Future Work

In this work, we addressed the problem of placing toponymy on a map, and specifically city maps, using a Tabu Search algorithm. The initial results are promising and worth a more in depth study along with new challenging experiments. As a future research line, it would also be interesting to tackle the problem using a different population based metaheuristic, such as Ant Colony Optimization. Some preliminary experiments we have carried show that with an adequate calibration of the parameters, an ACO algorithm may actually produce good results, in a relatively short time and in a small number of iterations. Another interesting research topic is dynamic map labeling. The wide-spread use of smartphones and location-based internet services (e.g., Google Maps) gives the opportunity to collect user locations, social interactions [6], and cultural interests (see [3–5]), a city map could be labelled following the user interests. Thus, some labels could, dinamically, be temporarily hidden while others highlighted according to specific criteria. As a consequence, an effective labeling of a city map could be simplified while users could still be clearly read labels of points of interests.

**Acknowledgements.** This research is supported by the project Future Artificial Intelligence Research (FAIR) - PNRR MUR Cod. PE0000013 - CUP: E63C22001940006

## References

1. Abe, N., Kuroda, K., Kamata, Y., Midoritani, S.: Implementation and evaluation of a fast area feature labeling method using auxiliary lines. ISPRS Int. J. Geo-Inf. **9**(9), 529 (2020)
2. Bae, W.D., Alkobaisi, S., Narayanappa, S., Vojtechovsky, P., Bae, K.Y.: Optimizing map labeling of point features based on an onion peeling approach. J. Spat. Inf. Sci. **2**, 3–28 (2011)
3. Cavallaro, C., Verga, G., Tramontana, E., Muscato, O.: Multi-agent architecture for point of interest detection and recommendation. In: CEUR Workshop Proceedings, vol. 2404, pp. 98–104 (2019)
4. Cavallaro, C., Verga, G., Tramontana, E., Muscato, O.: Eliciting cities points of interest from people movements and suggesting effective itineraries. Intelligenza Artificiale **14**(1), 75–87 (2020). https://doi.org/10.3233/IA-190040
5. Cavallaro, C., Vizzari, G.: A novel spatial–temporal analysis approach to pedestrian groups detection. Proc. Comput. Sci. **207**, 2364–2373 (2022). https://doi.org/10.1016/j.procs.2022.09.295
6. Cutello, V., Fargetta, G., Pavone, M., Scollo, R.A.: Optimization algorithms for detection of social interactions. Algorithms **13**(6), 139 (2020)
7. Formann, M., Wagner, F.: A packing problem with applications to lettering of maps. In: Proceedings of the Seventh Annual Symposium on Computational Geometry, pp. 281–288. SCG 1991, ACM Press, New York, NY, USA (1991)
8. Harrie, L., Stigmar, H., Koivula, T., Lehto, L.: An algorithm for icon labelling on a real-time map. In: Developments in Spatial Data Handling, pp. 493–507. Springer, Heidelberg (2005). https://doi.org/10.1007/3-540-26772-7_38
9. Kakoulis, K.G., Tollis, I.G.: On the complexity of the edge label placement problem. Comput. Geom. **18**(1), 1–17 (2001)
10. Lhuillier, A., van Garderen, M., Weiskopf, D.: Density-based label placement. Vis. Comput. **35**(6–8), 1041–1052 (2019)
11. Lu, F., Deng, J., Li, S., Deng, H.: A hybrid of differential evolution and genetic algorithm for the multiple geographical feature label placement problem. ISPRS Int. J. Geo-Inf. **8**(5), 237 (2019)
12. Openstreetmap. Planet dump (2017). https://planet.osm.org/
13. Schrijver, A.: Combinatorial Optimization - Polyhedra and Efficiency. Springer, Heidelberg (2003)
14. Stracquadanio, G., Greco, O., Conca, P., Cutello, V., Pavone, M., Nicosia, G.: Packing equal disks in a unit square: an immunological optimization approach. In: 2015 International Workshop on Artificial Immune Systems (AIS), pp. 1–5 (2015)
15. Wagner, F., Wolff, A.: A combinatorial framework for map labeling. In: Whitesides, S.H. (ed.) GD 1998. LNCS, vol. 1547, pp. 316–331. Springer, Heidelberg (1998). https://doi.org/10.1007/3-540-37623-2_24
16. Xie, M.K., Huang, S.J.: Partial label learning with noisy label identification. IEEE Trans. Pattern Anal. Mach. Intell. **44**(7), 3676–3687 (2022)

17. Zito, F., Cutello, V., Pavone, M.: Deep learning and metaheuristic for multivariate time-series forecasting. In: García Bringas, P., et al. (eds.) SOCO 2023. LNNS, vol. 749, pp. 249–258. Springer, Cham (2023). https://doi.org/10.1007/978-3-031-42529-5_24
18. Zito, F., Cutello, V., Pavone, M.: A machine learning approach to simulate gene expression and infer gene regulatory networks. Entropy **25**(8), 1214 (2023). https://doi.org/10.3390/e25081214

# How to Turn a Leaky Learner into a Sealed One

Ch. Zaugg[1,3](✉), R. Ingold[3], R. M. Füchslin[1,2], and A. Fischer[3,4]

[1] ZHAW, ACSS, Technikumstrasse 9, 8400 Winterthur, Switzerland
{christoph.zaugg,rudolf.fuechslin}@zhaw.ch
[2] European Centre for Living Technology, Ca' Bottacin, Dorsoduro 3911, Calle Crosera, 30123 Venice, Italy
[3] Department of Informatics, DIVA, Fribourg University, Boulevard de Pérolles 90, 1700 Fribourg, Switzerland
{christoph.zaugg,rolf.ingold,andreas.fischer}@unifr.ch
[4] iCoSys, HES-SO, Boulevard de Pérolles 80, 1700 Fribourg, Switzerland

**Abstract.** The outstanding performance of deep neural networks often comes at the expense of a lack of explainability of the results. In this paper, we investigate mutual information between network layers as an information-theoretic means to understand the learning process better. When considering network layers as high-dimensional continuous random variables, the computation of mutual information is a challenging problem. We focus on an approximation method provided by Rényi's matrix-based entropy functional and evaluate it in a classification task with a multi-layer perceptron. We validate the approximation by checking the data processing inequalities. Furthermore, we use mutual information to detect data leaks, i.e., a loss of information between layers. Sealing such leaks at the beginning of training improves the network's performance in our classification experiments.

**Keywords:** neural network · Rényi's matrix-based entropy functional · data processing inequality · information bottleneck principle · minimal sufficient statistic

## 1 Introduction

More and more deep neural networks (DNNs) of various architectures solve tasks equally well or even better than humans. In [3], Huber et al. compare the performance of humans of four age classes versus state-of-the-art DNNs regarding object recognition tasks with distorted images. The authors estimate that the number of images humans refer to in training is three orders of magnitudes less than in the case of the DNNs examined. How do DNNs learn, and why is human learning more data-efficient than machine learning? We present a tool to observe the learning dynamics of a network in training. Instead of a state-of-the-art DNN, we restrict to a multilayer perceptron (MLP) with an architecture

Supported by organization Swiss Federal Office of Civil Aviation FOCA.


M. Villani et al. (Eds.): WIVACE 2023, CCIS 1977, pp. 29–40, 2024.
https://doi.org/10.1007/978-3-031-57430-6_3

similar to Tishby and Schwartz introduced in [6]. These authors carefully design discrete synthetical data and track mutual information (MI) through their network. We replace the discrete and synthetical data of Tishby and Schwartz with limited training, validation, and test sets we assemble from the MNIST data. We regard each layer as a continuous random variable in computing the MI values between all network layers. The dimensionalities are much higher than in Tishby and Schwartz's example, and Rényi's matrix-based entropy functional, introduced by Giraldo et al. in [2], will provide an excellent approximation.

Relying on an approximation method requires an assessment of its quality. The fact that an MLP can not recover information in a higher layer once lost in a lower layer gives rise to a data processing inequality (DPI). We will check all possible DPIs and adjust the approximated MI values such that the approximation of the entropy in the target is close to the entropy in the sense of Shannon. Displaying epoch-wise the MI values with the cross-entropy objective illustrates the learning dynamics of our MLP that we can interpret in a consistent information-theoretic environment due to the validity of the DPIs. We quantify the leaks in the hidden layers between the information available for training and the information our MLP captures epoch-wise in training, and the information bottleneck (IB) principle seals the leaks between the target and the hidden layers.

In [8], the authors promote the same approximation method to quantify the non-linear projection spaces' behavior in training an autoencoder. Due to the mirror symmetry at the bottleneck layer, they established two types of DPIs. The first compares the MI values between the input and the last hidden layer in the encoder and has a dual version in the decoder. The second compares MI values between symmetric layers of the en- and decoder. To compensate for the lack of symmetry in our MLP, we check the validity of all possible DPIs, including the one involving the target.

We organize the paper as follows. We first motivate and summarize both methods on which we base our study, i.e., Rényi's matrix-based entropy functional, the IB principle, and its role in providing approximately minimal sufficient statistics. Next, we describe the experiments, i.e., the data sets, the model, and the initialization schemes that mimic the leaky and sealed learner. Then, we present the results, comprising the information-theoretic consistency by fulfilling the DPIs, sealing the leaks between the target and the hidden layers, propagating the sealing to other layers, and comparing the learning dynamics between the leaky and the sealed learner. In the last section, we discuss our work, i.e., we point out numerical issues and asses our result as a starting point for further investigation.

## 2 Methods

### 2.1 Motivation

Our overall goal is to display the learning dynamics of an MLP at the beginning of its training in a consistent information-theoretical environment. Since our layers are continuous random variables, the computation of the MI between its layers is a non-trivial task, i.e., we have to rely on an approximation method. We decided to validate Rényi's matrix-based entropy functional presented by Giraldo et al. in [2] because it offers two striking features. Firstly, it avoids estimating the data's probability distribution function (PDF), and secondly, the convergence does not depend on the dimensionality of the random variables. However, the convergence rate might depend on it. Hence, we must check that the convergence is uniform in the range of dimensionalities we face in our MLP cf. Fig. 1. In the rest of the paper, we will refer to the authors' method in [2] by the term "approximation."

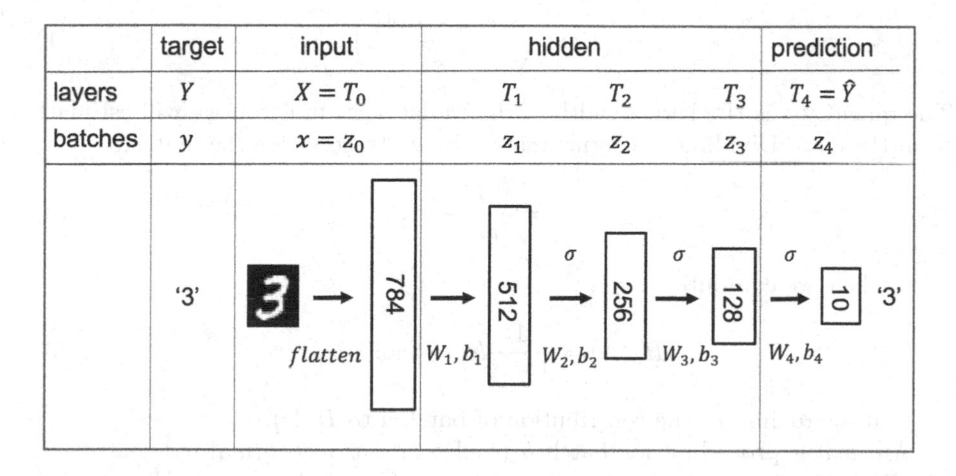


**Fig. 1.** The input is a picture with (28 × 28) pixels of the figure '3'. It is flattened to a vector of 784 components and passed through the layers (number of neurons indicated), followed by a sigmoid activation function. The prediction layer contains the pseudo-probabilities (softmax), and the prediction corresponds to the label of the highest value.

Using the IB principle between the target and the hidden layers not only seals the desired leaks. It also turns the hidden layers into maximally compressed input representations under the constraint of sealing the leaks. The principle allows the network to focus on relevant features present in the data while ignoring noise or features built into its architecture.

### 2.2 Rényi's Matrix-Based Entropy Functional

Alfred Rényi generalizes in [5] the axioms that determine Shannon's entropy and presents a continuous family of real-valued functions that generate entropy-like quantities converging to Shannon's entropy as the parameter $\alpha$ of the family tends to one, cf. [4], Sect. 2.2. The authors of [2] turn Rényi's generalization into a matrix-based functional that assigns approximation values to entropy, joint entropy, and mutual information in the case of high dimensional continuous random variables.

We represent the target layer as $Y$, the input layer as $X = T_0$, the three hidden layers as $T_i$ for $i = 1, 2, 3$, and the prediction layer as $T_4 = \hat{Y}$ with corresponding batches $y$, $x = z_0$, $z_i$ for $i = 1, 2, 3$ and $z_4$, cf. Fig. 1 for the notation of layers and corresponding batches. Next, we outline how to compute the approximations in a loop over pairs of input-target batches. We exemplify the procedure by approximating the MI $I(Y;T_2)$. We set $a = y$ and $b = z_2$ to avoid a cluttered notation. We denote the i-th row vector of $a$ by $a_i$ and the square of its Euclidean norm by $\|a_i\|^2$. From $a$, we compute the kernel matrix:

$$K_{ij} = \exp\left(-\frac{\|a_i - a_j\|^2}{s^2}\right) \tag{1}$$

The quantity $s$ is the kernel width of the radial basis function we will estimate from the data. Dividing the kernel matrix by its trace yields the matrix:

$$A = \frac{K}{tr(K)} \tag{2}$$

from where we compute:

$$H_\alpha(A) = \frac{1}{1-\alpha} log_2\left(tr(A^\alpha)\right) \tag{3}$$

that approximates the contribution of batch $a$ to $H(Y)$.

An analog procedure for batch $b$ results in a trace normalized matrix $B$. Finally, we normalize the matrices $A$ and $B$ such that the uniform distribution results on their diagonal. The last procedure ensures the DPIs, cf. [2], formula (34). With the Hadamard product, $A \circ B$, the contributions to the joint entropy $H(A, B)$ and the mutual information $I(A; B)$ become:

$$H(A, B) = H_\alpha\left(\frac{A \circ B}{tr(A \circ B)}\right) \tag{4}$$

$$I(A; B) = H_\alpha(A) + H_\alpha(B) - H_\alpha(A, B) \tag{5}$$

Averaging the contributions to $I(A; B)$ over the number of loop passes yields the approximation of $I(Y; T_2)$. Replacing $a$ and $b$ by any two different batches displayed in Fig. 1, we get the collection of 14 possible approximations of MI values.

### 2.3 Information Bottleneck Principle

For given random variables $Y$ and $X$, the IB principle provides a maximally compressed representation $\hat{X}$ of $X$ that is as informative about $Y$ as possible.

The IB principle minimizes $I(X;\hat{X})$ under the constraint that $I(Y;\hat{X})$ stays above a threshold. Since the random variables $Y$, $X$, and $\hat{X}$, given in that order, form a Markov chain, the following DPI is valid:

$$I(Y;X) \geq I(Y;\hat{X}) \tag{6}$$

In the case of equality, $\hat{X}$ is a sufficient statistic of $X$ w.r.t. $Y$, and if in addition $I(X;\hat{X})$ is minimal, $\hat{X}$ is a minimal sufficient statistic of $X$ w.r.t. $Y$. In [7], Tishby and Zaslavsky apply the IB principle to approximate a minimal sufficient statistic by minimizing the Lagrangian:

$$\mathcal{L} = I(X;\hat{X}) - \beta \cdot I(Y;\hat{X}) \tag{7}$$

The positive Lagrange multiplier $\beta$ governs the bottleneck tradeoff between the complexity $I(X;\hat{X})$ and the predictivity $I(Y;\hat{X})$, i.e., the amount of information about the target $Y$ preserved in the compression $\hat{X}$. To get the sealed learner, we will sequentially turn each hidden layer of our MLP into an approximately minimal sufficient statistic of its previous layer w.r.t. the target by minimizing suitably modified Lagrangians on the validation data. The procedure ensures that the MLP does not lose information about the target in the hidden layers.

## 3 Experiments

### 3.1 Data

We assemble stratified training, validation, and test sets of sizes 1000, 1000, and 200 from the MNIST data. The examples are handwritten digits from zip codes containing labels. To set the stage, Fig. 2(a) presents a sample of 100 input examples, and Fig. 2(b) shows the details of a specific input example, i.e., an array of shape $(28, 28)$ whose entries are greyscale values ranging from 0 to 255. The number 0 corresponds to the color black, and the number 255 encodes white. Since the approximation $H_\alpha(Y)$ tends to Shannon's entropy $H(Y)$ as $\alpha$ tends to one, we chose the hyper-parameter $\alpha = 1.01$. We must be careful since we divide by $1 - \alpha$ in the formula (3). A value of $\alpha$ too close to one results in a division by a small number. With batch size 128, we can estimate the kernel width $s_{est}$ from the validation data such that the relative error between the approximation $H_\alpha(Y)$ and the Shannon entropy $H(Y)$ computed from the validation data is one percent in absolute value.

### 3.2 Model

Our model is a fully connected MLP, as shown in cf. Fig. 1. We flatten the input to obtain a vector of 784 components and pass it through the layers by multiplying with the associated connectivity matrix containing the weights $W_i$, adding a bias vector $b_i$, for $i = 1, 2, 3$ and applying the sigmoid activation function $\sigma$. The parameters $W_4$ and $b_4$ govern the prediction layer that assigns the label with the highest pseudo-probability.

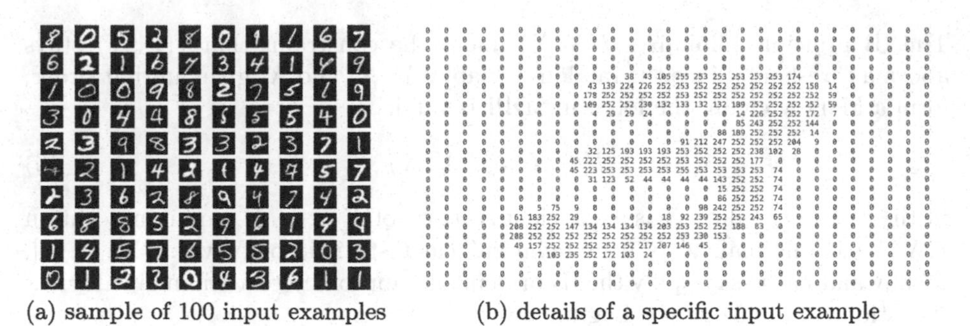

(a) sample of 100 input examples (b) details of a specific input example

**Fig. 2.** MNIST: handwritten digits from zip codes

### 3.3 Initialization Schemes

We model the leaky and the sealed learner with two initialization schemes for our MLP. Scheme $S_1$ is the leaky learner we randomly initialize. To obtain scheme $S_2$ of the sealed learner we modify formula (7) for $i = 1, 2, 3$:

$$\mathcal{L}_i = I(T_{i-1}; T_i) - \beta \cdot I(Y; T_i) \tag{8}$$

We first minimize $\mathcal{L}_1$ and freeze the network parameters up to $T_1$, then we minimize $\mathcal{L}_2$ and freeze the network parameters up to $T_2$. Finally, we minimize $\mathcal{L}_3$ and randomly initialize the network parameters governing the prediction layer. We evaluate the ratio between successive mutual information values to assess how close to equality we come, cf. the four biggest terms in equation (10). E.g., to illustrate how close $I(Y; T_2)$ comes to $I(Y; T_1)$ we compute the ratio $r_2$:

$$r_2 = \frac{I(Y; T_2)}{I(Y; T_1)} \leq 1 \tag{9}$$

Figure 3 shows the sensitivity of $r_2$ in the bottleneck trade-off parameter $\beta$. The choice $\beta = 3.5$ yields $r_2 = 0.99$ after performing 200 iterations to minimize the corresponding Lagrangian on the validation set. We train both learners with the cross-entropy objective.

## 4 Results

### 4.1 Consistency

We aim to check the totality of the possible DPIs arranged in four chains. The first chain starts with the target layer $Y$, the second with the input layer $X = T_0$, the third with the first hidden layer $T_1$ and the fourth with the second hidden layer $T_2$:

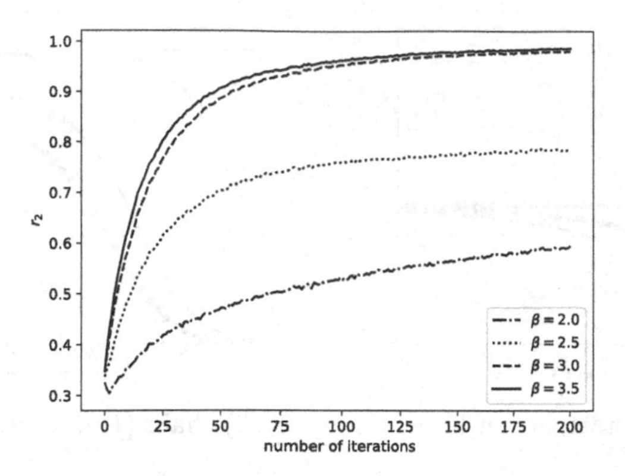


**Fig. 3.** Sensitivity of $r_2$ in $\beta$: the closer its value is to one the less information about the target is lost in passing from the first to the second hidden layer.

$$I(Y;T_0) \geq I(Y;T_1) \geq I(Y;T_2) \geq I(Y;T_3) \geq I(Y;T_4) \tag{10}$$

$$I(T_0;T_1) \geq I(T_0;T_2) \geq I(T_0;T_3) \geq I(T_0;T_4) \tag{11}$$

$$I(T_1;T_2) \geq I(T_1;T_3) \geq I(T_1;T_4) \tag{12}$$

$$I(T_2;T_3) \geq I(T_2;T_4) \tag{13}$$

Figure 4 and Fig. 5 depict the learning dynamics of the leaky and the sealed learner during fifty training epochs, i.e., the curves corresponding to the MI values in the chains (10), (11), (12) and (13). From the arrangement of the curves in both figures, we observe that all possible DPIs are valid throughout the training of the learners. In Fig. 5(a) the curves of $I(Y;T_0)$ and $I(Y;T_1)$ almost coincide. Figure 6(a) is a zoom of Fig. 5(a), where we drop the MI between the target and the prediction layer. The figure shows that even if the curves only vary in a band of small width, the approximation respects the DPIs. The fact indicates that the approximation is excellent, although the number of nodes in the layers ranges from 10 to 784. The choice of the kernel width such that $H_\alpha(Y)$ is close to $H(Y)$ justifies the unit *bit* on the vertical axes. We can interpret the learning dynamics in a consistent information-theoretic setting. Our next goal is to analyze the learning dynamics of both learners.

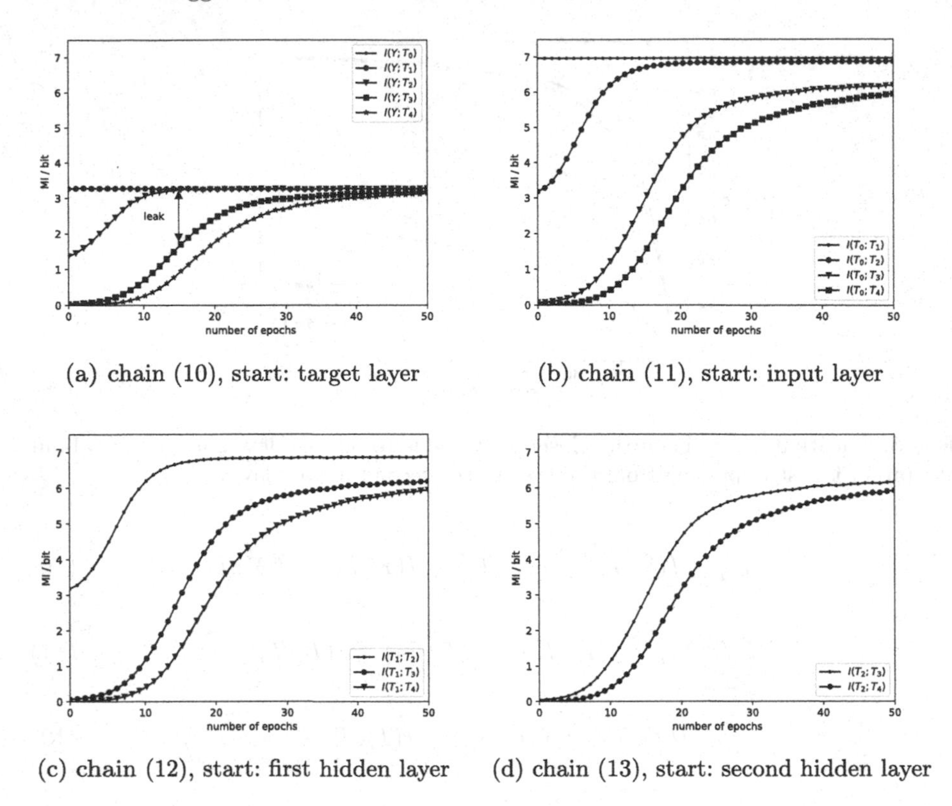


(a) chain (10), start: target layer

(b) chain (11), start: input layer

(c) chain (12), start: first hidden layer

(d) chain (13), start: second hidden layer

**Fig. 4.** Learning dynamics of $S_1$ (leaky learner): captured by all possible DPIs, ordered by the starting layer.

## 4.2 Learning Dynamics of Both Learners

All curves in Fig. 4 monotonically increase until they saturate. Saturation becomes a natural stopping criterion for further training. Figure 4(a) depicts a leak between the target and the second hidden layer. The second hidden layer captures less than 95 percent of the information about the target, i.e., the MLP loses information. Similarly, we can define leaks between the input and its higher layers, between the first hidden and higher layers, and between the second and third hidden layers. Figure 5(a) shows that the optimization procedure affects the learning dynamics as expected, i.e., it seals the leaks between the target and the hidden layers. A computation shows that none of the curves in 5(a) corresponding to $I(Y;T_1)$, $I(Y;T_2)$ and $I(Y;T_3)$ loses more than 5 percent of $I(Y;T_0)$. Although we only seal the leaks between the target and the hidden layers, we observe that the effect propagates to Fig. 5(b) and 5(c). Except for the prediction layer, no curve of these figures falls below 5 percent of the curve's value at the top. In addition, the test accuracy of the sealed learner outperforms the leaky learner after a few training epochs, cf. Fig. 6(b).

# 5 Discussion

## 5.1 Numerical Issues

The kernel matrices corresponding to all batches have the universal shape (128,128). To compare the MI values between all layers of our MLP, we choose a kernel width estimate that is also universal to all kernel matrices. The fact that $H(Y)$ is as close to $H_\alpha(Y)$ as possible guides us. We assess our choice by checking the strict validity of all possible DPIs. To perform the assessment, we have to avoid two numerical issues. Firstly, if the kernel estimate is too small, some kernel matrices may be ill-conditioned. In that case, the computation of the eigenvalues is not possible. Any choice of the kernel width estimate has to take care of it. Secondly, even if the kernel matrices are positive definite symmetric, the algorithm may compute negative real eigenvalues of small absolute values. According to the formula (3), we add the eigenvalues. Neglecting small negative eigenvalues in the sum solves the issue.

## 5.2 Assessment and Further Investigation

Up to 50 training epochs, we can interpret the learning dynamics of a simple MLP performing a classification task in a consistent information-theoretic setting, and we can influence the learning behavior. At about 80 training epochs, the MI curves in Fig. 4 and Fig. 5 begin to cross. To observe the learning dynamics of a fully trained MLP, we need to refine the approximation. We also have to analyze the influence of the learning rate. We trained all our networks with the learning rate $10^{-4}$. Since we scale the MI values such that the approximation of the entropy in the target is closest to Shannon's entropy, we can even compare the dynamics of MLPs that differ in architecture. The result is promising to investigate the influence of architecture on learning dynamics. We also tested our tool by analyzing an MLP Alemi et al. present in [1]. It has three hidden layers of nodes $1024 - 1024 - 256$ with *relu* activation functions. If we train it with the cross-entropy objective on the same training set and compute the top four MI curves of chain (10), Fig. 7(a) shows that it is well sealed, and Fig. 7(b) shows that its training and test accuracies vastly outperform our MLP. The sealing in the chain (10) becomes a tool to assess the performance of competing MLPs. Although we can represent the learning dynamics of ours and Alemi's MLP, the tool does not yet explain the reason that makes Alemi's MLP such an efficient learner.

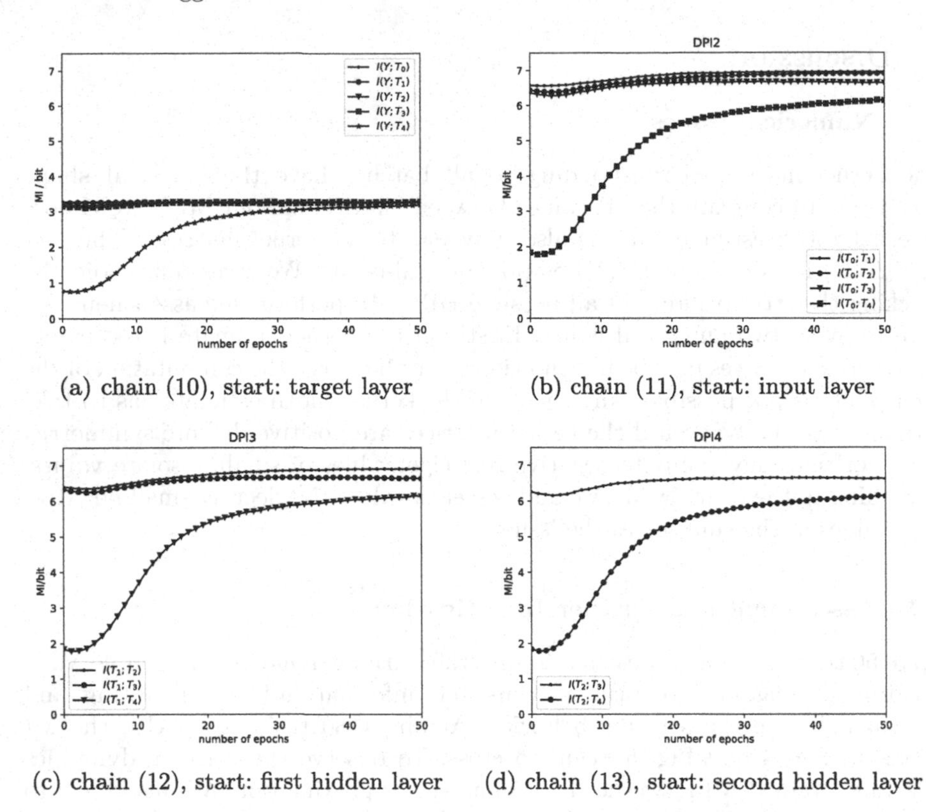


(a) chain (10), start: target layer (b) chain (11), start: input layer

(c) chain (12), start: first hidden layer (d) chain (13), start: second hidden layer

**Fig. 5.** Learning dynamics of $S_2$(sealed learner): captured by all possible DPIs, ordered by the starting layer.

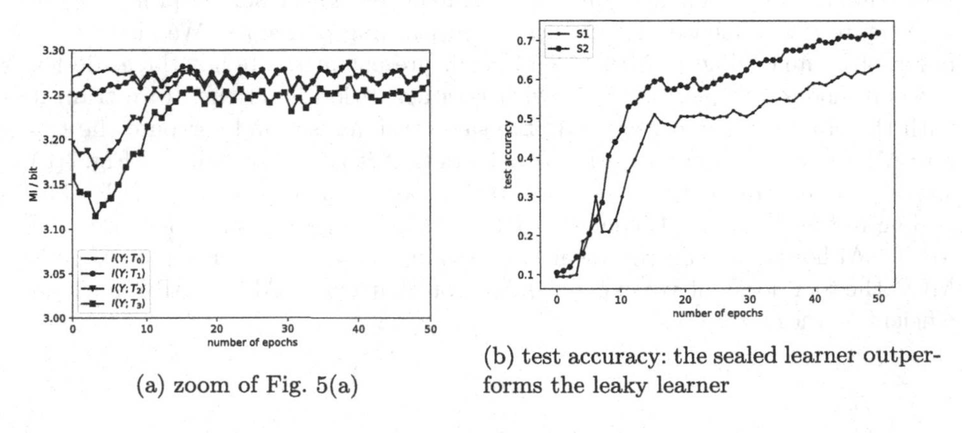


(a) zoom of Fig. 5(a)

(b) test accuracy: the sealed learner outperforms the leaky learner

**Fig. 6.** zoom and comparison of test accuracies

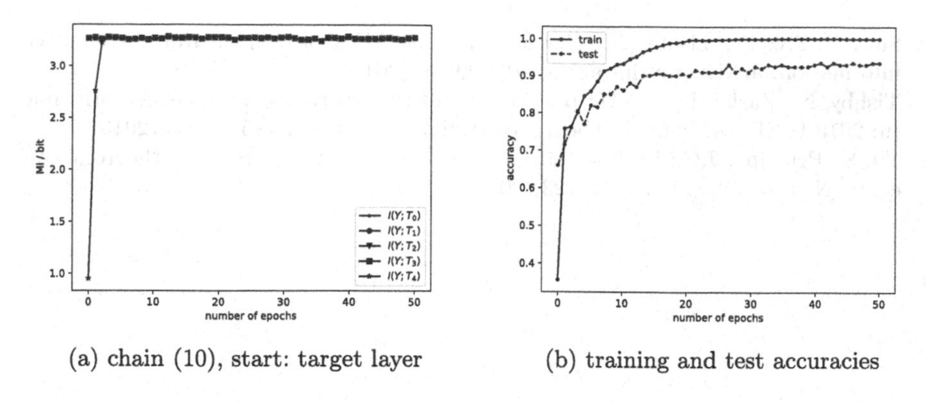


(a) chain (10), start: target layer

(b) training and test accuracies

**Fig. 7.** Alemi's MLP: well sealed and high accuracies

# 6 Conclusion

Overall, Rényi's matrix-based entropy functional has provided a promising approximation for computing mutual information between neural network layers, especially at the beginning of the learning process. For a classification task with a multi-layer perceptron, all data processing inequalities held over 50 epochs, allowing the investigation of the learning dynamics with a sound information-theoretic interpretation. The mutual information allowed us to identify information leaks in the network. Sealing the leaks led to an improved classification accuracy.

In future work, we aim to reduce the approximation error further and address potential numerical problems in later stages of the training process. Another line of investigation focuses on the extension of our experiments to more network architectures and tasks, including, in particular, convolutional and recurrent neural networks.

# References


1. Alemi, A.A., Fischer, I., Dillon, J.V., Murphy, K.: Deep variational information bottleneck. arXiv preprint arXiv:1612.00410 (2016)
2. Giraldo, L.G.S., Rao, M., Principe, J.C.: Measures of entropy from data using infinitely divisible kernels. IEEE Trans. Inf. Theory **61**(1), 535–548 (2014)
3. Huber, L.S., Geirhos, R., Wichmann, F.A.: The developmental trajectory of object recognition robustness: children are like small adults but unlike big deep neural networks. J. Vis. **23**(7), 4–4 (2023)
4. Principe, J.C.: Information Theoretic Learning. ISS, Springer, New York (2010). https://doi.org/10.1007/978-1-4419-1570-2
5. Rényi, A.: On measures of entropy and information. In: Proceedings of the Fourth Berkeley Symposium on Mathematical Statistics and Probability, Volume 1: Contributions to the Theory of Statistics, vol. 4, pp. 547–562. University of California Press (1961)

6. Shwartz-Ziv, R., Tishby, N.: Opening the black box of deep neural networks via information. arXiv preprint arXiv:1703.00810 (2017)
7. Tishby, N., Zaslavsky, N.: Deep learning and the information bottleneck principle. In: 2015 IEEE Information Theory Workshop (ITW), pp. 1–5. IEEE (2015)
8. Yu, S., Principe, J.C.: Understanding autoencoders with information theoretic concepts. Neural Netw. **117**, 104–123 (2019)

# Capturing Emerging Complexity in Lenia

Sanyam Jain[1], Aarati Shrestha[1], and Stefano Nichele[1,2](✉)

[1] Østfold University College, Halden, Norway
{sanyamj,aaratis,stefano.nichele}@hiof.no
[2] Oslo Metropolitan University, Oslo, Norway

**Abstract.** This work investigates the emergent complexity in Lenia, an artificial life platform that simulates ecosystems of digital creatures. Lenia's ecosystem consists of a continuous cellular automaton where simple artificial organisms can move, grow, and reproduce. Measuring long-term complex emerging behavior in Lenia is an open problem. Here we utilize evolutionary computation where Lenia kernels are used as genotypes while keeping other Lenia parameters, such as the growth function, fixed. First, we use Variation over Time as a fitness function where higher variance between the frames is rewarded. Second, we use Auto-encoder based fitness where variation of the list of reconstruction loss for the frames is rewarded. Third, we perform a combined fitness where higher variation of the pixel density of reconstructed frames is rewarded. Finally, after performing several experiments for each fitness function for 500 generations, we select interesting runs for an extended evolutionary time of 2500 generations. Results indicate that the kernel's center of mass increases with a specific set of pixels and the overall complexity measures also increase. We also utilize our evolutionary method initialized from known handcrafted kernels. Overall, this project aims at investigating the potential of Lenia as ecosystem for emergent complexity in open-ended artificial intelligence systems.

**Keywords:** Continuous CA · Lenia · Evolution · Artificial Life · Complexity

## 1 Introduction

Open-endedness is considered an important feature in Artificial Intelligence (AI) [1,2], because it enables the development of more flexible, creative, and autonomous systems that can solve a wider range of tasks. It also facilitates the emergence of unexpected and potentially useful behaviors that may not have been anticipated by human designers. In addition, open-endedness may allow AI systems to continually learn and adapt to changing environments. Lenia is an artificial life [3,4] where simple rules can give rise to complex behavior in a digital system. It is a digital simulation that uses a continuous cellular automaton (CA) to generate complex patterns. Lenia differs from traditional CA in that its cells are not limited to discrete states like "on" or "off". Instead, each cell can


M. Villani et al. (Eds.): WIVACE 2023, CCIS 1977, pp. 41–53, 2024.
https://doi.org/10.1007/978-3-031-57430-6_4

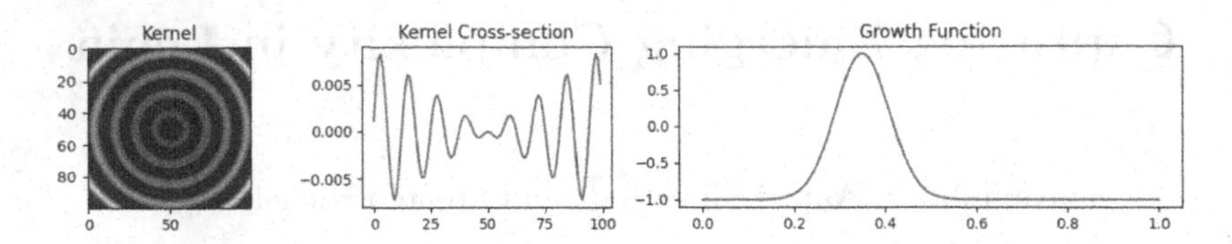


**Fig. 1.** Visualising Kernel, Gaussian Kernel Cross-Section Function and Gaussian Growth Function

take on a continuous range of values, which allows for more fluid and organic patterns to emerge. Lenia is also unique in that it allows for a wide range of parameters to be adjusted, such as the size and shape of the grid, the rules for how cells interact (kernel), and the speed at which the simulation runs (growth function). More in general, cellular automata (CA) have been recently shown to be well suited for AI tasks [13,14]. In [5], a variant of CA called Neural Cellular Automata (NCA) has been used to develop a control system for a cart-pole agent, demonstrating stable behavior over many iterations and exhibiting regeneration and robustness to disruptions. Further, extensions of classical Lenia have been proposed, such as Sensorimotor Lenia [6], Flow Lenia [7], and Energy based Particle Lenia [8], which show advanced physics, chemistry and biology of the matter. In this paper, we investigate the emergent complexity in Lenia using an evolutionary approach.

## 2 Lenia

Lenia [3,4] is a digital life simulation software that utilizes a unique approach to simulate various patterns and behaviors through the use of an abstract concept called the "growth function." The growth function in Lenia represents the process of growth and decay of various shapes and structures, which is governed by two important parameters: $\mu$ and $\sigma$.

The $\mu$ parameter in Lenia's growth function is the mean of the function, which determines the overall behavior of the shapes in the simulation. The $\sigma$ parameter, on the other hand, represents the standard deviation of the function. It controls the degree of randomness in the growth and decay process. Together, the $\mu$ and $\sigma$ parameters of the growth function enable Lenia to generate a wide range of complex patterns. Over 500 species have been discovered so far [12], with each residing in a unique combination of these parameters.

Similarly, the Gaussian kernel is used for smoothing and blurring the simulation board, and it is a circular matrix that is centered around the current cell. The number of smooth rings or "shells" in the Gaussian kernel is determined by the "peaks" parameter, which specifies the amplitude of the peaks for each shell. A detailed view of kernel and growth function can be seen in Fig. 1. For more details on Lenia, please refer to [3,4].

# 3 Methodology

We propose modified versions of two known approaches called compression based metric and deviation based approach to evaluate cellular automata (CA) [11]. Previously these approaches have been studied on classical CA [9,10], we propose novel modified approaches namely, AutoEncoder (AE) based compression metric, Variation over Time (VoT) based deviation metric and a completely novel technique called as Auto Encoder Variation Over Time (AEVoT) which is built on top of known AE and VoT approaches. We describe each of them as follows:

## 3.1 Variation over Time (VoT) Fitness

One way to quantify complexity in Lenia is to measure the variation in the state of the board over time. One approach is to measure the number of active cells on the board at each time step. The number of active cells can provide a rough estimate of the complexity of the system, as more active cells typically indicate a higher degree of complexity. However, it is important to note that the number of active cells can vary significantly between different time steps, and may not be a reliable measure of complexity in all cases. A threshold (a hyper-parameter) is used to determine the alive state of the grid. For pixel values smaller than threshold, cells are considered dead. Count from each grid timestep is used to calculate the standard deviation. Deviation is a measure of the randomness or differences in states of a system, and can be used to quantify the variation of complexity in a system over time. The VoT approach is particularly useful in identifying temporal patterns and trends in the system's behavior, which can be indicative of its underlying complexity. An overview is shown in Fig. 2.

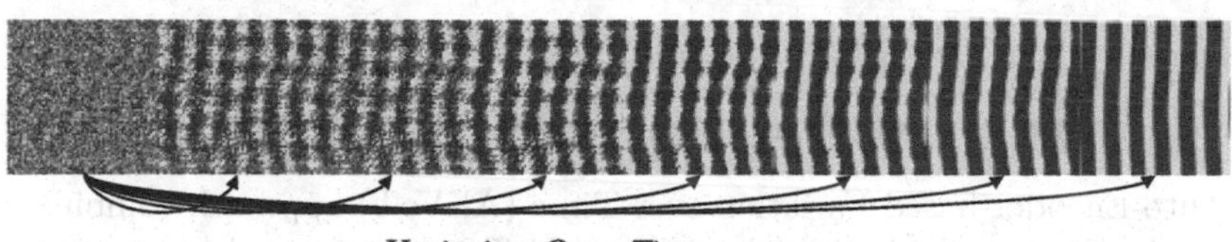


**Fig. 2.** Variation over time approach to measure complexity, where Lenia states (indicated by arrows) change over time

## 3.2 Autoencoder (AE) Based Fitness

Another approach, which uses an encoder-decoder network to compress the input images, has been proposed in [11] as a way to measure the complexity of CA. In this approach, the encoder network is trained to encode the current state of the Lenia board into a lower-dimensional feature space, while the decoder network is trained to decode the features back to the original state. The key idea is that the amount of information that is lost during the encoding and decoding process (reconstruction loss) can be used as a measure of the complexity of the Lenia board. Our hypothesis is that if the input is complex or chaotic, it is more challenging for the decoder to reconstruct the input (high reconstruction loss), while in the ordered behaviour the reconstruction loss would be small. The input frames are reconstructed using the trained AE and the Mean Squared Error (MSE) loss is stored. The MSE is calculated between original frames and reconstructed frames. After that, the standard deviation is calculated between those errors. If the list of errors has a higher deviation, then there should be a larger variance between the frames generated using the evolved kernel. To train the encoder-decoder network, a dataset of Lenia board states is first generated. The encoder network is then trained to map each board state to a set of features that capture the key patterns and structures in the board. The decoder network is trained to reconstruct the board state from the features, with the goal of minimizing the reconstruction error. An overview is shown in Fig. 3.

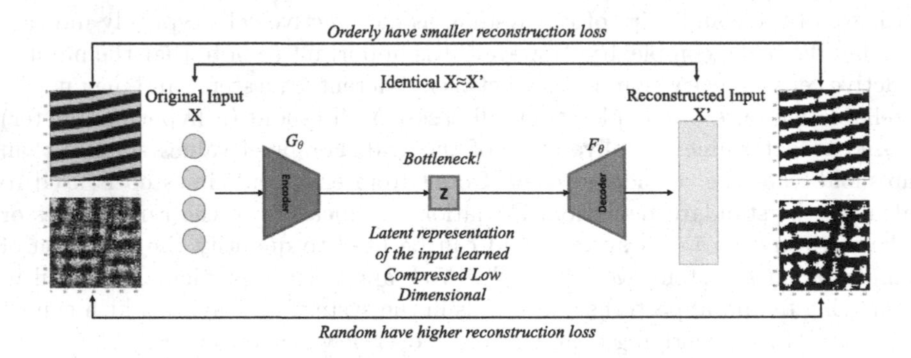


**Fig. 3.** Auto-encoder approach to measure complexity

### 3.3 Auto Encoder Variation over Time

The Auto-Encoder based Variation over Time (AEVoT) approach combines the two previously described methods, AE and VoT, to measure the complexity of the Lenia patterns over time on reconstructed frames. It is very similar to the AE and VoT approaches, with a difference that VoT is calculated on reconstructed images from the AE. To use an AE for time series data, we can treat each time step as a separate input and train the AE to reconstruct the original sequence of data. The reconstruction loss, which is the difference between the input and the reconstructed output, can then be used to measure the complexity of the time series data. By analyzing the variation in the reconstruction loss over time, we can gain insights into the complexity of the time series. For example, if the reconstruction loss is relatively constant over time, it may indicate that the time series is relatively simple or predictable. However, if the reconstruction loss varies widely over time, it may indicate that the time series is more complex and difficult to predict.

## 4 Experimental Setup and Results

A high level overview of experimental setup is provided in Fig. 4.

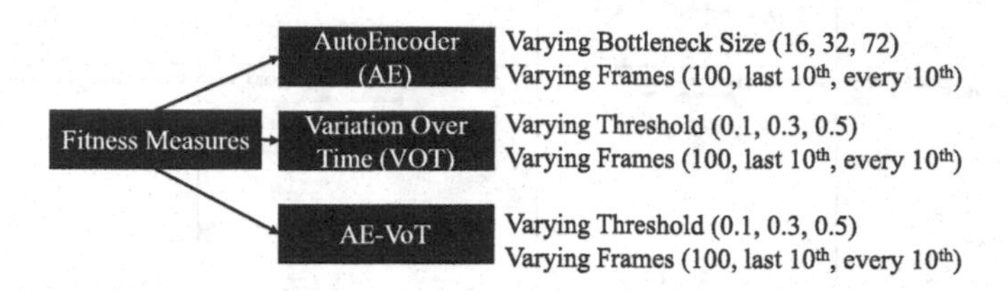


**Fig. 4.** Overview of experiments performed with three different fitness measures and relative parameters

We developed an evolutionary algorithm where Lenia kernels are evolved with the three fitness functions outlined above. The initial population comprises randomly initialized kernels that are applied from a randomly initialized Lenia board. Additionally, the Lenia frames are processed in three different ways, namely "all produced frames for $t$ time steps", "every 10th frame" and "last 10 frames". The idea is to capture long term emerging complexity so the behavior should be present not only at initial frames but in later frames as well.

The selection function used in the code is a Roulette Wheel Selection method, which selects individuals based on their fitness proportionate to the sum of the fitness of all individuals in the population. The number of elites in the population is set to one.

The mutation operator mutates an individual's genes by randomly altering each gene with a probability equal to the *mutation_rate* value. The *mutation_rate* value is set to 0.02, which means that each gene has a 2% chance of being mutated. If the random value generated for a gene is less than *mutation_rate*, the gene is randomly reassigned a value between 0 and 1, rounded to 3 decimal places.

For each of the experiments, a system configuration of 64 GB of RAM, 250 GB Hard Disk space, 8vCPU, in a AWS Sagemaker ML.R5.2xLarge instance is used. It takes almost 24 h to finish one run of 500 generations with our configuration. For the AE training dataset, the Lenia parameters are given in Table 1. The configuration for AE is provided in Table 2 and the overall configuration for the Evolutionary Algorithm is provided in Table 3.

All experimental results for the different combinations of parameters are displayed for 500 generations. In addition best fitness and average fitness are also plotted in the Fig. 5 for VoT, Fig. 6 for AE and Fig. 7 for AEVoT.

Once all the experiments in Figs. 5, 6 and 7 were finished, we analysed and selected one experiment from AEVoT and one from VoT on the basis of their promising performance. We also found that AE itself is not able to perform well as per the experiments shown in Fig. 6. AE alone could not capture the complex of Lenia behaviours because it merely reconstructs the frames and tries

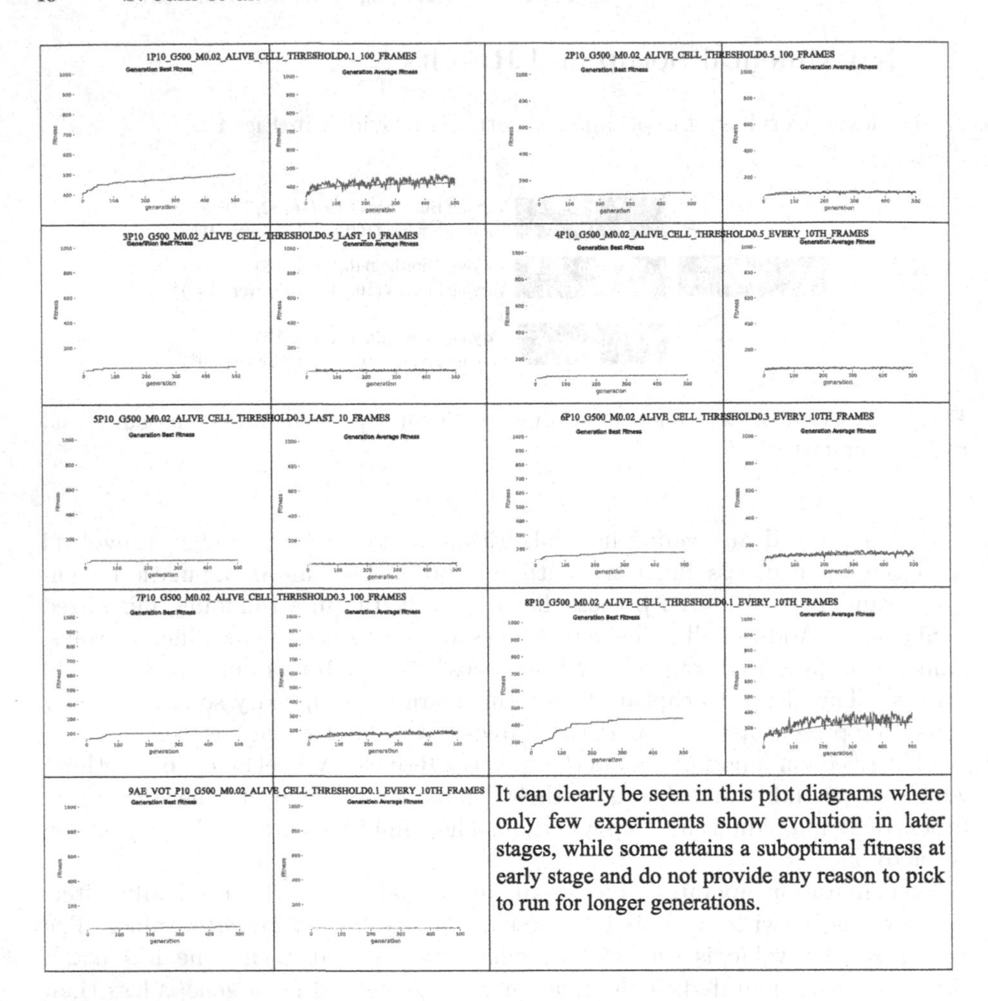


**Fig. 5.** Plot data for Variation over Time fitness. Configuration should be read as for example 1AE_36bottleneck_allframe Autoencoder fitness for bottleneck size of 36 ran for all 100 frames. Y-axis has upper limit of 1000.

to measure MSE loss over the reconstruction. AEVoT achieved the best fitness performance when ran for 2500 generations. We found, for AEVoT, Active cell threshold 0.5 and every $10^{th}$ frame is working better than other configurations (with fitness value of 500.27). Moreover in VOT, Active cell threshold 0.1 for every $10^{th}$ frame is working particularly well (with fitness value of 435.91).

It may be observed that for variation over time, active cell threshold should be small and has to be calculated for a time delta of a significant value over span of total time steps. In our case, we have a total time step of 100 where time delta is 10 and hence fitness is being calculated for every $10^{th}$ frame. Hence, we provide a detailed analysis for the two selected experiments with kernel visuals from generation 1 to generation 2500 along with their long term best and average

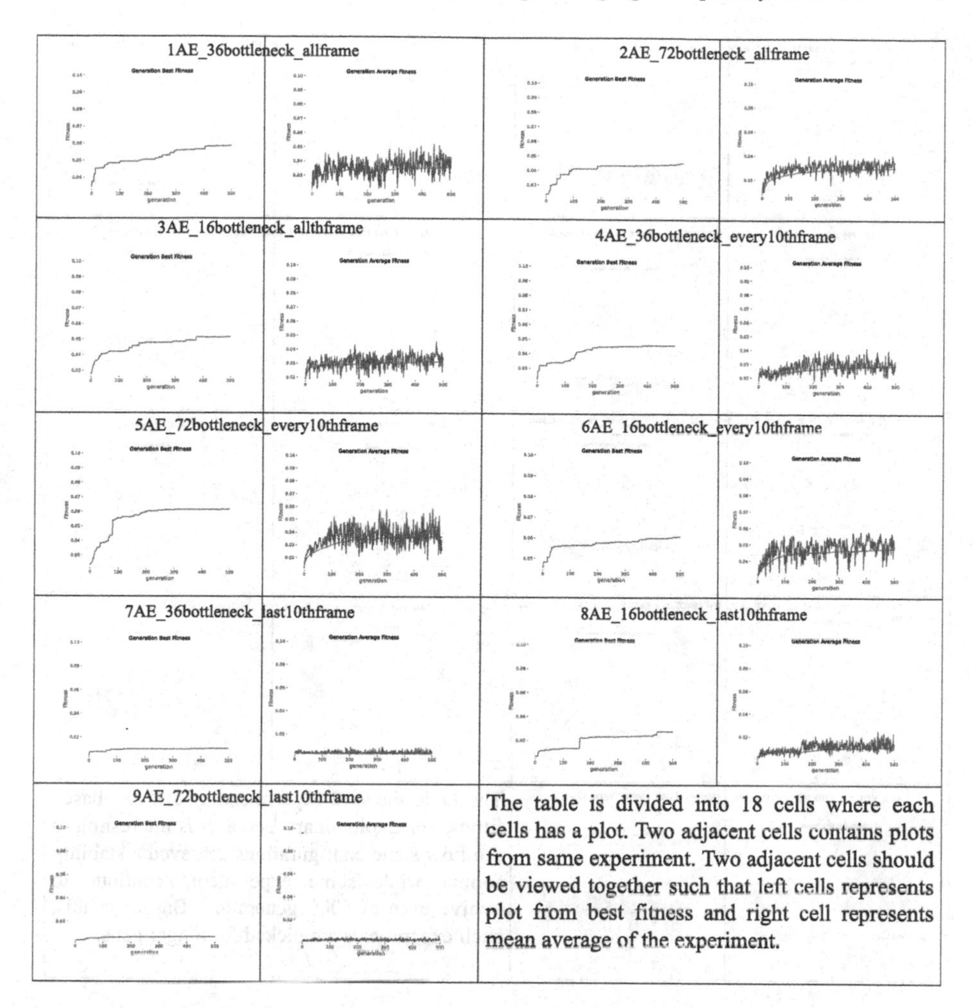


**Fig. 6.** Plot data for AutoEncoder fitness. Configuration should be read as for example 1P10_G500_M0.02_ALIVE_CELL_THRESHOLD 0.1_100_FRAMES should be read as population size (P) as 10 ran for 500 generations (G), with mutation rate (M) of 0.02, active cell threshold 0.1 for 100 frames. Y-axis has upper limit of 0.1.

fitness plots in Fig. 10. It is important to note that after 2500 generations, the kernel centre of mass should have more density of similar pixels (can be seen in "ellowish" color map in the kernel center, however all the dark pixels with dark bluc color are outlined towards boundary in the plot). Finally we plot five-fold-average for best and average fitness values of the same experiments shown in Fig. 10 ran on five different run-times (with randomly initialized initial kernel populations), which is shown in Fig. 8.

To compare our experiments with handcrafted kernels or known kernel, we perform six experiments keeping growth function and the rest of the configu-

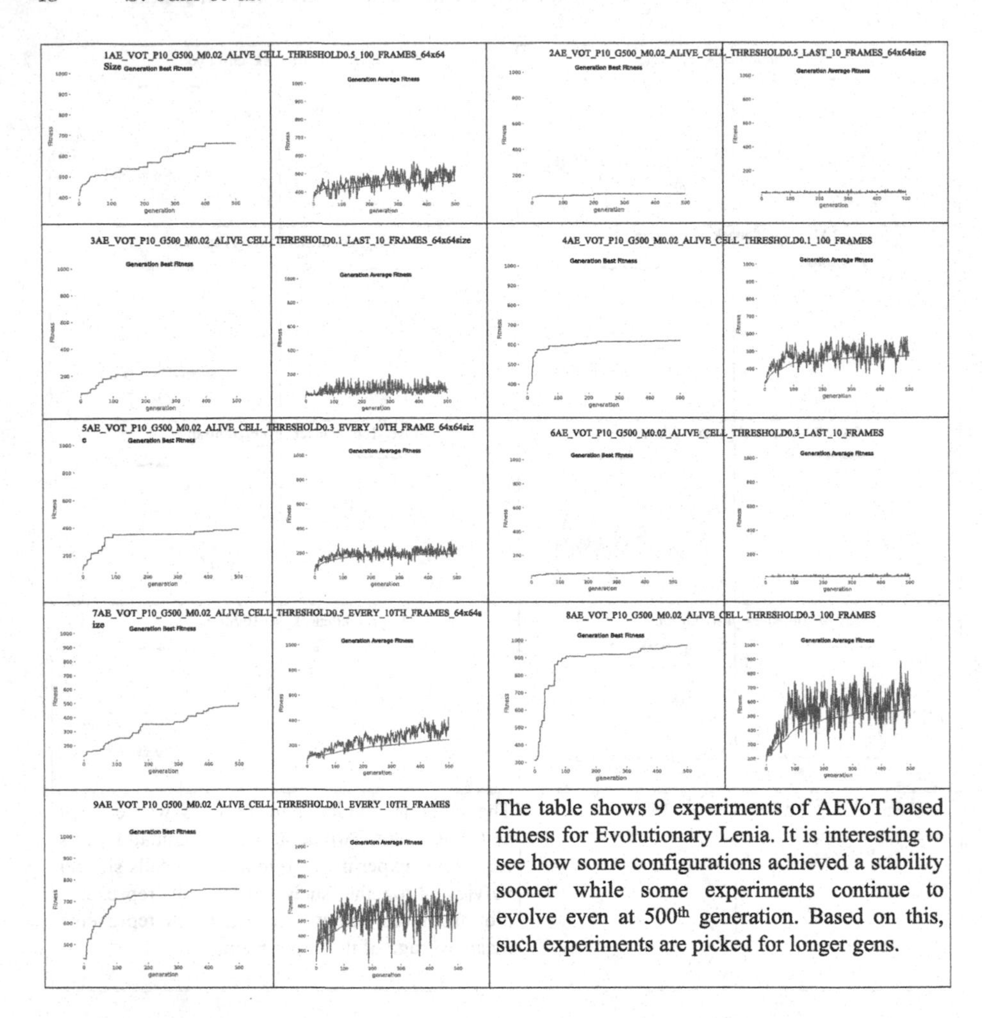


**Fig. 7.** Plot data for AutoEncoder with Variation over Time fitness. Configuration should be read as for example 1AE_VOT_P10_G500_M0.02_ALIVE_CELL_THRESHOLD0.5_100_FRAMES should be read as population size (P) as 10 ran for 500 generations (G), with mutation rate (M) of 0.02, active cell threshold 0.5 for 100 frames with 64 as input image size. Y-axis has upper limit of 1000.

ration the same. Known kernels are very robust to perturbations. Even after doing mutations for these kernels they show similar behaviour and complexity measurement remains high, which can be shown by the generation fitness curve depicted in Fig. 11. At the first place, even after mutations, there are very less changes happened in the fitness measurement value. In other words, we let evolution run for 500 generations and each generation has mutation that affects 2 pixels. However, the growth of fitness happens slowly in those generations, which allows the behavior to remain robust even after perturbations. For example, with

**Table 1.** Dataset Specifications

| Parameter | Value |
|---|---|
| Dataset Size | 3000 frames |
| DPI | 50 |
| $Mu_G$ | 0.31 |
| $Sigma_G$ | 0.057 |
| Dt | 0.1 |
| FPS | 30 |
| Kernel Size | 16 |
| Board Size | 64 |

**Table 2.** Training Configuration for AE

| Parameter | Value |
|---|---|
| Image Size | $64 \times 64$ |
| AE test size | 0.30 |
| AE Input Size | $64 \times 64$ |
| AE Hidden Size | 36 |
| AE Ouptut Size | $64 \times 64$ |
| Epochs | 300 |
| Batch Size | 128 |

**Table 3.** Configuration for Genetic Algorithm

| Parameter | Value |
|---|---|
| Kernel size | 16 |
| Board size | 64 |
| Mutation rate | 0.02 |
| Population size | 10 |
| Generation | 500 |
| Elites | 1 |

the kernel configuration and best plots in the $3^{rd}$ column shown in Fig. 11, the growth has happened only at first few generations and then it became stable very soon. Finally, after handcrafting a kernel with a smoothed spherical Gaussian pixel distribution which is shown in Fig. 9, along with the popular known growth function, shows particularly complex and self-organising behaviour. Metaphorically, it shows a ring like bacteria pattern emerging from a random grid. All the results from such a kernel are shown in Fig. 9. The configuration used here is the same used in Fig. 11.

Full code with results is openly available at this link: https://s4nyam.github.io/evolenia. All experimental visualisations and animations were programmatically processed to produce an overall summary shown on our YouTube channel: www.bit.ly/leniaonyt. The GitHub repository is www.github.com/s4nyam/APCSP.

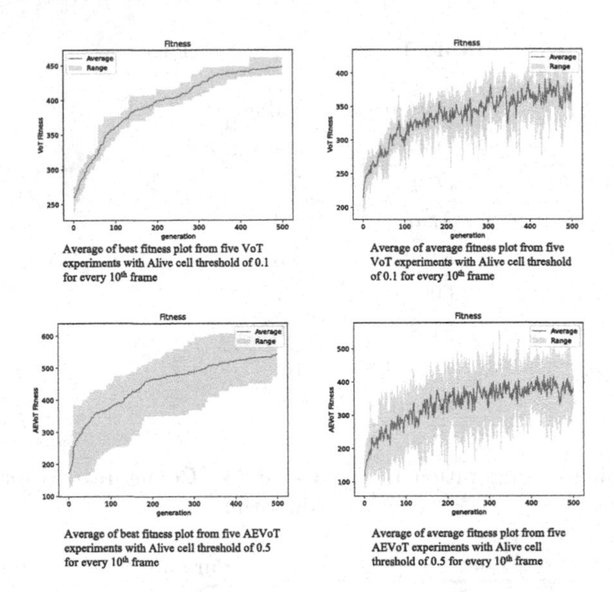


**Fig. 8.** Five-fold-averaging of best and average fitness values of five experiments with same configurations shown in Fig. 10. First row shows VoT run over 500 generations (5 times) using Alive threshold of 0.1 and measuring over every tenth frame. Second row shows AEVoT run over 500 generations (5 times) using Alive threshold of 0.5 and measuring over every tenth frame.

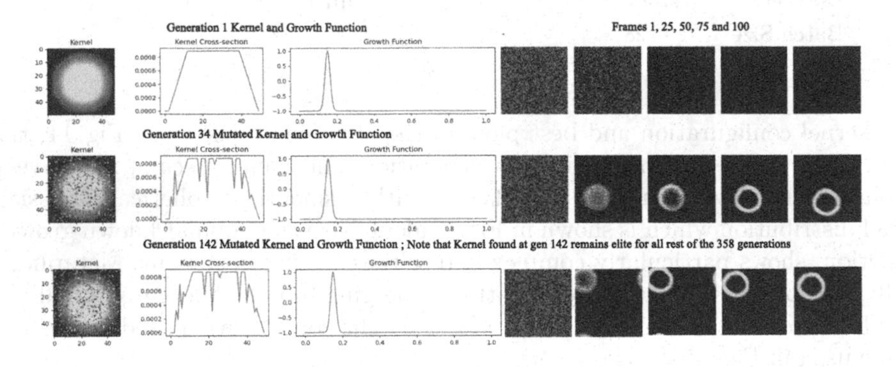


**Fig. 9.** Complex behaviour emerging in Evolutionary Lenia with the known kernel and its mutation for 500 generations

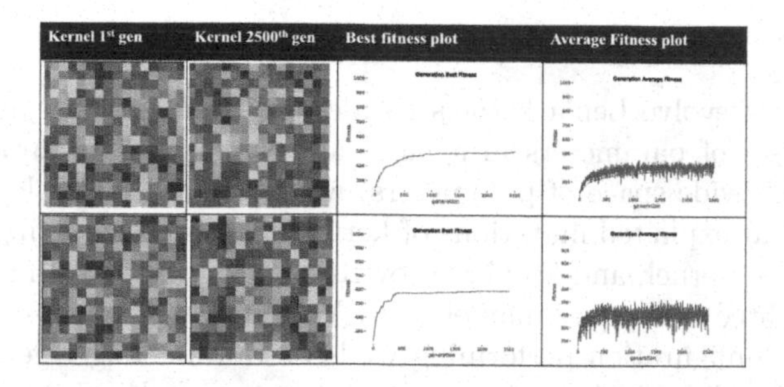


**Fig. 10.** Best and average fitness plots with kernel visualisations for the selected configuration running for 2500 generations. First row is VoT based simulation with configuration picked from Fig. 5 $8^{th}$ row and second row is AEVoT based simulation with configuration picked from Fig. 7 $7^{th}$ row

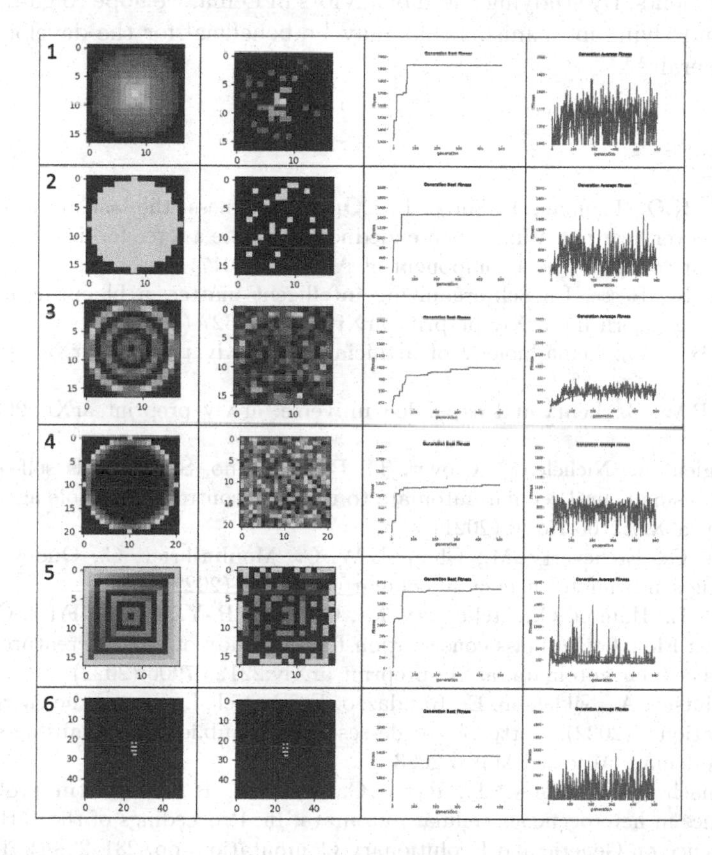


**Fig. 11.** Running 500 generation based simulation for known kernels. The configuration is kept same as the simulation shown in Fig. 10 second row. First column shows the initial known kernel, second column shows mutated kernel after 500 generations. Additionally, third column shows best fitness plot and fourth column shows average fitness plot with cumulative mean-average line in between. (Plots have a y-limit of 2000)

## 5 Conclusions

In this work, we evolve Lenia kernels to identify complex emerging behaviour for a specific set of parameters in a large and wide existing parameter-space. To explore such wide space of parameters, we performed multiple evolutionary experiments that exploited mutations of kernel pixels while starting from a randomly initialised kernel, and keeping growth function fixed. We measured complexity using three techniques, namely AE, VoT and AEVOT. Once we found a specific set of configuration performing well, we ran for longer generations and achieved a rather different behaviour emerging in Lenia when compared with the initial random behaviour. Lenia, as a computational system, provides a platform for exploring emergent behaviors that mimics some of the properties of living systems, such as self-organization, pattern formation, and adaptation to changing environments. By studying these behaviors in Lenia, we hope to gain insights into key underlying mechanisms that may be beneficial for the development of a more general AI.

## References

1. Stanley, K.O., Lehman, J., Soros, L. : Open-endedness: the last grand challenge you've never heard of, While open-endedness could be a force for discovering intelligence, it could also be a component of AI itself (2017)
2. Gregor, K., Besse, F.: Self-organizing intelligent matter: a blueprint for an AI generating algorithm. arXiv preprint arXiv:2101.07627 (2021)
3. Chan, B.W.-C.: Lenia-biology of artificial life. arXiv preprint arXiv:1812.05433 (2018)
4. Chan, B.W.-C.: Lenia and expanded universe. arXiv preprint arXiv:2005.03742 (2020)
5. Variengien, A., Nichele, S., Glover, T., Pontes-Filho, S.: Towards self-organized control: using neural cellular automata to robustly control a cart-pole agent. arXiv preprint arXiv:2106.15240 (2021)
6. Hamon, G., Etcheverry, M., Chan, B.W.-C., Moulin-Frier, C., Oudeyer, P.-Y.: Learning sensorimotor agency in cellular automata (2022)
7. Plantec, E., Hamon, G., Etcheverry, M., Oudeyer, P.-Y., Moulin-Frier, C., Chan, B.W.-C.: Flow Lenia: mass conservation for the study of virtual creatures in continuous cellular automata. arXiv preprint arXiv:2212.07906 (2022)
8. Mordvintsev, A., Niklasson, E., Randazzo, E.: Particle Lenia and the energy-based formulation (2022). https://google-research.github.io/self-organising-systems/particle-lenia/. Accessed Mar 6 2023
9. Medernach, D., Kowaliw, T., Ryan, C., Doursat, R.: Long-term evolutionary dynamics in heterogeneous cellular automata. In: Proceedings of the 15th Annual Conference on Genetic and Evolutionary Computation, pp. 231–238 (2013)
10. Cisneros, H., Sivic, J., Mikolov, T.: Evolving structures in complex systems. In: IEEE Symposium Series on Computational Intelligence (SSCI), pp. 230–237 (2019)
11. Cisneros, H., Sivic, J., Mikolov, T.: Visualizing computation in large-scale cellular automata. arXiv preprint arXiv:2104.01008 (2021)
12. Chan, B.W.C.: Towards Large-Scale Simulations of Open-Ended Evolution in Continuous Cellular Automata. arXiv preprint arXiv:2304.05639 (2023)

13. Nichele, S., Ose, M.B., Risi, S., Tufte, G.: CA-NEAT: evolved compositional pattern producing networks for cellular automata morphogenesis and replication. IEEE Trans. Cogn. Dev. Syst. **10**(3), 687–700 (2017)
14. Mordvintsev, A., Randazzo, E., Niklasson, E., Levin, M.: Growing neural cellular automata. Distill **5**(2), e23 (2020)

# On the Detection of Significant Pairwise Interactions in Complex Systems

Giada Fini[1], Gianluca D'Addese[1], Luca La Rocca[1], and Marco Villani[1,2](✉)

[1] Department of Physics, Informatics and Mathematics, University of Modena and Reggio Emilia, 41121 Modena, Italy
marco.villani@unimore.it
[2] European Centre for Living Technology, 30123 Venice, Italy

**Abstract.** Many systems in nature, society and technology are complex systems, i.e., they are composed of numerous parts that interact in a non-linear way giving rise to positive and negative feedback. The dynamic organization of these systems often allows the emergence of intermediate structures that once formed profoundly influence the system and therefore play a key role in understanding its behavior. In the recent past our group has devised an effective method for identifying groups of interacting variables within a system, based on their observation. The result is a set of entities, each of which connects two or more nodes of the system: this result can therefore be represented by a hypergraph, which can be of considerable use for understanding the system under consideration. In particular, we use an index that allows us to evaluate the level of integration of a group of variables. In order for a group to be identified as significant, the value of this index must exceed a threshold that corresponds (under appropriate hypotheses) to a level of statistical significance decided by the user. In this work we propose a more elaborate approach to determining the significance threshold, which is (i) in itself theoretically interesting and (ii) of considerable practical utility. We use the new approach to determine collections of pairwise relationships in meaningful cases, such as relationships in gene regulatory networks.

**Keywords:** regulatory networks · graph reconstruction · relevance index · dynamic organization · integration

## 1 Introduction

Many systems in nature, society and technology are composed of numerous nonlinearly interacting parts. The dynamic organization of these systems often allows the emergence of intermediate structures that once formed deeply affect the system, and therefore play a key role in understanding its behavior [1, 2].

In order to study complex systems, it is useful (if not essential) to represent them using a unifying mathematical language. In recent decades, researchers focused their attention on two frameworks potentially able to represent the systems under study: graphs and hypergraphs [3]. Indeed, a complex system is essentially a collection of units and their


M. Villani et al. (Eds.): WIVACE 2023, CCIS 1977, pp. 54–64, 2024.
https://doi.org/10.1007/978-3-031-57430-6_5

relations, and therefore we need representations able to mirror this composition. The units of both frameworks are called nodes (or vertices). Graphs include a collection of pairwise relations between nodes (called edges), whereas hypergraphs include a collection of relationships linking an arbitrary number of nodes (hyperedges); graphs are therefore a special case of hypergraphs with links only between two nodes at a time.

In the recent past our group has devised an effective method for the identification of groups of interacting variables within a system, based on their repeated observation [4–8]. The result is a collection of entities, each linking two or more nodes of the system: this result can therefore be represented by a hypergraph, which can be of considerable use in understanding the system under examination. In particular, we use an index (called *zI* in [5, 7]) that allows the evaluation of the integration level of a group of variables. In order for a group to be identified as significant, the value of this index must exceed a threshold $\theta$ corresponding (under appropriate hypotheses) to a level of statistical significance decided by the user. A simple semi-empirical rule, based on the ubiquitous normal distribution, consists in setting this threshold equal to 3.0 (or 5.0 in case of need for greater selectivity).

In this work, we propose a more elaborate approach to the determination of the significance threshold, which is (i) in itself theoretically interesting and (ii) of considerable practical utility. In a nutshell, we go beyond statistical significance and also consider practical significance. We use the new approach for determining collections of pairwise relationships and so, in the following, we will discuss systems representable within the graph framework.

The determination of a threshold for our method (or in general for any kind of methodology) leads to a more interesting discussion concerning the possibility of identifying relationships starting from observational data [9, 10]. In this work we assume that we do not always have time-ordered data series available: we therefore exclude the analysis of causal relationships and focus on relationships identified by measures of correlation or entropies.

## 2 The zI Index

Considering a system composed of $m$ random variables $X_1$, $X_2$, …,$X_m$ we suppose that $S_k$ is a subset composed of $k$ elements, with $0 < k < m$. Each variable $X_i = \{X_i^0, X_i^1, \ldots, X_i^n\}$ is present in $n$ observations. Our purpose is to identify subsets of variables that behave in a somehow coordinated way. To do this we make use of the so-called Integration.

$$I(S_k) = \sum\nolimits_{s \in S_k} H(s) - H(S_k) \tag{1}$$

where $H(S_k)$ denotes the joint entropy of the variables in group $S_k$ and $H(s)$ is the marginal entropy of $X_s$.[1]

The index we use has the form:

$$zI(S_k) = \frac{2nI(S_k) - \langle 2nI(S_k) \rangle}{\sigma(2nI(S_k))} = \frac{2nI(S_k) - d_k}{\sqrt{2d_k}} \tag{2}$$

[1] In the case of pairs of variables, this measure can be declined as mutual information.

where $n$ is the number of observations, and $< 2nI(S_k) >$ and $\sigma(2nI(S_k))$ are, respectively, the average and the standard deviation of $2nI(S_k)$ for a matching homogeneous system (a system of the same size of the system under examination, whose variables are mutually independent). These averages can be effectively approximated through a Chi Square distribution, whose $d_k$ degrees of freedom depend on the size of the subset and on the cardinality of its alphabet [6, 7]. Interestingly, the integration is related to the identification of dynamical criticality in complex systems [11].

The *zI* index has thus the form of a distance from the independence case, measured in standard deviations, whence the semi-empirical rule of using the threshold 3.0 (or 5.0, or even 7.0, in case of need for greater safety) as the lower value [12, 13].

## 3 The New Approach

Given the formula of the zI index, in which there is an explicit dependence on the number of observations involved, a fixed threshold is not completely satisfactory. Typically, more observations allow to identify relationships (of a given strength) at a higher level of statistical significance, or to detect as statistically significant (at a given level) weaker relationships, which would not have been detected with fewer observations. Since a fixed threshold corresponds to a fixed level of statistical significance, the fixed threshold approach detects relationships of decreasing strength as the number of observations increases. However, this could pose a problem, because the practical significance of a relationship is linked to its strength, and we are not really interested in detecting weak interactions, whose identification would only introduce confusion (sometimes to the point of making the reconstructed graph unusable). Furthermore, series of observations from the same system, but having different lengths, could lead to different reconstructions (sometimes a disturbing scenario for a researcher). An approach based on the strength of interesting interactions thus becomes noteworthy.

The proposed approach is based on a simple stochastic model of interaction. In this work, we focus on interaction between two variables: we are interested in identifying collections of connections between two nodes, and therefore in reconstructing graphs (an approach already used in [8]). Let us consider the binary (for simplicity) variables $A$ and $B$ and suppose that: $A$ takes value "1" with probability $P_a$ (value "0" with probability $1-P_a$), then $B$ either copies $A$ with probability $P_c$ or takes value "1" with probability $P_b$ (value "0" with probability $1-P_b$). In this case it is possible to directly calculate the entropy of the pair $(A, B)$ as

$$H(AB) = -(P_{00}ln(P_{00}) + P_{01}ln(P_{01}) + P_{10}ln(P_{10}) + P_{11}ln(P_{11})) \tag{3}$$

Where

$$\begin{cases} P_{00} = (1 - P_a)P_c + (1 - P_a)(1 - P_c)(1 - P_b) \\ P_{01} = (1 - P_a)(1 - P_c)P_b \\ P_{10} = (1 - P_b)(1 - P_c)P_a \\ P_{11} = P_aP_c + P_aP_c(1 - P_c) \end{cases} \tag{4}$$

and similarly calculate $H(A)$ from the marginal probabilities $P_{1+} = P_{10} + P_{11} = P_a$ and $P_{0+} = P_{00} + P_{01} = 1 - P_a$, as well as $H(B)$ from $P_{+1} = P_{01} + P_{11} =$

$P_cP_a + (1 - P_c)P_b$ and $P_{+0} = P_{00} + P_{10} = P_c(1 - P_a) + (1 - P_c)(1 - P_b)$. It is then possible to directly calculate the integration $I(S_{AB}(P_a, P_b, P_c))$ of the pair $(A, B)$ and therefore the value of the $zI$ index in case of $n$ observations.

We consider the above described variables $A$ and $B$ as representative of the kind of interaction we are interested in and, for a suitable (application specific) choice of $P_c$, of the minimum level of interaction that is practically significant, that is, of the weakest relationship that is worth detecting. This leads us to the new threshold:

$$\theta_{zI}(P_a, P_b, P_c) = max\left(3.0, \frac{2nI(S_{AB}(P_a, P_b, P_c)) - d_k}{\sqrt{2d_k}}\right) \tag{5}$$

where $n$ is the number of observations, and the probabilities $P_a$ and $P_b$ can be estimated from data (thus making the threshold specific for the case under examination). Notice that the threshold (5) cannot be less than the aforementioned normal value (in this work equal to 3.0) to avoid confusing noise with signal.[2] The resulting behaviour (when the threshold is above its minimum allowed value) increases linearly with the number of observations, and shows a significant dependence on the values actually assumed by the probabilities $P_a$, $P_b$ and $P_c$ (Fig. 1).

A very similar approach can also be carried out in the case of variables having 3 levels (conventionally, "0", "1" and "2"). The procedure for calculating the threshold in this case is shown in Appendix A.

If we wish to use the network obtained by thresholding pairwise integrations to reconstruct the underlying structure of the system under examination not all the detected relationships are really necessary, or "primitive": a large fraction of these relations is indeed indirect in nature. If variable $A$ affects variable $B$, and if variable $B$ in turn affects variable $C$, any correlation measure will also show a relationship between variable $A$ and variable $C$, even in the absence of a direct causal link.

However, it is known that the evidence of this "epiphenomenal" link will be less than or equal to the evidence of correlation between variable $A$ and variable $B$, or between variable $B$ and variable $C$ (an information theoretic property called Data Processing Inequality—in short, DPI) [14]. It is therefore possible to carry out a pruning operation by eliminating the link with lower evidence in each closed path. In most practical cases, it is enough—and faster—to eliminate the link with lower evidence in each clique of size three, a procedure already adopted in other contexts [8, 14, 15]. In case of tree-like relationship structures, it is shown that this procedure leads to the correct underlying network [15].

[2] It can be noted that the proposed approach is asymmetric with respect to the "roles" of $A$ (copied variable) and $B$ (variable that can copy $A$ with probability $P_c$). Leaving the possible exploitation of this asymmetry to further works, in this paper for each analyzed pair we will test both roles, and we will use the threshold with the highest resulting value.

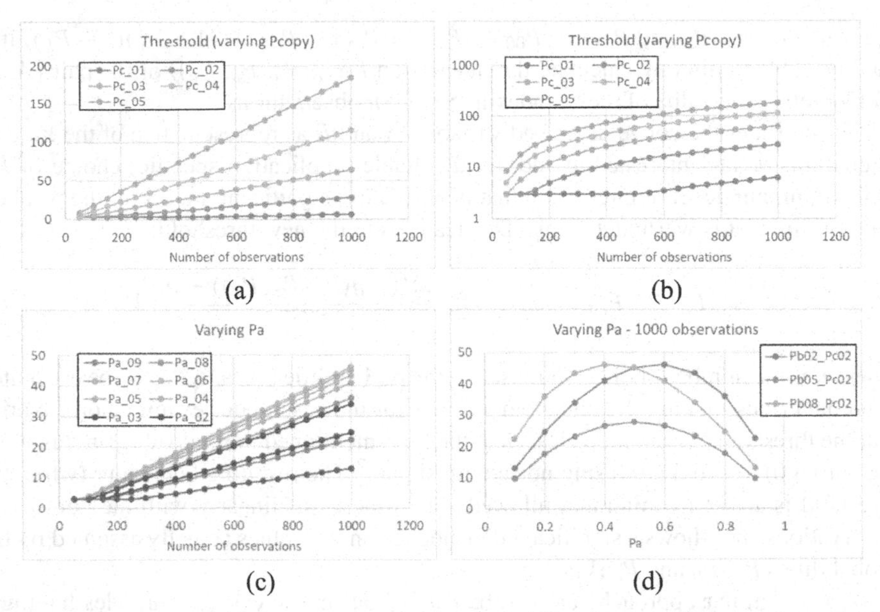


**Fig. 1.** (a) The value of the threshold as the copy probability $P_c$ varies between 0.1 and 0.5, with $P_a = 0.5$ and $P_b = 0.5$. (b) The same as in (a), with the Y axis in logarithmic scale which highlights the behavior of the function in correspondence with the minimum threshold value (in this case equal to 3.0). (c) The value of the threshold as the probability $P_a$ varies (between 0.1 and 0.9), with $P_b = 0.5$ and $P_c = 0.2$. (d) The threshold value as $P_a$ varies, with the number of observations set to 1000 ($P_b = 0.2$ and $P_c = 0.2$).

## 4 Results

We evaluated the accuracy of our approach on two datasets with a known dynamic structure, both already used to evaluate ARACNE [15, 16], an algorithm that is currently one of the most used and performing in the field of genetic regulatory network reconstruction.

Specifically, regarding the analysis of single-cell data [17, 18] we used the BEELINE data [19], which provide a framework for benchmarking algorithms that infer graphs from observational data (in particular, genetic regulatory networks starting from single cell gene expression data). BEELINE provides datasets with various levels of reconstruction difficulty, six different architectures and various motif shapes. The data series have different lengths; for each length and for each topology there are ten different series. In order to analyze the data, we discretized them by using two levels (arbitrarily labelled as "0" and "1") depending on the fact that each individual value is lower or higher than the overall mean of the specific gene in the observations. We used these data to highlight the performance differences between the *zI* algorithm with a fixed threshold (hereinafter set at 3.0) and the zI algorithm with a variable threshold (given in Eq. 5).

We then examined data resulting from the simulation of knock-out experiments (silencing of a single gene, followed by counting which and how many genes consequently changed their expression) [20], presented by [21] as a platform for comparison of reverse engineering algorithms; these are known as Century data. These networks consist of 100 genes and 200 links organized in one of three topologies: Erdös-Rényi (random network – RND in the following) [22], scale-free (SF) [23] or "small worlds" (SW) [24]. In the first topology (RND), each vertex of a graph is equally likely to be connected to any other vertex; in the second one (SF), the distribution of the number of connections associated with each vertex follows a power law and large interactions hubs are present; the third topology (SW) allows a fast diffusion of information, while maintaining a high degree of compactness. Many biological systems appear to exhibit organizations like the last two [25]. In order to analyze the data, we discretized them by using three levels (arbitrarily labelled as "0", "1", and "2") depending on whether the gene has decreased, maintained, or increased its normal activity during the KO, a procedure already used in [6]. ARACNE instead has its own ways to discretize continuous data, so we have directly passed to it the data matrix. As for ARACNE, all the genes are considered transcription factors. The reconstructed networks are dynamic correlation networks, and it is not possible to define a correct reference network (a ground truth). As usual in the field, we therefore use as approximation of the reference the graph formed by the structural links present in the equations of the model originating the data [16, 21].

In order to evaluate the effectiveness of the reconstruction we employed the precision and recall performance metrics [26], comparing the reconstructed graphs and the original ones. In our context, precision and recall are defined as follows: precision = $TP/(TP + FP)$ and recall = $TP/(TP + FN)$, where $TP$ are the true positives (number of correctly inferred true relations), $FP$ are the false positives (number of spurious relations inferred) and $FN$ are the false negatives (number of true relations that are not inferred). The closer both precision and recall are to 1, the better. We remark that biologists are mainly interested in precision, which allows them to be more confident about the acquired knowledge. The introduction of a variable threshold in the new approach has resulted in significant improvements compared to the approach with a fixed threshold. The extent of these improvements varies depending on the underlying architecture of the data. Our results are reported in Fig. 2.

The BEELINE data structure also allows other types of comparison, in particular:

- between the reconstructed graphs (concerning the same topology) starting from different series having the same length;
- between the reconstructed graphs (concerning the same topology) starting from different series having different lengths.

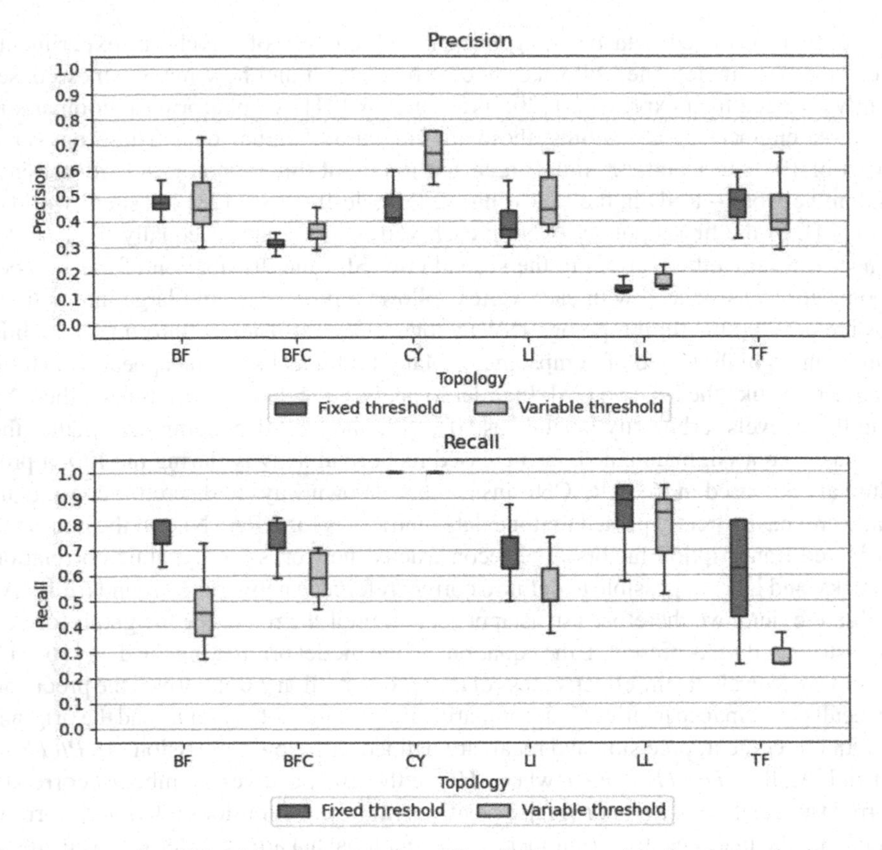


**Fig. 2.** A Boxplot of Precision (upper row) and Recall (lower row) in the six topologies present in BEELINE. In four out of six cases, the approach with variable threshold has significantly better precision (the most relevant index in the cases of interest).

In order to compare two independent graph reconstructions concerning the same system it is interesting to know how many links have been identified in the two reconstructions, and in particular how many of them are present in both. We therefore evaluated the *intersection* between all the possible pairs in the two cases mentioned above (defined as the ratio of the number of common edges to the minimal cardinality of the two), both using the fixed threshold approach and using the new approach with variable threshold. The higher the intersection of a pair of reconstructions, the greater their consistency (and consequently the confidence placed in the reconstruction method).

We measured the intersections before applying the DPI, in order to highlight the only advantage of the variable threshold, compared to the fixed threshold. The intersection values obtained using the variable threshold method are in most cases equal to or greater than those obtained using the fixed threshold method (Fig. 3). It should be highlighted that comparable graphs have been obtained by using data series of significantly different lengths, varying even by an order of magnitude.

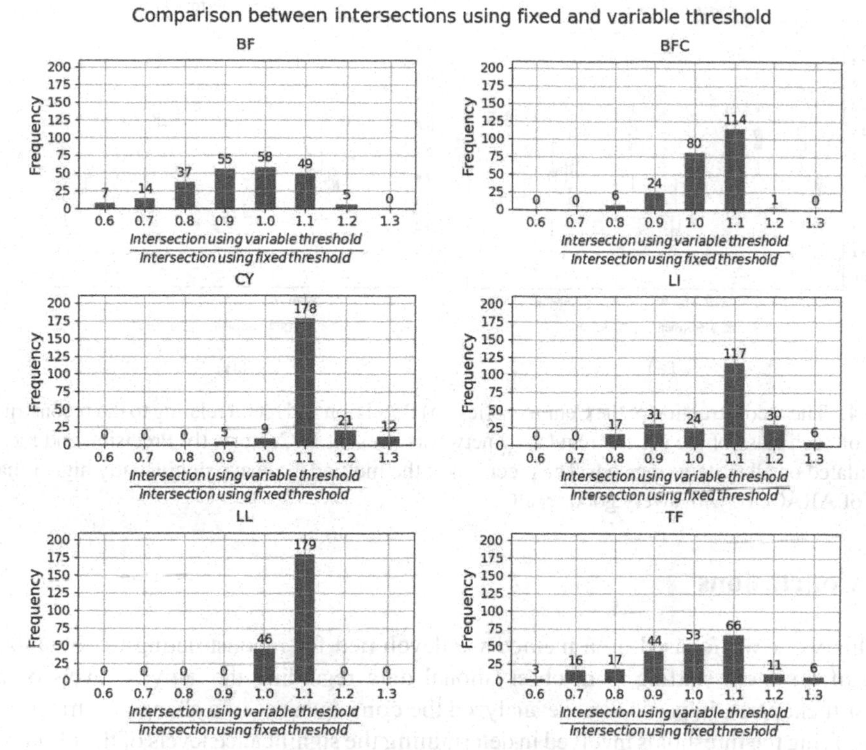


**Fig. 3.** For every possible pair of reconstructions of the same system obtained from different series (it does not matter whether of the same length or not), the graphs show the ratio between the intersection obtained using a variable threshold and the intersection obtained using the fixed threshold. Ratios equal to or greater than 1.0 indicate better intersections by the variable threshold approach. The ratios settle in the majority of cases close to 1 or exceed it, which means that in general the use of a variable threshold keeps the coherence of the algorithm almost unchanged or even better increases it.

Finally, we applied the variable threshold method to the Century data. The reconstruction of the Century series of artificial regulatory networks by our method is very good. In particular, the precision of the method is always significantly higher than that of ARACNE, with a very good recall—in SF and SW system a recall higher than that of ARACNE (Fig. 4).

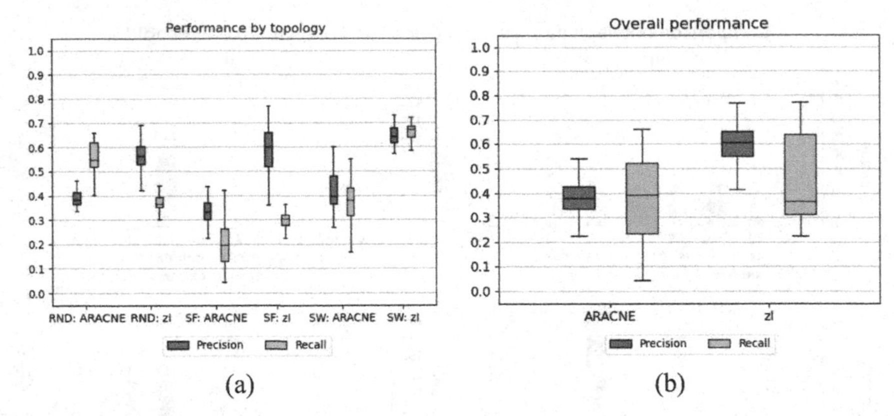


(a) (b)

**Fig. 4.** The reconstruction of the Century series. (a) Precision and recall relating to the reconstruction of each class of the genetic regulatory networks present in Century. (b) Precision and recall calculated on all systems together. The precision of the method is always significantly higher than that of ARACNE, with a very good recall.

## 5 Conclusions

In this work we focused on a method we developed for reconstructing the organization of a system starting from observational data regarding the activity levels of its constituent parts. In particular, we analyzed the consequences of a change in strategy in identifying the thresholds involved in determining the significance levels of the observed correlations.

Interestingly, the considered approach turned out to be stimulating from a theoretical point of view and of substantial practical usefulness. We applied the method to the significant case of reconstructing the organization of genetic regulatory networks, usually represented by collections of couples, that is, by graphs, starting from single-cell data and knock-out experiments.

We are currently using the new method with variable threshold to analyze data from real biological situations, and we are evaluating the extension of the proposed approach to the identification of (hyper) links connecting more than two entities at a time (hypergraphs).

**Funding.** This research was funded by Università degli Studi di Modena e Reggio Emilia (FAR2023 project of the Department of Physics, Informatics and Mathematics). Financial support was provided also by the MUR-PRIN grant 2022 SMNNKY, CUP B53D23009470006.

## Appendix A

The stochastic model presented in the paper can also be used to calculate the threshold in the case of three-level variables. To this aim, let us consider the ternary variables $A$ and $B$ and suppose that $A$ takes value "0" with probability $P_{a0}$, "1" with probability $P_{a1}$ and "2" with probability $P_{a2}$, then $B$ either copies $A$ with probability $P_c$ or takes value "0" with probability $P_{b0}$, "1" with probability $P_{b1}$ and "2" with probability $P_{b2}$.

In this case it is possible to directly calculate the entropy of the pair (A, B) as:

$$H(AB) = -(P_{00}ln(P_{00}) + P_{01}ln(P_{01}) + P_{02}ln(P_{02}) + P_{10}ln(P_{10}) + P_{11}ln(P_{11}) + P_{12}ln(P_{12}) + P_{20}ln(P_{20}) + P_{21}ln(P_{21}) + P_{22}ln(P_{22}))$$

where:

$$\begin{cases} P_{00} = P_{a0}P_c + P_{a0}(1 - P_c)P_{b0} \\ P_{01} = P_{a0}(1 - P_c)P_{b1} \\ P_{02} = P_{a0}(1 - P_c)P_{b2} \\ P_{10} = P_{a1}(1 - P_c)P_{b0} \\ P_{11} = P_{a1}P_c + P_{a1}(1 - P_c)P_{b1} \\ P_{12} = P_{a1}(1 - P_c)P_{b2} \\ P_{20} = P_{a2}(1 - P_c)P_{b0} \\ P_{21} = P_{a2}(1 - P_c)P_{b1} \\ P_{22} = P_{a2}P_c + P_{a2}(1 - P_c)P_{b2} \end{cases}$$

As before, we can calculate *H(A)* and *H(B)* from the marginal probabilities. It is then possible to directly calculate the integration $I(S_{AB}(P_{a0}, P_{a1}, P_{a2}, P_{b0}, P_{b1}, P_{b2}, P_c))$ of the pair *(A, B)* and therefore the value of the *zI* index in case of *n* observations.

This leads us to the new threshold:

$$\theta_{zI}(P_{a0}, P_{a1}, P_{a2}, P_{b0}, P_{b1}, P_{b2}, P_c) = max\left(3.0, \frac{2nI(S_{AB}(P_{a0}, P_{a1}, P_{a2}, P_{b0}, P_{b1}, P_{b2}, P_c)) - d_k}{\sqrt{2d_k}}\right)$$

where *n* is the number of observations, and the probabilities $P_{aj}$ and $P_{bj}(j = 0,1,2)$ can be estimated from data (thus making the threshold specific for the case under examination). Notice again that the threshold cannot be less than the normal value of 3.0, to avoid confusing noise with signal.

## References


1. Bar-Yam, Y., McKay, S.R., Christian, W.: Dynamics of complex systems (studies in nonlinearity). Comput. Phys. **12**, 335–336 (1988)
2. Lane, D.: Hierarchy, complexity, society. In: Hierarchy in Natural and Social Sciences, vol. 3, pp. 81–119, Springer, Dordrecht (2006). https://doi.org/10.1007/1-4020-4127-6_5
3. Torres, L., Blevins, A.S., Bassett, D., Eliassi-Rad, T.: The why, how, and when of representations for complex systems. SIAM Rev. **63**(3), 435–485 (2021)
4. Filisetti, A., Villani, M., Roli, A., Fiorucci, M., Serra, R.: Exploring the organisation of complex systems through the dynamical interactions among their relevant subsets. In: Proceedings of the ECAL 2015: the 13th European Conference on Artificial Life. ECAL 2015: the 13th European Conference on Artificial Life. York, UK, pp. 286–293 (2015)
5. Sani, L., D'Addese, G., Pecori, R., Mordonini, M., Villani, M., Cagnoni, S.: An integration-based approach to pattern clustering and classification. In: Ghidini, C., Magnini, B., Passerini, A., Traverso, P. (eds.) AI*IA 2018. LNCS (LNAI), vol. 11298, pp. 362–374. Springer, Cham (2018). https://doi.org/10.1007/978-3-030-03840-3_27
6. Villani, M., et al.: An iterative information-theoretic approach to the detection of structures in complex systems. Complexity **2018**, 1–15 (2018)

7. D'Addese, G., Sani, L., La Rocca, L., Serra, R., Villani, M.: Asymptotic information-theoretic detection of dynamical organization in complex systems. Entropy **23**(4), 398 (2021)
8. D'Addese, G., Casari, M., Serra, R., Villani, M.: A fast and effective method to identify relevant sets of variables in complex systems. Mathematics **9**(9), 1022 (2021)
9. Mooij, J.M., Peters, J., Janzing, D., Zscheischler, J., Schölkopf, B.: Distinguishing cause from effect using observational data: methods and benchmarks. J. Mach. Learn. Res. **17**(1), 1103–1204 (2016)
10. Igelström, E., Craig, P., Lewsey, J., Lynch, J., Pearce, A., Katikireddi, S.V.: Causal inference and effect estimation using observational data. J. Epidemiol. Community Health **76**(11), 960–966 (2022)
11. Roli, A., Villani, M., Caprari, R., Serra, R.: Identifying critical states through the relevance index. Entropy **19**(2), 73 (2017)
12. Wheeler, D.J., Chambers, D.S.: Understanding Statistical Process Control. SPC Press, Knoxville (1992)
13. Pukelsheim, F.: The three sigma rule. Am. Stat. **48**(2), 88–91 (1994)
14. Cover, T.M., Thomas, J.A.: Elements of Information Theory. Wiley, New York (1991)
15. Lachmann, A., Giorgi, F.M., Lopez, G., Califano, A.: ARACNe-AP: gene network reverse engineering through adaptive partitioning inference of mutual information. Bioinformatics **32**, 2233–2235 (2016)
16. Margolin, A.A., et al.: ARACNE: an algorithm for the reconstruction of gene regulatory networks in a mammalian cellular context. BMC Bioinform **7**, 1–15 (2006)
17. Eberwine, J., Sul, J.Y., Bartfai, T., Kim, J.: The promise of single-cell sequencing. Nat. Methods **11**(1), 25–27 (2014)
18. Gawad, C., Koh, W., Quake, S.R.: Single-cell genome sequencing: current state of the science. Nat. Rev. Genet. **17**(3), 175–188 (2016)
19. Pratapa, A., Jalihal, A.P., Law, J.N., Bharadwaj A., Murali. T.M.: Benchmarking algorithms for gene regulatory network inference from single-cell transcriptomic data. Nat. Methods **17**(2), 147–154 (2020)
20. Griffiths, A.J., Miller, J.H., Suzuki, D.T., Lewontin, W.C., Gelbart, W.M.: An Introduction to Genetic Analysis, 7th edn. W. H. Freeman, New York (2000)
21. Mendes, P., Sha, W., Ye, K.: Artificial gene networks for objective comparison of analysis algorithms. Bioinformatics **19**(Suppl. S2), ii122–ii129 (2003)
22. Erdos, P., Renyi, A.: On random graphs. Publicationes Mathematicae **6**, 290–297 (1959)
23. Barabasi, A.L., Albert, R.: Emergence of scaling in random networks. Science **286**(5439), 509–512 (1999)
24. Watts, D.J., Strogatz, S.H.: Collective dynamics of 'small-world' networks. Nature **393**(6684), 440–442 (1998)
25. Newman, M.E.J.: The structure and function of complex networks. SIAM Rev. **45**(2), 167–256 (2003)
26. Powers, D.M.: Evaluation: from precision, recall and F-measure to ROC, informedness, markedness and correlation. Int. J. Mach. Learn. Technol. **2**, 37–63 (2011)

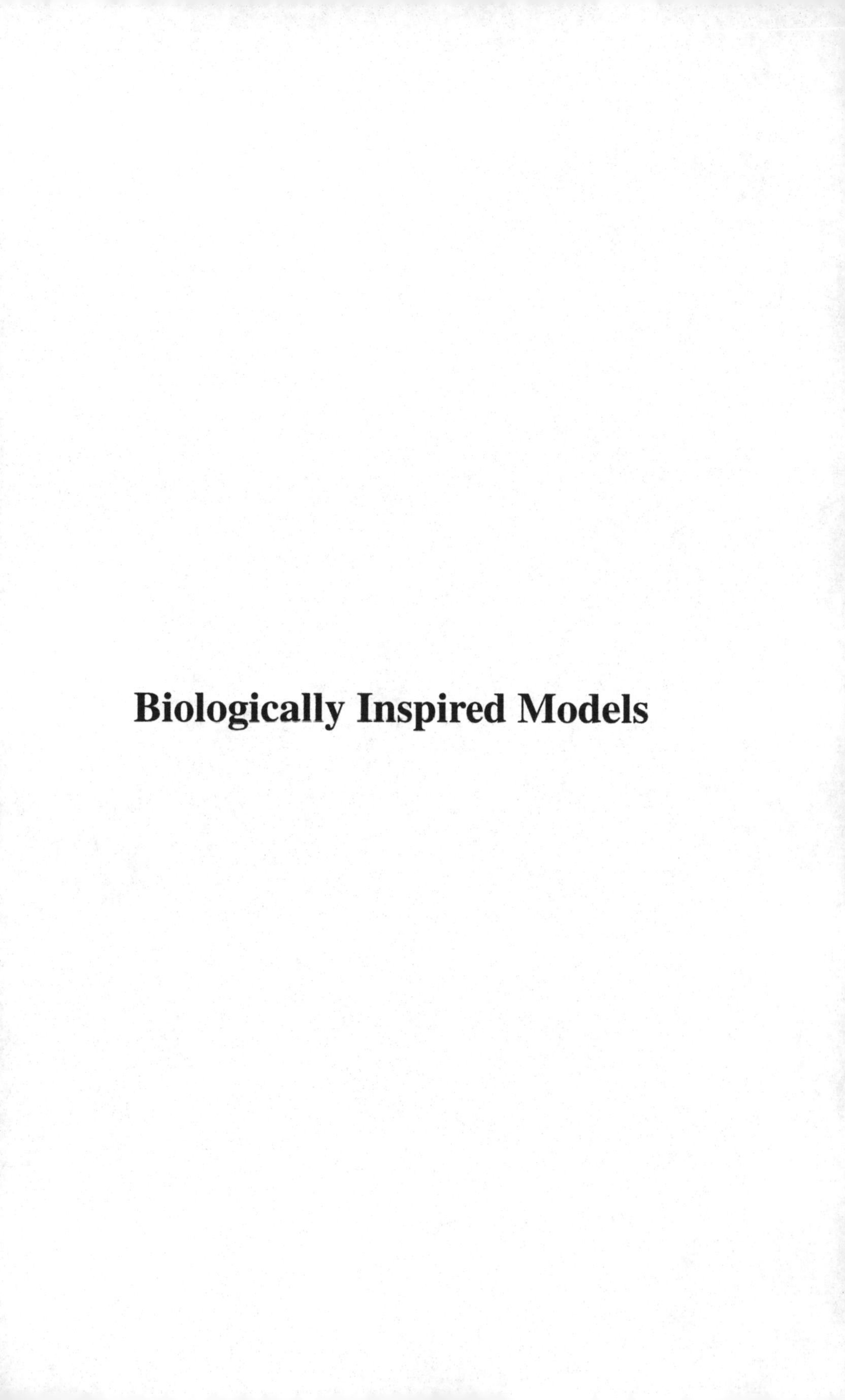

# Biologically Inspired Models

# The Properties of Pseudo-Attractors in Random Boolean Networks

Marco Villani[1,2(✉)], Matteo Balugani[1], and Roberto Serra[1,2,3]

[1] Department of Physics, Informatics and Mathematics, Modena and Reggio Emilia University, Modena, Italy
marco.villani@unimore.it
[2] European Centre for Living Technology, Venice, Italy
[3] Institute for Advanced Study, Amsterdam University, Amsterdam, The Netherlands

**Abstract.** Random Boolean Networks are dissipative dynamical models of gene regulatory networks, which are older than fifty years but still raise considerable interest. In this paper we will rely on two key concepts which had been introduced in previous works, namely those of pseudo-attractors (which are obtained by projecting true dynamical attractors onto constant vectors) and of the "common sea" (de-fined as the set of nodes which take the same value in every pseudo-attractor of a given network realization). In particular, we will study the dependence of the number of pseudo-attractors and of the relative size of the common sea upon the values of some key parameters, like the average number of connections per node and the so-called bias of the set of Boolean functions, paying particular attention to dynamically critical networks. We will also comment on the relationship of these models with measured gene expression values in single-cell observations.

**Keywords:** gene regulatory networks · random Boolean networks · models · common sea · pseudo-attractors · dynamical regime

## 1 Introduction

Random Boolean Networks (RBNs for short) are extremely interesting models of gene regulatory networks, which are older than fifty years [1] but still raise considerable interest (see e.g. [2]).

Many books and papers describe the model (including inter alia [3–5]), so we limit here to summarize the main features of its original and still most widely used (synchronous) version: a RBN is a $N$-dimensional dynamical system with Boolean variables $X = (X_1 ... X_N)$, which changes in discrete time steps according to well-defined, deterministic rules. $X_i$ is meant to represent the activation of the $i$-th gene (a node of the network). A fixed Boolean function $F_i$ is associated to each variable, and the next state $X_i(t+1)$ is obtained by applying $F_i$ to a set of k input variables. This set does not change in time, and all the nodes are simultaneously updated at each time step.

The most original feature of the model is that connections are drawn at random and Boolean functions are also chosen at random. This makes it unsuitable to describe a


M. Villani et al. (Eds.): WIVACE 2023, CCIS 1977, pp. 67–74, 2024.
https://doi.org/10.1007/978-3-031-57430-6_6

specific genetic circuit or an organism, but also well-suited to study the generic properties of families of circuits or organisms. When Kauffman introduced RBNs very little was known about gene regulation. Nowadays many more information have become available, however the search for generic properties of families of networks is still an important research topic, since (i) the specific information are not complete, so important parts of the regulatory systems are still unknown and (ii) knowing the properties of families of networks is important per se, as it allows generalizations beyond the study of single organisms.

This is particularly relevant since theory is now lagging behind data in biology, due to the recent "data deluge". And RBNs are among the few models that can contribute to the development of wide-ranging theories. Our research is inspired by the idea is that these networks can nowadays be improved, by introducing more precise constraints based upon increased biological knowledge.

Among others, a particularly important class of experimental data are the measurements of gene activation values, related with the amount of mRNA in single cells, which allows one in principle to observe a single gene network at work [6]. A comparison of the RBN model with these data requires considerable care. Besides the fairly obvious remark that the Boolean approximation may be a poor representation of the existing levels of gene expression, it is important to stress that the attractors (stable asymptotic states, which can be reached by some initial condition) of a finite deterministic RBN are bound to be either constant (fixed points) or oscillating (limit cycles) states. Fixed points can be directly compared with measurements on single cells, but how shall we interpret cycles?

One possibility might be that of comparing them with time courses, but these cannot come from a single cell, which is destroyed by the measurement procedure. Moreover, the properties, and the very existence of cyclic states are strongly related to the choice of synchronously updating all the nodes, which is manifestly unrealistic since (i) the various genes are not subject to any "central clock" and (ii) each node has its own "time step", whose length is determined by the possibility of forgetting the state at time $t - 1$ when updating the state at time $t + 1$; it is therefore related to the life-time of the molecule which is synthesized by the corresponding gene (often a protein), and these lifetimes can differ widely among various genes [7].

Should we then simply forget about RBNs? We do not think that this is the case, since they provide useful indications, and they stand among the very few models that maintain a certain degree of understandability and can aim at some generality. A promising approach is that of introducing a gene-protein model, where different proteins may have different decay times. This is a well-founded strategy, which we have introduced in the past [8–10], which however pays the price of increasing the (already quite large) number of model parameters. An alternative approach is that of projecting the time-varying expression values of oscillating genes onto fixed points, which might then be directly compared with single-cell measurements.

A possible way to perform this projection has been proposed and discussed in depth in [11], where we introduced some new variables, including pseudo-attractors. For any true dynamical attractor $A_d$, a constant Boolean pseudo-attractor $A_p$ is defined as follows:

- If the corresponding component of $A_d$ is constant, that component of $A_p$ takes that same value
- If the corresponding component of $A_d$ oscillates, that component of $A_p$ takes the value 0 (resp., 1) if the time average of the oscillating values is larger than (resp., smaller than or equal to) a given threshold $\theta$

The pseudo-attractors thus represent a sort of coarse-graining of the set of dynamical attractors, and their values can be compared with those of single-cell data, if the latter are binarized using another threshold $\chi$. Moreover, since single-cell expression data are very noisy and can easily provide false negatives, it is convenient to take for comparison average (or median) values on a number of different single-cell measurements, coming from supposedly identical cell types in identical conditions [11]. A possible alternative might be that of using samples which are directly taken from several cells belonging to the same type, provided that the presence of cells of different types can be excluded.

The usual approach in applying RBNs to multicellular organisms is that of associating the various attractors to cell types, since they represent a "way of functioning" of the gene regulatory network which is coherent and sustainable. Since pseudo-attractors are obtained from dynamical attractors by freezing out their oscillations (which we regard as artifacts due to synchronous updating), we will interpret here pseudo-attractors as corresponding to cell types. This allows one to make some comparisons with experimental data, provided that the latter are binarized in some way (e.g., by comparison with some threshold value of gene expression). In this way, it is possible for ex-ample to interpret the number of "1"s in a pseudo-attractor as the number of active genes in the corresponding cell type. Of course, since we deal with random models, we cannot associate a specific pseudo-attractor to a specific cell types but, as we shall see, we can get useful information from statistical properties of these values.

Another interesting concept is that of a "common sea" CS of genes: for a given network realization, it is defined as the set of all the nodes which take the same value in every pseudo-attractor. Its relative size $\phi_{CS}$, defined as the ratio between the number of nodes in CS to the total number of nodes, can also be compared to the corresponding quantity which can be estimated from experimental data.

In this paper we will deal with models, and we refer the reader interested in how experimental data can be obtained from single cell measurements to [11]. Here we will shed light on the behavior of pseudo-attractors, comparing it to that of dynamical attractors, for different sets of parameter values. The main parameters of a family of RBNs are the number of connections per node $k$ and the properties of the al-lowed Boolean functions, which are summarized by the parameter $b$ (the bias) which measures the probability that a set of input values turns the target node on [3, 4]: for every node, for every set of input values, the corresponding output "0" or "1" is drawn at random, $b$ being the probability that "1" is chosen. We will mainly concentrate on so-called critical networks, which are particularly interesting for a number of reasons, widely discussed in the literature (see [2, 3, 12, 13]). Critical behaviors correspond to specific pairs of values $(k,b)$ and our analysis will largely (but not exclusively) focus on these pairs.

The outcome of various simulations will be described in Sect. 2, where it will be observed that RBNs tend to give rise to very large common seas, which do not seem to correspond to those which are actually found in data. While this can provide useful hints to modify the basic model, here (in Sect. 3) we limit to show that Boolean networks with a not-so-large common sea can actually exist, although they are no longer fully random but obtained by evolving a population of networks with a genetic algorithm. The final Sect. 4 will be dedicated to summarize the main conclusions and to provide indications for further work.

## 2 Simulation Results

In this section we will show different simulation results concerning different interesting quantities. Unless otherwise stated, the networks comprise 100 nodes, the Boolean functions are drawn at random with bias *b*. The attractors are identified by starting from 10.000 fully random initial states; the search for cyclic attractors is limited to those whose periods do not exceed 1000 steps. The maximum allowed length of the transients never exceeds 50.000 steps. The number *k* is actually the number of input connections per node, which is the same for every node. Since the input nodes are drawn at random with uniform probability, the number of output connections per node approximately follows a Poisson distribution, whose average is of course *k*. Averages and medians are computed on ensembles of 100 networks.

Figure 1 shows the behavior of the number of pseudo-attractors vs the number of dynamical attractors for different sets of critical networks (with three different values of the number k of connections per node). Since the number of pseudo-attractors cannot exceed that of attractors, all the points lie below the bisector.

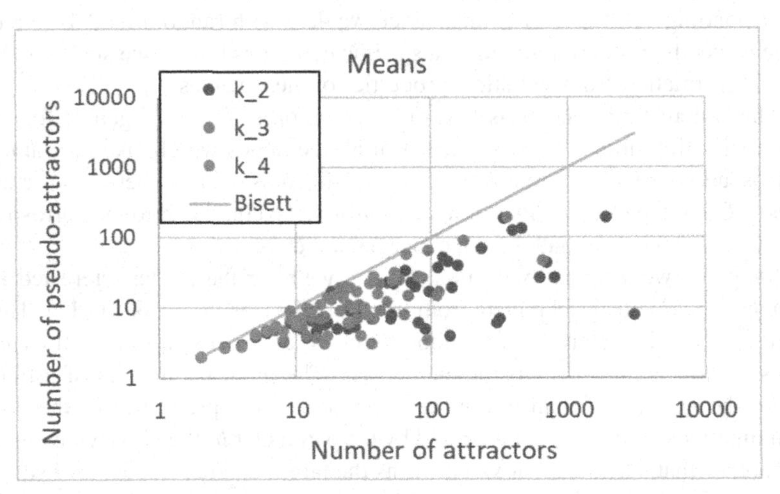


**Fig. 1.** Number of pseudo-attractors vs. number of attractors, in three critical ensembles with different *(k,b)* pairs (log-log scale): values of 3*100 = 300 RBNs are shown. The green line is the bisector. The averages are taken by counting, for every value of the parameter *k*, the average number of networks with the corresponding value of the number of pseudo-attractors (Color figure online).

It might be noticed that, although alle the networks are all critical, their distributions look somewhat different. Figure 2 shows the relative size of the CS for different k values and for different values of the number of nodes. It should be observed that $\phi_{CS}$ is an increasing function of both the total number of nodes and the number of connections per node.

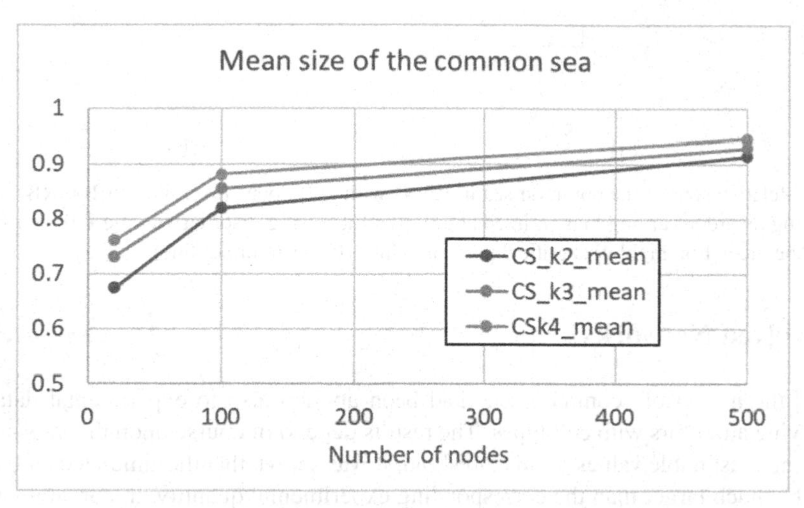


**Fig. 2.** Common part vs. number of nodes, critical ensembles each composed by 100 RBNs. Note that (i) the greater the average connectivity *k*, the greater the size of the common part and that (ii) the greater the number of nodes, the greater the size of the common part.

So far, all the networks which have been considered are critical. Let us now look at the size of the CS for different ensembles. Figure 3 shows the relative size of the CS for various ordered, critical and chaotic networks. It is interesting to observe (Fig. 3a) that $\phi_{CS}$ seems to decrease slowly as the Derrida parameter increases, in the case of a fixed number of connections per node. The Derrida parameter measures in a sense the "degree of chaoticity" of a network, it takes the value 1 for critical networks, and larger values for chaotic networks. In the case of Fig. 3a, higher values of the Derrida parameter are associated to higher values of the bias. On the other hand (Fig. 3b), when the size of the bias is fixed, the size of the CS seems insensible to changes in Derrida parameter (which increases as $k$ increases).

The analysis of the behavior of chaotic nets requires considerable care, since it may happen that no attractor is found before the time limits of the simulation are reached. While this is a well-known problem in dealing with chaotic networks, a further problem is encountered here in analyzing ordered networks, where one sometimes finds only a single attractor; in these cases it makes no sense to study the common sea, so one needs to also disregard them (they are of course not well-suited to describe multicellular organisms with different cell types).

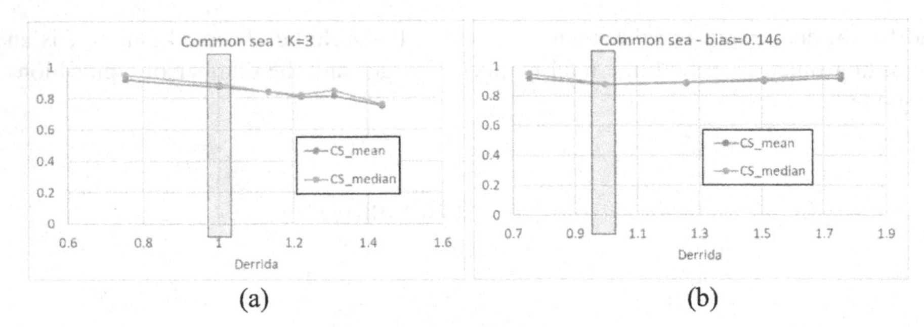


(a) (b)

**Fig. 3.** Relative size of the common sea in RBNs with 100 nodes (averages on 100 RBNs). (a) Increasing disorder varying bias (k fixed at 3.0). (b) Increasing disorder varying k (bias fixed at 0.146 The green box highlights critical pairs of values (Color figure online).

## 3 Evolved Networks

In [11] the notion of "common sea" had been applied also to experimental data, by identifying attractors with cell types. The results depend of course upon the threshold $\theta$ but, using reasonable values of this threshold, it was shown that the simulated $\phi_{CS}$ turns out to be much larger than the corresponding experimental quantity. In our simulations for this work we have tested a wide set of *(k,b)* values, but we continued to find quite large common seas.

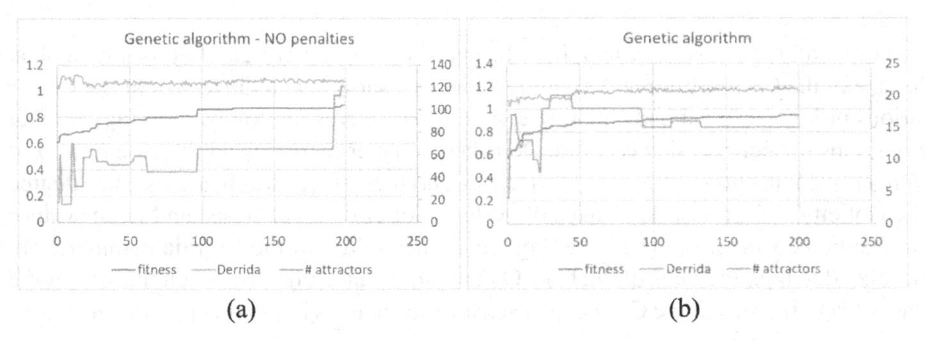


(a) (b)

**Fig. 4.** Genetic algorithm applied to the search for a limited CS. X-axis: generation number; y-axis: red line, fitness; grey line, number of dynamical attractors; yellow line, Derrida parameter. (a) Simulation carried out without limiting the number of attractors (b) Simulation carried out by limiting the maximum number of attractors to 20. It should be noted that the number of final attractors is lower than this threshold (Color figure online).

One is therefore led to wander whether it is actually possible to obtain a "reasonable" size of the common sea with a RBN. We tentatively used a Genetic Algorithm with fitness function equal to $1\text{-}\phi_{CS}$ to evolve networks with a smaller CS. Interestingly, the answer turned out to be positive (Fig. 3a), although of course the evolved networks were no longer fully random. It is also interesting to observe that the GA used a fairly understandable "strategy" to achieve that goal, by increasing the number of attractors, which involves an increase in the number of pseudo-attractors and a shrinking of the CS. But, since the number of pseudo-attractors should be related to that of different cell types, their proliferation might lead to unwanted consequences. We therefore modified the GA in order to exclude networks with more than a predefined number of attractors, and also in this case the system found networks with a limited number of attractors and a CS of reasonable size (Fig. 4b). This proves that BNs with those properties do actually exist.

## 4 Conclusions

A general observation is that the notions of pseudo-attractor and of common sea makes sense when there are several different attractors which can be identified, therefore these concepts are particularly useful for networks which are neither too chaotic (this would make it very difficult to find the attractors in time-limited simulations) nor too ordered (which sometimes leads to a single attractor).

Critical networks, which have attracted considerable attention and which seem able to account for interesting experimental data [12, 13] belong to this region, so they are amenable to be analyzed using these notions. It is interesting to observe that different ensembles of networks, which are all critical but differ in number of nodes (and, correspondingly, in bias of the Boolean functions), also show different sizes of the common sea (while their Derrida parameters are all close to 1). One is therefore led to conclude that the size of the common sea is not determined uniquely by the dynamical regime, when measured according to the usual Derrida parameter.

The networks of Sect. 3 are not fully random, but they have been evolved to match the typical values of the CS observed in experiments. It is well-known that the dynamics of such networks may differ from those of the initial ones [15], but it is interesting to observe (Fig. 4) that the evolved networks, discovered by the GA, seem to be slightly chaotic, since their Derrida parameter $\lambda$ is slightly larger than one. This could make sense, since chaotic networks have a larger number of attractors, and therefore (typically) a smaller CS. However, it has already been observed elsewhere [14] that the classical way to compute $\lambda$, by using fully random initial conditions, may be misleading when applied to strongly non-ergodic systems like RBNs. Therefore we plan to use in a future work some different measures of the dynamical regime, like e.g. the sensitivity on attractors [14].

One might also consider hybrid networks, mixing with some known circuits (like e.g. the TCA cycle [16]) with some random parts; while interpreting those results might be really challenging, it would be interesting to observe which kind of phenomena may show up.

The results shown above need to be complemented by further simulations, aimed in particular at exploring the chaotic region. In spite of the difficulties of identifying attractors in these simulations, they may provide a route to finding networks with a more reasonable size of the CS than that of critical RBNs. Moreover, it might of course be interesting to analyze further sets of single-cell experiments, to confirm the results of the analysis in [11] and to get more accurate data.

**Funding.** This research was funded by Università degli Studi di Modena e Reggio Emilia (FAR2023 project of the Department of Physics, Informatics and Mathematics).

## References

1. Kauffman, S.A: Metabolic stability and epigenesis in randomly constructed genetic nets. J. Theor. Biol. **22**(3), 437–467 (1969)
2. Kauffman, S.A., Serra, R., Schmulevich, I., Huang, S. (eds.): Entropy, Special Issue on "The Principle of Dynamical Criticality" (2023)
3. Kauffman, S.A: Origins of Order. Oxford University Press, Oxford (1993)
4. Aldana, M., Coppersmith, S., Kadanoff, L.P.: Boolean dynamics with random couplings. In: Kaplan, E., Marsden, J., Sreenivasan, K.R. (eds.) Perspectives and Problems in Nonlinear Science. Springer, New York (2003). https://doi.org/10.1007/978-0-387-21789-5_2
5. Bastolla, U., Parisi, G.: Relevant elements, magnetization and dynamical properties in Kauffman networks: a numerical study. Phys. D: Nonlinear Phenom. **115**(3–4), 203–218 (1998)
6. Wang, D., Bodovitz, S.: Single cell analysis: the new frontier in 'omics.' Trends Biotechnol. **28**(6), 281–290 (2010)
7. Fornasiero, E.F., Savas, J.N.: Determining and interpreting protein lifetimes in mammalian tissues. Trends Biochem. Sci. **48**(2), 106–118 (2023)
8. Graudenzi, A., Serra, R., Villani, M., Colacci, A., Kauffman, S.A.: Robustness analysis of a boolean model of gene regulatory network with memory. J. Comput. Biol. **18**, 559–577 (2011)
9. Graudenzi, A., Serra, R., Villani, M., Damiani, C., Colacci, A., Kauffman, S.A.: Dynamical properties of a boolean model of gene regulatory network with memory. J. Comput. Biol. **18**, 1291–1303 (2011)
10. Sapienza, D., Villani, M., Serra, R.: Dynamical properties of a gene-protein model. In: Pelillo, M., Poli, I., Roli, A., Serra, R., Slanzi, D., Villani, M. (eds.) WIVACE 2017. CCIS, vol. 830, pp. 142–152. Springer, Cham (2018). https://doi.org/10.1007/978-3-319-78658-2_11
11. Villani, M., D'Addese, G., Kauffman, S.A., Serra, R,: Attractor-specific and common expression values in random Boolean network models (with a preliminary look at single-cell data). Entropy **24**, 331 (2022)
12. Muñoz, M.: Colloquium: criticality and dynamical scaling in living systems. Rev. Mod. Phys. **90**, 031001 (2018)
13. Roli, A., Villani, M., Filisetti, A., Serra, R.: Dynamical criticality: overview and open questions. J. Syst. Sci. Complex. **31**, 647–663 (2018)
14. Villani, M., Campioli, D., Damiani, C., Roli, A., Filisetti, A., Serra, R.: Dynamical regimes in non-ergodic random Boolean networks. Nat. Comput. **16**, 353–363 (2017)
15. Benedettini, S., et al.: Dynamical regimes and learning properties of evolved Boolean networks. Neurocomputing **99**, 111–123 (2013)
16. Kay, J., Weitzman, P.D.J.: Krebs' Citric Acid Cycle: Half a Century and Still Turning: Biochemical Society Symposium No. 54 Held at University of Leicester, April 1987. Biochemical Society (Great Britain) Symposia, Leicester, England (1987)

# Analysing the Expressiveness of Metabolic Networks Representations

Irene García[1], Bessem Chouaia[2], Mercè Llabrés[1], Pere Palmer-Rodríguez[1], and Marta Simeoni[2,3](✉)

[1] Mathematics and Computer Science Department, University of the Balearic Islands, Palma, Spain

[2] Department of Environmental Science, Computer Science and Statistics, University Ca' Foscari of Venice, Venice, Italy
simeoni@unive.it

[3] European Centre for Living Technology (ECLT), Venice, Italy

**Abstract.** In this work, we explore the expressiveness of three graph-based representations of metabolic networks. We consider Abstract Metabolic Networks (AMNs), metabolic-Directed Acyclic Graphs (m-DAGs) and Reaction Graphs (RGs). These representations form a hierarchical view of the metabolism, AMNs being the most abstract, m-DAGs serving as the intermediate, and RGs being the most detailed. We evaluate their expressiveness for a case study comprising 331 Vertebrates and by using the Weisfeiler-Lehman graph kernel to perform the comparison. The results show that AMNs are not able to discern the various taxonomic groups at the Class level, while m-DAGs and RGs clearly distinguish Mammals, Fishes and Birds. When focusing on Mammals at the Order level, only m-DAGs are partially able to identify some of the taxonomic groups. Moreover, m-DAGs are able to distinguish Primates at the Infraorder level of taxonomy. Based on the obtained results, it emerges that m-DAGs are a good compromise between the amount of network information and the computational effort needed to obtain reliable patterns on the taxonomic clustering of the different organisms.

**Keywords:** Metabolic Networks · Abstract Metabolic Networks · Metabolic DAGs · Reaction Graphs

## 1 Background

Metabolism is the organism's machinery that breaks down complex organic molecules to produce the energy and building blocks needed for the organism's normal functioning. It comprises all chemical and physical processes that occur within the cells of living organisms and that allow for maintaining life. These

This work was partially supported by DAIS - Ca' Foscari University of Venice within the IRIDE program and Grant PID2021-126114NB-C44 funded by MCIN/AEI/ 10.13039/501100011033 and by "ERDF A way of making Europe".


M. Villani et al. (Eds.): WIVACE 2023, CCIS 1977, pp. 75–87, 2024.
https://doi.org/10.1007/978-3-031-57430-6_7

processes interact with one another, creating a complex network structure. Each process, in turn, is made up of a pathway of chemical or physical reactions, and all the pathways together operate as a highly integrated network [6].

Literature on metabolomics in the last two decades ranges from the analysis of single pathways [10,15], to the comparative analysis of a set of pathways [1], together with the metabolic networks' dynamics [3,17]. Automatic metabolic network reconstruction and comparison can be performed thanks to the knowledge stored in metabolic databases such as BioCyc [4], BioModels [5] and KEGG (Kyoto Encyclopedia of Genes and Genomes) [11]. However, this reconstruction generally requires human intervention since data repositories can be incomplete, heterogeneous and incoherent. Furthermore, the comparison of metabolic networks is computationally challenging due to the huge number of chemical reactions involved in the metabolism.

A crucial choice that greatly influences automatic metabolic network reconstruction and comparison is how the network is represented. In [9], we employed *abstract metabolic networks* (AMNs) to compare the metabolic networks of different species. An AMN is a graph that represents metabolic pathways as nodes, and there is an edge between two nodes if their corresponding pathways share one or more compounds. No information about chemical reactions is included. The AMN of a given organism results in a small network that requires low computational power to be analysed and makes it a suitable model to perform a large-scale comparison of organisms' metabolism. AMNs were shown to be able to discriminate macroevolutionary events, indicating that they are expressive enough to capture key steps in metabolism evolution. However, they are clearly unsuitable for fine-grain metabolic network comparison. Therefore, further research questions are the following: What are the benefits of refining the representation of the metabolism? To what extent is the addition of information useful to discriminate between different species? Is there a trade-off between representation expressiveness and computational cost in studying metabolism?

To address these questions, we aim to extend the analysis in [9] to two other representations of metabolic networks, namely m-DAGs and reaction graphs. A *reaction graph* (RG) represents a metabolic network under study as a directed graph where chemical reactions are nodes, and there is a directed edge from reaction A to reaction B if a metabolite produced by A is consumed by B. A *m-DAG* [2] is obtained from an RG by collapsing into a single node each strongly connected component, which is called a metabolic building block in this context. A metabolic building block corresponds to an autonomous subsystem, a minimal set of reactions that can operate at a steady state. Contracting them into single nodes allows for considerably reducing the size of the resulting graph while keeping the interesting information. Note that AMNs, m-DAGs and RGs form a hierarchical view of the metabolism, AMNs being the most abstract (only the topological organisation of the metabolic pathways is considered) and RGs the most detailed (all the reactions are included). For instance, for Homo sapiens, the corresponding AMN representation is a graph with 84 nodes, the m-DAG has 1006 nodes, and the reaction graph has 2218 nodes.

The aim of the paper is to analyse the different contributions of AMNs, m-DAGs and RGs to discerning the various taxonomic groups and species. As a case study, we consider a set of 331 Vertebrates stored in the KEGG database, an extensive, reliable and widely used source of metabolic data. We derive their AMN, m-DAG and RG representations and study their expressiveness. For each representation technique, we perform the pairwise comparison of the obtained graphs and then analyse the results through Multi Dimensional Scaling (MDS) [8] and clustering techniques. Since comparing graphs is computationally impractical, we resort to the use of graph kernels as they allow for a feasible, although not exact, comparison. Specifically, we employ the Weisfeiler-Lehman (WL) graph kernel [18], which is based on subtrees comparison. In this way, the pairwise comparison of the 331 considered Vertebrates can be done in a few seconds for each of the considered graph-based representations. We remark that the whole pipeline is performed in an automatic way and by using just the information provided by KEGG.

The results demonstrate that AMNs are unable to differentiate between various taxonomic groups at the Class level, whereas m-DAGs and RGs successfully distinguish between Mammals, Fishes, and Birds. Focusing on Mammals at the Order level, only m-DAGs are able to partially identify some taxonomic groups. Additionally, m-DAGs can also distinguish Primates at the Infraorder level of taxonomy. Based on the results, it becomes evident that, for the considered case study, m-DAGs offer a balanced compromise between the amount of network information and the computational effort required to obtain reliable patterns for the taxonomic clustering of different organisms.

The paper is organised as follows: Sect. 2 introduces all the preliminary definitions and techniques needed to develop our study. Section 3 presents the performed analyses and discuss the results. Finally, Sect. 4 draws some conclusion and delineates some directions for future work.

## 2 Methods

In this section, we introduce the preliminary notions and methods needed to develop our case study. First, we illustrate KEGG as a source of metabolic data, and then we introduce AMNs, m-DAGs and RGs. Afterwards, we briefly introduce the Weisfeiler-Lehman graph kernel and the methods employed for evaluating the comparisons results.

### 2.1 KEGG as a Source of Metabolic Data

The first step in automatically creating a metabolic network representation of a specific organism is automatic data retrieval. We used KEGG as the sole source of metabolic data. Our approach strictly depends on the data representation and the knowledge available in KEGG, including data incompleteness, inconsistency and biases that could reflect negatively on the subsequent comparison and analysis phases. It is worth highlighting that biases exist in all metabolic databases and cannot be avoided while performing large-scale automated processes. Nonetheless, we can count on the fact that KEGG is a widely known database, whose content is constantly updated based on new knowledge.

The metabolic activity in KEGG is divided into various categories, which are shown in Table 1. Each category is then composed of various metabolic pathways. In order to consider the whole metabolic network, all categories and all the corresponding pathways should be contemplated.

**Table 1.** List of KEGG metabolic categories: each one is composed by many metabolic pathways.

| Metabolic category |
|---|
| Carbohydrate metabolism |
| Energy metabolism |
| Lipid metabolism |
| Nucleotide metabolism |
| Amino-acid metabolism |
| Metabolism of other amino-acids |
| Glycan biosynthesis and metabolism |
| Metabolism of cofactors and vitamines |
| Metabolism of Terpenoids and polyketides |
| Biosynthesis of other secondary metabolites |
| Xenobiotics biodegradation and metabolism |

Since the metabolic pathways are quite preserved among organisms, KEGG associates to each metabolic function, a unique *reference pathway*, which corresponds to the union of the corresponding pathways in all the organisms included in the database. A pathway of a specific organism can be obtained from the corresponding reference pathway. This standardised and modular representation of pathways plays an important role in avoiding incoherence in metabolism comparison.

KEGG supplies two related representations for each pathway in its repository: a graphical representation (*pathway map*), showing the network of chemical reactions composing the pathway, and a textual one written in an XML format, a *KGML file*, where KGML stands for *KEGG Markup Language*. Such a file contains the information represented in the corresponding map. To automatically represent the metabolism of a specific organism, it is necessary to download the KGML files of each pathway of the organism through the public KEGG's APIs, and to parse each KGML file to extract the relevant information according to the chosen representation (in our case the compounds and the reactions of each pathway).

## 2.2 Abstract Metabolic Networks, Reaction Graphs and Metabolic DAGs

We present the three graph-based representations of metabolic networks under study, namely AMNs, m-DAGs and RGs.

An *abstract metabolic network* (AMN for short) is an undirected graph in which each node identifies a metabolic pathway, and an edge between two nodes indicates that the two corresponding pathways share one or more compounds among the ones defining their reactions network. Nodes are labelled with the KEGG pathway identifiers, and edges are not labelled. Note that a shared compound $C$ between two pathways may represent different situations:

- $C$ is produced by one pathway and consumed by the other;
- $C$ is a compound used in the shared part of two overlapping pathways;
- $C$ is a compound present in both pathways, even though they express unrelated functions or work in different environments or conditions.

Ubiquitous compounds such as $H_2O$, phosphate, ATP and ADP are not considered since, though they are needed and assumed to be present everywhere, they are not "representative" compounds of any pathway, and their consideration would only add noise to our representation.

In this representation, all the reactions involved in the various pathways are omitted, and just the minimal information about the presence of pathways and their interconnections is taken into consideration. This clearly results in an abstract and coarse-grain view of the metabolism.

A *reaction graph* representation of a metabolism is a directed graph $G_R = (R, E)$ whose set of nodes is the set $R$ of chemical reactions present in the metabolism, and its set of arcs, $E$, is defined as follows: there is an arc from $R_i$ to $R_j$ if, and only if, there exists at least one metabolite produced by $R_i$ that is consumed by $R_j$. Reversible reactions are modelled by two different nodes, one for the forward reaction and the other for the backward reaction.

When a directed graph has no cycles, it is called a *directed acyclic graph*, DAG for short. A *path* from a node $u$ to a node $v$ in a directed graph $G$ is a sequence of nodes $\{u_0, u_1, ...u_k\}$ such that $u_0 = u$, $u_k = v$ and $(u_i, u_{i+1})$ is an arc in $G$ for $i = 0, ..., k-1$. Two nodes $u, v$ are said to be biconnected if there is a path in each direction between them. A *strongly connected component* of a directed graph $G$ is a subgraph such that every pair of nodes in it are biconnected, and it is maximal under inclusion with this property [7,20]. Since biconnectivity is an equivalence relation, the collection of strongly connected components forms a partition of the set of nodes of $G$. If each strongly connected component is contracted to a single vertex, the resulting quotient graph is a DAG, the *condensation* of G. Notice that there is an arc from a node $s_i$ to a node $s_j$ in the condensation of a directed graph $G$ if, and only if, there is an arc in $G$ from a node $u \in s_i$ to a node $v \in s_j$. Thus, for every reaction graph $G_R$, we can consider its collection of strongly connected components and compute its condensation, which will be a DAG. We call *metabolic building blocks* (MBBs for short) the strongly connected components in the reaction graph $G_R$, and we call *metabolic DAG*, (m-DAG for short), the condensation of a reaction graph $G_R$. In this representation, all the reactions involved in the MBBs are omitted, and just the information about the presence of cycles and their interconnections is taken into consideration. This clearly results in a simplified view of the reaction graph topology.

### 2.3 Graph Kernels

Graph kernels can be intuitively understood as functions measuring the similarity of pairs of graphs. Given two graphs in input, a graph kernel calculates a real number that measures their similarity according to some graph's features. More in general, given a set of $n$ graphs in input, the kernel's output is a $n \times n$ similarity matrix where each cell $(i, j)$ contains the result of the pairwise comparison between graphs $i$ and $j$. In our case the matrix is normalised, that is, all similarity values are real numbers in the interval [0, 1], where 0 and 1 indicate the minimum and maximum similarity values, respectively. The matrix is symmetric, and the main diagonal contains only 1 values, each cell being the comparison of a graph with itself.

A formal introduction to graph kernels can be found, e.g., in [12–14]. In this paper we used the Weisfeiler-Lehman graph kernel [18] as implemented in the GraKel Python library [19]. Roughly speaking, the Weisfeiler-Lehman (WL) subtree kernel compares the subtrees rooted on each node of the two graphs under consideration, taking into account their node labels: the more the subtrees structures and labels coincide, the higher their similarity. The fundamental idea of the Weisfeiler-Lehman algorithm is to replace the label of each node with a multiset of labels consisting of the original label of the node and the sorted set of labels of its neighbours. The resultant multiset is then compressed into a new short label, which reflects the knowledge of the node and its neighbourhood. This relabelling process is then repeated for $h$ iterations. By performing this procedure simultaneously on a set of input graphs, it follows that two nodes from different graphs will get an identical new label if and only if they have identical multisets of labels. The kernel function, in this case, compares the node labels of the graphs resulting after $h$ iterations and summarises the comparison with a real number. It can be shown that this is equivalent to comparing the number of shared subtrees between the two input graphs (the kernel considers all subtrees up to height $h$).

### 2.4 Data Visualisation and Analysis

We performed exploratory data analysis using different techniques, such as Multi-dimensional scaling to visualise the comparison results, and clustering analysis, specifically k-means and hierarchical clustering, to classify the considered organisms and compare the obtained clusters with the real taxonomic groups. All the analyses were performed using the $R$ package [16], a language and environment for statistical computing.

Multi-dimensional Scaling (MDS) is a classic multivariate data analysis technique that allows for obtaining a low-dimensional representation of the observed similarities [8]. Given a normalised similarity matrix $\mathbf{Q}$, we implemented a metric MDS by transforming $\mathbf{Q}$ into a distance matrix with elements.

$$d_{ij}^2 = q_{ii} + q_{jj} - 2q_{ij} = 2(1 - q_{ij}),$$

where $q_{ij}$ is the element in row $i$ and column $j$ in $\mathbf{Q}$ and $q_{ii} = q_{jj} = 1$.

The objective of the MDS is to represent the observed distances using a set of variables $y_1, \ldots, y_k$, where $k < n$, such that the Euclidean distances between the coordinates of the elements for these variables are close to the original distances. In this way, the graphical representation in $k$ dimensions will faithfully reproduce the observed structure.

K-means clustering divides the dataset into $k$ subsets, where $k$ is pre-specified. This algorithm finds $k$ groups with the smallest within the sum of squares (WSS) and the largest between the sum of squares (SS). Each cluster is represented by the mean of the data points belonging to the cluster itself. We specified the number of clusters according to the phylogeny and using the so-called Elbow method [8]. The Elbow method consists in plotting the value of the WSS produced by different values of $k$. The bend (elbow) location in the plot is generally considered an indicator of the appropriate number of clusters.

Agglomerative hierarchical clustering is an alternative approach that does not require specifying the number of clusters to be generated. We used an average linkage clustering, i.e. the distance between two clusters is calculated as the average of distances between all pairs of organisms from each group. The formula for calculating the distance is $d_{rs} = T_{rs}/(N_r \times N_s)$, where $T_{rs}$ is the sum of all pairwise distances between cluster $r$ and cluster $s$, $N_r$ is the number of organisms in cluster $r$, and $N_s$ is the number of organisms in cluster $s$.

The result of hierarchical clustering is a tree-based representation of the objects; the visualisation is shown through a graph known as a dendrogram.

## 3 Results and Discussion

In this section, we analyse the expressiveness of the three different representations of metabolic networks for the considered case study. We first consider the whole group of KEGG Vertebrates and test the ability of the three representations to distinguish the various taxonomic groups at the Class level, that is, Mammals, Birds, Fishes, Reptiles and Amphibians. Subsequently, we focus on Mammals at the Order taxonomic level and, finally, we consider only the Primates at Infraorder level. For each experiment, we show the comparison results and discuss them in detail.

### 3.1 Vertebrates Analysis

In this section we consider whole group of KEGG Vertebrates. Figure 1 shows the MDS plots of the comparison results for AMNs (plot on the left), m-DAGs (plot in the center) and RGs (plot on the right). We notice that, while m-DAGs

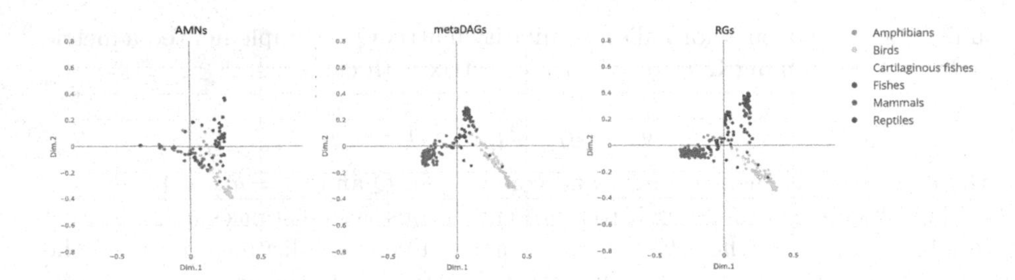


**Fig. 1.** MDS plots of Vertebrates compared using the WL graph kernel: AMNs (left), m-DAGs (center) and RGs (right). The colors reflect the class each organism belongs to (see the legend on the right side).

and RGs clearly distinguish Mammals, Fishes and Birds, AMNs are not really able to separate them.

This is also evidentiated by the k-means analysis in Table 2, which reports the results for AMNs, m-DAGs and RGs considering 4 clusters, as suggested by the elbow method. We observe that AMNs do not exhibit homogeneous clusters, and Mammals, Birds and Fishes are always mixed with other species. On the other hand, m-DAGs and RGs present similar results. With the m-DAG model, Fishes and Mammals fall into homogeneous clusters. With RG, the Mammals cluster also contains 4 outliers and Fishes are split into two clusters, one of which is homogeneous. Birds are equally classified in the two models: most of them fall into a single cluster, together with 3 outliers. This may reflect that m-DAGs are indeed a suitable reduction of RGs for studying and comparing metabolic networks. They are able not only to highlight the relevant topological structure under the RG, but also to capture the key metabolic information to discern the different taxonomy groups as RGs.

**Table 2.** k-means clustering of Vertebrates for AMNs, m-DAGs and RGs

| Class level | AMNs Clusters | | | | m-DAGs Clusters | | | | RGs Clusters | | | |
|---|---|---|---|---|---|---|---|---|---|---|---|---|
| | 1 | 2 | 3 | 4 | 1 | 2 | 3 | 4 | 1 | 2 | 3 | 4 |
| Amphibians | 1 | 0 | 0 | 5 | 0 | 6 | 0 | 0 | 5 | 0 | 0 | 1 |
| Birds | 0 | 43 | 7 | 20 | 0 | 9 | 0 | 61 | 8 | 62 | 0 | 0 |
| Cartilaginous fishes | 0 | 0 | 0 | 2 | 0 | 2 | 0 | 1 | 2 | 1 | 0 | 0 |
| Fishes | 75 | 0 | 0 | 14 | 86 | 3 | 0 | 0 | 13 | 0 | 76 | 0 |
| Mammals | 3 | 1 | 93 | 43 | 0 | 2 | 137 | 0 | 3 | 0 | 0 | 136 |
| Reptiles | 8 | 3 | 3 | 10 | 0 | 22 | 0 | 2 | 19 | 2 | 0 | 3 |

Tests performed on Vertebrates show that m-DAGs and RGs are able to distinguish three main groups, roughly corresponding to Fishes, Birds and Mammals. Fishes usually live in entirely different habitats from other animals (i.e.

water and land, respectively), which impact their respective water metabolism and the physiological mechanisms they have evolved to maintain water homeostasis. One crucial feature is oxygen intake, for which fish use gills to "extract" oxygen from the surrounding water environment. On the other hand, land animals developed new organs (namely lungs) to "extract" oxygen from the air. Moreover, Mammals, in addition to Birds, are the only endotherms. The latter physiological feature requires the evolution of the pathways to metabolically produce heat, which is absent in fishes. Instead, Amphibians and Reptiles are not well separated, probably because they are more closely related than other groups. Moreover, amphibians and reptiles possess several physiological features that are similar (e.g., ectodermy, similar immune mechanisms).

### 3.2 Mammals Analysis

Since Mammals are well discriminated by m-DAGs and RGs, we further evaluate their expressiveness by focusing on Mammals only and descending at the Order level, see Fig. 2. This analysis shows that both m-DAGs and RGs are unable to distinguish the various orders. However, m-DAGs clearly separate Primates from the rest of Mammals, while RGs fail to separate many of them.

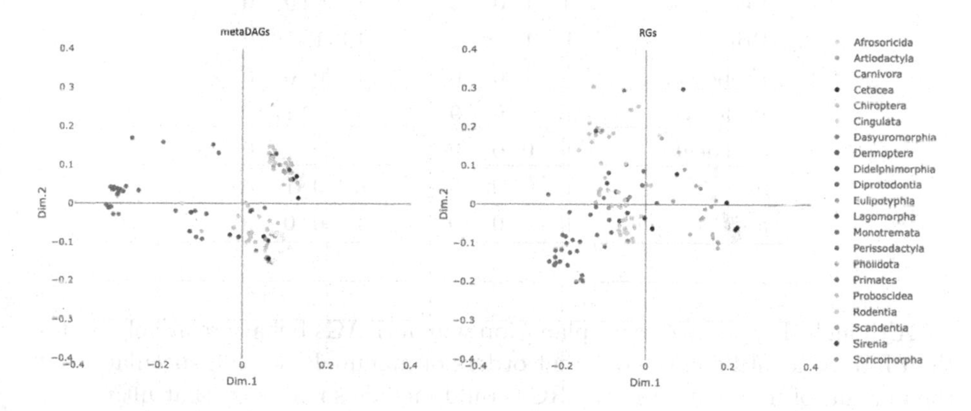


**Fig. 2.** MDS plots for Mammals with m-DAGs (left) and RGs (right). The colors reflect the order each organism belongs to (see the legend on the right).

The MDS results are further supported by the k-means clustering, see Table 3. For this analysis we used k = 4, as suggested by the elbow method. The results show that while RGs do not separate the different clusters based on their taxonomical affiliation, m-DAGs are at least able to identify the Primates cluster. Namely, cluster 3 contains most of Primates and two outliers, but similarly to RG, all the other orders fall in the various clusters without a clear separation. One possible reason that could explain these results is that most of the orders are represented by a very low number of individuals.

**Table 3.** k-means clustering of Mammals for m-DAGs and RGs.

| Order level | m-DAGs Clusters | | | | RGs Clusters | | | |
|---|---|---|---|---|---|---|---|---|
| | 1 | 2 | 3 | 4 | 1 | 2 | 3 | 4 |
| Afrosoricida | 1 | 0 | 0 | 0 | 1 | 0 | 0 | 0 |
| Artiodactyla | 2 | 2 | 0 | 12 | 2 | 2 | 0 | 12 |
| Carnivora | 13 | 2 | 0 | 13 | 8 | 1 | 0 | 19 |
| Cetacea | 3 | 0 | 0 | 3 | 2 | 0 | 1 | 3 |
| Chiroptera | 5 | 8 | 0 | 5 | 10 | 3 | 5 | 0 |
| Cingulata | 0 | 0 | 1 | 0 | 0 | 1 | 0 | 0 |
| Dasyuromorphia | 0 | 2 | 0 | 0 | 1 | 1 | 0 | 0 |
| Dermoptera | 0 | 0 | 1 | 0 | 1 | 0 | 0 | 0 |
| Didelphimorphia | 0 | 2 | 0 | 0 | 2 | 0 | 0 | 0 |
| Diprotodontia | 1 | 0 | 0 | 0 | 1 | 0 | 0 | 0 |
| Eulipotyphla | 0 | 1 | 0 | 0 | 1 | 0 | 0 | 0 |
| Lagomorpha | 0 | 2 | 0 | 0 | 1 | 1 | 0 | 0 |
| Monotremata | 0 | 0 | 0 | 1 | 1 | 0 | 0 | 0 |
| Perissodactyla | 0 | 1 | 0 | 2 | 1 | 0 | 2 | 0 |
| Pholidota | 0 | 1 | 0 | 0 | 1 | 0 | 0 | 0 |
| Primates | 1 | 1 | 26 | 1 | 10 | 1 | 18 | 0 |
| Proboscidea | 0 | 1 | 0 | 0 | 1 | 0 | 0 | 0 |
| Rodentia | 13 | 0 | 0 | 9 | 4 | 1 | 17 | 0 |
| Scandentia | 1 | 0 | 0 | 0 | 0 | 0 | 1 | 0 |
| Sirenia | 1 | 0 | 0 | 0 | 0 | 0 | 1 | 0 |
| Soricomorpha | 1 | 0 | 0 | 0 | 1 | 0 | 0 | 0 |

Although there is no clear explanation why m-DAGs behave relatively better than RGs in identifying the different orders of mammals, we can speculate that the amount of information in the RGs could include some noise that ultimately may result in blurring the identification of the clusters. It may also be possible that the analysis carried out with the WL graph kernel could provide a better result using the m-DAGs w.r.t RGs when considering lower taxonomic levels (i.e. when comparing very similar species).

### 3.3 Primates Analysis

As a last experiment, we test the expressiveness of m-DAGs on Primates at the Infraorder level. As we are now comparing fewer species, namely 29 Primates, we opted for a dendrogram visualisation to show the results, see Fig. 3. We observe that all *Simiiformes* are clustered together except for *Piliocolobus tephrosceles* (KEGG code `pteh`), the Ugandan red Colobus, which falls in the other cluster

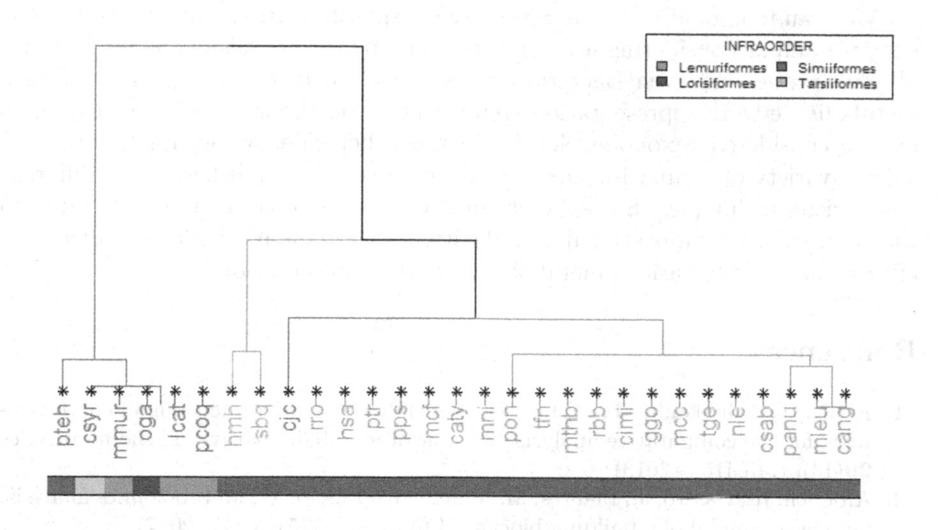


**Fig. 3.** Hierarchical clustering of Primates with m-DAG representation and WL kernel. Colors in the tree identify the different branches. Colors in the horizontal bar are in agreement with the Infraorder classification (see the legend on the top right box).

containing the rest of Primates. Conversely, the other taxa are grouped together with no clear cluster. The difficulty of m-DAGs to identify clustering patterns among the Lemuriformes, Lorisiformes, and Tarsiformes could potentially be attributed to the low number of individuals in each infraorder. We also performed the same experiment at a lower level of taxonomy, i.e., at the Family level, but m-DAGs were unable to clearly distinguish the various taxonomic groups.

## 4 Conclusion

In this paper we explored the expressiveness of AMNs, m-DAGs and RGs for a case study comprising all current KEGG Vertebates. We used the Weisfeiler-Lehman graph kernel as a comparison method and exploratory data analysis techniques to evaluate the results.

Representation and analyses of metabolic networks are not a substitute for sequence-based phylogenetic studies to shed light on the evolutionary and taxonomic relationships between the different organisms. Nonetheless, the different representations of the metabolic network are a fast tool to observe and detect patterns that could later be the focus of more detailed studies. In this context, it emerges that among the three possibilities of metabolic network representation, m-DAG provides a good compromise between the amount of information to use and the reliability of the clustering patterns that we observed. Such an approach could represent a starting point to identify clusters that could be studied further using more traditional biological and evolutionary tools.

We plan to extend our exploration by considering various directions. On one side, we aim at considering a much larger number of organisms, as for instance all Eukaryotes and even Bacteria and Archaea. In fact, the effectiveness of a metabolic network representation could depend on the species under exam, or on the considered taxonomic level. On the other side, we would like to consider a variety of comparison methods, such as other graph kernels or different comparison techniques. Since exact comparison methods are unfeasible due to the impractical computational complexity, different heuristics could affect the effectiveness of the various metabolic network representations.

## References

1. Abaka, G., Bıyıkoğlu, T., Erten, C.: CAMPways: constrained alignment framework for the comparative analysis of a pair of metabolic pathways. Bioinformatics **29**(13), i145–i153 (2013)
2. Alberich, R., Castro, J., Llabrés, M., Palmer-Rodríguez, P.: Metabolomics analysis: finding out metabolic building blocks. PLOS One **12**(5), 1–25 (2017)
3. Beguerisse-Díaz, M., Bosque, G., Oyarzún, D., Picó, J., Mauricio Barahona, M.: Flux-dependent graphs for metabolic networks. NPJ Syst. Biol. Appl. **4**, 32 (2018)
4. Biocyc Database Collection. https://biocyc.org/
5. Biomodels database. https://www.ebi.ac.uk/biomodels/
6. Caetano, R.A., Ispolatov, Y., Doebeli, M.: Evolution of diversity in metabolic strategies. eLife **10**, 1–20 (2021)
7. Chartrand, G., Lesniak, L., Zhang, P.: Graphs & Digraphs, 5th edn. Chapman & Hall/CRC (2010)
8. Cox, M., Cox, T.: Multidimensional scaling. In: Chen, C., Härdle, W., Unwin, A. (eds.) Handbook of Data Visualization. Springer Handbooks Comp.Statistics, pp. 315–347. Springer, Heidelberg (2008). https://doi.org/10.1007/978-3-540-33037-0_14
9. García, I., Chouaia, B., Llabrés, M., Simeoni, M.: Exploring the expressiveness of abstract metabolic networks. PLoS One **18**(2) (2023)
10. Kauffman, K.J., Prakash, P., Edwards, J.: Advances in flux balance analysis. Curr. Opin. Biotechnol. **14**(5), 491–496 (2003)
11. Kyoto Encyclopedia of Genes and Genomes. https://www.genome.jp/kegg/
12. Kriege, N.M., Johansson, F.D., Morris, C.: A survey on graph kernels. Appl. Netw. Sci. **5**(1), 6 (2020). https://doi.org/10.1007/s41109-019-0195-3
13. Nikolentzos, G., Meladianos, P., Vazirgiannis, M.: Matching node embeddings for graph similarity. In: Proceedings of the Thirty-First AAAI Conference on Artificial Intelligence, AAAI 2017, pp. 2429–2435. AAAI Press (2017)
14. Nikolentzos, G., Siglidis, G., Vazirgiannis, M.: Graph kernels: a survey. arXiv e-print (2019)
15. Price, N.D., Reed, J.L., Palsson, B.: Genome-scale models of microbial cells: evaluating the consequences of constraints. Nat. Rev. Microbiol. **2**(11), 886–897 (2004)
16. R Core Team: R: A Language and Environment for Statistical Computing. R Foundation for Statistical Computing, Vienna, Austria (2023). https://www.R-project.org/
17. Rawls, K.D., Dougherty, B.V., Blais, E.M., Stancliffe, E., et al.: A simplified metabolic network reconstruction to promote understanding and development of flux balance analysis tools. Comput. Biol. Med. **105**, 64–71 (2019)

18. Shervashidze, N., Schweitzer, P., Leeuwen, E., Mehlhorn, K., Borgwardt, K.: Weisfeiler-Lehman graph kernels. J. Mach. Learn. Res. **12**(Sep), 2539–2561 (2011)
19. Siglidis, G., Nikolentzos, G., Limnios, S., Giatsidis, C., Skianis, K., Vazirgiannis, M.: GraKeL: a graph kernel library in python. ArXiv abs/1806.02193 (2018)
20. Valiente, G.: Algorithms on Trees and Graphs. Springer, Berlin (2002). https://doi.org/10.1007/978-3-662-04921-1

# scFBApy: A Python Framework for Super-Network Flux Balance Analysis

Bruno G. Galuzzi[1,2,3] and Chiara Damiani[1,3(✉)]

[1] Department of Biotechnology and Biosciences, University of Milano-Bicocca, Milan, Italy
{bruno.galuzzi,chiara.damiani}@unimib.it
[2] Institute of Molecular Bioimaging and Physiology (IBFM), Segrate, Italy
[3] SYSBIO Centre of Systems Biology/ISBE.IT, Milan, Italy

**Abstract.** Constraint-based modelling (CBM) is a computational method used in systems biology to predict metabolic fluxes. However, modelling metabolic fluxes with CBM remains challenging due to the complexity of metabolism and the need for omics data integration. This study introduces scFBApy, a Python-based tool for simulating CBM and the metabolic cooperation between cells. It allows the flux simulation of a population of networks for a target objective, such as biomass production, with or without cooperation. The tool integrates single-cell transcriptomics data using Reaction Activity Scores and uses a denoising algorithm for pre-processing scRNA-seq data. Five real-world scRNA-seq datasets were used to demonstrate the applicability of the pipeline. Results showed that cooperation between cells increased biomass production compared to independent cell simulations. The scFBApy package provides an open-source alternative to MATLAB-based CBM tools.

**Keywords:** Metabolic networks · Flux Balance Analysis · Constraint-based modelling · super-network based modelling

## 1 Introduction

Constraint-based modelling (CBM) is a computational method commonly used in systems biology to predict metabolic fluxes, with many possible applications in health, wellness, and bio-transformations [7]. It is based on the principle that cellular processes operate under certain constraints, such as the availability of nutrients and energy. By mathematically representing these constraints and the metabolic reactions within a cell, CBM can predict how cells will behave under different conditions and identify optimal strategies for cellular functions.

Modelling metabolic fluxes with CBM remains challenging for many reasons. First, creating a feasible and plausible metabolic network is not simple. Toy models made by a few reactions are not able to capture the complexity of metabolism, and large genome-wide models can have many numerical artefacts such as unfeasible loops or hide the essentiality of such reactions. Second,


M. Villani et al. (Eds.): WIVACE 2023, CCIS 1977, pp. 88–97, 2024.
https://doi.org/10.1007/978-3-031-57430-6_8

a correct simulation of metabolic fluxes requires the integration of omics data (proteomics, transcriptomics, and metabolomics) into a metabolic network, and, even if several methods have been provided to do this [1,10,18], there is no general consensus on the best way to integrate such data. In particular, the integration of transcriptomics data at the single-cell (sc) level represents an important challenge due to the possible generation of numerical artefacts due to the high number of false zero values in the sc-RNA seq data [6,12].

### 1.1 State of the Art

CBM is usually used to simulate an individual metabolic network. This network can either represent an entire organism, such as Escherichia Coli, or a part of a multi-cellular organism, as in tissue-specific models [2], or a single-cell.

Recently, some works have also explored the possibility of using a super-network model, in which a set of cells are modelled as a set of networks, one for each cell, to simulate a possible environment in which cells interact with each other and cooperate for a common purpose like the growth or the ATP maintenance. Such type of model could be useful to describe, for instance, the tumour microenvironment (TME) in which interaction/cooperation phenomena between cells have been reported in the literature [14]. In [4], the authors provided an extension of the classical FBA, called popFBA, to explore how metabolic heterogeneity and cooperation phenomena affect the overall growth of cancer cell populations. They showed that a population of cells may follow several different metabolic paths and cooperate to maximize the growth of the total population. In [6], the authors proposed single-cell Flux Balance Analysis (scFBA), a computational tool to translate single-cell transcriptomes into single-cell fluxes. In this framework, the growth of a population of cells is simulated as in popFBA, but also cell-specific constraints from sc-RNA seq data with the use of Reaction Activity Scores(RAS) [13].

All these frameworks have been developed in MATLAB, currently limiting the application of the pipeline, because MATLAB is proprietary software and many life scientists do not have the software licence. Moreover, the optimization of metabolic fluxes can be performed on the entire super-network, which is time-consuming for large single-cell datasets.

### 1.2 Our Contribution

To the best of our knowledge, no Python software has ever been implemented to simulate CBM and cooperation between cells. Here, we present scFBApy (single-cell Flux Balance Analysis in Python), a Python-based COBRApy extension that provides a simple and effective method for the simulation of a population of networks for a specific purpose like biomass production. The simulation can be done assuming cooperation between all the cells, or without cooperation. In this last case, the optimization problem is split into $n$ sub-problem, where $n$ is the number of cells. This strategy is less time-consuming and parallelizable.

We also added the possibility to pre-process the RNA-seq data with a denoising algorithm, called MAGIC [8], to remove the possible false zero values in single-cell RNA-seq data.

Finally, we demonstrate the applicability of the pipeline to five real-world scRNA-seq datasets, where we computed the optimal flux configuration of the metabolic super-network with and without the interactions between cells.

## 2 Material and Methods

### 2.1 Constraint-Based Modelling

A metabolic network can be represented by a stoichiometric matrix $S$ of dimensions $M \times R$, where $M$ is the number of metabolites and $R$ is the number of reactions. A steady-state condition is imposed, meaning that the total production of any metabolite must equal the total consumption of that metabolite. Therefore, any possible configuration of metabolic fluxes is represented by a vector $\vec{v}$ such that $S\vec{v} = 0$.

Among the numerous feasible steady-state flux distributions, Flux Balance Analysis (FBA) [16] calculates a single feasible flux distribution that maximizes the flux through a target function by solving the following linear programming (LP) problem:

$$\max z = f(\vec{v}), \qquad (1)$$
$$S \cdot \vec{v} = \vec{0},$$
$$\vec{v_L} \leq \vec{v} \leq \vec{v_U},$$

where $\vec{v_L}$ and $\vec{v_U}$ represent the possible flux bounds, and $f(\vec{v})$ in an objective function to maximize.

### 2.2 From a Single-Network to a Super-Network

To pass from a single-cell metabolic network model to a super-network one, we follow an approach similar to that in Damiani et al. [6].

We consider a generic input network of $R$ reactions per $M$ metabolites. From this network, we extract all the internal reactions $R_{int}$ and the transportation reactions $R_t$, i.e. the reactions which transport one or more metabolites from the internal to the extra-cellular compartment.

We create a new network formed by $C \times (R_{int} + R_t)$ reactions where we label the reactions with the suffix "*_cell_*" $+c$ where $c$ is one of the $C$ cells. We do the same thing for the metabolites except for the external metabolites. In this way, we are able to simulate a situation in which any cell can uptake/secret any metabolite from any other cell if there is a transportation reaction from the internal to the external compartment, and if this reaction is reversible. Therefore, any cell can cooperate with any other cell of the tissue even if cells are far from each other. We added the $R_{exc} = R - R_{int} - R_t$ exchange reactions of the original network, to simulate the exchange of metabolites with an external compartment, such as a growth medium or blood. The final network has $C \times (R_{int} + R_t) + R_{exc}$ reactions. Finally, we create a new objective function, that corresponds to the sum of all the biomass reactions of all the cells.

### 2.3 Transcriptomics-Derived Constraints to Metabolic Fluxes

Starting from the RNA-seq count matrix and the reactions involved in the metabolic network, we computed the Reaction Activity Scores(RAS) matrix as an $\overline{R} \times C$ matrix, where $\overline{R} \leq R$ represents the number of reactions associated with a Gene Protein Reaction (GPR) rule, and $C$ is the total number of cells. The entries in the matrix are calculated by substituting the mRNA abundances into the corresponding GPRs, as done in [13]. To solve the logical expressions, the minimum transcript value is taken when multiple genes are joined by an AND operator, and the sum of their values is taken when multiple genes are joined by an OR operator.

After the RAS computation, specific constraints on internal fluxes of the network are built following the approach adapted in [10,12]. In a nutshell, for each reaction $j = 1, \ldots, \overline{R}$ and cell $c = 1, \ldots, C$, an upper bound $U_j^c$ and a lower bound $L_j^c$ to the flux capacity are defined, based on the following formulas:

$$U_j^c = F_j^u \times \frac{RAS_j^c}{\max_c RAS_j^c}, \tag{2}$$

$$L_j^c = F_j^l \times \frac{RAS_j^c}{\max_c RAS_j^c}, \tag{3}$$

where $F_j^u$ and $F_j^l$ represent the maximum and the minimum flux that reaction $j$ might carry, obtained by Flux Variability Analysis [15], and $RAS_j^c$ is the RAS value for cell $c$ and reaction $j$. These constraints are used to map RNAseq data, with a one-to-one correspondence between the single cell transcriptomics profile and the corresponding sub-network of the super-network. Therefore, each sub-network has specific constraints derived from the transcriptomics and defined in Eq. 2 and 3.

### 2.4 Data Pre-processing

The integration of scRNA-seq data in a metabolic network requires special attention. Indeed, the high number of possible false zeros in a scRNA-seq count matrix causes a high number of fluxes constrained to be zero using Eqs. 2 and 3. So many zero fluxes can cause feasibility problems for FBA solutions.

To mitigate the presence of false zero values in the transcriptomics data, we applied a denoising algorithm, called MAGIC [8], on the count matrix using the default values of the algorithm, except for the numbers of nearest neighbours that we set to 3 instead of 5.

### 2.5 The ScFBApy Package

The core code of the tool is a Python function that takes as input a metabolic network model and a scRNA-seq dataset. The metabolic network is loaded from SBML or JSON format using the COBRApy library [11], while the scRNA-seq data is provided as an annotated data matrix (AnnData), an efficient data structure provided by the Scanpy library [19]. The output is an AnnData object in which the optimal flux matrix of the entire super-network is provided. The integration with Scanpy offers the possibility to apply several methods, e.g., quality check, denoising, clustering, and differential expression testing on the flux matrix.

### 2.6 Datasets

To demonstrate the applicability of our strategy to real-world datasets, we considered five different single-cell datasets:

**LCPT45** Composed of 34 cells acquired from a xenograft, obtained by sub-renal implantation in mice of a surgical resection of a 37-mm irregular primary lung lesion in the right middle lobea of a 60-year-old untreated male patient.

**H358** Composed of 50 cells from NCI-H358 bronchioalveolar carcinoma cell line.

**LCMBT15** Composed of 49 cells acquired from a xenograft, obtained by sub-renal implantation in mice of a surgical resection of a metachronous brain metastasis acquired from a 57-year-old female after standard chemotherapy and erlotinib treatments.

**BC04** [3] Composed of 59 human epidermal growth factor receptor 2 positive (HER2+) cells.

**BC03LN** [3] Composed of 55 lymph node metastases of human estrogen receptor-positive (ER+) and human epidermal growth factor receptor 2 positive (HER2+) cells.

The first three datasets are obtained from the NCBI Gene Expression Omnibus (GEO) data repository under GEO accession number GSE69405. The last two datasets are a breast cancer dataset of scRNA-seq under GEO accession number GSE75688. Each of the 5 datasets includes the gene expression level in the form of Transcript Per Kilobase Milion (TPM). For each dataset, we retained only the genes included in the metabolic network.

### 2.7 The Metabolic Network Model

We used the recently published network model ENGRO2 [10] of the human central carbon and essential amino acids metabolism. It contains 484 reactions, 403 metabolites, and 497 genes, and represents a follow-up of the core model of human central metabolism ENGRO1 [5]. 337 model reactions are associated with a GPR rule. Biomass pseudo-reaction is the biomass reaction of the Recon3D model, in terms of the set of metabolites considered and corresponding stoichiometric coefficients. The only difference is that, given that ENGRO2 does not explicitly include lipid synthesis, the biomass pseudo-reaction assigns the sum of the stoichiometric coefficients of 1-Phosphatidyl-1D-Myo-Inositol, Phosphatidylcholine, Phosphatidylethanolamine, Phosphatidylglycerol, Cardiolipin, Phosphatidylserine, and Sphingomyelin to palmitate [10].

### 2.8 Experimental Setting

We simulated a growth medium condition in which a specific set of metabolites was abundantly available. This set included glucose, glutamine, arginine, glycine, cystine, oxygen, water, hydrogen, folic acid, and all essential amino acids. We examined two scenarios: one without any cooperation, where each cell's network operated independently from others (NO-COOP), and one with cooperation (COOP), where metabolites secreted by one cell could be utilized by other cells. In the COOP scenario, cells had the potential to secrete several metabolites, including all non-essential amino acids, glucose, lactate, palmitate, and pyruvate, while maintaining a significant biomass production. The production of Pyruvate in particular can occur in multiple ways in the ENGRO2 model, e.g. from the glycolysis pathway, lactate dehydrogenase and the degradation of alanine, serine, or cysteine.

## 3 Experimental Results

### 3.1 Cooperation Between Cells Increases the Biomass Production

In Table 1, we report, for each dataset, the total biomass production per cell and the percentage of cells having a non-negligible biomass production (% feasibility) in case of no cooperation (NO-COOP) and cooperation (COOP) between cells. We can note that the cooperation increases the biomass production across all five datasets. This improvement varies between 32% for the BC03LN to 216% for the BC04 dataset. For this last dataset, we have also an increase in the percentage of feasible cells, which passes from 0.78 to 1.

**Table 1.** Biomass production per cell and the percentage of cells having a non-negligible biomass production in case of no cooperation (NO-COOP) and cooperation (COOP) between cells.

| Dataset | NO-COOP | | COOP | |
|---|---|---|---|---|
| | Biomass per cell | Feasibility | Biomass per cell | Feasibility |
| BC04 | 2.42858 | 0.784314 | 5.268854 | 1 |
| BC03LN | 27.83925 | 1 | 36.81522 | 1 |
| LCPT45_SC | 21.5471 | 1 | 28.72713 | 1 |
| H358_SC | 34.0013 | 1 | 44.97986 | 1 |
| LCMBT15_SC | 44.0021 | 1 | 58.18925 | 1 |

### 3.2 Cells Exchange Specific Metabolites to Increase the Biomass Production

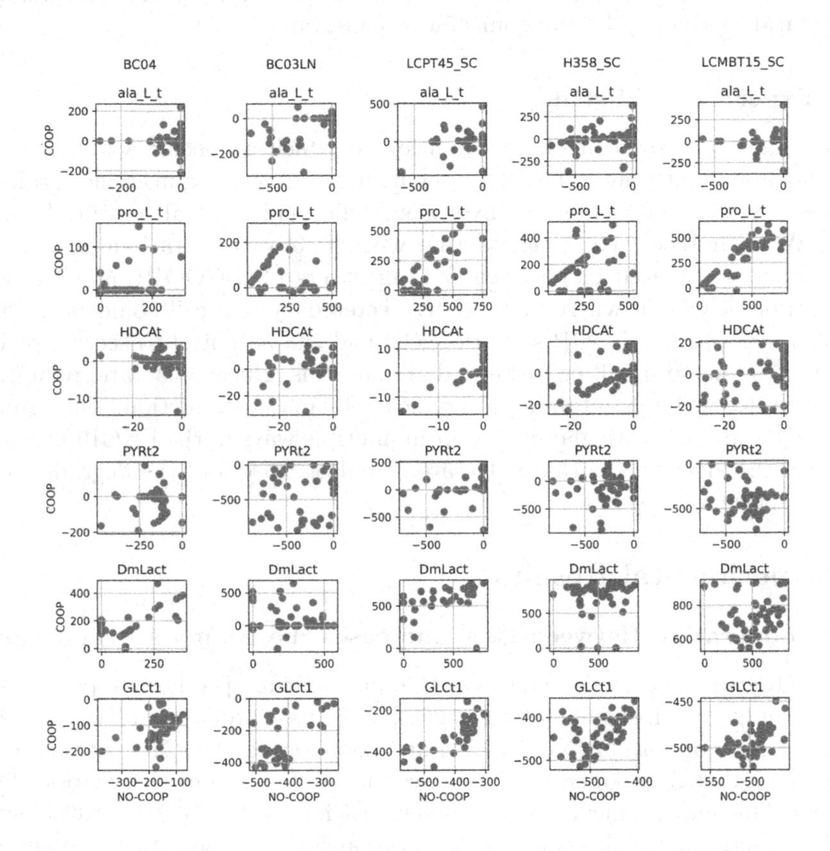


**Fig. 1.** Scatter-plots illustrating the impact of cooperation. Each subplot represents a different simulation of FBA on the network, with the columns representing different datasets, and the rows representing different reactions. The scatter-plot in each subplot compares the values obtained with (COOP) and without cooperation (NO-COOP).

We examined the level of cooperation between cells for the five different datasets. More in detail, we verified that for the super-network models, some cells secret metabolites that are up-taken by other cells, when cooperation is present. In Fig. 1, we reported the comparison of the fluxes for the alanine (Ala_L_t), proline (pro_L_t), palmitate (HDCAt), pyruvate (PYRt2), lactate (DmLact), glucose (GLCt1) exchange, in the no cooperation case (NO-COOP, horizontal axis) and the cooperation one (COOP, vertical axis). Since the growth medium does not contain any of these metabolites, in case of no cooperation, such reactions are used only to secret the metabolites. In the case of cooperation, the flux can be positive or negative as a function of the secretion or consumption of these metabolites, respectively. In particular, it can be observed that a consistent group of cells consumes the alanine, pyruvate, and palmitate that are secreted by the other cells. On the contrary, there is no significant lactate exchange between cells for all the datasets except for BC03LN one. The proline exchange is moderate but non-negligible in all the datasets. No cell in any dataset secretes glucose to be consumed by other cells, reasonably because the medium is rich in glucose, hence the cells do not need to exchange glucose with each other. However, it is interesting to note that the uptake of glucose inside the cells in the cooperative case is generally lower than in the non-cooperative one. We speculate that it could mean that the general glucose consumption of the super-network is optimized for biomass growth.

Finally, to demonstrate the significance of specific metabolite exchange between cells, in Table 2, we reported the relative variation in biomass production when the exchange of a particular metabolite is restricted on the super-network. It can be noted that the removal of proline exchange strongly affects biomass production for the BC04 dataset only. We observed also that the palmitate exchange between cells affects biomass production for three of these datasets. On the contrary, lactate exchange is not strictly required for maximal growth for any cells of any dataset.

**Table 2.** Relative variation in biomass production per cell when the uptake of a particular metabolite is restricted for all cells.

| Reaction | BC04 | BC03LN | LCPT45 | H358 | LCMBT15 |
|---|---|---|---|---|---|
| Palmitate | 1 | 1 | 0.96 | 0.96 | 0.99 |
| Pyruvate | 1 | 1 | 1 | 1 | 1 |
| Lactate | 1 | 1 | 1 | 1 | 1 |
| Alanine | 1 | 1 | 1 | 0.99 | 1 |
| Asparagine | 1 | 1 | 1 | 0.99 | 1 |
| Aspartic acid | 1 | 1 | 1 | 1 | 1 |
| Serine | 1 | 1 | 1 | 0.98 | 1 |
| Proline | 0.60 | 1 | 1 | 1 | 1 |

### 3.3 Software Availability and Computational Architecture

All computations were performed on an Intel(R)@3 GHz 32 GB, using GLPK as solver and one single CPU. The source code and documentation are available at https://github.com/CompBtBs/scFBApy.

## 4 Discussion and Conclusions

Constraint-based modelling to calculate feasible metabolic fluxes is gaining growing significance in comprehending the mechanisms associated with the physiological and pathological conditions of cells or organisms. In this work, we considered

Flux Balance Analysis for a super-network model. Starting from a previously published framework in MATLAB, we provide a Python tool for single-cell Flux Balance Analysis which integrates the information coming from transcriptomics data and allows cells to cooperate to maximize the total biomass production.

The possible limitations of this work are the following. First, since the definition of a biomass reaction is a hard task, this can be a factor limiting our ability to describe metabolism using FBA [9]. This is because the composition of biomass can be tissue and physiological condition-dependent. Another limitation of the work is related to the assumption of biomass production as a common FBA objective of the cell population of the four datasets. Defining an objective function of FBA is a hard task, even in the simplest case of a single cell. Indeed, defining one or multiple objective function(s) intrinsically introduces an observer bias as to what the main "goal" of the cell is, in the context of the analysis. However, biomass production can be a realistic assumption for proliferative conditions like in a tumour microenvironment [17]. Moreover, we remark that in our implementation we do not assume that each cell must grow, but that the overall population must. To this aim, another possible limitation of our approach is that any cell can cooperate with any other cell in the tissue to increase the total biomass production, even if they are far from each other. Finally, one last limitation is related to the denoising strategy used to remove false zero values. Since the application of denoising alters the RNA profiles, it might be necessary to check the robustness of the results using different parameter configurations or denoising algorithms.

As a further work, we plan to include spatial information on the super-network, to have possible cooperation between cells as a function of their proximity. This could be useful for the analysis, e.g. of datasets of single-cell spatial transcriptomics. Indeed, with this tool, one could simulate the single-cell spatial fluxomics from spatial scRNAseq data and perform flux cluster analysis to study the possible metabolic differences between sub-populations of cells, along the space, and compare the clusters with the ones obtained from transcriptomics.

## References

1. Agren, R., Bordel, S., Mardinoglu, A., Pornputtapong, N., Nookaew, I., Nielsen, J.: Reconstruction of genome-scale active metabolic networks for 69 human cell types and 16 cancer types using INIT. PLoS Comput. Biol. **8**, e1002518 (2012)
2. Bordbar, A., Feist, A., Usaite-Black, R., Woodcock, J., Palsson, B., Famili, I.: A super-tissue type genome-scale metabolic network for analysis of whole-body systems physiology. BMC Syst. Biol. **5**, 1–17 (2011)
3. Chung, W., et al.: Single-cell RNA-seq enables comprehensive tumour and immune cell profiling in primary breast cancer. Nat. Commun. **8**, 1–12 (2017)
4. Damiani, C., Di Filippo, M., Pescini, D., Maspero, D., Colombo, R., Mauri, G.: popFBA: tackling intratumour heterogeneity with Flux Balance Analysis. Bioinformatics **33**, i311–i318 (2017)
5. Damiani, C., et al.: A metabolic core model elucidates how enhanced utilization of glucose and glutamine, with enhanced glutamine-dependent lactate production,

promotes cancer cell growth: the WarburQ effect. PLoS Comput. Biol. **13**, e1005758 (2017)
6. Damiani, C., et al.: Integration of single-cell RNA-seq data into population models to characterize cancer metabolism. PLoS Comput. Biol. **15**, e1006733 (2019)
7. Damiani, C., Gaglio, D., Sacco, E., Alberghina, L., Vanoni, M.: Systems metabolomics: from metabolomic snapshots to design principles. Curr. Opin. Biotechnol. **63**, 190–199 (2020)
8. Dijk, D., et al.: MAGIC: a diffusion-based imputation method reveals gene-gene interactions in single-cell RNA-sequencing data. BioRxiv, p. 111591 (2017)
9. Dikicioglu, D., Kírdar, B., Oliver, S.: Biomass composition: the "elephant in the room" of metabolic modelling. Metabolomics **11**, 1690–1701 (2015)
10. Di Filippo, M., et al.: INTEGRATE: model-based super-omics data integration to characterize super-level metabolic regulation. PLoS Comput. Biol. **18**, e1009337 (2022)
11. Ebrahim, A., Lerman, J., Palsson, B., Hyduke, D.: COBRApy: constraints-based reconstruction and analysis for python. BMC Syst. Biol. **7**, 1–6 (2013)
12. Galuzzi, B., Vanoni, M., Damiani, C.: Combining denoising of RNA-seq data and flux balance analysis for cluster analysis of single cells. BMC Bioinform. **23**, 1–21 (2022)
13. Graudenzi, A., et al.: Integration of transcriptomic data and metabolic networks in cancer samples reveals highly significant prognostic power. J. Biomed. Inform. **87**, 37–49 (2018)
14. Junttila, M., De Sauvage, F.: Influence of tumour micro-environment heterogeneity on therapeutic response. Nature **501**, 346–354 (2013)
15. Mahadevan, R., Schilling, C.: The effects of alternate optimal solutions in constraint-based genome-scale metabolic models. Metab. Eng. **5**, 264–276 (2003)
16. Orth, J., Thiele, I., Palsson, B.: What is flux balance analysis? Nat. Biotechnol. **28**, 245–248 (2010)
17. Santi, A., et al.: Cancer associated fibroblasts transfer lipids and proteins to cancer cells through cargo vesicles supporting tumor growth. Biochimica Et Biophysica Acta (BBA)-Mol. Cell Res. **1853**, 3211–3223 (2015)
18. Wagner, A., et al.: Metabolic modeling of single Th17 cells reveals regulators of autoimmunity. Cell **184**, 4168–4185 (2021)
19. Wolf, F., Angerer, P., Theis, F.: SCANPY: large-scale single-cell gene expression data analysis. Genome Biol. **19**, 1–5 (2018)

# Semantic Information as a Measure of Synthetic Cells' Knowledge of the Environment

Lorenzo Del Moro[1], Maurizio Magarini[1(✉)], and Pasquale Stano[2(✉)]

[1] Dipartimento di Elettronica, Informazione e Bioingegneria (DEIB), Politecnico di Milano, Milan, Italy
lorenzo.delmoro@mail.polimi.it, maurizio.magarino@polimi.it

[2] Department of Biological and Environmental Sciences and Technologies (DiSTeBA), University of Salento, Lecce, Italy
pasquale.stano@unisalento.it

**Abstract.** The concept of semantic information refers to the type of information that has some "significance" or "meaning" for a given system. Its use to describe how precisely the desired meaning is conveyed makes possible to characterize systems in terms of autonomous agents that are able to achieve an intrinsic goal or to accomplish a specific task. Two different types of semantic information are well recognized and used in the literature: i. 'stored' semantic information, which refers to information exchanged between a system and its environment in its initial distribution, and ii. 'observed' semantic information, which denotes the information that is dynamically acquired by a system to maintain its own existence. Both the concepts of stored and observed semantic information were first introduced by Kolchinsky and Wolpert in 2018.

In this paper we present an approach to measure observed semantic information. Its quantitative measure is obtained for a smart drug delivery scenario where synthetic cells sense an environment made up of cancerous cells. These release a signal molecule that triggers the production of a cytotoxic drug by the synthetic cell. For the same scenario, the stored semantic information has already been computed. The main novel contribution compared to the evaluation of stored semantic information consists in a measure of the minimal perception of the environment [in bits] that allows a system to maintain its own functionality (as a proxy of its own existence) during its joint dynamic evolution with the environment, i.e. not decreasing its viability compared to full environment perception. Moreover, we provide a preliminary discussion about how the quantification of semantic information can contribute to better define what is meaningful to an agent. With this result we emphasize once again the role that "synthetic cells" have as new (bio)technological platform for theoretical and applied investigations of semantic information in biological systems.

**Keywords:** Semantic Information · Synthetic Cells · Molecular Communications


M. Villani et al. (Eds.): WIVACE 2023, CCIS 1977, pp. 98–110, 2024.
https://doi.org/10.1007/978-3-031-57430-6_9

## 1 Introduction

Recent studies in biological systems [8] have acknowledged the importance of examining living systems and their components from an information-centric perspective. The classical measures from Shannon information theory, however, seem not to be suitable to fully characterize all aspects of the transfer of information in living systems, as they focus on syntactic aspects, disregarding the concept of what the information means, i.e., its semantic/pragmatic aspects, its "meaning" for the sender and the receiver (considered, each one or both of them, as agents). Terms as semantic information and meaning need more specifications. As evidenced in [17] (and references therein), the definition and usage of the information concept (a mathematical one), is not connected to its meaning. Developed mainly as a theory for the transmission of digital signals, the Shannon information theory does not deal with what messages mean. Messages, of course, typically have meaning, but their meaning is not necessary for the Shannon information theory. This approach was considered incomplete by other scholars (e.g., MacKay [2,18]), who wanted to highlight semantic aspects of information too, according to the common usage of the word information in human communication. Actually, a theory of semantic information was developed in the 1950s [5] for logical propositions (for a review of this and other theories, see [10]). The concept of meaning refers, directly or indirectly, to the purpose for which the information-bearing data was sent and/or what changes the information-bearing data generate once received. Transmitted data should "make sense" for the sender and/or for the receiver in order to be meaningful, otherwise they represent just noise. The meaning of information, in a communication process, can be identified by the observers, but the process that gave rise to its origin is more complex. In biological agents, it can be argued that it is linked to their functioning mechanisms, their specific way of interacting with the environment – in other words to their evolution [3,9,21–23]. However, the mechanism of meaning origination is gaining major relevance also in the sciences of artificial, because artifacts normally do not recognize meaning in their input/output data [13]. Understanding meaning and semantic information becomes, therefore, a timely scientific question. For these reasons, the interest in the semantic aspects of information and meaning has increased recently.

Although the importance of adopting a measure of information that takes into account the semantic aspects is of fundamental importance in the context of biological systems, up to now only few works in the literature have addressed the problem of its quantitative evaluation. Among the approaches to measure semantic information, that proposed by Kolchinksy and Wolpert (KW) [14] seems to be one of most promising for application into the biological domain because of its operative definition. Its introduction allows for the definition of new quantitative metrics to measure semantic information that are more suitable to describe the interaction between an agent and its surrounding environment. The authors of [19] first conceived the idea of applying the KW approach to model the behavior of a synthetic cell (SC). The idea was then developed from a operational point of view in [24], where a (bio)technological platform was proposed to describe how

SCs extract semantic information from the environment with the intrinsic goal of releasing a cytotoxic drug to kill cancerous cells. An alternative to the KW approach was proposed in [4], where a computational state machine was introduced to evaluate the "subjective information." Subjective information refers to the ability of a living system to optimize its growth and survival over time and its capacity to maximize information acquisition from its surroundings. Reported simulation results demonstrates that a strategy that maximizes information efficiency grows at a slower rate than a strategy that gets less information but has a greater survival value. It is worth noting that, while the subjective information plays an important role in characterizing the ability of an individual organism to opportunistically manipulate the information channel between itself and its environment, our interest is more towards the cost of acquiring or processing more "relevant" information from the environment. We therefore keep the focus on the quantitative evaluation of the semantic information to gain a deepen insights on how it affects the agent's viability.

The main goal of this article is to delve further the application of the KW approach to characterize the impact of coarse-grained joint dynamics induced by interventions of the SC and its surrounding environment. Differently from [7,24], where the goal is to evaluate the amount of *stored semantic information* between an autonomous agent and its environment, in this work we present a computational approach to evaluate the *observed semantic information* in the same scenario as [24], i.e., a "smart" drug delivery system (an SC) engaging semantic molecular communication with its environment.

The rest of the article is organized as follows. In Sect. 2 we introduce the concept of observed semantic information, its mathematical formalism, and give its interpretation for the considered SC scenario. Numerical results showing the temporal behaviour of the viability for different intervened joint agent-environment dynamics are reported in Sect. 3. Section 4 presents a preliminary discussion on how semantic information can be associated with the degree of "knowledge" SCs possess about their environment. Finally, conclusions are reported in Sect. 5, where the main properties of semantic information are summarized and interpreted from the point of view of meaningfulness for an SC.

## 2 Observed Semantic Information

Observed semantic information refers to that amount of syntactic information that, differently from the stored one, is acquired during the autonomous agent-in-the-environment joint dynamics, and that causally contributes to maintain the autonomous agent in existence. Joint dynamics describe the interaction and coordination of an agent with its surrounding environment, wherein the environment and the agent both influence and are influenced by the generated, transmitted, and updated semantic information. In the context of the present analysis, joint dynamics are defined by the joint conditional probabilities of the states of the agent and of the environment in a time instant given those at the previous time. To evaluate observed semantic information, interventions are applied to the joint dynamics of the autonomous agent and of the environment with the objective of perturbing the flow of information between them and then evaluate the part of such a flow that affects the autonomous agent existence. The temporal flow of syntactic information is evaluated by computing the concept of transfer entropy from the state description of the environment $Y_t$ at time $t$ to the state of the system $X_{t+1}$ at time $t+1$ as [25]

$$T_p(Y_t \rightarrow X_{t+1}) = I_p(X_{t+1}; Y_t|X_t) = H(X_{t+1}|X_t) - H(X_{t+1}|X_t, Y_t), \tag{1}$$

where $I_p(X_{t+1}|X_t, Y_t)$ is the conditional mutual information, i.e. the difference of conditional Shannon entropies $H(\cdot|\cdot)$, for a given joint distribution $p_{X_{t+1},X_t,Y_t}$ [14]. The above equation defines how much uncertainty is reduced about the state of the autonomous agent $X_{t+1}$ at time $t+1$ by the knowledge of the state of $Y_t$ at time $t$, conditioned the full knowledge of the state $X_t$ at time $t$.

Furthermore, since the goal is to compute the transfer entropy over time from 1 to $\tau$, that is the time interval of the simulation, the overall syntactic information flow is simply the sum over time of each dynamic flow from an instant to another, so that we compute:

$$\sum_{t=0}^{\tau-1} T_p(Y_t \rightarrow X_{t+1}) = \sum_{t=0}^{\tau-1} I_p(X_{t+1}; Y_t|X_t). \tag{2}$$

Interventions are defined by coarse-grain functions that are applied to the overall dynamic of the system. This latter is given by a first order Markov described by the following conditional probability $p_{X_{t+1},Y_{t+1}|X_t,Y_t} = p_{X_{t+1}|X_t,Y_t} \cdot p_{Y_{t+1}|X_{t+1},X_t,Y_t}$, which is factored by applying the chain of conditional probability. To evaluate observed semantic information, interventions are applied only to the first term on the right-hand side since, in this manner, only the information flow between the environment and the system is scrambled and not the opposite flow of information.

The intervened conditional probability is then computed as reported by KW in the following way

$$\hat{p}^{\phi}(x_{t+1}|x_t, y_t) \triangleq p(x_{t+1}|x_t, \phi(y_t)) = \frac{\sum_{y_t':\phi(y_t')=\phi(y_t)} p(x_{t+1}|x_t, y_t')\hat{p}^{\phi}(x_t, y_t')}{\sum_{y_t':\phi(y_t')=\phi(y_t)} \hat{p}^{\phi}(x_t, y_t')}, \tag{3}$$

where $\phi(\cdot)$ denotes the coarse-graining function that is applied to the discrete state space of the environment. The sum over $y_t'$ means that the intervention coarse-grains the environment state and the system cannot recognize that granularity.

Clearly, there is the need to understand which of the different interventions destroy only that part of the overall syntactic information that is meaningful (that causally contributes to the system existence), and therefore an *optimal* intervention must be found. This optimal intervention is defined as the one that preserves the viability function of the system (that describes the autonomous agent existence), while reducing as much as possible the transfer entropy. In this case the remaining part of transfer entropy is the part of information that is meaningful for the system existence. Eventually, the optimal intervention is defined as

$$\hat{p}^{opt}_{X_{0,\dots,\tau},Y_{0,\dots,\tau}} \in \arg\min_{\hat{p}^{\phi}_{X_{0,\dots,\tau},Y_{0,\dots,\tau}}:\phi\in\Phi} \sum_{t=0}^{\tau-1} T_{\hat{p}^{\phi}}(Y_t \rightarrow X_{t+1})$$
$$s.t \quad V(\hat{p}^{\phi}_{X_\tau}) = V(p_{X_\tau}), \tag{4}$$

where $\hat{p}^{\phi}_{X_{0,\dots,\tau},Y_{0,\dots,\tau}}$ describes the joint distributions over time, from 0 to $\tau$ and $\Phi$ represents the whole set of possible interventions.

Once the optimal intervention $\hat{p}^{opt}_{X_{0..\tau},Y_{0..\tau}}$ is obtained, it is straightforward to find the amount of observed semantic information as

$$S_{obs} \triangleq \sum_{t=0}^{\tau-1} T_{\hat{p}^{opt}}(Y_t \rightarrow X_{t+1}). \tag{5}$$

From the perspective of more traditional digital communication systems, interventions play the same role as that of the semantic source encoder, whose role is to encode noisy data gathered by a sensor and deliver the pertinent and important piece of information [11]. Interventions consists in a randomization of the environment with the goal of breaking the (possible) correlations it has with the agent. Here, the randomization method suggested by KW is used, where interventions are generated by applying permutations with repetition to give the coarse-graining functions merging the discrete states of the environment. Randomization ensures that any potential irrelevant factor, known or unknown, is similarly distributed in each of the intervened groups. Therefore, the measure of the observed semantic information can be associated with the maximum degree of randomization of the environment that does not affect the viability of the agent. It is measured in bits and the resulting value gives the power of 2 that corresponds to the minimal number of states that allow for not decreasing the agent's viability and maintains it in the desired state of staying alive.

## 3 Numerical Results

As detailed in [7,24], the scenario consists of an SC that perceives signaling molecules from cancerous cells (CCs) in the surrounding and releases, based on an internal mechanism, cytotoxic drug molecules with the goal of killing CCs. In this manner, we sought to devise and model a realistic scenario, inspired to published reports [15,16]. In particular, it has been shown that SCs can produce a toxin called *Pseudomonas* exotoxin A, capable of killing 4T1 breast cancer cells in culture [15]. Such a result, combined with other investigations showing that SCs can start protein synthesis upon receiving a chemical signal [1], makes our scenario convincing, although its actual realization is still lacking. In the following, we will use the term "toxin", but the general case would correspond to SCs capable of producing any type of cytotoxic drug.

**Table 1.** Initial parameters of the simulation.

| Parameters | | |
|---|---|---|
| Environment and SC variable states | Possible levels | Initial probabilities |
| $X_{S_{per}}$ | $\{0, 1, 2, 3, 4, 5\}$ | $\{0, \frac{1}{5}, \frac{1}{5}, \frac{1}{5}, \frac{1}{5}, \frac{1}{5}\}$ |
| $X_{S_{in}}$ | $\{0, 1, 2, 3, 4, 5\}$ | $\{1, 0, 0, 0, 0, 0\}$ |
| $X_{ptox}$ | $\{0, 1, 2, 3, 4, 5, 6, 7, 8\}$ | $\{0, 0, 0, 0, 1, 0, 0, 0, 0\}$ |
| $Y_S$ | $\{0, 1, 2, 3, 4, 5\}$ | $\{0, \frac{1}{5}, \frac{1}{5}, \frac{1}{5}, \frac{1}{5}, \frac{1}{5}\}$ |

The environment $\mathcal{Y}$ is characterized by the level of signalling molecules released by the CCs and described by a single state variable $Y_S = \{0, \ldots, 5\}$, while the autonomous agent $\mathcal{X}$, i.e. the SC, is described by a state variable $X$ with three degree of freedom, such that $X = X_{S_{in}} \times X_{S_{per}} \times X_{ptox}$. In particular $X_{S_{per}} = \{0, \ldots, 5\}$ is the variable associated with the perceived level of the signaling molecules, supposed to be always equal to $Y_S$. $X_{S_{in}} = \{0, \ldots, 5\}$ is instead the internalization level of $Y_S$ and it is directly related to $X_{S_{per}}$. In addition, $X_{ptox} = \{0, \ldots, 8\}$ is the amount of toxin molecules inside the SC. At time $t = 0$ the signaling molecules levels are considered to be uniformly distributed over $1, \ldots, 5$ which means that at least a signaling molecules is always present. Furthermore, we supposed to have perfect correlation between $Y_S$ and $X_{S_{per}}$, i.e. $Y_S = X_{S_{per}}$, while initially the internalization level and the amount of toxin molecules inside the SC are $X_{S_{in}} = 0$ and $X_{ptox} = 5$. Table 1 summarize the initial parameters of the simulation.

As in [7,24], the joint dynamics are based on internal SC mechanisms and on the interaction with the signalling molecules. Even though, the aim is the evaluation of the meaningful flow of syntactic information exchanged during the dynamics, the scenario and the dynamics are exactly the same.

### 3.1 Evaluation of Viability and of Semantic Information

Figure 1A shows the viability over time for different interventions. The reference curve for the viability is the non-intervened one, which is that with the highest values in each time step. The interventions that do not lead to a decrease of the viability are those for which there is not any degradation in the SC's perception of the environment. These can be obtained only by evaluating the temporal evolution of the viability considering all the possible interventions. Therefore, in Fig. 1A are reported the different temporal evolution's of the viability obtained considering all the possible interventions for the considered scenario. As it can be observed, only six different curves are distinguishable because the overlap among the obtained curves. The differences in the viability are due to the reduction of SC's perception of the environment that arises due by the merging of the states induced by permutations with repetitions used for intervention. Some of these interventions leads to the same reduction of the SC's perception. The dashed line in the figure defines the time instant $\tau = 6$, which will be considered in the following to evaluate the observed semantic information. From the figure it is possible to note how some interventions can decrease the viability over time, reducing consequently the meaningful flow of information during the dynamics. Here, a 'coarse-grained' intervention has been used to scramble the channel between the SC and the environment. This type of intervention is quite a conventional approach in the evaluation of information-theoretic metrics when applied to a discrete state variable. A decrease of the viability is associated with coarse-grained interventions where the merging of the states lead to a reduced perception of the environment, and therefore destroy part of the information flowing across the channel.

An issue can be noted. There should not be a steady state viability, after a certain time instant, different for each intervention. Viability in the steady state regime has to be roughly the same for each intervention, due to the fact that the autonomous agent must be dead after a certain time instant (we consider the SC dead when all the toxin are released) and thus it must have a viability of $-100$ bits at least, which is the dead entropy value we set. Therefore, it is needed to understand why, even though interventions affect viability, the steady state regime is not the same for each of them.

Another issue, related to the simulation itself, comes from the highly variable number of possible states we have introduced for the autonomous agent and the environment. In fact, in this particular simulation, a reduced states space is used. However, in case we want to increase it, we may face some computational problems due to the limited size of the computer memory.

Back to the first issue, we can say it is related to the probability distribution $p_{Y_{t+1}|X_{t+1},X_t,Y_t}$, in fact this probability is not normalized due to our particular joint dynamics. This means that even though we have a defined probability for a possible joint state $X_{t+1}, X_t, Y_t$, the probability of being in state $Y_{t+1}$ is not always defined. In order to solve this problem, we introduce four different approaches to normalize $p_{Y_{t+1}|X_{t+1},X_t,Y_t}$. These are explained in the following section.

In addition, Fig. 2 depicts the viability curve at time $\tau = 6$ (vertical red dot line in Fig. 1) both for the normalized and not normalized case. Applying interventions as previously explained leads to the decrease of the viability. In our simulation, there are more optimal interventions and these lead to an observed semantic information $S_{obs} = 3.91$ bits both in absence of normalization and by applying a normalization as explained in the next section.

### 3.2 Normalization of $p_{Y_t+1|X_t+1,X_t,Y_t}$

The first approaches consists in having a stochastic matrix of the dynamics that is strictly positive and consequently the matrix $p_{Y_{t+1}|X_{t+1},X_t,Y_t}$ is fully specified. The second one, also proposed in [14], consists instead in the definition of a Bayesian network, so that we can specify the matrix $p_{Y_{t+1}|X_{t+1},X_t,Y_t}$. By the way, the first approach depends on the dynamics we want to apply, while the second one is sometimes hard to implement due to a complex dynamics, therefore they are not always feasible. A third approach could be that of approximating

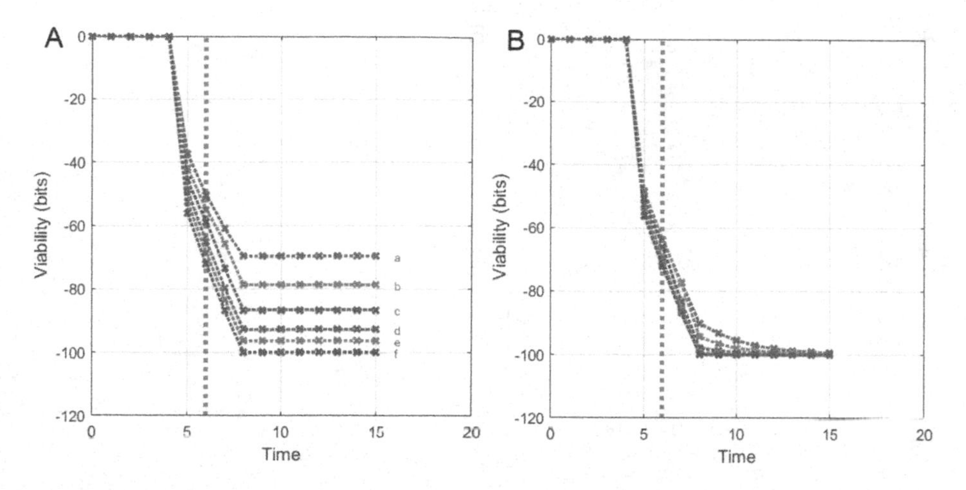


**Fig. 1.** (A) Viability over time without normalization of $p_{Y_{t+1}|X_{t+1},X_t,Y_t}$. Different viabilities under different interventions are associated to letters from 'a' to 'f'. (B) Viability over time with normalization of $p_{Y_{t+1}|X_{t+1},X_t,Y_t}$. The figure has been presented as a poster at the NANOCOM 2023 conference, see Acknowledgements and [6].

the matrix $p_{Y_{t+1}|X_{t+1},X_t,Y_t}$ by specifically adding some values, when it is not specified with the aim to normalize it, but of course we cannot master the degree of approximation. A fourth approach, instead, requires the a-priori knowledge of the intervened matrix of the dynamics. In this way, from the already known intervened matrix we can obtain the matrix $p_{Y_{t+1}|X_{t+1},X_t,Y_t}$ that is normalized where we are interested. As an example, Fig. 1B shows the viability over time for different interventions, when the third approach is used. We choose the third method because it is easy to implement and because it gives the same semantic information as the non-normalized case.

## 4 Interpreting Semantic Information Values as a Measure of "Knowledge" SCs Have About Their Environment: A Preliminary Discussion

We have shown how to calculate the values (in bits) of observed semantic information referred to a SC situated in a very simple environment (E), constituted by CCs that secrete a chemical signal capable of activating a SC response (Sect. 3). In a previous study, we have calculated, for the same system, the stored semantic information [24]). The dynamics of the SC/E have been described, in both cases, by a set of rules that, based on SC and E variables, allow the calculation of the viability $V$. By applying coarse-graining functions to the distribution of variables in E, the SC viability in several intervened E can be calculated, and semantic information finally obtained by a ranking process. As metrics for measuring stored and, respectively, observed semantic information, we have used

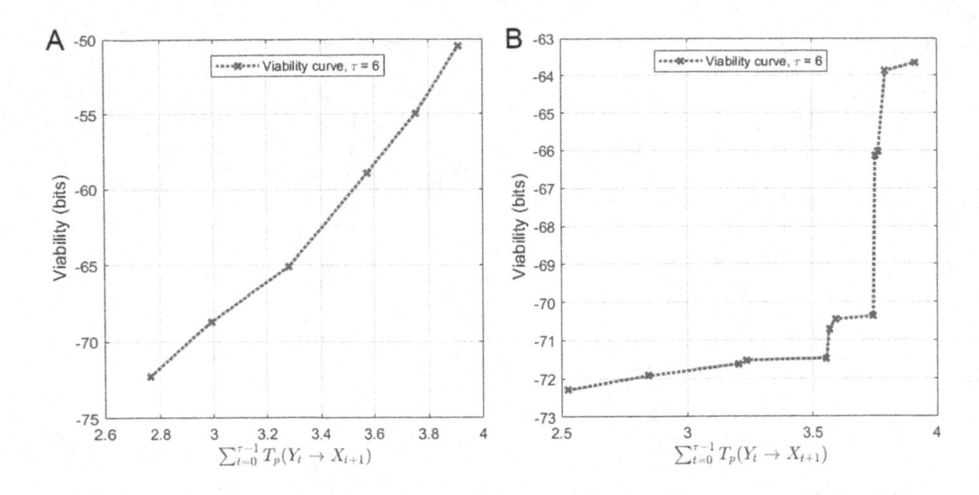


**Fig. 2.** (A) Viability curve over time without normalization of $p_{Y_{t+1}|X_{t+1},X_t,Y_t}$. (B) Viability curve at time instant 6 with normalization of $p_{Y_{t+1}|X_{t+1},X_t,Y_t}$.

mutual information and transfer entropy, respectively (in principle, other metrics could be used [14]). The calculated values are 1.92 bits (stored) and 3.91 bits (observed). The next crucial question is the following: does semantic information values correlate with the knowledge SCs have of their environment, and how?

Two additional considerations are needed to start a discussion, even if preliminary as in the current case. First, we need to make a step back and recall that the viability function $V$, in our specific scenario, has been defined based on the SC capacity of producing a toxin that would kill CCs. In other words, our scenario is referred to a system whose goal is performing a useful action (killing CCs). At this aim, SCs have internal mechanisms that are functional to merely achieve this goal. SCs are intended as *chemical machines*, designed and constructed for a goal which is different than the production of themselves. By production of themselves we mean the production of their own constitutive components, necessary to their very existence as dynamical systems. In other words, they are *allopoietic* machines. On the other hand, KW devised their semantic information theory to investigate systems (in particular, living ones although the analysis can be applied to not living ones too) whose viability function specifically mirrors their capacity of maintaining their own structure and their functioning, in order to continue to exist. These are self-maintaining systems. It is important to recall, indeed, that the systems we are interested in are dynamical systems existing in non-equilibrium conditions. In the context of SC research, when systems exist because they maintain their structure via a continuous production of their components, which otherwise disappear because of disruptive transformations, they are called *autopoietic* machines, and are alive according to Maturana and Varela [20]. Therefore, although the results of our numerical simulations refer to allopoietic systems, a similar approach could have been applied to autopoietic ones.

Second, we should consider that the specific SC we have devised, whose behavior is determined by a certain set of rules, it is just a particular case of a hypothetical variety of SCs whose individual behavior differs owing to different internal mechanisms. We can call a "set of rules", for short, an "organization". Regardless of whether SCs under scrutiny are allopoietic or autopoietic machines, we can imagine a variety of SCs, and therefore a variety of different organizations, that will generate different SC/E dynamics, different values of the viability function, and different semantic information values. Clearly, the scenario we have studied is particularly simple, and probably the allowed variability of organization/behavior/viability function is, for all practical purposes, rather limited. On the other hand, if E would have been much more complex (e.g., made of several variables such as nutrients, signals, inhibitors, etc., each one with its own peculiar distribution), the diversity among all possible organizations that allow SCs cope with the environment (and the set of joint dynamics) would be much more evident.

Based on these consideration, we can extrapolate a hypothetical scenario that is useful to propose the following interpretation of semantic information. Stored or observed semantic information, measured in bits, is a measure of the "knowledge" SCs have about their environment. Let us imagine have a population of different autopoietic SCs situated in a certain environment E. The goal of autopoietic SCs is maintaining their own structure in a dynamical homeostatic state, i.e., maintain their internal processes of self-production so that their existence is assured (SCs with organizations not compatible with E will disappear, and they are of no interest). SCs, whose organization leads to viability function values that allow self-maintenance, will survive. However, because of their diversity, it is expected that their viability function values will be different, and consequently the semantic information associated to the SC/E dynamics will differ too. High values of semantic information, in general, would mean that SCs lose viability even when minor coarse-graining interventions are applied to the environmental variables. Low values of semantic information instead would mean the opposite, i.e., that SC lose viability only when extensive coarse-graining interventions are applied. In the latter case, SCs internal organization, although assuring high-enough viability values, cannot distinguish the fine details of the environment: coarse-graining interventions does not damage their viability. For these SCs, not all details of the environmental variables' distributions "make sense" or "have meaning" (e.g., they can be scrambled, with no - or minor - effects on SCs functioning). Vice versa, SCs characterized by high values of semantic information have internal organizations that match well and with a major *resolution* with a certain E, exploiting the environmental patters at a full extent. For these SCs, environmental variables' distribution and their details do make sense, or have meaning. It can be said that SCs, which "extract" more semantic information from the environment, know better their environment, or are better observers of it.

In this artificial scenario, whereby SCs have been constructed in the laboratory, their inner organization is given by the designer. In a natural evolution scenario, the SC inner organization would instead result from a process of progressive adaptation of the SC to the environment, in a process of reciprocal influence (in language of autopoietic theory: due to the structural coupling).

The discussion on the link between semantic information, meaning, and its origin in natural and artificial systems, clearly, is highly speculative at this stage. But we believe that the tools provided by the KW strategy are powerful ones and very apt to face these investigations. Simulations specifically devised to explore the behavior of varieties of SCs have not been performed yet. Moreover, the best models would be autopoietic systems, not allopoietic ones. One critical hypothesis in our discussion is that SCs with different organizations, placed in the same environment, all display high, but essentially comparable, value of viability function. The second (implicit) hypothesis is that the maximum value of the metric used to quantify information (e.g., mutual information or transfer entropy) is at least approximately constant for the joint SC/E dynamics for different SCs. Future work will be devoted to answer these questions.

The final remark is that the meaning and its emergence have a *relational* nature. It depends on the SC/E interplay, in particular on how SCs literally bring environmental perturbations *into its organization*, augmenting its probability of existence. This is, actually, the basis of the concept of cognition in the autopoiesis theory.

## 5 Conclusion

In conclusion, here we have presented results about the quantification of KW observed semantic information based on the dynamic exchange of information across a channel that characterizes a SC in an environment, where it operates according to a perception-action dynamics. This is achieved by resorting to the concepts and the methodologies firstly introduced by [14] and applied by us to a realistic SC scenario. In particular, while Kolchinsky and Wolpert in their seminal work have defined ideas, terms, and mathematical approaches to compute the stored and semantic information, here we have applied their strategy to a specific case. The key elements for applying the KW approach are the "interventions" that consist in coarse-graining, randomizing, scrambling the distributions of environmental variables in order to quantify the part of syntactic information that has a causal role in maintaining the SC functionality (or, in the more general case, the SC existence), measured by a viability function.

Next, we have provided a preliminary discussion about whether, and at what extent, it is possible to interpret semantic information values, calculated *à la* KW, as a measure of knowledge that agents have about their environment: i.e., with respect to what can be considered as meaningful to them. The question refers to a possible connection to the more general concepts of meaning, origin of meaning [3], and the mechanisms of "assignment" of meaning to those environmental perturbations affecting the agent dynamics (this is akin, in AI language, to the symbol grounding problem [12]). While SC are constructed in the laboratory and, for them, what is – or is not – meaningful has been decided by the designer/constructor, in biological cells meaning has emerged by evolutionary processes. Measures of semantic information are, probably, the best tools for quantitatively addressing the meaningful and the meaningless parts of the information flow impinging on an agent. Our discussion is largely speculative and here we intended to just sketched some hypotheses. Rather than providing a definitive answers to these questions, which are of great relevance for all artificial systems, the proposed discussion (Sect. 4) should motivate further enquiries, possibly carried out thanks to the application of KW approach.

**Acknowledgements.** Part of this research has been carried out within the PRIN-2022 funding program, supported by the Ministero dell'Università e della Ricerca (MUR), grant number 20222HHXAX, "Org(SB-EAI) – An Organizational Approach to the Synthetic Modeling of Cognition based on Synthetic Biology and Embodied AI". The synthetic cell model on observed semantic information have been also presented, as a poster [6], at the NANOCOM 2023 conference (20–22 September 2023, Coventry, United Kingdom).

## References

1. Adamala, K.P., Martin-Alarcon, D.A., Guthrie-Honea, K.R., Boyden, E.S.: Engineering genetic circuit interactions within and between synthetic minimal cells. Nat. Chem. **9**(5), 431–439 (2017)
2. Anta, J.: Information, meaning and physics: the intellectual evolution of the English School of Information Theory during 1946–1956. Sci. Context **34**(3), 357–373 (2021)
3. Atlan, H.: Self creation of meaning. Phys. Scr. **36**, 563–576 (1987)
4. Barker, T.S., Pierobon, M., Thomas, P.J.: Subjective information and survival in a simulated biological system. Entropy **24**(5), 639 (2022)
5. Carnap, R., Bar-Hillel, Y.: An outline of a theory of semantic information. Massachusetts Institute of Technology, Research Laboratory Electronics Technical Report No. 247 (1952)
6. Del Moro, L., Magarini, M., Stano, P.: On the evaluation of observed semantic information in synthetic cells. In: Proceedings of the 10th ACM International Conference on Nanoscale Computing and Communication, NANOCOM 2023, pp. 156–157. Association for Computing Machinery, New York (2023)
7. Del Moro, L., et al.: Chemical neural networks and semantic information investigated through synthetic cells. In: De Stefano, C., Fontanella, F., Vanneschi, L. (eds.) WIVACE 2022. CCIS, vol. 1780, pp. 27–39. Springer, Cham (2023). https://doi.org/10.1007/978-3-031-31183-3_3
8. Farnsworth, K.D., Nelson, J., Gershenson, C.: Living is information processing: from molecules to global systems. Acta. Biotheor. **61**, 203–222 (2013)
9. Fields, C., Levin, M.: How do living systems create meaning? Philosophies **5**(4), 36 (2020)
10. Floridi, L.: The Philosophy of Information. Oxford University Press, Oxford (2011)
11. Getu, T.M., Kaddoum, G., Bennis, M.: Tutorial-cum-survey on semantic and goal-oriented communication: research landscape, challenges, and future directions. arXiv preprint arXiv:2308.01913 (2023)
12. Harnad, S.: The symbol grounding problem. Physica D **42**(1), 335–346 (1990)

13. Kiverstein, J., Kirchhoff, M.D., Froese, T.: The problem of meaning: the free energy principle and artificial agency. Front. Neurorobot. **16**, 844773 (2022)
14. Kolchinsky, A., Wolpert, D.H.: Semantic information, autonomous agency and non-equilibrium statistical physics. Interface Focus **8**, 20180041 (2018)
15. Krinsky, N., et al.: Synthetic cells synthesize therapeutic proteins inside tumors. Adv. Healthc. Mater. **7**(9), e1701163 (2018)
16. Leduc, P.R., et al.: Towards an in vivo biologically inspired nanofactory. Nat. Nanotechnol. **2**(1), 3–7 (2007)
17. Logan, R.K.: What is information?: why is it relativistic and what is its relationship to materiality, meaning and organization. Information **3**(1), 68–91 (2012)
18. MacKay, D.M.: Information, Mechanism and Meaning. MIT Press, Cambridge (1969)
19. Magarini, M., Stano, P.: Synthetic cells engaged in molecular communication: an opportunity for modelling Shannon-and semantic-information in the chemical domain. Front. Commun. Netw. **2**, 724597 (2021)
20. Maturana, H.R., Varela, F.J.: Autopoiesis and Cognition: The Realization of the Living, 1st edn. D. Reidel Publishing Company (1980)
21. Nauta, D.: The Meaning of Information. Approaches to Semiotics [AS}, Mouton (De Grutyer), The Hague (1972)
22. Newman, J.: Some observations on the semantics of "information". Inf. Syst. Front. **3**(2), 155–167 (2001)
23. Rovelli, C.: Meaning and intentionality = information + evolution. In: Aguirre, A., Foster, B., Merali, Z. (eds.) Wandering Towards a Goal. FRONTCOLL, pp. 17–27. Springer, Cham (2018). https://doi.org/10.1007/978-3-319-75726-1_3
24. Ruzzante, B., Del Moro, L., Magarini, M., Stano, P.: Synthetic cells extract semantic information from their environment. IEEE Trans. Mol. Biol. Multi-Scale Commun. **9**(1), 23–27 (2023)
25. Schreiber, T.: Measuring information transfer. Phys. Rev. Lett. **85**(2), 461 (2000)

# General Lines, Routes and Perspectives of Wetware Embodied AI. From Its Organizational Bases to a Glimpse on Social Chemical Robotics

Luisa Damiano[1](✉) and Pasquale Stano[2](✉)

[1] RG-ESA (Research Group on the Epistemology of the Sciences of the Artificial), Department of Communication, Arts and Media, IULM University, Milan, Italy
luisa.damiano@iulm.it

[2] Department of Biological and Environmental Sciences and Technologies (DiSTeBA), University of Salento, Lecce, Italy
pasquale.stano@unisalento.it

**Abstract.** In this contribution we would like to put forward a proposal about a novel form of AI, called "Wetware Embodied AI", based on the construction of bio-chemical dynamical systems intended as models of living and cognitive systems. The ambition is complementing common approaches in robotics and AI, mainly characterized by behavioral imitation of the cognitive processes under inquiry, with more radical approaches, aiming at creating artificial models reproducing (aspects of) the organizational mechanisms underlying the target process in nature. To this aim, we will first recapitulate some aspects of frontier research lines in AI (in particular, Embodied AI and related approaches), and then present a wetware version of it, together with a brief plan for theoretical and experimental investigations. The role of synthetic biology and systems chemistry to accomplish these goals will be highlighted. Considering the connection of this program with the current experimental trends about communicating synthetic cells, a final brief comment on the "social chemical robotics" perspectives closes the article.

**Keywords:** Autopoiesis · Artificial Life · Cognition · Embodied AI · Synthetic Biology · Synthetic (Artificial) Cells · Philosophy of Artificial Intelligence · Wetware Embodied AI

## 1 Exorcizing the "Ghost" in the Machine

In line with the programmatic ambition of releasing the cognitive mind from the "ghostly" status it had assumed in computationalism [1–3], pioneering Embodied AI (EAI) programs worked on synthetically modeling natural cognition not through computer programs, but through "embodied agents" [4–6]. The idea was that of "biological-like robots", which learn about their environment and accomplish cognitive tasks through their bodies, as living systems do. By modeling in these robots the whole range of natural cognitive processes through an "emergent design", the goal was the generation of


M. Villani et al. (Eds.): WIVACE 2023, CCIS 1977, pp. 111–122, 2024.
https://doi.org/10.1007/978-3-031-57430-6_10

increasingly complex cognitive processes by means of combining basic sensory-motor processes [6]. The novel programmatic idea, we would like to discuss here, makes one step forward and beyond EAI. It starts from the consideration that hardware robots, even biological-like ones, miss a fundamental life-like mechanism, i.e., the continuous regeneration of their material body via a continuous destruction and reconstruction of its parts. In accordance with the theory of autopoiesis [7], taken here as the reference theory, we propose the Wetware EAI: a synthetic study of natural cognitive processes based on the construction and experimental exploration of wetware – i.e., chemical – implementations of the model of the biological organization.

## 2 Wetware EAI

### 2.1 From EAI to "Organismically-Inspired Robotics", "Enactive AI" and Beyond: A Recap

EAI emerged in the early 90s to overcome the "crisis" of computationalist or classic AI by focusing on the role of the body in cognition [8, 9]. In the late 90s, while EAI was already producing effective "adaptive robots", it became evident that this goal was out of reach. Pioneers such as Brooks announced that a "fundamental change" was needed, as EAI's robots lacked "organizing principles of biological systems" and could not aspire to their cognitive performances [10]. Related analyses in philosophy of AI prospected the needed turn as the transition from "organismoid" to "organismic" robots. The referred shift points to move from artifacts sharing superficial features of natural organisms, such as anatomical structures, to artifacts sharing the living form of organization, viewed as the network of functional relations generating living systems' self-production – their capability of producing their material identity by themselves, based on metabolism [11, 12]. Dedicated philosophical debate can be divided into two camps: a pessimistic camp, affirming the impossibility of recreating the biological organization, e.g., [13], and a proactive camp, engaged in filling EAI's gap. The latter generated some of the most interesting EAI's programs, such as "Organismically-inspired robotics" [14] and "Enactive AI" [15]. Both are (*i*) inspired by a well-defined theoretical model of the biological organization, life, and cognition, namely, autopoiesis [7]; (*ii*) limited to hardware and software models that implement autopoiesis and its filiation [2, 16], at an acceptable level of abstraction. Despite their theoretical value, more than ten years after their definition, these programs have not produced concrete organizational approaches in EAI, that is, approaches grounding their ambition of modeling natural cognition in systematic attempts of implementing artificially the biological organization. Indeed, the most significant recent results in EAI are extensive advances in building organismoid robots.

### 2.2 Wetware EAI: The General Lines

Our theoretical-experimental research program intends to join the proactive camp of Philosophy of AI to lay the grounds of a new organizational approach for EAI. As mentioned, the programmatic idea (wetware EAI) refers to the synthetic study of natural cognitive processes based on the construction and experimental exploration of wetware – i.e., chemical – implementations of the autopoietic model of the biological organization

[7]. It can be properly identified as wetware autopoietic EAI. The program will be possible thanks to the experimental "platform" provided by Synthetic Biology, Systems Chemistry, and related approaches such as Molecular Communication, Neuromorphic Chemical Networks, intended as a set of frontier chemical approaches of reference, with a specific attention for those dedicated to the construction of "artificial cells" (ACs).

Wetware autopoietic EAI will fill the gap constituted by the missing *effective* organizational approach in other forms of EAI. Attempts of grounding modeling of the biological organization in hardware and software systems cannot generate the required dynamics we hold as fundamental for actual modeling life and cognition in the physical (material) domain. Our plan for theoretical and experimental investigations includes: (i) a theoretical framework of reference, (ii) technical and experimental specifications for its implementation, (iii) epistemological criteria for assessment, (iv) a detailed description of potential applications, and (v) ethical indications. Current theories of biological organization have to be used to generate theoretical models implementable in wetware models. In view of such an implementation, theoretical models have to be used as references to determine the technical general lines of their appropriate translations into chemical models. Moreover, epistemological criteria need to be developed in order to evaluate the relevance, for the scientific understanding of life and cognition, of the wetware models to be implemented. On these bases, wetware models can be ideated, analyzed, implemented, and empirically explored for scientific purposes. At the same time, a library of potential applications should be developed, and the elaboration of related ethical guidelines to warrant the sustainability of the applicative endeavor should be included.[1] The wetware autopoietic EAI adopts (E)AI's "understanding-by-building" method [9, 12], through which it engages in the (bio)chemical fabrication of non-trivial chemical systems that display at least some of the living or living-like properties of biological organisms.

## 2.3 Wetware Modeling of Life and Cognition

Historically, the first attempt of building chemical autopoietic systems has been due to Humberto Maturana and Francisco Varela, the authors of the theory of autopoiesis, by means of a project called "molecular protobionts". This program was based on coacervates made of cedar oil and proteinoids in presence of salts, sand and sugars [19], but without any significant dynamics (i.e., it just showed the formation of coacervates). Interestingly, in that report Gloria Guiloff mentioned very clearly that "a molecular protobiont should not include a permanent component" (p. 119), highlighting the ultimate feature of the chemical networks that generate autopoietic systems: the precariousness of the components, which is continuously counteracted by processes for their production. Said in another way, it is not the structure that counts, even if it is the first thing we might note, but the underlying processes which keep the system out-of-equilibrium (from the viewpoint of chemical energy fluxes). However, the processes must also generate a structure that is "permanent", *despite and also thanks to* the underlying (anabolic/catabolic) processes.

[1] More exhaustive descriptions of the theoretical, technical and applicative aspects of the wetware autopoietic EAI program can be found in, e.g., [12,17]. The epistemological aspects are thoroughly discussed in [18].

Some years later, Varela and Luisi proposed that fatty acid-based reverse micelles – while undergoing some chemical transformations – could be a viable example of chemical autopoiesis [20]. It was a prolific intuition as it has been followed by a series of seminal studies on micelles, vesicles, and solute filled vesicles [21–23], generally referred to as synthetic cells, minimal cells, artificial cells (ACs). It should be remarked that these names can be occasionally misleading because AC complexity is not comparable to the complexity of biological cells, and ACs, even the most advanced ones which have been recently built, are not yet alive. Ultimately, however, the research on autopoietic solute-filled vesicles kicked-off the bottom-up branch of Synthetic Biology in early 2000s, and some current research lines in Systems Chemistry, as well as the most common branch of wetware Artificial Life. Around these themes a mature movement involving several dozens of highly skilled researchers worldwide has gathered.

Two comments come immediately in mind when chemical systems of this sort are employed and conceived as models of living and cognitive systems. The first refers to the unique features of chemical transformation, chemical interactions (which are "informationally open" [24]) and intrinsic self-regulative dynamics (e.g., think, for example, to the Gibbs free energy principle of minimization in chemical equilibria and to the Le Chatelier's principle). With respect to this first aspect, we note that the chemical domain, made of molecules, supramolecular systems, colloidal structures and reaction networks, provides an opportunity to construct models of life and cognition in a way that hardware and software approaches cannot achieve. In particular, wetware models, being physically embodied and thermodynamically constrained, can undergo physical interactions with the world in a very permissive manner (only partially achieved in robotics). Wetware models are open to all sorts of interactions/perturbations. Elements of the environment can freely alter the course of chemical reactions and, consequently, can elicit adaptive behavior. Chemical systems (networks of reactions) are intrinsically open to those interactions and can find, possibly, a new dynamical organization in order to accommodate these perturbations, in an autonomous manner (i.e., plasticity). These features, in turn, allow the emergence of meanings and suggest how to overcome the symbol-grounding problem [25, 26], in artificial systems – a well-known fundamental issue in Artificial Intelligence (AI), and similarly important in AL.

In the virtual domain (e.g., software models), all sorts of interactions are possible, and they are pre-determined by the designer, even when the outcome of the simulation is not foreseeable. Constraints, if present, have been set by the designer. Hardware systems have a body, and for this reason they can have physical interactions with the surrounding world (e.g., with the ground, with gravity) and with themselves (a robotic arm and a robotic leg cannot occupy the same space), and must deal with energy issues to allow movement. However, the macroscopic size of a robot and the materials used for its construction set limits to the kind of physical interactions the robot can experience. On the contrary, wetware (chemical) AL models are able to experience all sorts of physical and chemical stimuli because of their size and to the allowed energy exchanges at the nanoscale. It has been emphasized that molecular machines are privileged objects because various forms of energy (thermal, chemical, mechanical, and electrostatic) have the same order of magnitude [27, 28].

The second comment is instead about the type of processes that should occur in the designed and built chemical network. Essentially all current studies focus on 'anabolic' reactions, i.e., processes of production of the system components. As highlighted in the very first report on chemical autopoiesis [19], this constitutes only one part of the story. The two Janus-like faces of autopoiesis include necessarily anabolism and catabolism, which must coexist to generate a circularly closed organization. Moreover, only in such precarious systems it is true that the structure and the out-of-equilibrium state are permanent *thanks to and despite* a continuous component turnover. Moreover, this dynamic state allows (and needs) self-regulation as a result of inner organization rules (autonomy). Such conception of simultaneous anabolic/catabolic co-existence (and its major implication: circular metabolism) is generally not present in current experimental research. Many studies focus on the capacity of synthesizing some component of the system, such as DNA, or lipids, or energy-rich molecules. This is acceptable, because research in the field is still at its infancy, and achieving these syntheses in artificial systems is still a challenge. However, the core idea behind the autopoietic organization should not be forgotten.

A notable exception to the generalized interest toward anabolic reactions comes from a noteworthy 2001 study, not often mentioned, where the synthetic and disruptive reactions related to fatty acids simultaneously occurred on fatty acid vesicles [29]. The two opposing reactions were finely tuned in order to run at the same rate, so that the vesicles could remain as they are (in a homeostatic state) despite a continuous turnover of their components. Despite its simplicity (and the need of re-investigating this interesting system, furthering the study with more accurate analyses), this experimental approach to artificial homeostasis is a beautiful example of structural stability at a higher hierarchical level (the vesicle) despite the instability of components at a lower hierarchical level. It is also an example of how out-of-equilibrium processes generate a "structure", which in turn allows for the existence of such processes.

## 2.4 Autopoiesis and Autonomy

While it is true that only a chemical approach matches the requirements for the autopoietic dynamics, contemporary research has also shown that building artificial autopoietic systems is quite challenging (and not yet achieved, to date). By keeping in mind the central idea of circularity, an alternative and not less interesting theoretical target is the concept of autonomy, as expressed and defined by Varela [16]. While in autopoiesis the processes that generate an autopoietic system are intended as "processes of production (transformation and destruction)" of the components participating into the autopoietic dynamics, in the definition of autonomy the specification of "processes of production" is dropped off, in favor of processes which "recursively depend on each other in the generation and realization of the processes themselves" (p. 55). However, the processes still need to "constitute the system as a unity recognizable in the space (domain) in which the processes exist" (p. 55). A somewhat analogous shift – from autopoiesis to autonomy – can be found in recent discussions on artificial agency in robotics ("Instead of building robots that instantiate metabolic processes that self-organize to form autonomous networks, the strategy has been to build robots whose sensorimotor processes self-organize to form autonomous networks" [30]). The analogous goal in wetware autopoietic EAI would be minimal forms of chemical autonomy by exploiting AC systems that include, in their chemical network, multiple entailing steps for a closure organization pattern.

This perspective can suggest new and original directions to generate autonomous ACs, the latter being potentially less demanding than autopoietic ACs. For example, it would be possible to refer to processes that entail each other, in closed loops. We can imagine developing one or more autonomous module(s) to be engrafted in non-autopoietic ACs. We imagine having sets of reactions (modules) that satisfy the first condition for minimal autonomy (recursive dependence from each other in order to generate the processes themselves), while the second condition is met at the containment-property level (the processes occur inside a system that is physically distinct from the environment, like an AC). However, a link should be provided to causally join the set of reactions and the containment-property, e.g., the production of the boundary molecules, as in early chemical autopoiesis reports from Luisi and collaborators [20–23, 31]. The autonomous module, in this simplified and tolerant design, could be surrounded and sustained by other supporting pathways that most follow logics of non-cyclic transformations. If the demanding conditions for realizing full-fledged autopoiesis are provisionally eliminated, opportunities arise to focus on how to make autonomous moduli adaptable, plastic, self-regulative.

Examples of "bricks" that could be assembled as individual entailing steps in more complex networks can be identified in the literature of bottom-up ACs. Now "classic" investigations, such as the "bioreactor" of Noireaux and Libchaber [32], the Yomo's self-encoded Qβ-replicase [33], the Kuruma's protein synthesis that self-promote its realization via light-induced ATP generation [34], the Rossi's enzyme-generated self-perturbation of membrane curvature and vesicle division [35]; the Sugawara's enzyme-generated cooperative DNA binding on membrane and the resultant vesicle division [36], and others can be mentioned. The common traits of these studies are the series of entailing processes, sometimes spanning multiple organizational levels, that can serve as bricks for constructing autonomous ACs or as inspiration sources for self-regulation and adaptive behavior. They can surely be starting points for further explorations, modifications, and engineering.

Another clue might come from the translation of "sensorimotor" dynamics from the hardware to wetware domain. The sensorium layer of ACs can be made of a single receptor on their membrane, or by an array of receptors. The motoric part corresponds to the chemical processes that are activated upon receiving signals. These considerations have inspired us to propose scenarios based on the integration of Synthetic Biology with neuromorphic engineering: the inclusion of neural network-like circuitries in ACs [37–39]. Here, practical importance can refer to classification problems (how to behave, given an environment). A more intriguing option, related to the concept of autonomy, can be identified when the results of a neural network computation feedback in the network (i.e., a recurrent chemical neural network [40]). The latter dynamic is possible because in chemical systems, as we have emphasized elsewhere [41, 42], the difference between *computer and computed* substantially blurs, so that the result of a computation can well be one of the computing elements itself. A concentration changes for one element in the chemical neural network, for example, would correspond to a change of the organization (it is a trivial change, but it does have consequences on the network dynamics).

## 3 Social Robotics in the Chemical Domain

One of the most interesting applicative directions of AC research is related to the capacity of communication [43, 44] and the construction of artificial or hybrid (artificial/natural) networks of interacting agents, for example to realize application scenarios such as the "smart" drug delivery system ones [45]. This is a perspective that displays intriguing implications for the Philosophy of AI in particular for the similarities they have with the emerging techno-scientific area called *Social Robotics*, born at the dawn of the new millennium from the convergence of frontier fields in Robotics such as *Human-Robot Interaction* (HRI), *Cognitive Robotics* and *Social Embodied AI*, among others [46].

The application of ACs in the development of hybrid (artificial/natural) networks of interacting agents offers to Social Robotics an unprecedented opportunity to extend their groundbreaking lines of development, prolonging them into the chemical domain. The new scenario that today emerges is that of a *Chemical Social Robotics*, engaged in the creation of synthetic and hybrid "nano-cognitive-ecologies".

In particular, here we would like to offer a preliminary epistemological consideration (see also [47]) of these hybrid ecologies emerging from cross-fertilization between fields as synthetic biology, systems chemistry, biology, neurobiology, cognitive sciences, unconventional computation, AI and artificial life.

### 3.1 Social Robotics

Since its birth, Social Robotics has been engaged in the design and construction of electromechanical robots capable of interacting through social signals—e.g., gaze, gestures, emotional expressions, verbal communication. Research on Social Robotics currently focuses in several directions.

A first one is the creation of robotic agents that communicate with each other through shared signals and, on this basis, generate robotic social aggregates endowed with novel cognitive capacities of a collective nature, as occurs in paradigmatic cases of social insects and, more generally, in social forms of collective intelligence. Another one is the implementation of social interactions between natural and robotic agents. This is realized either (i) by creating robotic "social partners" for humans, capable of communicating with humans through social signals compatible with human ones; (ii) by introducing more intuitive and effective ways of interacting with robotic platforms destined for a variety of operational uses. The latter finds a major implementation for the construction of a "social presence" for robots, that is, appearance and behavioral features apt to stimulate humans to recognize these machines as agents with whom they can interact socially.

But a conceptually relevant implication of projects and approaches in social robotics lies in the development of mixed (human–robot) "social ecologies", prospected as hybrid social contexts in which social collaboration can be established by humans with a variety of robotic agents, so that human performance is improved. Enhancing the human capacity to solve cognitive tasks based on relations of social coordination with robotic artifacts is an empowerment often described through wide notions of the cognitive mind (e.g., [48]). The concept of *distributed mind*, directly stemming from the above-mentioned scenarios, can be defined as a co-evolutionary complex network of human and robotic agents that collaboratively coordinate their behaviors based on effective social signaling to generate (augmented) cognitive performance [49]. To maximize the success of a hybrid (human-robot) distributed mind, social robotics must be grounded in synthetic modeling: the incorporation of scientific hypotheses about human social processes into these robotic artifacts and their behavioral patterns. This adds a significant scientific interest to social robots, making them synthetic tools to deepen the scientific knowledge of human sociality based on a specific way of implementing the understanding-by-building method [50]. In other words, social robots have to be conceived of not only as artifacts to improve our daily lives but also, and inseparably, as tools serving our self-knowledge. This latter consideration brings us to extend such a vision to a chemical Social Robotics and its potentials.

### 3.2 Chemical Social Robotics

Considering ACs as chemical robot-like agents, among other implications, has the consequence that it is possible, although still in a speculative manner, to conceive them in a Chemical Social Robotics scenario.

To this aim, it is firstly necessary to imagine a hybrid artificial/natural scenario that parallels the human/robot one described in Sect. 3.1. The growing interest in developing *communicating ACs* perfectly suits this need as it implies the capacity of engineering ACs that exchange chemical signals between each other or with biological cells. Seen from this perspective, ACs actually can be conceived as chemical robot-like agents that can be placed in scenarios as those described above. In particular, ACs that communicate to each other represent multi-agent systems that generate an overall behavior thanks to the coordination of individual ACs; while ACs that communicate with biological cells represent hybrid multi-agent systems where the actions of ACs affect (and are affected by) biological cells. Moreover, ACs can be conceived, in some cases, as a cognitive extension of biological cells (think, for example, to ACs that, thanks to specifically designed mechanisms, can "translate" toxic xenobiotic signals to biochemical signals, allowing biological cells perceiving, the xenobiotic ones without being directly exposed to them).

The subject of communicating ACs is one of the most recent ones in synthetic biology, currently under intense investigation. In 2012, based on previous reports on ACs that could send chemical signals to bacteria [51], we highlighted the perspective of exploiting gene expression mechanisms to build communicating ACs [52], and indeed several studies have confirmed this capability. Similar ideas have been developed by scholars in the community of "Molecular Communication" (MC), a sub-field of communication engineering initiated by Tadashi Nakano and collaborators in the early 2000s [53]. In the latter field, high relevance is given to the application of information and communication theories, *à la* Shannon, to chemical signaling, and therefore to ACs. The interest of the communication engineering community for application has led to a concept that fits well with the hybrid artificial/natural network we have envisioned here. In particular, it has been imagined that a network of "nano-machines" (artificial and biological) will come into existence when (and if) it will be possible to finely control molecular communication inside the human body – for medical applications. The evocative term for such a network is "Internet of Bio-Nano Things" (IoBNT, [54]), and represents a long-term goal of bioengineering, based on these new (unconventional) mechanisms of information transfer and information processing.

From our viewpoint, we recognize the importance of practical applications and the utility, in some contexts, of adopting the usual engineering perspective, the IoBNT is first of all a platform that offers AC research an unprecedented opportunity: extending the theoretical approaches currently developed (and under development) of (electromechanical) Social Robotics into the chemical domain (chemical Social Robotics). The new scenario we are referring to is based on the creation of synthetic and hybrid *nano-cognitive ecologies* grounded in effective forms of signaling—communication. The wide range of potential and incipient applications of ACs in the IoBNT, as described elsewhere [47], expresses well the need of a vast, programmatic work of developing new or augmented technological functionalities.

The cross-fertilization between ACs and a IoBNT scenario opens the possibility of exploring, at the nano-level, natural cognition and, more specifically, social cognition based on a new application of the understanding-by-building method. Deepening our understanding of social cognition, down to the cellular/artificial cellular level, will be possible by synthetically reconstructing and studying the involved nano-mechanisms so as to identify the transition from non-cognitive to minimally cognitive processes, and from non-social to minimally social processes. These open questions will be faced by crucial attempts, like those aiming at nano-forms of distributed organizations capable of accomplishing social cognitive tasks based on behavior coordination grounded in effective signaling. As in electromechanical Social Robotics, chemical Social Robotics also requires an active effort to orient its development toward sustainability involving the constitution of partnerships in the form of effective communications and collaborations between disciplines, aiming to create a positive interplay between the processes of knowledge generation through these new techno-scientific developments.

## 4 Concluding Remarks

The wide range of potential and incipient applications of ACs for EAI, Artificial Life, Synthetic Biology suggest the emerging option of a vast, programmatic work of design of forms of chemical systems as models of natural processes (life, cognition). Mostly, these perspectives are generated by focusing on reference organization theories such as the autopoiesis, and via a productive cooperation between artificial, or artificial and natural, interactive systems. The employment of ACs in basic and applied fields (such as the IoBNT) opens the possibility of exploring, at the nano-level, natural cognition, and, more specifically, social cognition, based on a new application of the understanding-by-building method. The range of potential scientific developments is broad, including (i) the transition from non-cognitive to minimally cognitive processes, and from non-social to minimally social processes, and (ii) the construction of wetware autonomous systems in an Artificial Life context, (iii) the reconstruction and the study of nano-forms of distributed organizations capable of accomplishing cognitive tasks based on behavior coordination grounded in effective signaling.

**Acknowledgments.** Part of this research has been carried out within the project "Org(SB-EAI) – An Organizational Approach to the Synthetic Modeling of Cognition based on Synthetic Biology and Embodied AI" (PRIN-2022, grant number 20222HHXAX), funded by the Ministero dell'Università e della Ricerca (MUR). The discussion presented in Sect. 3.1 has been adapted from our recently published article [47].

## References

1. Ryle, G.: The Concept of Mind. University of Chicago, Chicago (1949)
2. Varela, F.J., Thompson, E.T., Rosch, E.: The Embodied Mind: Cognitive Science and Human Experience. MIT Press, Cambridge (1991)
3. Ceruti, M., Damiano, L.: Plural embodiment(s) of mind. Genealogy and guidelines for a radically embodied approach to mind and consciousness. Front. Psychol. **9**, 2204 (2018)
4. Brooks, R.: Intelligence without representation. Artif. Intell. **47**(1–3), 139–159 (1991)
5. Steels, L., Brooks, R. (eds.): The Artificial Life Route to Artificial Intelligence. Routledge, London (1995)
6. Pfeifer, R., Bongard, J.: How the Body Shapes the Way We Think. MIT Press, Cambridge (2006)
7. Maturana, H.R., Varela, F.J.: Autopoiesis and Cognition: The Realization of the Living, 1st edn. D. Reidel Publishing Company, Dordrecht (1980)
8. Clark, A.: Being There: Putting Brain, Body, and World Together Again. MIT Press, Cambridge (1997)
9. Pfeifer, R., Scheier, C.: Understanding Intelligence. MIT Press, Cambridge (1999)
10. Brooks, R.A.: From earwigs to humans. Robot. Auton. Syst. **20**(2), 291–304 (1997)
11. Ziemke T.: Are robots embodied? In: Balkenius, C., Zlatev, J., Kozima, H., Dautenhahn, K., Breazeal, C. (eds.) Proceedings of the First International Workshop on Epigenetic Robotics. Modeling Cognitive Development in Robotic Systems (17–18 September 2001; Lund, Sweden), Lund University Cognitive Studies, vol. 85, pp. 75–83 (2001)
12. Damiano, L., Stano, P.: Synthetic biology and artificial intelligence. Grounding a cross-disciplinary approach to the synthetic exploration of (embodied) cognition. Complex Syst. **27**, 199–228 (2018)
13. Dreyfus, H.L.: Why Heideggerian AI failed and how fixing it would require making it more Heideggerian. Artif. Intell. **171**(18), 1137–1160 (2007)
14. Di Paolo, E.A.: Organismically-inspired robotics: homeostatic adaptation and teleology beyond the closed sensorimotor loop. In: Murase K., Asakura T. (eds.) Dynamical Systems Approach to Embodiment and Sociality, pp. 19–42. Advanced Knowledge International, Adelaide, Australia (2003)
15. Froese, T., Ziemke, T.: Enactive artificial intelligence: investigating the systemic organization of life and mind. Artif. Intell. **173**(3), 466–500 (2009)
16. Varela, F.J.: Principles of Biological Autonomy. Elsevier North-Holland Inc, New York (1979)
17. Stano, P., Damiano, L.: Synthetic cell research: is technical progress leaving theoretical and epistemological investigations one step behind? Front. Robot. AI **10**, 1143196 (2023)
18. Damiano, L., Stano, P.: Explorative synthetic biology in AI. Criteria of relevance and a taxonomy for synthetic models of living and cognitive processes. Artif. Life **29**, 367–387 (2023)
19. Guiloff, G.D.: Autopoiesis and neobiogenesis. In: Zeleny, M. (ed.) Autopoiesis. A Theory of Living Organization, pp. 118–124. North Holland, New York (1981)
20. Luisi, P.L., Varela, F.J.: Self-replicating micelles – a chemical version of a minimal autopoietic system. Orig. Life Evol. Biosph. **19**(6), 633–643 (1989)

21. Oberholzer, T., Wick, R., Luisi, P.L., Biebricher, C.K.: Enzymatic RNA replication in self-reproducing vesicles: an approach to a minimal cell. Biochem. Biophys. Res. Commun. **207**(1), 250–257 (1995)
22. Oberholzer, T., Albrizio, M., Luisi, P.L.: Polymerase chain-reaction in liposomes. Chem. Biol. **2**(10), 677–682 (1995)
23. Oberholzer, T., Nierhaus, K.H., Luisi, P.L.: Protein expression in liposomes. Biochem. Biophys. Res. Commun. **261**(2), 238–241 (1999)
24. Cariani, P.: To evolve an ear. Epistemological implications of Gordon Pask's electrochemical devices. Syst. Res. **10**, 19–33 (1993)
25. Harnad, S.: The symbol-grounding problem. Phys. D Nonlinear Phenom. **42**, 335–346 (1990)
26. Floridi, L.: The Philosophy of Information. Oxford University Press, Oxford (2011)
27. Hoffmann, P.M.: Life's Ratchet: How Molecular Machines Extract Order from Chaos. Basic Books. A Member of the Perseus Books Group, New York (2012)
28. Phillips, R., Quake, S.R.: The biological frontier of physics. Phys. Today **59**, 38–43 (2006)
29. Zepik, H.H., Blochliger, E., Luisi, P.L.: A chemical model of homeostasis. Angew. Chem. Int. Edit. **40**, 199–202 (2001)
30. Kiverstein, J., Kirchhoff, M.D., Froese, T.: The problem of meaning: the free energy principle and artificial agency. Front. Neurorobot. **16**, 844773 (2022)
31. Walde, P., Wick, R., Fresta, M., Mangone, A., Luisi, P.L.: Autopoietic self-reproduction of fatty-acid vesicles. J. Am. Chem. Soc. **116**, 11649–11654 (1994)
32. Noireaux, V., Libchaber, A.: A vesicle bioreactor as a step toward an artificial cell assembly. Proc. Natl. Acad. Sci. U.S.A. **101**(51), 17669–17674 (2004)
33. Kita, H., et al.: Replication of genetic information with self-encoded replicase in liposomes. ChemBioChem **9**(15), 2403–2410 (2008)
34. Berhanu, S., Ueda, T., Kuruma, Y.: Artificial photosynthetic cell producing energy for protein synthesis. Nat. Commun. **10**(1), 1325 (2019)
35. Miele, Y., et al.: Self-division of giant vesicles driven by an internal enzymatic reaction. Chem. Sci. **11**, 3228–3235 (2020)
36. Kurihara, K., Tamura, M., Shohda, K.-I., Toyota, T., Suzuki, K., Sugawara, T.: Self-reproduction of supramolecular giant vesicles combined with the amplification of encapsulated DNA. Nat. Chem. **3**(10), 775–781 (2011)
37. Stano, P.: Chemical neural networks and synthetic cell biotechnology: preludes to chemical AI. In: Chicco, D., et al. (eds.) Computational Intelligence Methods for Bioinformatics and Biostatistics. CIBB 2021. Lecture Notes in Computer Science, vol. 13483, pp. 1–12. Springer, Cham (2022). https://doi.org/10.1007/978-3-031-20837-9_1
38. Stano, P.: Sketching how synthetic cells can function as a platform to investigate chemical AI and information theories in the wetware domain. In: Bindhu, V., Tavares, J.M.R.S., Vuppalapati, C. (eds.) Proceedings of Fourth International Conference on Communication, Computing and Electronics Systems. ICCCES 2022. Lecture Notes in Electrical Engineering, vol. 977, pp. 571–584. Springer, Singapore (2023). https://doi.org/10.1007/978-981-19-7753-4_43
39. Gentili, P.L., Stano, P.: Chemical neural networks inside synthetic cells? A proposal for their realization and modeling. Front. Bioeng. Biotechnol. **10**, 927110 (2022)
40. Braccini, M., Collison, E., Roli, A., Fellermann, H., Stano, P.: Recurrent neural networks in synthetic cells: a route to autonomous molecular agents? Front. Bioeng. Biotechnol. **11**, 1210334 (2023)
41. Gentili, P.L., Stano, P.: Tracing a new path in the field of AI and robotics: mimicking human intelligence through chemistry. Part I: molecular and supramolecular chemistry. Front. Robot. AI **10**, 1238492 (2023)
42. Gentili, P.L., Stano, P.: Tracing a new path in the field of AI and robotics: mimicking human intelligence through chemistry. Part II: systems chemistry. Front. Robot. AI **10**, 1266011 (2023)

43. Lentini, R., et al.: Two-way chemical communication between artificial and natural cells. ACS Cent. Sci. **3**(2), 117–123 (2017)
44. Rampioni, G., et al.: Synthetic cells produce a quorum sensing chemical signal perceived by pseudomonas aeruginosa. Chem. Commun. **54**, 2090–2093 (2018)
45. Leduc, P.R., et al.: Towards an in vivo biologically inspired nanofactory. Nat. Nanotechnol. **2**(1), 3–7 (2007)
46. Damiano, L., Dumouchel, P.G.: Emotions in relation. Epistemological and ethical scaffolding for mixed human-robot social ecologies. HUMANA.MENTE J. Philos. Stud. **13**(37), 181–206 (2020)
47. Stano, P., Gentili, P.L., Damiano, L., Magarini, M.: A role for bottom-up synthetic cells in the internet of bio-nano things? Molecules **28**, 5564 (2023)
48. Miłkowski, M., et al.: From wide cognition to mechanisms: a silent revolution. Front. Psychol. **9**, 2393 (2018)
49. Damiano, L.: Homes as human–robot ecologies: an epistemological inquiry on the "domestication" of robots. In: The Home in the Digital Age; Routledge, Abingdon, UK (2021)
50. Dumouchel, P., Damiano, L.: Living with Robots. Harvard University Press, Cambridge (2017)
51. Gardner, P.M., Winzer, K., Davis, B.G.: Sugar synthesis in a protocellular model leads to a cell signalling response in bacteria. Nat. Chem. **1**, 377–383 (2009)
52. Stano, P., Rampioni, G., Carrara, P., Damiano, L., Leoni, L., Luisi, P.L.: (2012) Semi-synthetic minimal cells as a tool for biochemical ICT. BioSystems **109**, 24–34 (2012)
53. Nakano, T.: Molecular communication: a 10 year retrospective. IEEE Trans. Mol. Biol. Multi-Scale Comm. **3**, 71–78 (2017)
54. Akyildiz, I.F., Pierobon, M., Balasubramaniam, S., Koucheryavy, Y.: The internet of bio-nano things. IEEE Commun. Mag. **53**(3), 32–40 (2015)

# A Proposed Mechanism for *in vivo* Programming Transmembrane Receptors

Roger D. Jones[1,2,3,4(✉)] and Alan M. Jones[2,3,4]

[1] European Centre for Living Technology, Ca' Foscari, University of Venice, Venice, Italy
RogerDJonesPhD@gmail.com
[2] Department of Biology, University of North Carolina at Chapel Hill, Chapel Hill, NC, USA
[3] Systems Engineering and Research Center, Stevens Institute of Technology, Hoboken, NJ, USA
[4] Department of Pharmacology, University of North Carolina at Chapel Hill, Chapel Hill, NC, USA

**Abstract.** Transmembrane G-protein coupled receptors (GPCRs) are ideal drug targets because they resemble, in function, molecular microprocessors for which outcomes (e.g. disease pathways) can be controlled by inputs (extracellular ligands). The inputs here are ligands in the extracellular fluid and possibly chemical signals from other sources in the cellular environment that modify the states of molecular switches, such as phosphorylation sites, on the intracellular domains of the receptor. Like in an engineered microprocessor, these inputs control the configuration of output switch states that control the generation of downstream responses to the inputs.

Many diseases with heterogeneous prognoses including, for example, cancer and diabetic kidney disease, require precise individualized treatment. The success of precision medicine to treat and cure disease is through its ability to alter the microprocessor outputs in a manner to improve disease outcomes. We previously established *ab initio* a model based on maximal information transmission and rate of entropy production that agrees with experimental data on GPCR performance and provides insight into the GPCR process. We use this model to suggest new and possibly more precise ways to target GPCRs with potential new drugs.

We find, within the context of the model, that responses downstream of the GPCRs can be controlled, in part, by drug ligand concentration, not just whether the ligand is bound to the receptor. Specifically, the GPCRs encode the maximum ligand concentration the GPCR experiences in the number of active phosphorylation or other switch sites on the intracellular domains of the GPCR. This process generates a memory in the GPCR of the maximum ligand concentration seen by the GPCR. Each configuration of switch sites can generate a distinct downstream

---

**Supplementary Information** The online version contains supplementary material available at https://doi.org/10.1007/978-3-031-57430-6_11.


M. Villani et al. (Eds.): WIVACE 2023, CCIS 1977, pp. 123–137, 2024.
https://doi.org/10.1007/978-3-031-57430-6_11

response bias. This implies that cellular response to a ligand may be programmable by controlling drug concentration. The model addresses the observation paradox that the amount of information appearing in the intracellular region is greater than amount of information stored in whether the ligand binds to the receptor. This study suggests that at least some of the missing information can be generated by the ligand concentration. We show the model is consistent with assay and information-flow experiments.

In contrast to the current view of switch behavior in GPCR signaling, we find that switches exist in three distinct states: inactive (neither off nor on), actively on, or actively off. Unlike the inactive state, the active state supports a chemical flux of receptor configurations through the switch, even when the switch state is actively off. Switches are activated one at a time as ligand concentration reaches threshold values and does not reset because the ligand concentration drops below the thresholds. These results have clinical relevance. Treatment with drugs that target GPCR-mediated pathways can have increased precision for outputs by controlling switch configurations. The model suggests that, to see the full response spectrum, fully native receptors should be used in assay experiments rather than chimera receptors.

Inactive states allow the possibility for novel adaptations. This expands the search space for natural selection beyond the space determined by pre-specified active switches.

## 1 Introduction

Protein receptors that span the cell membrane are molecular microprocessors [20]. They gather information from outside the cell and process and transmit the information to the intracellular space, where it is directed to chemical pathways that lead to cellular response to extracellular conditions. The molecular microprocessors are programmed by natural selection and chemical conditions within the organism. For disease control, the programming is achieved mostly through the directed application of drugs. It requires higher-resolution drug targets and increased understanding of the effects of high-resolution drug targeting [2].

The most important class of these transmembrane receptors, both scientifically and medically, are the G-protein coupled receptors (GPCRs). More than 30% of all prescription drugs target GPCRs [18]. The advent of precision medicine has increased the importance of understanding how information can be controlled at higher levels of resolution. As an example, the hormone angiotensin II increases blood pressure and prolonged hypertension drives diabetic kidney disease (DKD). Renin angiotensin system inhibitors (RASi), such as angiotensin converting enzyme inhibitors (ACEis) and angiotensin II receptor blockers (ARBs) block the formation and action of angiotensin II and lower systemic blood pressure. Interestingly, when compared with other antihypertensive agents, ACEIs and ARBs stabilize kidney function at the same level of achieved blood pressure better than conventional antihypertensive therapy [22].

This suggests that angiotensin II also operates in other processes [2,13]. Indeed, the angiotensin receptor and other G-protein coupled receptors can trigger distinct multiple downstream responses that depend on the cellular environment [6,11,17] and thereby may lead to heterogeneous disease progression and effect of therapy.

As the angiotensin example illustrates, the era of precision medicine requires much higher resolution drug targeting to reduce the heterogeneity of disease progression. This requires a better understanding of the details of how the information is processed by GPCRs. Recent attempts [6] at making progress on this ambitious goal have relied on teleonomy, the imparting of goals to natural selection. In this case, a model was developed that imparted goals of maximal rate of energy production and information transmission. In this study, we use the model developed in [6] to improve the resolution of GPCR information processing and increase our understanding of how this resolution can be used to program the molecular microprocessors to individualize patient treatments for improved prognosis.

Information is a type of entropy [5,7,10,21]. Systems in equilibrium are those in which entropy is maximized and entropy flow is zero. Systems in equilibrium are not alive. Living systems are those in which entropy, and hence information, flows. Entropy flow can occur, for instance, when a chemical concentration is far from its equilibrium value as is the case with adenosine triphosphate (ATP) in biological systems. Entropy and information flow can also occur in the presence of spatial gradients such as those found at cell membranes. Both types of entropy flow are important in this study. Figure 1A illustrates a case in which information flows across a cell membrane. Figure 1C illustrates an active phosphorylation switch in which a chemical flux is driven by excess of ATP.

Information contained in the extracellular space is transmitted to the intracellular space by receptors (Fig. 1A). Ligands, such as hormones, auxins, nutrients, neurotransmitters, and many other molecules in the extracellular fluid, announce their presence to the cell by binding to extracellular domains of the receptor and allosterically altering the intracellular properties of the receptor. The intracellular changes in the receptor effect the cell's response to the ligand stimulus. In many cases, the extracellular changes take the form of phosphorylated and unphosphorylated sites on the C tail of the receptor protein (Figs. 1B and C) [11]. The sites form a barcode that is read by intracellular processes that respond to the information in the code [3,4,11,23–26]. Other receptor conformations are possible but the barcode process described here is a good representative exemplar.

A number of questions emerge. How is the barcode programmed? In other words, which sites are phosphorylated and which are unphosphorylated? How many active sites, or switches, are there? How is the phosphorylation state of the barcode changed? What determines the number of active phosphorylation sites and how is that number changed? The issue is that any set of molecules that store information, as does the barcode, must be stable to thermal and other fluctuations [19]. Yet, rearrangement of the barcode must be possible. Since many phosphorylation sites may be present, this requires a significant amount of energy to effect a global change. Typically sites are phosphorylated and dephosphorylated by catalyst kinases and phosphatases, respectively. How can both kinases and

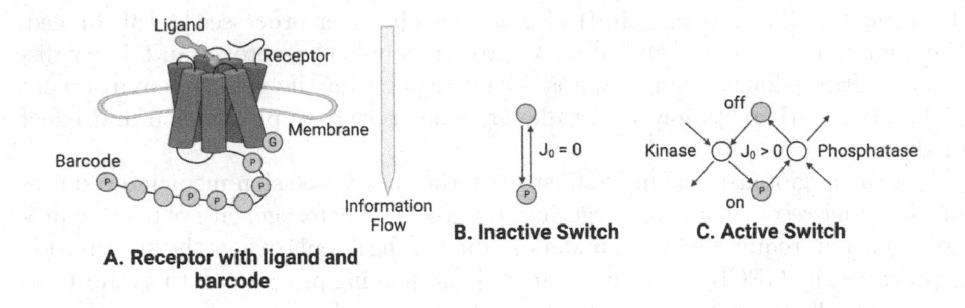


**Fig. 1. A.** A ligand binds to the extracellular domains of a receptor that spans the cell membrane. A number of phosphorylation sites occupy the C tail of the receptor. Some of the sites are inactive (not shown); some are active and phosphorylated (designated by P); and some are active and unphosphorylated. Phosphorylated sites are designated on, while unphosphorylated states are designated off. The site labeled G is a GTPase switch found in G-protein coupled receptors (GPCRs). **B.** An inactive PdPC switch. Every physical path from unphosphorylated (off) to phosphorylated (on) has a microscopically reversed path. The chemical flux in the forward path is equal to the flux in the microscopically reversed path. The flux is in detailed balance and the reaction is in equilibrium. **C.** An active PdPC switch. The reactions in a forward circular direction have significant chemical flux $J_0$ and are catalyzed by kinases and phosphatases. The microscopically reversed pathways in the opposite direction have negligible flux and are not displayed. The switch is driven far from equilibrium by the flux $J_0$.

phosphatases act in global concert to change the information content across an entire barcode of several phosphorylation switches?

This study attempts to address some of these questions. Our approach is grounded in recent theoretical work [6] fitting biochemical data [17], site-directed spectroscopy observations [11], molecular dynamic simulations [11] and information-flow measurements [9]. Our approach differs from the mass-action-driven modeling approaches on signaling bias, e.g. [1], thus it may be unfamiliar to some due to its new approach and its heavy reliance on non-equilibrium thermodynamic principles. However, the fundamentals of the original model are well described by [6]. Therefore, we only introduce here some of the relevant features of the approach in Figs. 1 and 2. The new findings are illustrated and supported by data in Table 1 and Figs. 3 and 4. Briefly, we show that in addition to information for output (e.g. effector coupling, MAPK cascading, gene expression) induced by ligand binding to the GPCR, the dynamics of ligand concentration presented to this receptor, both concentration and time, provide information that is transmitted to downstream responses. Consequently, new drug presentation strategies may provide an additional tool in targeting GPCRs for disease control.

## 2 Approach and Results

### 2.1 Relation Between Ligand Concentration and Number of Active Phosphorylation Sites

We build on the study in [6]. That work used maximum entropy production along with maximum information storage and transfer to generate the bag-of-independent-switches (BOIS) model picture, which imposes a number of constraints on the barcode. The model predicts that three switch configurations are possible, on, off, and inactive and that the phosphorylation sites (switches) are effectively uniform, except for a small number of differences that divide the switches into the three configurations. Active switches are distinguished from inactive switches in that active switches support a finite chemical flux through the switches [16], while inactive switches have zero flux Fig. 1B and C). Active switches are divided into phosphorylated (on) and unphosphorylated (off) sites. The chemical flux in active switches serves as a local energy/heat source that drives the process. The basic exemplar switch is a phosphorylation-dephosphorylation cycle (PdPC) [16] displayed in Fig. 1B for the inactive switch and Figure C for an active switch. The number of switches $N$, in this case, is equal to the number of phosphorylation sites on the C tail. The receptor concentration is designated $R_T$.

For a receptor that has never been exposed to a particular ligand, all switches in the barcode are inactive with no chemical flux. As ligand concentration increases, the switches activate one at a time (Fig. 2). The first switch activates when the receptor concentration reaches an experimentally accessible [17] reference concentration of $R_{ref} = R_T/N$ that is independent of the type of ligand or the ligand concentration [6]. As ligand concentration increases, the switches activate one-by-one until all $N$ switches are activated. If the ligand concentration is then reduced, the switches that have been activated do not deactivate. This is conceptually similar to a ratchet and pawl that ratchets to a higher number of activated states (Supplement A.1).

According to the BOIS model, the maximum concentration $L^*$ that the ligand reaches, the total number $N$ of switches, and the number $M$ of activated switches obey the relationship (Supplement A.2)

$$\frac{R_L}{R_{ref}} = \frac{L^*}{K_D + L^*}\,(N - M) < 1 \tag{1}$$

where $K_D$ is the ligand/receptor dissociation constant. Equation 1 is a constraint on the number of activated switches $M$. If the ligand concentration is very large, $L_{max} \gg K_D$, then all the phosphorylation sites are activated, $M = N$. If the receptor has never encountered the ligand, $L^* = 0$, then the arguments of Supplement A.2 indicate that no switches are activated and $M = 0$. The predicted value for $M$ is the minimum value of $M$ for which the inequality Eq. 1 is true. This can be written

$$M = \text{floor}\left[(N+1)\frac{L^*}{K_D + L^*}\right] \tag{2}$$

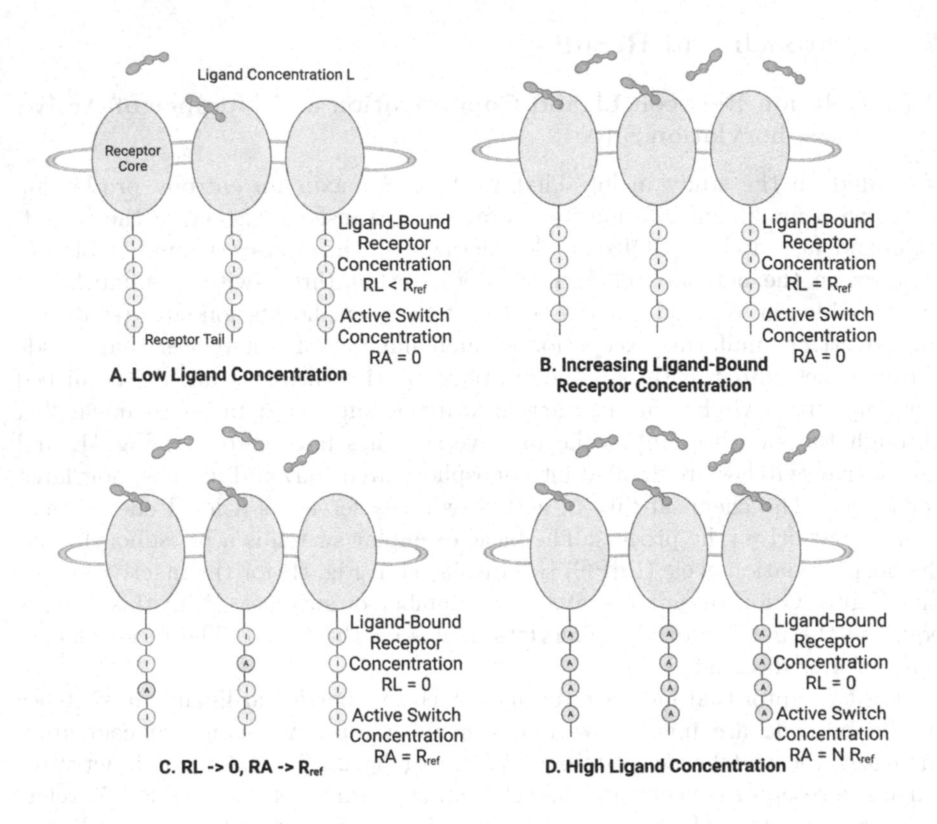


**Fig. 2.** ratchet. **A.** Early time, low ligand concentration. The ligand-bound core concentration $R_L = 0$. The concentration of active switches $R_A = 0$ All switches are inactive (I). **B.** As ligand concentration increases, the ligand-bound core concentration $R_L$ approaches the reference concentration $R_{ref}$. **C.** When the ligand-bound core concentration $R_L$ equals $R_{ref}$, the core concentration $R_L$ goes to zero and the concentration $R_A$ of active switches goes to $R_{ref}$. **D.** The process repeats until the number $M$ of active switches is equal to the number pf phosphorylation sites $N$. At any given time the concentration of active switches is $R_A = M\ R_{ref}$. The number of active switches $M$ does not decrease when the ligand concentration decreases.

where floor chooses the largest integer less than the argument. When the maximum ligand concentration is zero, $M = 0$. When the argument of floor becomes slightly greater than one, then $M = 1$. When the maximum ligand concentration becomes very large, then $M = N$. The ligand concentrations $L(M)$ at which switch $M$ turns on is determined by

$$M = N\frac{L^*(M)}{K_D + L^*(M)} \tag{3}$$

The quantity $L^*(M)$ can be approximated by the half-maximal effective concentration $\text{EC}_{50}(M)$ for switch $M$. This prediction is consistent with observations that significant downstream response can be observed even when a significant fraction of receptors are not engaged causing $EC_{50}$ values to be potentially much smaller than the value of the dissociation constant $K_D$ [27].

We see from Eq. 2 that an approximation of the value of the largest ligand concentration the receptor has encountered is coded into the number of active switches, according to the BOIS model. The amount of information per bit $I_M$ stored in this manner is (Supplement A.2)

$$I_M = \log_2 \frac{N!}{(N-M)!} \tag{4}$$

where $M$ is given by Eq. 2. The BOIS model thus provides a mechanism for encoding and transmitting information related to extracellular ligand concentration. The maximum information $I_B$ stored in the binding of ligand with the receptor is

$$I_B = \log_2 2 = 1 \text{ bit} \tag{5}$$

Therefore, the amount of information we have identified in the extracellular fluid that can be transmitted to the intracellular fluid is the sum of Eqs. 4 and 5. This extra information about the maximum ligand concentration may contribute to the observed excess transmitted information [9].

### 2.2 Application to G-Protein Coupled Receptors

An important GPCR feature, in the context of this study, is that, in addition to the barcode on the C tail, there is a GTPase switch [16] that activates particular downstream responses to ligand stimuli [14].

The BOIS model predictions of the previous section can be compared with GPCR assay observations [17]. The results are summarized in Table 1 and Fig. 3. The assays examined the response in two receptors, adrenergic and angiotensin, when tested with several ligands. The responses in two downstream pathways were measured, the response to the $G_\alpha$ GTPase switch turning on and the response of the recruitment of $\beta$ arrestin ($\beta$arr) to the C tail barcode of the receptor GPCR [12]. The $\beta$arr is thought to be a scaffold for other responses mediated by the barcode [12]. To reduce noise, the C tails of the light receptors in the assay were replaced with C tails of the vasopressin. Receptors with this alternate tail are known as chimeric. While the $\beta$arr recruitment site was preserved, the remaining phosphorylation sites on the C tail may not have been preserved on the chimeric-receptor tail.

The ligand concentration was slowly increased from 0 molar to a concentration at which the response was saturated. It was found that the maximum response was approximately equal to a common reference concentration $R_{ref}$ for each receptor in agreement with the BOIS predictions [6]. This can be seen in Fig. 3C and D where the maximum values of the response concentrations normalized to $R_{ref}$ are found at the corners, (00), (10), (01), and (11), of a hypercube. Moreover, the switches turned on one at a time also as indicated by the BOIS

**Table 1. Summary of Assay Data.** Two receptors were tested with several ligands. The detailed results and the ligands are given in the Supplement and reference [17]. The purple row is for those outcomes in which both the $G_\alpha$ and the $\beta$arr recruitment switches are turned on by the ligands. The red cells indicate assays in which the $G_\alpha$ switch is turned on by the ligands, but the $\beta$arr recruitment switch is off. The blue cells indicate assays in which the $G_\alpha$ switch is turned off by the ligands, but the $\beta$arr recruitment switch is on. The yellow cell indicates a ligand that did not turn on either the $G_\alpha$ or the $\beta$arr switch. The $G_\alpha$ column indicates the mean logarithm of the molar ligand concentration at which the $G_\alpha$ switch turns on. The $\beta$arr column indicates the concentration at which the $\beta$arr switch turns on. An $X$ indicates that the switch did not turn on. The Order columns indicate the order in which the switch turns on as ligand concentration increases. We see, for balanced ligands, that the second switch turns on at a ligand concentration approximately one order of magnitude higher than the concentration at which the first switch turns on. For the biased ligands, the order of the switch turning on is determined by comparison with the concentrations of the balanced ligands. For example, the biased $G_\alpha$ ligands are determined to be the second switch turning on by noting that the concentration -7.76 for the biased turn on is approximately equal to -7.95, the concentration of the second switch to turn on for the balanced ligands.

| **Adrenergic** | | | | | **Angiotensin II** | | | | |
|---|---|---|---|---|---|---|---|---|---|
| | Order | $G_\alpha$ | Order | $\beta$arr | | Order | $G_\alpha$ | Order | $\beta$arr |
| **Bal** | $1^{st}$ | **-9.44** | $2^{nd}$ | **-7.95** | **Bal** | $1^{st}$ | **-8.06** | $2^{nd}$ | **-7.14** |
| **Bias** | $2^{nd}$ | **-7.76** | | X | **Bias** | | X | $2^{nd}$ | **-6.86** |
| None | | X | | X | | | | | |

model (Table 1, Fig. 3A and B). In the case we have here in which the ligand concentration of the second switch is much greater than the ligand concentration of the first switch, then, from Eq. 2, the dissociation constant $K_D$ is approximately the $L^*(1) = EC_{50}(1)$ of the first switch, which is given in Table 1 as $10^{-9.44}$ M for the adrenergic receptor and $10^{-8.06}$ M for the angiotensin II receptor. This implies, from Eq. 3, that the total number $N$ of switches is

$$N = 2 \tag{6}$$

This means that the experimenters [17] observed all the states in the two receptors that were affected by the increasing ligand concentration. Moreover, it implies that the replacement of the adrenergic and angiotensin II C tails with vasopressin tails did not preserve the ability of the barcode to trigger responses other than $\beta$arr recruitment.

From Eq. 4, we see that two bits ($2 \log_2 2$) of information is transmitted across the cell membrane to the intracellular space.

The BOIS model predicts that these receptors are able to detect if a ligand is attached and whether the ligand concentration is greater than or less than the half maximal effective concentration $EC_{50}(1)$ for the first switch.

Four distinct responses were observed, a balanced response to the bound ligand in which both the $G_\alpha$ switch and the $\beta$arr recruitment switch turn on

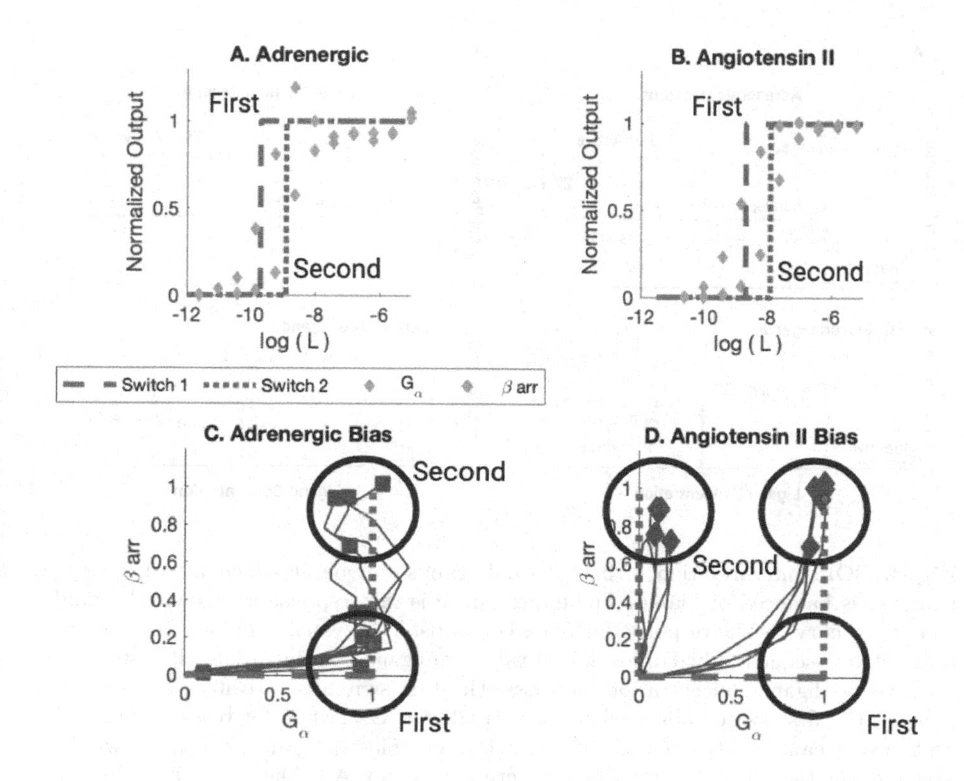


**Fig. 3. A.** Adrenergic Receptor with Formoterol as Ligand. Here, $L$ is the ligand concentration. The theory is displayed in red. The dashed line is the simulation of the activation of the first switch. The dotted curve is the activation of the second switch. The yellow markers are the observed assay dose response for the $G_\alpha$ pathway. The cyan markers are the observed assay dose response for $\beta$ arr. **B.** Angiotensin II Receptor with Angiotensin II as Ligand. **C.** Bias Plot for All Ligands for Adrenergic Receptor (see Table 1). The simulation results are displayed in red. Note that some ligands are $G_\alpha$ biased; their endpoints lie close to the $G_\alpha$ axis. Other ligands are balanced; their endpoints lie at (1,1). No $\beta$ arr bias is seen in this set of ligands. **D.** Bias Plot for All Ligands for angiotensin II Receptor. This plot illustrates balanced bias and $\beta$ arr bias. For balanced bias the first switch is the GTPC and the second switch is the PdPC that activates arrestin recruitment. For $\beta$ arr bias, the GTPC is not activated. The BOIS model predicts that the first ACTIVATED is a PdPC that is not observed. Here, $G_\alpha$ bias is only seen for ligand concentrations that are associated with sub-maximal response.

(purple row in Table 1), a situation in which just the $G_\alpha$ switch turned on (red cells), a situation in which just the $\beta$arr switch turned on (blue), and a situation in which no switches turned on (yellow). For the balanced ligands, both switches are seen to activate, first the $G_\alpha$ switch and then the $\beta$arr recruitment switch (Fig. 4). The first switches to activate in the case of biased ligands is now

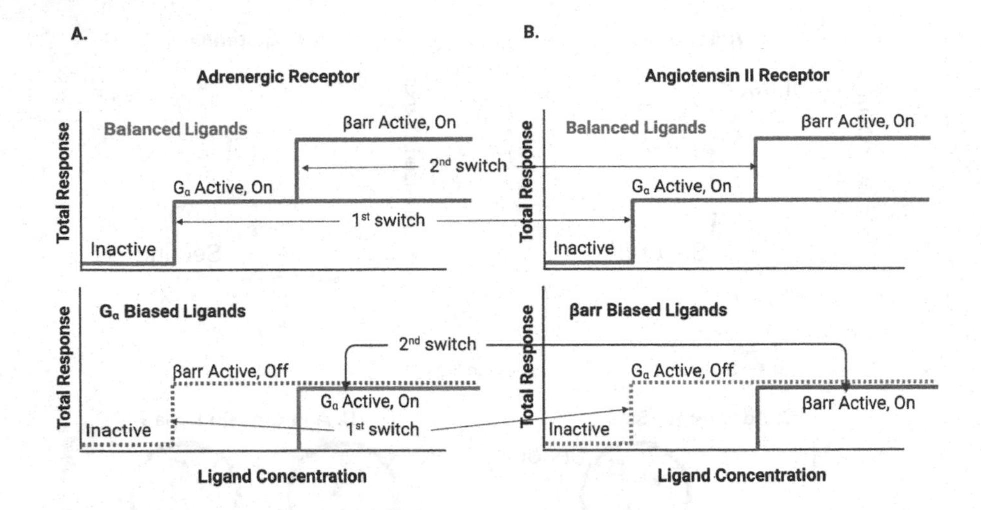


**Fig. 4.** BOIS interpretation of bias data. Ligands can be classified by whether the response is balanced or biased. The upper curve is the response for balanced ligands. The lower curve is the response for biased ligands. The system supports two switches that activate sequentially. For low initial values of ligand concentration, all switches are inactive. As ligand concentration increases, the first switch is activated. For balanced ligands, the first switch observed to be activated is $G_\alpha$, while for biased ligands the first switch can be either $G_\alpha$ **A.** or $\beta$arr **B.**. For biased ligands, the first switch is activated in the off state. For the adrenergic receptor **A.**, The $G_\alpha$ switch does not need to be activated in order to activate the $\beta$arr switch into the off state for biased ligands. After the first switch activates, but before the second switch activates, half the switches are active and half are inactive. As ligand concentration increases further, the second switch activates activating the $\beta$arr recruitment pathway for balanced ligands and the remaining inactive pathways for the biased ligands. Note that the BOIS model predicts that if the ligand concentration is lowered from its maximum that the activated switches do not deactivate. Therefore, the number of active switches is a measure of the maximum ligand concentration.

observed, however. The adrenergic receptor responds with $G_\alpha$ response when the second switch activates, while the angiotensin II receptor responds with the $\beta$arr recruitment response when the second switch activates.

The BOIS model predicts that, for biased ligands in these observations, the first switch is activated, but set to the off state. In other words, the switch sites are absorbing energy and dissipating heat but they are in the off position. This means, for example, that if the switch is a phosphorylation site, it is active and generating chemical flux through the site but the site spends most of its time unphosphorylated and not activating downstream response.

The BOIS model predicts that, for the one ligand that displayed no response, the switches may have been activated but they were in the off state.

## 3 Discussion

This study starts from an abstract model, the BOIS model [6] that was built on the assumptions of maximum rate of entropy production and maximum storage and flow of information. To validate the model with experiment, the BOIS model was specialized in [6] to address information flow in GPCRs. The process of this specialization was continued in this study and further hypotheses were generated that were compared with assay observations of [17] and information-flow experiments [28].

Specifically, switches can exist as phosphorylation sites on the C tail or on the GPCR or they can exist as GTPase switches on the core of the receptor. The switches can exist in three distinct states [6], inactive, active/on, and active/off. Active states are distinguished from inactive states in that active states support a chemical flux of receptor states through the switches while the inactive states have zero flux. For switches in the on state the receptor spends most of the time in the phosphorylated state in the case of phosphorylation switches. In the off state, the receptor spends most of the time in the unphosphorylated state.

Each active switch associates with a quantum of the total receptor concentration equal to $R_{ref} = R_T/N$ where $R_T$ is the receptor concentration and $N$ is the number of switches. This is true for both inactive and active switches [6]. This is a collective process. Each switch has a local property $R_{ref}$ affected by a global quantity $N$, which is non-local. It is not clear how an individual switch gains access to how many switches, active and inactive, there are.

The switches are activated one at a time in units of $R_{ref}$ as ligand concentration increases from an initial zero value and assuming all switches in the receptor are inactive. Therefore $N$ is the total number of switches than can be activated. Whether a switch is in the on state or to the off state seems to be determined at activation. If a switch becomes activated, then it remains active if the ligand concentration drops. Therefore, the number of active switches is a measure of the maximum concentration the receptor has experienced. The expression for this is given in Eqs. 3 and 4.

These predictions are consistent with experimental observations that significant response can occur when only a small number of total receptors are engaged with active switches [27]. They also suggests that observations that do not use chimeric-receptor C tails may be of importance. Chimeric-receptor tails were used to reduce noise, but the model predicts that the only switches available in assay experiments [17] were the $\beta$arr recruitment switch and the $G_\alpha$ switch when chimeric-receptor tails were employed. One expects many more phosphorylation switches [11]. This hints that the native C tails had relevant phosphorylation sites that did not appear on the chimeric-receptor tails. This suggests that the some of the noise that was eliminated by the chimeric-receptor tails was actually signal. If one redoes the assay experiments with native tails and measures more than two downstream responses, then the model predicts a series of signal switches that turn on sequentially and that the $EC_{50}$s of those switches would be closer together.

The amount of information transmitted through the membrane, Eq. 4, from the ligand concentration can account for some of the information found missing in information flow measurements [28]. In the observations, two bits or more of information was observed transmitting through the membrane, but if the only source of the information was whether the ligand was bound or not to the receptor, then only one bit, at most, was available to be transmitted. This suggests that there might be information other than simple binding that is associated with the ligand. At least some of the missing information seems to originate with the value of the ligand concentration.

Comparison of the BOIS model with assay experiments [17] indicates that the first switch that is activated may be either a $G_\alpha$ switch or a $\beta$arr switch for biased ligands (Fig. 4). For the two receptors observed, the first switch was $G_\alpha$ for balanced ligands and the second switch was for $\beta$arr recruitment. Bias may be determined at the time of switch activation. Our small observational sample [17] suggests that the first state activated is activated to the off state and determines the bias, at least in the case of a small number of switches ($N = 2$).

This picture differs from the canonical model of GPCR activation [15]. In that model, the $G_\alpha$ switches activates and turns on first. Subsequently, the $\beta$arr switch activates and turns on. Somewhat surprisingly, for the adrenergic receptor in this study coupled with the assay study [17], the $G_\alpha$ switch does not need to be activated in order to activate the $\beta$arr switch into the off state for biased ligands. Within the context of the BOIS model and the assay experiments [17], the $\beta$arr switch is first to activate, but it is turned off. The $G_\alpha$ switch may be turned on after the $\beta$arr recruitment switch, but the mechanism for this is currently unclear. These observations may also be affected by the information lost in the chimeric-receptor tail.

These results suggest some options for programming the receptor microprocessors with drugs for application in precision medicine. The new information in this study relates to the importance of ligand concentration in determining the downstream response.

In addition to ligand type, downstream response seems to be determined by the ligand concentration. Ligand concentration determines the number of active switches and the downstream response. Presumably there is a different downstream set of responses for each value of the number of active switches. Bias is determined by first switch activated. Ligand selection indicates that ligands determine the $G_\alpha$ pathway and the $\beta$arr pathway.

This study indicates that bias can be determined at the time that switches are activated. It is still not clear, however, if or how switch states can be altered between on and off after the switch is activated. Altering switch states at this level is a competition between the longevity and stability requirements for useful information and the need to be able to change the states for adaptability [19].

The observation that a biological switch can have three possible states rather than the two in a Boolean world leads to implications for adaptability. The switches in inactive states are switches that can be either on or off when activated. They represent the possibility of future information encoding for unknown future cell responses. The possible information in the barcode is no longer constrained by the number of active switches that are definitely in on and off states. The number of potential barcode configurations has increased significantly over the number encoded by on and off. This concept of increasing the size of the available search space to respond to future unknown events has been dubbed adjacent possibility [8] and has been suggested to be a necessary requirement for natural selection.

**Acknowledgements.** Biological images created with BioRender.com.

**Funding.** Partially funded by the European Union's Horizon 2020 research and innovation programme under grant agreement No 848011". Views and opinions expressed are however those of the author(s) only and do not necessarily reflect those of the European Union or Innovative Medicines Initiative 2 Joint Undertaking (Grant No. 115974, 2015). Neither the European Union nor the granting authority can be held responsible for them. Research in the laboratory of Alan M. Jones is supported by the National Science Foundation (MCB-0718202) and the National Institute of General Medical Sciences (R01GM065989).

## References

1. Black, J.W., Leff, P.: Operational models of pharmacological agonism. Proc. R. Soc. Lond. Ser. B. Biol. Sci. **220**(1219), 141–162 (1983)
2. Cardon, L.R., Harris, T.: Precision medicine, genomics and drug discovery. Hum. Mol. Genet. **25**(R2), R166–R172 (2016)
3. Chakravorty, D., Assmann, S.M.: G protein subunit phosphorylation as a regulatory mechanism in heterotrimeric G protein signaling in mammals, yeast, and plants. Biochem. J. **475**(21), 3331–3357 (2018)
4. Chen, H., Zhang, S., Zhang, X., Liu, H.: QR code model: a new possibility for GPCR phosphorylation recognition. Cell Commun. Signaling **20**(1), 1–16 (2022)
5. Ebeling, W.: Physical basis of information and the relation to entropy*. Eur. Phys. J. Spec. Top. **226**, 161–176 (2017)
6. Jones, R.D., Jones, A.M.: Model of ligand-triggered information transmission in G-protein coupled receptor complexes. Front. Endocrinol. **14**, 879 (2023)
7. Jones, R.D., Redsun, S.G., Frye, R.E.: Entropy generation by a Maxwell demon in the sequential sorting of the particles in an ideal gas. arXiv preprint physics/0311023 (2003). https://arxiv.org/abs/physics/0311023
8. Kauffman, S.: Is there a fourth law for non-ergodic systems that do work to construct their expanding phase space? Entropy **24**(10), 1383 (2022)
9. Keshelava, A., et al.: High capacity in G protein-coupled receptor signaling. Nat. Commun. **9**(1), 1–8 (2018)
10. Landauer, R.: The physical nature of information. Phys. Lett. A **217**(4–5), 188–193 (1996)
11. Latorraca, N.R., et al.: How GPCR phosphorylation patterns orchestrate arrestin-mediated signaling. Cell **183**(7), 1813–1825 (2020)
12. Latorraca, N.R., et al.: Molecular mechanism of GPCR-mediated arrestin activation. Nature **557**(7705), 452–456 (2018)

13. Leehey, D.J., Singh, A.K., Alavi, N., Singh, R.: Role of angiotensin II in diabetic nephropathy. Kidney Int. **58**, S93–S98 (2000)
14. Lefkowitz, R.: Nobel lecture: a brief history of G protein coupled receptors. Nobel Lect. **52**, 6366–6378 (2012)
15. Lefkowitz, R.J.: A brief history of G-protein coupled receptors (Nobel lecture). Angew. Chem. Int. Ed. **52**(25), 6366–6378 (2013)
16. Qian, H.: Thermodynamic and kinetic analysis of sensitivity amplification in biological signal transduction. Biophys. Chem. **105**(2–3), 585–593 (2003)
17. Rajagopal, S., et al.: Quantifying ligand bias at seven-transmembrane receptors. Mol. Pharmacol. **80**(3), 367–377 (2011)
18. Santos, R., et al.: A comprehensive map of molecular drug targets. Nat. Rev. Drug Discov. **16**(1), 19–34 (2017)
19. Schrodinger, R., Schrödinger, E., Dinger, E.S.: What is Life?: With Mind and Matter and Autobiographical Sketches. Cambridge University Press, Cambridge (1992)
20. Smith, J.S., Lefkowitz, R.J., Rajagopal, S.: Biased signalling: from simple switches to allosteric microprocessors. Nat. Rev. Drug Discov. **17**(4), 243–260 (2018)
21. Szilard, L.: On the decrease of entropy in a thermodynamic system by the intervention of intelligent beings. Behav. Sci. **9**(4), 301–310 (1964)
22. Thomas, M.C., Atkins, R.C.: Blood pressure lowering for the prevention and treatment of diabetic kidney disease. Drugs **66**, 2213–2234 (2006). https://doi.org/10.2165/00003495-200666170-00005. PMID: 17137404
23. Tobin, A.: G-protein-coupled receptor phosphorylation: where, when and by whom. Br. J. Pharmacol. **153**(S1), S167–S176 (2008)
24. Tobin, A.B., Butcher, A.J., Kong, K.C.: Location, location, location... site-specific GPCR phosphorylation offers a mechanism for cell-type-specific signalling. Trends Pharmacol. Sci. **29**(8), 413–420 (2008)
25. Tunc-Ozdemir, M., Li, B., Jaiswal, D.K., Urano, D., Jones, A.M., Torres, M.P.: Predicted functional implications of phosphorylation of regulator of g protein signaling protein in plants. Front. Plant Sci. **8**, 1456 (2017)
26. Yang, Z., et al.: Phosphorylation of g protein-coupled receptors: from the barcode hypothesis to the flute model. Mol. Pharmacol. **92**(3), 201–210 (2017)
27. Zhu, B.T.: The competitive and noncompetitive antagonism of receptor-mediated drug actions in the presence of spare receptors. J. Pharmacol. Toxicol. Methods **29**(2), 85–91 (1993)
28. Zielińska, K., Katanaev, V.: Information theory: new look at oncogenic signaling pathways. Trends Cell Biol. **29**(11), 862–875 (2019)

# Complex Chemical Systems

# Kauffman Model with Spatially Separated Ligation and Cleavage Reactions

Johannes Josef Schneider[1,2(✉)], Peter Eggenberger Hotz[1], William David Jamieson[3], Alessia Faggian[4], Jin Li[2], Hans-Georg Matuttis[5], Adriano Caliari[4], Mathias Sebastian Weyland[1], Dandolo Flumini[1], Aitor Patiño Diaz[4], Silvia Holler[4], Federica Casiraghi[4], Lorena Cebolla Sanahuja[4], Martin Michael Hanczyc[4,6], David Anthony Barrow[2], Pantelitsa Dimitriou[2], Oliver Castell[3], and Rudolf Marcel Füchslin[1,7]

[1] Institute of Applied Mathematics and Physics, School of Engineering, Zurich University of Applied Sciences, Technikumstr. 9, 8401 Winterthur, Switzerland
johannesjosefschneider@googlemail.com, {scnj,eggg,weyl,flum,furu}@zhaw.ch

[2] School of Engineering, Cardiff University, Queen's Buildings, 14-17 The Parade, Cardiff, Wales CF24 3AA, UK
{LiJ40,Barrow,dimitrioup}@cardiff.ac.uk

[3] Welsh School of Pharmacy and Pharmaceutical Science, Cardiff University, Redwood Building, King Edward VII Ave, Cardiff, Wales CF10 3NB, UK
{jamiesonw,CastellO}@cardiff.ac.uk

[4] Laboratory for Artificial Biology, Department of Cellular, Computational and Integrative Biology (CIBIO), University of Trento, 38123 Trento, Italy
{alessia.faggian,adriano.caliari,aitor.patino,silvia.holler, federica.casiraghi,lorena.cebolla,martin.hanczyc}@unitn.it

[5] Department of Mechanical Engineering and Intelligent Systems, The University of Electrocommunications, Chofu Chofugaoka 1-5-1, Tokyo 182-8585, Japan
hg@mce.uec.ac.jp

[6] Chemical and Biological Engineering, University of New Mexico, MSC01 1120, Albuquerque, NM 87131-0001, USA

[7] European Centre for Living Technology, Ca' Bottacin, Dorsoduro 3911, 30123 Venice, Italy
https://www.zhaw.ch/en/about-us/person/scnj/

**Abstract.** One of the open questions regarding the origin of life is the problem how macromolecules could be created. One possible answer is the existence of autocatalytic sets in which some macromolecules mutually catalyze each other's formation. This mechanism is theoretically described in the Kauffman model. We introduce and simulate an extension of the Kauffman model, in which ligation and cleavage reactions are spatially separated in different containers connected by diffusion, and provide computational results for instances with and without autocatalytic sets, focusing on the time evolution of the densities of the vari-

This work has been partially financially supported by the European Horizon 2020 project *ACDC – Artificial Cells with Distributed Cores to Decipher Protein Function* under project number 824060.


M. Villani et al. (Eds.): WIVACE 2023, CCIS 1977, pp. 141–160, 2024.
https://doi.org/10.1007/978-3-031-57430-6_12

ous molecules. Furthermore, we study the rich behavior of a randomly generated instance containing an autocatalytic metabolism, in which molecules are created by ligation processes and destroyed by cleavage processes and vice versa or generated and destroyed both by ligation processes.

**Keywords:** Kauffman model · origin of life · chemical evolution · Fick diffusion · autocatalytic set

# 1 Introduction

Over the past decades, the Kauffman model [11–13] has been intensively studied [7,9,10,21]. It deals with one of the basic questions of the origin of life [14] how macromolecules could be created via chemical evolution. As a possible answer, it proposes the emergence of autocatalytic sets in which some molecules are able to mutually catalyze each other's formation and which are self-sustaining if some food source in the form of monomers or small oligomers is provided. The basic condition for the production of macromolecules from an autocatalytic set is that the framework of catalyzed ligation and cleavage reactions forms a graph which in principle allows the production of the desired macromolecules [8]. This condition is necessary but not sufficient. Also the dynamics has to be considered as e.g. in the work of Bagley and Farmer [1]: They define an autocatalytic metabolism (ACM) as a coupled set of catalyzed reactions which lead to permanent concentrations $p_i(t)$ for the various molecules $i$ that significantly depart from values one would obtain without catalysis. Füchslin et al. [5] simulated the Kauffman model in one container: They chose appropriate values for the occurrence of catalyzed cleavage and ligation reactions, started off with $p_i(t = 0) = 1$ for all molecules, allowed only an inflow of two constituent monomeric molecules, and measured probabilities for the occurrence of an ACM and the sizes of the ACM by having a look at the final values of $p_i$ for the non-monomeric molecules. If at least one of them was larger than a proposed threshold, an ACM existed in their system.

In nature, one will find that catalyzed reactions are often only performed under some specific conditions, as e.g. enzymes only work in specific pH ranges. We assume that these specific conditions which change spatially might increase the probability for the existence of autocatalytic sets leading to macromolecules required for more complex forms of life. In order to make a first step in investigating this assumption, we extend the work by Füchslin et al. [5] to a system comprised of two containers: In one container, only catalyzed cleavage reactions shall be performed, in the other container, only catalyzed ligation reactions. Both containers are connected, such that molecules can diffuse into the other container depending on the concentration difference and the diffusion constant. This paper is organized as follows: We describe in general our extended model with spatially separated ligation and cleavage reactions in Sect. 2 and provide the simulation details for its application to a system of copolymers in Sect. 3.

Computational results for two randomly generated instances, one of them containing an ACM and the other displaying no ACM, are discussed in Sect. 4, before a conclusion and an outlook to future work is given in Sect. 5.

## 2 Extension of the Kauffmann Model

Based on the statements above, we now extend the model to a system with multiple containers. We thus deal with unnormalized densities $p_{i,j}$ annotated with two indices, where the first index $i$ denotes as before the number of the corresponding molecule and the new second index $j$ denotes the number of the container. The time evolution of $p_{i,j}$ is described by a set of differential equations. The total derivative $\mathrm{d}p_{i,j}/\mathrm{d}t$ subsummizes the various temporal changes of $p_{i,j}(t)$ imposed by different processes.

### 2.1 "In-Out" Processes

As most basic processes, we assume a constant inflow $k_{i,j,\text{in}}$ and an outflow which depends linearly on the density $p_{i,j}$ with a factor $k_{i,j,\text{out}}$ in each container:

$$\left(\frac{\mathrm{d}p_{i,j}}{\mathrm{d}t}\right)_{\text{in}-\text{out}} = k_{i,j,\text{in}} - k_{i,j,\text{out}} \times p_{i,j} \tag{1}$$

For these "in-out" processes, we consider the system of various containers as homogeneous, i.e., we set the $k$-parameters to the same values for all containers. Furthermore, we set all outflow-parameters for the various molecules to the same value. Second, we want to have the same amount of inflow for two constituing molecules with indices $i = 1$ and $i = 2$ only, all other molecules shall be created through ligation and cleavage processes. Thus, we have

$$k_{i,j,\text{out}} \equiv k_{\text{out}} > 0 \text{ and } k_{i,j,\text{in}} \equiv k_{i,\text{in}} = \begin{cases} k_{\text{in}} > 0 & \text{for } i = 1, 2 \\ 0 & \text{otherwise} \end{cases} . \tag{2}$$

In the absence of other processes, these in-out processes would converge to an equilibrium in which

$$(p_{i,j})_{\text{in}-\text{out}-\text{equ}} = \begin{cases} \dfrac{k_{\text{in}}}{k_{\text{out}}} & \text{for } i = 1, 2 \\ 0 & \text{otherwise} \end{cases} \tag{3}$$

in all containers, to which also systems with reactions but no ACM converge.

## 2.2 Cleavage and Ligation Processes

A ligation process is simply given by

$\alpha + \beta \longrightarrow \gamma$

and the corresponding cleavage process by

$\gamma \longrightarrow \alpha + \beta$ .

As already mentioned above, reactions enabled by catalyst molecules are considered within the Kauffman model. Let $\mathcal{R}$ be the set of possible reactions and $R = |\mathcal{R}|$ be their number. Then, for cleavage reaction $r$, there is a set $\mathcal{K}(r)$ containing $K(r) = |\mathcal{K}(r)|$ catalyst molecules $\kappa_k(r)$, $k = 1, \ldots, K(r)$, each of which is able to catalyze the reaction

$\gamma(r) \xrightarrow{\kappa_k(r)} \alpha(r) + \beta(r)$ .

Thus, for the cleavage reactions, we get the addend

$$\left(\frac{\mathrm{d}p_{i,j}}{\mathrm{d}t}\right)_{\text{cleavage}} = k_{j,C} \times \sum_{r=1}^{R} p_{\gamma(r),j} \left(-\delta_{i,\gamma(r)} + \delta_{i,\alpha(r)} + \delta_{i,\beta(r)}\right) \times \sum_{k=1}^{K(r)} p_{\kappa_k(r),j} \tag{4}$$

with the Kronecker symbol

$$\delta_{i,x} = \begin{cases} 1 & \text{if } i = x \\ 0 & \text{otherwise} \end{cases} \tag{5}$$

and the cleavage parameters $k_{j,C}$ relating the cleavage processes to the in-out processes.

Note that in the special case that there are no catalyst molecules for some cleavage reaction $\tilde{r}$, such that the reaction cannot be performed, the set of catalyst molecules is empty, $K(\tilde{r}) = 0$, and $\sum_{k=1}^{0} \cdots = 0$, such that this reaction does not contribute to the derivatives of the densities.

Analogously to the cleavage reactions, for each ligation reaction $r$, there is a set $\mathcal{L}(r)$ containing $L(r) = |\mathcal{L}(r)|$ catalyst molecules $\lambda_l(r)$, $l = 1, \ldots, L(r)$, each of which is able to catalyze the reaction

$\alpha(r) + \beta(r) \xrightarrow{\lambda_l(r)} \gamma(r)$ .

Thus, for the ligation reactions, we get the addend

$$\left(\frac{\mathrm{d}p_{i,j}}{\mathrm{d}t}\right)_{\text{ligation}} = \sum_{r=1}^{R} k_{\alpha(r),\beta(r),j,L} \times p_{\alpha(r),j} \times p_{\beta(r),j} \left(\delta_{i,\gamma(r)} - \delta_{i,\alpha(r)} - \delta_{i,\beta(r)}\right) \times \sum_{l=1}^{L(r)} p_{\lambda_l(r),j} \tag{6}$$

with the ligation parameters $k_{\alpha(r),\beta(r),j,L}$. The reason behind making the ligation parameters depending on the molecules is the insight that if two different molecules get into close enough contact for a reaction, their corresponding end monomers must get into contact. For simplicity, we define

$$k_{\alpha,\beta,j,L} = \frac{k_{j,L}}{\Lambda(\alpha) \times \Lambda(\beta)} \tag{7}$$

with constant ligation parameters $k_{j,L}$ and $\Lambda(\alpha)$ and $\Lambda(\beta)$ being the lengths of molecules $\alpha$ and $\beta$, rsp., which we define as the numbers of monomers in the molecules.

The dependency of the cleavage and ligation parameters $k_{j,C}$ and $k_{j,L}$ on the index $j$ of the container allows us to separate cleavage and ligation processes spatially as intended:

- If we intend to have only cleavage processes in some specific container $\hat{\jmath}$, then we set $k_{\hat{\jmath},L} = 0$ and $k_{\hat{\jmath},C}$ to some non-vanishing value.
- Analogously, if we intend to have only ligation processes in some specific container $\tilde{\jmath}$, then we set $k_{\tilde{\jmath},C} = 0$ and $k_{\tilde{\jmath},L}$ to some non-vanishing value.

### 2.3 Consideration of Finite Energy Amounts

In [5], the authors extend the Kauffman model by introducing an energy consideration: For many cleavage and ligation reactions in nature, an activation energy is required. However, the available amount of energy is rather limited. So far, the formulas (4) and (6) assume an infinite amount of energy or at least a large and renewable amount of energy which provides no obstacle for the reactions to be executed.

In order to include energy restrictions, a further variable $\varepsilon_j$ is introduced for each container $j$, with $\varepsilon_j(t)$ denoting the amount of energy available at time $t$ in container $j$. In order to consider the energy, the right sides of Eqs. (4) and (6) need to be multiplied with $\varepsilon_j$ and a further differential equation has to be added,

$$\begin{aligned}\left(\frac{\mathrm{d}\varepsilon_j}{\mathrm{d}t}\right)_{\text{total}} &= k_{E,j} - k_{E,j,\text{out}} \times \varepsilon_j \\ &\quad - \varepsilon_j \times k_{j,C} \times \sum_{r=1}^{R} p_{\gamma(r),j} \times \sum_{k=1}^{K(r)} p_{\kappa_k(r),j} \\ &\quad - \varepsilon_j \times \sum_{r=1}^{R} \frac{k_{j,L}}{\Lambda(\alpha(r)) \times \Lambda(\beta(r))} \times p_{\alpha(r),j} \times p_{\beta(r),j} \times \sum_{l=1}^{L(r)} p_{\lambda_l(r),j}\end{aligned} \tag{8}$$

with the energy inflow $k_{E,j}$, which we choose identical for all containers, i.e., then $k_{E,j} \equiv k_E$, and the outflow rate $k_{E,j,\text{out}}$, which we set identical with the corresponding parameter for all molecules, i.e., $k_{E,j,\text{out}} \equiv k_{\text{out}}$.

### 2.4 Diffusion Processes

So far, we only considered processes taking place separately in each container, such that we only have a set of separate containers up to now. But now, we want to take diffusion between neighboring containers into account. For simplicity, we want to rely hereby on Fick's first law of diffusion [4], according to which the diffusion velocity $v_D$ is given by

$$v_D = k_D \frac{\Delta c \times T}{\Delta x \times r \times \eta}, \tag{9}$$

with some diffusion constant $k_D$, the concentration difference $\Delta c$, the temperature $T$, the path length $\Delta x$ which needs to be transversed, the radius $r$ of the particle, and the viscosity $\eta$ of the medium. In our case, the radius corresponds to the length $\Lambda(i)$ of molecule $i$ and the concentration difference corresponds to the difference $p_{i,j} - p_{i,n}$ of the densities of molecule $i$ in neighboring containers $j$ and $n$. As diffusion processes occur only between neighboring pairs of containers, we need to create neighborhood lists: let $\mathcal{N}(j)$ be the set containing $N(j) = |\mathcal{N}(j)|$ index numbers of containers being neighbor to container $j$. Then we can write the addend for the diffusion processes as

$$\left(\frac{\mathrm{d}p_{i,j}}{\mathrm{d}t}\right)_{\text{diffusion}} = \sum_{n\in\mathcal{N}(j)} k_D \frac{p_{i,n} - p_{i,j}}{\Lambda(i)} = \frac{k_D}{\Lambda(i)} \times \left(-N(j)p_{i,j} + \sum_{n\in\mathcal{N}(j)} p_{i,n}\right) \tag{10}$$

with the diffusion constant $k_D$. As in Fick's first law of diffusion, the diffusion is proportional to the difference of the densities and inverse proportional to the length of the corresponding molecule. For containers of equal shape and volume with the same distance to all of their neighbors, we can omit the dependency on the path length.

Please note that we consider here only passive diffusion, i.e., diffusion does not use up any energy. Furthermore, only molecules can move to neighboring droplets, but there is no diffusion of energy in our model.

Such an approach with a time-independent diffusion constant $k_D$ is at odds with the experimental reality: Experiments performed on the development of aHL pores opening channels in bilayers between droplets by William David Jamieson at Cardiff University clearly show that it takes a significant amount of time until the first pore is formed. Thereafter, the number of pores increases monotonously in time, with decreasing slope. Thus, we have to alter Eq. (10) in order to consider the increase of the number $m_{j,n}(t)$ of pores between the neighboring containers $j$ and $n$:

$$\left(\frac{\mathrm{d}p_{i,j}}{\mathrm{d}t}\right)_{\text{diff.incr.}} = \frac{k_D}{\Lambda(i)} \sum_{n\in\mathcal{N}(j)} m_{j,n}(t) \times (p_{i,n} - p_{i,j}) \tag{11}$$

As the pores themselves are created by ligation of polymers (for example, aHL pores are comprised of seven macromolecules), one might think of considering the energy required for the creation of the pores, but we will abstain here from such an approach and neglect the energy consumption for the creation of pores. Furthermore, in order to show more clearly the effect of the opening of an increasing number of pores, we omit the dependencies of $m_{j,n}(t)$ on the container numbers $j$ and $n$, i.e. $m_{j,n}(t) \equiv m(t)$.

## 3 Simulation Details

In this paper, we apply the Kauffman model to a set of reactions generating and splitting copolymers comprised of a linear sequence of two different monomers A and B. Thus, there are $2^{\Lambda}$ different molecules containing $\Lambda$ monomers. (Please note that we assume the molecules to be directed, e.g. the molecules A − B and B − A are different molecules.) We consider only polymers comprised of up to a maximum number $\Lambda_{\max}$ of monomers.

The total number $M$ of different molecule types is given by

$$M = \sum_{\Lambda=1}^{\Lambda_{\max}} 2^{\Lambda} = 2^{\Lambda_{\max}+1} - 2. \tag{12}$$

The number $R$ of cleavage reactions can be determined to

$$R = \sum_{\Lambda=2}^{\Lambda_{\max}} (\Lambda - 1)2^{\Lambda} = \Lambda_{\max} 2^{\Lambda_{\max}+1} - 2M. \tag{13}$$

In contrast to [5], we do not allow ligation reactions leading to molecules with more than $\Lambda_{\max}$ monomers, such that the number of ligation reactions equals the number of cleavage reactions.

In our simulations, we use $\Lambda_{\max} = 3$, such that we have 12 non-monomeric different molecule types in our system and a reaction framework containing 20 possible reactions. For the various parameters, we choose values already used in [5]: The two monomers are not allowed to serve as catalysts, each of the other molecules is chosen with probability $r_C = 0.05$ to serve as catalyst for a cleavage reaction and with probability $r_L = 0.1$ to serve as catalyst for a ligation reaction. The other parameters are set to

$$k_{i,\mathrm{in}} = \begin{cases} 1 & \text{for the monomers, i.e., for } i = 1, 2 \\ 0 & \text{otherwise} \end{cases},$$

$k_{\mathrm{out}} = 0.02$, $k_C = 1$, $k_L = 1$, $k_E = 1$, and $k_D = 0.05$.

We will first have a look at the original Kauffman system with only one container for which we set $k_C = 1$ and $k_L = 1$. In order to study the effect of spatial separation of cleavage and ligation processes without any further side-effects occurring in more complex systems, we consider the extended Kauffman system with two containers only, one container $j = 1$, in which only cleavage reactions take place, and a second container $j = 2$, in which only ligation reactions are performed. For this purpose, we set $k_{1,C} = k_{2,L} = 1$ and $k_{1,L} = k_{2,C} = 0$.

We will also consider both diffusion processes as given in Eqs. (10) and (11). For the function $m(t)$, we choose

$$m(t) = \begin{cases} 0 & \text{if } t < 1 \\ \lfloor \log(t) \rfloor & \text{otherwise} \end{cases}. \tag{14}$$

This simple function is still able to reflect the key properties of the time evolution of the number of pores as observed in the experiment, i.e., the significant amount of time before the first pore opens up, the monotonous increase, and the concave shape after the opening of the first pore. As the pores open up at the times $t = e, e^2, e^3, \ldots$ according to Eq. (14), the changes due to the opening of more pores are equally spaced out on a logarithmic time scale.

We use the Dormand-Prince method [3] for the numeric solution of the set of differential equations. This algorithm which belongs to the class of Runge-Kutta methods allows us to adaptively change the length of the time interval between successive time steps by determining two different solutions of fourth and fifth order and halving the length of the time interval if the deviation between them is too large. We redo the calculation with a halved time interval if the relative deviation exceeds a value of $10^{-8}$. We integrate over the time interval from $t = 0$ to $t = 10^4$. As initial conditions, we set $p_{i,j}(t = 0) = \varepsilon_j(t = 0) = 1$ as in [5].

# 4 Computational Results

## 4.1 Revisiting the Original Kauffman Model Within One Container Only

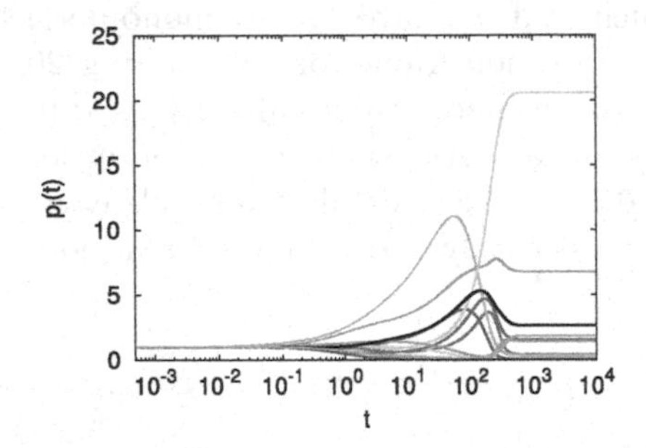


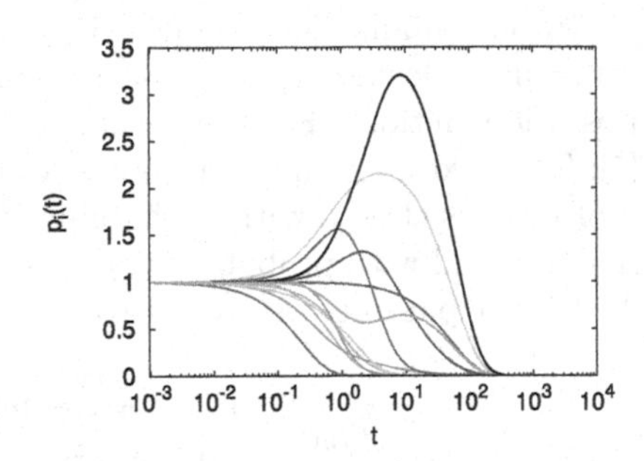


**Fig. 1.** Time evolution of the densities $p_i$ of the non-monomeric molecules in the original Kauffman model with one container and without energy consideration, for a randomly created instance displaying an ACM (left) and another randomly created instance displaying no ACM (right)

We start out simulating the original Kauffman model in one container only and present results for two randomly created catalyzed reaction instances, one of them displaying an ACM and another one displaying no ACM. For the instance displaying an ACM, we provide the list of reactions in Table 1. We will use these two instances with the same random choice of catalyst molecules for the various reactions also in the simulations for the next scenarios and will use the same colors for the same molecules. The time evolutions of the densities $p_i(t)$ of all 12 non-monomertic molecules are shown for both instances in Fig. 1. In order to better

**Table 1.** List of reactions in the instance containing an ACM: In this instance, we have 14 successively numbered molecules, with the monomers being denoted as Nos. 1 and 2, and 20 possible reactions. In the right two columns, only those cleavage and ligation reactions are noted for which there is at least one catalyst molecule. This instance was randomly generated.

| Reaction No. | reaction | cleavage | ligation |
|---|---|---|---|
| 1 | $3 \longleftrightarrow 1+1$ | | $1+1 \xrightarrow{5} 3$ |
| 2 | $4 \longleftrightarrow 1+2$ | | $1+2 \xrightarrow{3,6,10,11,12,14} 4$ |
| 3 | $5 \longleftrightarrow 2+1$ | | $2+1 \xrightarrow{4,6,8,12} 5$ |
| 4 | $6 \longleftrightarrow 2+2$ | | |
| 5 | $7 \longleftrightarrow 1+3$ | | $1+3 \xrightarrow{4,12} 7$ |
| 6 | $7 \longleftrightarrow 3+1$ | $7 \xrightarrow{10} 3+1$ | |
| 7 | $8 \longleftrightarrow 1+4$ | | |
| 8 | $8 \longleftrightarrow 3+2$ | | $3+2 \xrightarrow{12} 8$ |
| 9 | $9 \longleftrightarrow 1+5$ | | $1+5 \xrightarrow{6} 9$ |
| 10 | $9 \longleftrightarrow 4+1$ | $9 \xrightarrow{8} 4+1$ | $4+1 \xrightarrow{12,14} 9$ |
| 11 | $10 \longleftrightarrow 1+6$ | $10 \xrightarrow{5} 1+6$ | |
| 12 | $10 \longleftrightarrow 4+2$ | | $4+2 \xrightarrow{5,9} 10$ |
| 13 | $11 \longleftrightarrow 2+3$ | | $2+3 \xrightarrow{5,12,14} 11$ |
| 14 | $11 \longleftrightarrow 5+1$ | | |
| 15 | $12 \longleftrightarrow 2+4$ | $12 \xrightarrow{9} 2+4$ | |
| 16 | $12 \longleftrightarrow 5+2$ | | $5+2 \xrightarrow{4} 12$ |
| 17 | $13 \longleftrightarrow 2+5$ | | $2+5 \xrightarrow{7} 13$ |
| 18 | $13 \longleftrightarrow 6+1$ | $13 \xrightarrow{13} 6+1$ | $6+1 \xrightarrow{11,13} 13$ |
| 19 | $14 \longleftrightarrow 2+6$ | $14 \xrightarrow{13} 2+6$ | $2+6 \xrightarrow{7,8} 14$ |
| 20 | $14 \longleftrightarrow 6+2$ | | $6+2 \xrightarrow{9} 14$ |

display the developments at short time scales, we use a logarithmic time scale. After some intermediate increases and decreases of the various densities, the system converges to final values for the various densities between $t = 5 \times 10^2$ and $t = 10^3$. The non-vanishing final values for some densities in the left picture indicate that this instance displays an ACM, whereas the finally vanishing densities in the right picture show that the second instance does not contain an ACM. For the instance with an ACM, the Dormand-Prince method needs 194059 time steps, whereas only 1772 are needed for the instance without an ACM. In order to get results of equal quality for the time evolution of the densities, much more computing time is needed for an instance with an ACM.

Please note that each color in the curves in Fig. 1 corresponds to one distinct molecule. We will use the same color for the same distinct molecule in the curves of Figs. 2, 3, 4, 5, 6, 7 and 8 and in the boxes in Figs. 9, 10 and 11.

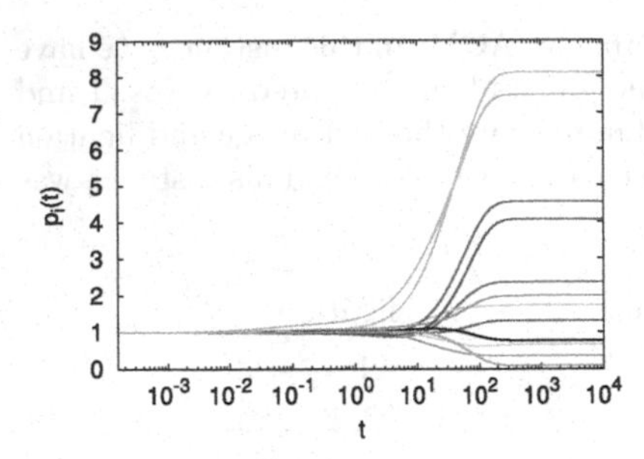


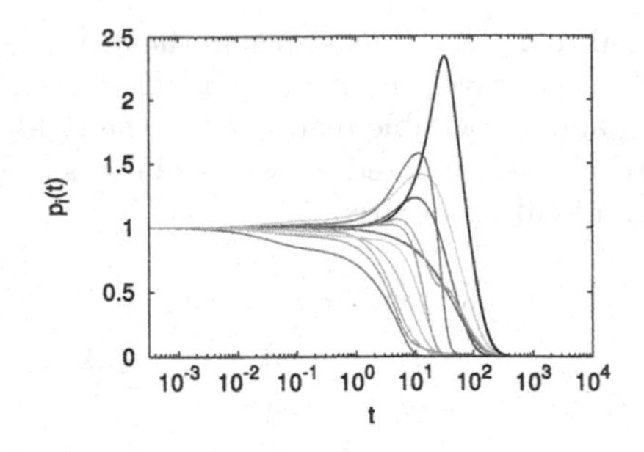


**Fig. 2.** Time evolution of the densities $p_i$ of the non-monomeric molecules as in Fig. 1, but now with energy consideration, both for the instance displaying an ACM (left) and the instance displaying no ACM (right)

In the next step, we consider the model extended with an energy consideration as introduced in [5]. We use the same instances as before. The computational results are shown in Fig. 2. The instance which displayed an ACM before again shows an ACM, but the resulting values of the densities for the various molecules differ strongly from those in Fig. 1. The other instance again contains no ACM. A further effect of considering the finite available energy is that it retards the dynamics, as the curves for the various molecules start to deviate from the original values significantly later. Furthermore, one can state that the consideration of the energy stabilizes the dynamics, as already mentioned in [5], as the intermediate maxima are much smaller than without energy consideration. The computing time required increases strongly when considering the energy: for the instance with an ACM, the Dornand-Price method requires more than 8.6 million time steps and it takes 8193 time steps for the instance without an ACM.

## 4.2 Two Separate Containers

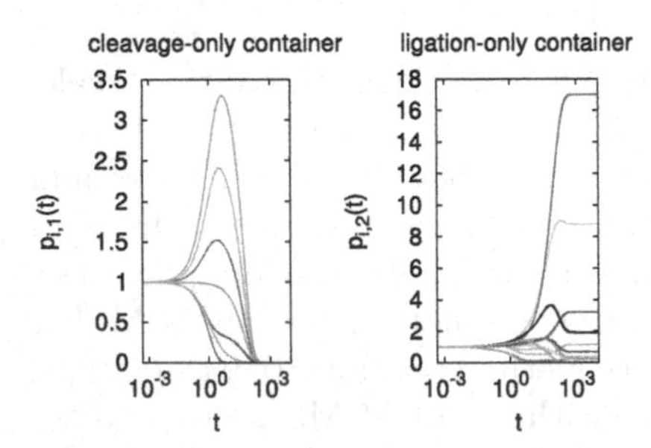


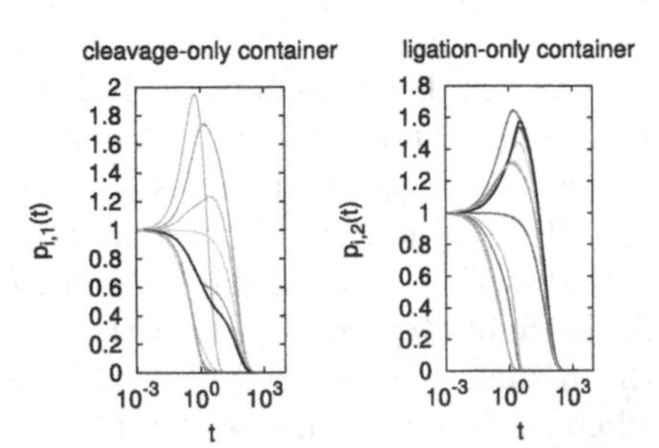


**Fig. 3.** Time evolution of the densities $p_{i,j}$ of the non-monomeric molecules as in Fig. 1, but now in two separate containers, one of them only allowing cleavage reactions (left part, $j = 1$), the other one only allowing ligation reactions (right part, $j = 2$), both for the instance displaying an ACM (left) and the instance displaying no ACM (right)

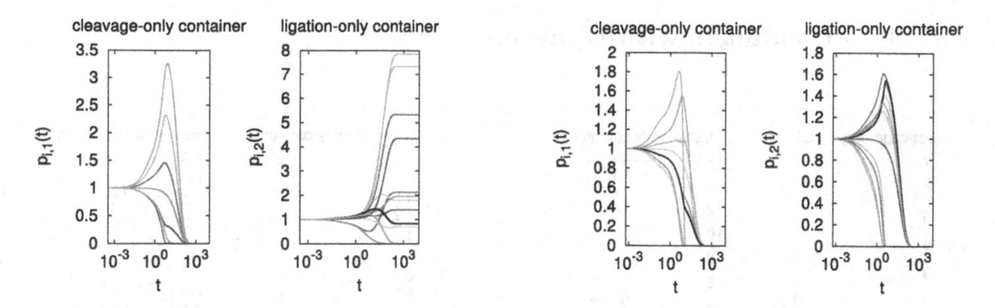


**Fig. 4.** Time evolution of the densities $p_{i,j}$ of the non-monomeric molecules as in Fig. 3, in two separate containers, one of them only allowing cleavage reactions (left part, $j = 1$), the other one only allowing ligation reactions (right part, $j = 2$), but now with energy consideration, both for the instance displaying an ACM (left) and the instance displaying no ACM (right)

The results discussed so far for one container can also be interpreted as the results obtained with two containers with spatially separated cleavage and ligation reactions if the diffusion takes place so fast that any density differences between the two containers are resolved immediately. Before considering the connected containers with diffusion, we would like to study the other extreme case first, in which the two containers are separate and in which no diffusion takes place. The results for the scenario without considering the energy are shown in Fig. 3, the results for the scenario including the energy in Fig. 4. For the instance, which contained an ACM before, we find that the densities vanish in the container with the cleavage processes as has to be expected (The original densities of the non-monomeric molecules decrease due to outflow and cleavage reactions and no new non-monomeric molecules are formed as there is no ligation. So, there is no ACM.), whereas an ACM can still be found in the container with the ligation processes. We also investigated other instances, for some of them which contained an ACM in one container only there is also no ACM in the container with only ligation processes.

### 4.3 Two Containers with Diffusion

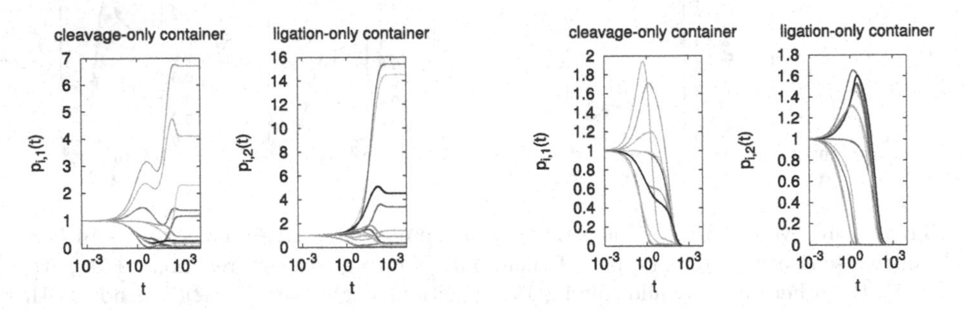


**Fig. 5.** Time evolution of the densities $p_{i,j}$ of the non-monomeric molecules as in Fig. 3, but now with diffusion between two connected containers, one of them only allowing cleavage reactions (left part, $j = 1$), the other one only allowing ligation reactions (right part, $j = 2$), both for the instance displaying an ACM (left) and the instance displaying no ACM (right)

Finally, we get to the main point of this paper, the Kauffman model implemented with spatially separated cleavage and ligation processes in two containers with diffusion between them. First, we consider diffusion according to formula (10) and present computational results without energy consideration in Fig. 5 and with energy consideration in Fig. 6. We find that the instance which contained an ACM before now contains ACMs in both containers despite the fact that the diffusion constant $k_D$ is set to a very small value. On the other hand, the instance which displayed no ACM before now also shows no ACMs. The molecules whose curves plotted in light blue and light green dominated the ACM in the left picture in Fig. 1 now also belong to the dominating molecules in the ACMs in the left pictures of Fig. 5, but there are now more significant contributions of other molecules as well. The same behavior is found for the molecules plotted in dark yellow and orange when comparing Figs. 2 and 6.

Then we have a look at the diffusion with an increasing number of pores according to formulas (11) and (14). The computational results are shown in Fig. 7 for the scenario without energy consideration and Fig. 8 with consideration of the energy. For the left instance, we again get an ACM, the steps in the curves reflect the opening of an increasing number of pores. These steps are smaller if the energy is considered. As the number of pores increases in time, equilibrium cannot be reached. For the instance without an ACM, we hardly see any steps, the breakdown of the densities dominates the behavior.

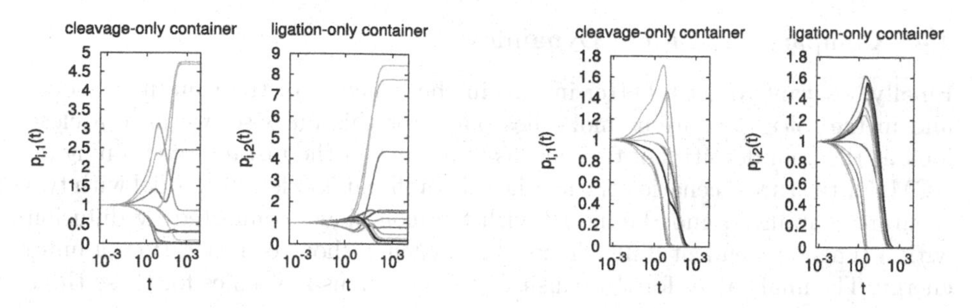


**Fig. 6.** Time evolution of the densities $p_{i,j}$ of the non-monomeric molecules as in Fig. 5, with diffusion between two connected containers, one of them only allowing cleavage reactions (left part, $j = 1$), the other one only allowing ligation reactions (right part, $j = 2$), but now with energy consideration, both for the instance displaying an ACM (left) and the instance displaying no ACM (right)

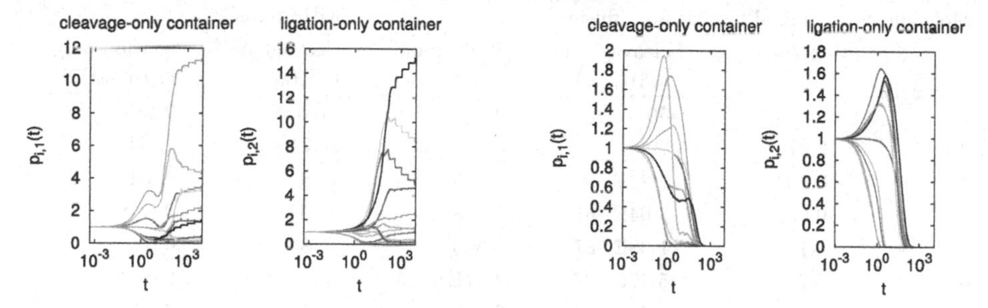


**Fig. 7.** Time evolution of the densities $p_{i,j}$ of the non-monomeric molecules as in Fig. 5, but now with diffusion through an increasing number of pores between two connected containers, one of them only allowing cleavage reactions (left part, $j = 1$), the other one only allowing ligation reactions (right part, $j = 2$), both for the instance displaying an ACM (left) and the instance displaying no ACM (right)

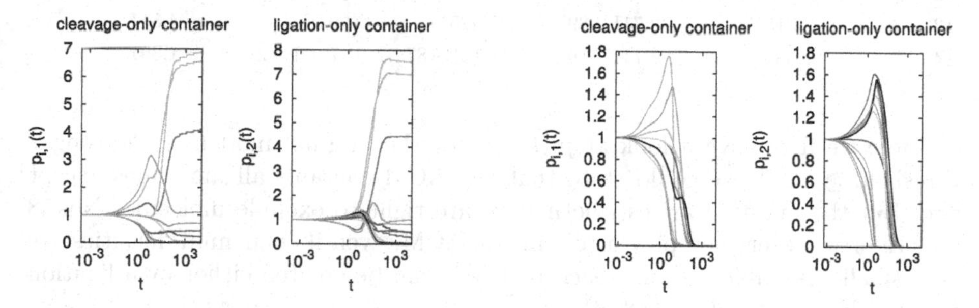


**Fig. 8.** Time evolution of the densities $p_{i,j}$ of the non-monomeric molecules as in Fig. 7, with diffusion through an increasing number of pores between two connected containers, one of them only allowing cleavage reactions (left part, $j = 1$), the other one only allowing ligation reactions (right part, $j = 2$), but now with energy consideration, both for the instance displaying an ACM (left) and the instance displaying no ACM (right)

### 4.4 Comparison of Final Dynamics

Finally, we want to get a better insight in the behavior of the Kauffman model and in the roles the various molecules play. For this purpose, we have a close look at the final densities of the various molecules in the instance containing an ACM, for the three scenarios of the original Kauffman model, the model with two separate containers, and the model with two containers connected by diffusion with a time-independent diffusion constant, each without consideration of finite energy. The final values for the densities of the various molecules for these three scenarios are provided in Table 2.

**Table 2.** Final values for the densities of the various molecules in the instance displaying an ACM for the original Kauffman model, for the extreme case without diffusion, and for the scenario with a constant diffusion constant, without energy consideration

| Molecule No. | original model | no diffusion | | with diffusion | |
|---|---|---|---|---|---|
| $i$ | $p_i(t_{\text{final}})$ | cleavage-only $p_{i,1}(t_{\text{final}})$ | ligation-only $p_{i,2}(t_{\text{final}})$ | cleavage-only $p_{i,1}(t_{\text{final}})$ | ligation-only $p_{i,2}(t_{\text{final}})$ |
| 1 | 0.505 | 50 | 0.457 | 21.76 | 0.635 |
| 2 | 0.070 | 50 | 6.52E−2 | 14.71 | 8.27E−2 |
| 3 | 0.404 | 2.25E−87 | 0.155 | 1.385 | 0.316 |
| 4 | 0.352 | 3.64E−87 | 1.138 | 4.108 | 0.889 |
| 5 | 0.145 | 1.38E−87 | 0.365 | 0.184 | 0.331 |
| 6 | 6.775 | 5.47E−87 | 2.47E−323 | 6.708 | 6.66E−2 |
| 7 | 0.023 | 5.08E−88 | 3.219 | 4.67E−2 | 3.369 |
| 8 | 6.15E−5 | 1.38E−87 | 0.171 | 6.80E−2 | 0.150 |
| 9 | 20.56 | 2.66E−109 | 8.803 | 2.309 | 14.51 |
| 10 | 1.553 | 2.66E−109 | 17.00 | 1.165 | 15.41 |
| 11 | 1.416 | 1.38E−87 | 0.264 | 0.420 | 0.924 |
| 12 | 8.69E−5 | 5.08E−88 | 0.677 | 2.36E−3 | 0.333 |
| 13 | 2.615 | 2.71E−89 | 1.915 | 0.256 | 4.511 |
| 14 | 1.856 | 2.71E−89 | 2.10E−87 | 7.91E−2 | 1.391 |

Here we first have a look at $p_i(t_{\text{final}})$ for the original model. If choosing a threshold $\geq 10^{-4}$, we could state that the ACM contains all molecules except two. But the question arises whether we are right to exclude molecules Nos. 8 and 12 or whether they play a role in the ACM, even if their final densities are very small. As Table 1 shows, each molecule can be created either by a ligation or by a cleavage reaction or both.

When having a look at the results for the two separate containers, we find as expected that there is no ACM in the cleavage-only container. All longer molecules are destroyed by the cleavage processes and by the outflow. No new longer molecules can be created, as the constant inflow of the two monomers cannot be used for ligation, as there are no ligation reactions. Thus, we get a value of 50 for the densities of the two monomers, which is just the ratio between the inflow and the outflow parameters, and a vanishing value for all other molecules. In the ligation-only container, we get final densities which partially slightly, but

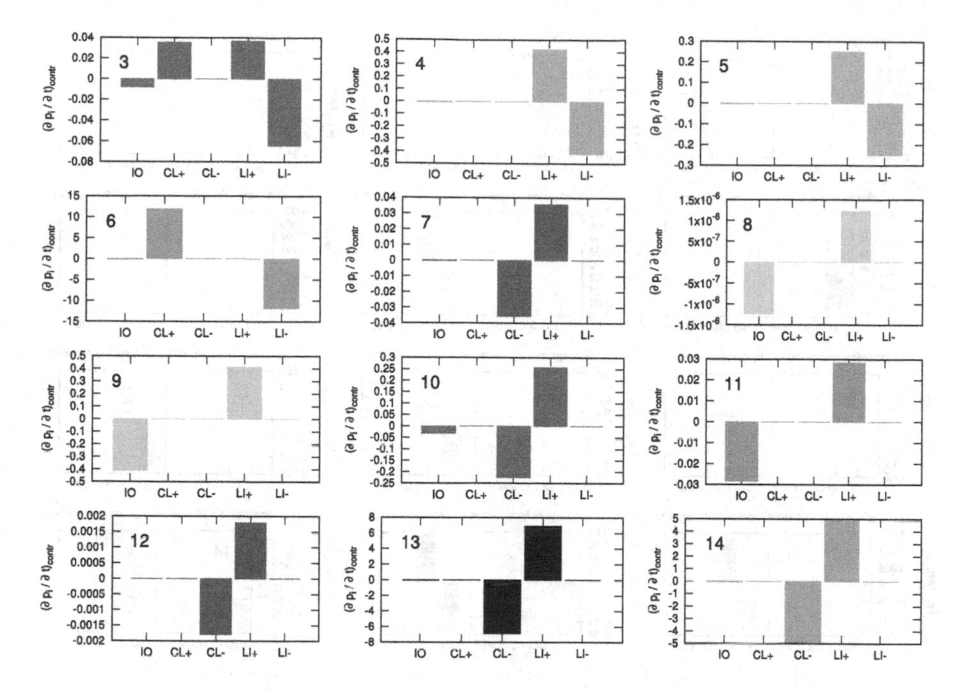


**Fig. 9.** Final dynamics: final values for the derivatives $(\mathrm{d}p_i/\mathrm{d}t)_{\mathrm{contr.}}$ of the densities for the various non-monomeric molecules for the instance containing an ACM in the original Kauffman model, for the various contributions: IO – inflow and outflow, CL+ – generated by cleavage processes, CL- – destroyed by cleavage processes, LI+ – generated by ligation processes, LI- – destroyed by ligation processes

most often strongly deviate from the final densities in the original Kauffman scenario. Obviously, the additional cleavage processes in the original Kauffman scenario lead to these large differences. Molecules Nos. 6 and 14 are obviously not part of the ACM anymore, their densities truly vanish.

Then we have a look at the results for two containers connected by diffusion. Here we see that the densities for the molecules Nos. 8 and 12 are much larger in both containers than in the original Kauffman model. Obviously, a spatial separation of reactions can lead to an enlargement of an ACM. But this result also raises questions to the approach of determining the size of an ACM, i.e., the number of molecules being part of an ACM, by excluding those molecules whose densities are smaller than some arbitrarily chosen threshold. Here one has to be very careful of how to choose the value of the threshold, in order to not exclude those molecules which contribute to the ACM, even if their contribution seems to be tiny.

In order to even better understand the final dynamics for these three scenarios, we have a look at the final values for the various contributions to the derivatives of the densities, which are shown for the three scenarios considered here in Figs. 9, 10, and 11. We consider separately the contributions by the

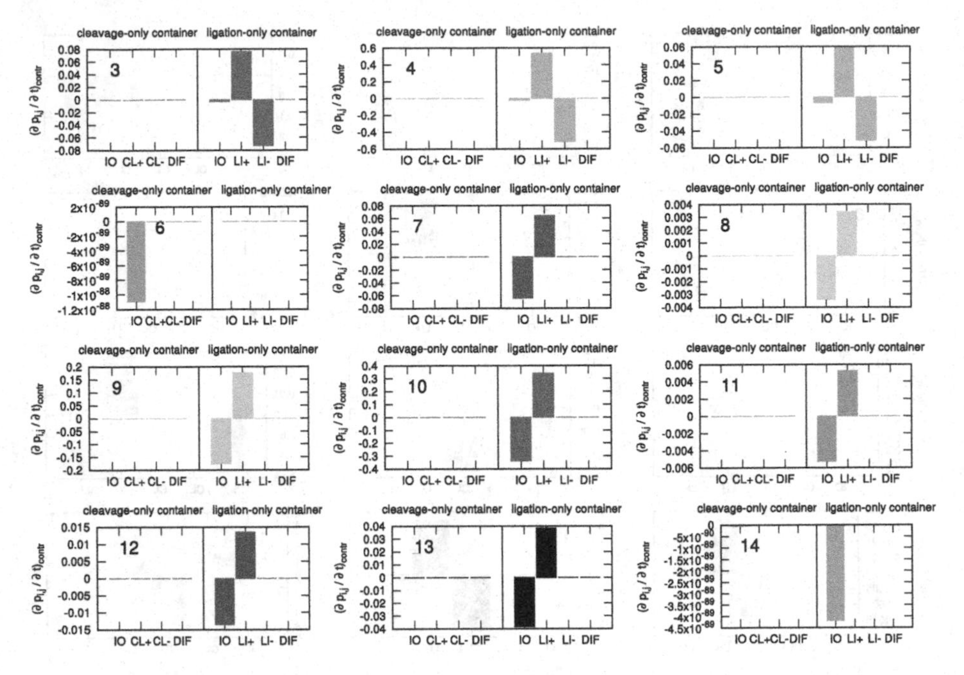


**Fig. 10.** Final dynamics: final values for the derivatives $(\mathrm{d}p_{i,j}/\mathrm{d}t)_{\mathrm{contr.}}$ of the densities for the various non-monomeric molecules in the cleavage-only container (left) and the ligation-only container (right) in the scenario with two separate containers, for the various contributions: IO – inflow and outflow, CL+ – generated by cleavage processes, CL- – destroyed by cleavage processes, LI+ – generated by ligation processes, LI- – destroyed by ligation processes, DIF – diffusion

various processes in-out-flow, cleavage, ligation, and diffusion. Already for the original Kauffman model, we see a rich behavior: Molecules 4 and 5 are created by ligation processes but also destroyed by ligation processes, molecules 7, 10, 12, 13, and 14 are created by ligation and destroyed by cleavage, while molecule 6 is created by cleavage and destroyed by ligation. Molecules 8, 9, and 11 are created by ligation, but destroyed by outflow. Molecule 3 exhibits the richest behavior, it is both created by cleavage and ligation and it is destroyed by ligation. Of course, for all of these molecules, also the outflow plays some role, but only for some of them the outflow provides the main contribution to decreasing their densities. Thus, we get a very rich variety of behaviors already here in Fig. 9 for the original Kauffman model. Please note that all the bars in the subgraphs for the various molecules add up to zero, as the final densities are constant such that the sum of all contributions to their derivatives has to vanish.

Figure 10 shows the results for the scenario with two separate containers. As expected, there is no ACM in the cleavage-only container, such that the derivatives vanish there. In the ligation-only container, the derivatives for molecules 6 and 14 vanish as well, they are not part of the ACM here. But all other molecules contribute to the ACM. Molecules 7–13 are created by ligation processes, their

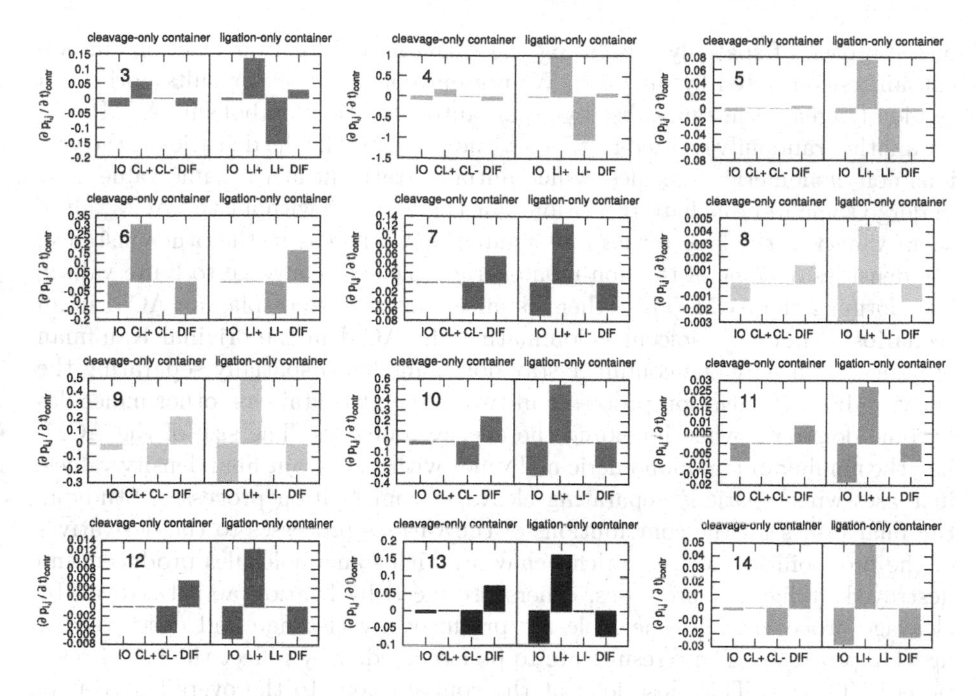


**Fig. 11.** Final dynamics: final values for the derivatives $(\mathrm{d}p_{i,j}/\mathrm{d}t)_{\text{contr.}}$ of the densities for the various non-monomeric molecules in the cleavage-only container (left) and the ligation-only container (right) in the scenario with two containers connected by diffusion, for the various contributions: IO – inflow and outflow, CL+ – generated by cleavage processes, CL- – destroyed by cleavage processes, LI+ – generated by ligation processes, LI- – destroyed by ligation processes, DIF – diffusion

densities are reduced by outflow only. Molecules 4 and 5 are both created and destroyed by ligation processes, additionally also outflow is reducing their densities.

Figure 11 shows the corresponding results for the scenario with two containers connected by diffusion, for which we again find interesting behaviors. Molecules 7 and 9–14 share the same behavior: they are produced in the ligatrion-only container. Part of the density diffuses to the cleavage-only container where the incoming density is destroyed by a cleavage process. In both containers, also the outflow reduces the densities. Like in Fig. 9, molecule 6 demonstrates just the opposite behavior, it is produced by a cleavage process in the cleavage-only container, part of its density diffuses into the ligation-only container, where it is destroyed by a ligation process and the outflow. Molecules 4 and 5 are again dominated by the production and destruction via ligation processes. Please note that also here the bars need to add up to zero, separately for both containers.

## 5 Conclusion and Outlook

For this paper, we performed simulations for the original Kauffman model, the Kauffman model extended with the consideration of finite energy amounts, and our approach of spatially separating the cleavage and ligation reactions in two containers connected by diffusion. We present computational results both for a randomly created instance displaying an autocatalytic metabolism (ACM) and for another randomly created instance without an ACM. The densities of the various non-monomeric molecules, which initially start out at the same value, first undergo some intermediate transition. For the instance without the ACM, all of them vanish in the long term in all simulations, whereas in the other instance, the densities of some of the non-monomeric molecules converge to finite values, thus forming the ACM. This other instance continues to display an ACM in all scenarios. While the molecules dominating the ACM in the original Kauffman approach with only one container stay dominant when spatially separating the cleavage from the ligation processes in two different containers, other molecules become dominant when including the energy finiteness. The size of the ACM, i.e., the number of non-monomeric molecules with significant final density values, increases when spatially separating cleavage from ligation processes. Studying the final values for the contributions of the various processes to the derivatives of the probabilities, we get a rich behavior, with some molecules produced and destroyed by ligation processes, others produced by ligation and destroyed by cleavage processes, and one molecule produced by cleavage and destroyed by ligation processes. These results are to be expected, they reflect the list of reactions in Table 1. This close look at the contributions to the overall derivative obviously provides a better insight whether a molecule is part of the ACM than the comparison of its density to an arbitrarily chosen threshold.

We intend to continue our investigations by creating larger statistics of these scenarios and by applying all of them to large networks of spatially connected droplets [15–19], in which either only ligation or only cleavage reactions are performed. Depending on the simulation parameters, we expect to be able to enlarge the probability for the occurrence of an ACM, for those instances in which it is possible from a graph-theoretical aspect, but in which the dynamics prevents the creation of an ACM in one container only. We also expect to get much larger average sizes of an ACM. These large networks also provide a further advantage: While it is impossible to change the kinetic parameters for the reactions as well as the diffusion and transport parameters for real systems, large networks of droplets with spatially separated reactions allow to put more emphasis on some reactions, e.g., to implement many more droplets with ligation than with cleavage reactions, thus increasing the probability for the occurrence of an ACM.

Furthermore, instead of using the approach to set up and numerically solve a set of differential equations, we alternatively intend to apply a stochastic simulation framework, e.g., to work with the Gillespie algorithm [2,6], which we recently applied to a minimum reaction system with one undesired side product [20]. The Gillespie algorithm is better suited for large numbers of small containers in which the number of the various molecules could be so small that the continuous density approach is no longer justified. This stochastic simulation framework offers new possibilities and new insights, but also leads to further difficulties: Here we would not have to depend on the value of a threshold for a density whether we would consider a specific molecule to be part of the ACM or not. When using the Gillespie algorithm with its integer numbers for various molecules, we can simply state that a molecule is not part of the ACM if its number is exactly zero over a sufficiently long time period. However, in another simulation run, it might be the case that this molecule does not vanish and thus is part of the ACM. Thus, the shape of the ACM achieved with the Gillespie algorithm might depend on whether some number of catalyst molecules necessary for some reactions become exactly zero due to a specific sequence of stochastic random choices of reactions or due to the late opening up of pores, or due to some other reasons, whereas in the continuous density approach, the density might become very small but then has the chance to increase again.

## References


1. Bagley, R., Farmer, J.: Spontaneous emergence of a metabolism. In: Langton, C., Taylor, C., Farmer, J., Rasmussen, S. (eds.) Artificial Life II, Santa Fe Institute Studies in the Sciences of Complexity, vol. X, pp. 93–140. Addison-Wesley, Redwood City (1991)
2. Cieslak, M., Prusinkiewicz, P.: Gillespie-lindenmayer systems for stochastic simulation of morphogenesis. in silico Plants **1**, diz009 (2019)
3. Dormand, J.R., Prince, P.J.: A family of embedded Runge-Kutta formulae. J. Comput. Appl. Math. **6**(1), 19–26 (1980)
4. Fick, A.: Ueber diffusion. Ann. Phys. **170**(1), 59–86 (1855). https://doi.org/10.1002/andp.18551700105
5. Füchslin, R.M., Filisetti, A., Serra, R., Villani, M., Lucrezia, D.D., Poli, I.: Dynamical stability of autocatalytic sets. In: Fellermann, H., et al. (eds.) Proceedings of the Twelfth International Conference on the Synthesis and Simulation of Living Systems, ALIFE 2010, Odense, Denmark, 19–23 August 2010, pp. 65–72. MIT Press (2010). https://digitalcollection.zhaw.ch/bitstream/11475/2722/1/2010_Fuechslin_Dynamical%20stability%20of%20autocatalytic%20sets_Artificial%20live%20XII.pdf
6. Gillespie, D.T.: A general method for numerically simulating the stochastic time evolution of coupled chemical reactions. J. Comput. Phys. **22**, 403–434 (1976)
7. Hordijk, W.: A history of autocatalytic sets: a tribute to Stuart Kauffman. Biol. Theory **14**, 224–246 (2019)
8. Hordijk, W., Steel, M.: Detecting autocatalytic, self-sustaining sets in chemical reaction systems. J. Theor. Biol. **227**, 451–461 (2004)
9. Hordijk, W., Steel, M., Kauffman, S.: Autocatalytic sets arising in a combinatorial model of chemical evolution. Life **12**(11), 1703 (2022)
10. Kauffman, S., Steel, M.: The expected number of viable autocatalytic sets in chemical reaction systems. Artif. Life **27**(1), 1–14 (2021)
11. Kauffman, S.A.: Cellular homeostasis, epigenesis and replication in randomly aggregated macromolecular systems. J. Cybern. **1**(1), 71–96 (1971)
12. Kauffman, S.A.: Autocatalytic sets of proteins. J. Theor. Biol. **119**(1), 1–24 (1986)
13. Kauffman, S.A.: The Origins of Order: Self-Organization and Selection in Evolution. Oxford University Press, New York (1993)
14. Oparin, A.: The Origin of Life on the Earth, 3rd edn. Academic Press, New York (1957)

15. Schneider, J.J., Weyland, M.S., Flumini, D., Füchslin, R.M.: Investigating three-dimensional arrangements of droplets. In: Cicirelli, F., Guerrieri, A., Pizzuti, C., Socievole, A., Spezzano, G., Vinci, A. (eds.) WIVACE 2019. CCIS, vol. 1200, pp. 171–184. Springer, Cham (2020). https://doi.org/10.1007/978-3-030-45016-8_17
16. Schneider, J.J., Weyland, M.S., Flumini, D., Matuttis, H.-G., Morgenstern, I., Füchslin, R.M.: Studying and simulating the three-dimensional arrangement of droplets. In: Cicirelli, F., Guerrieri, A., Pizzuti, C., Socievole, A., Spezzano, G., Vinci, A. (eds.) WIVACE 2019. CCIS, vol. 1200, pp. 158–170. Springer, Cham (2020). https://doi.org/10.1007/978-3-030-45016-8_16
17. Schneider, J.J., et al.: Network creation during agglomeration processes of polydisperse and monodisperse systems of droplets. In: De Stefano, C., Fontanella, F., Vanneschi, L. (eds.) WIVACE 2022. CCIS, vol. 1780, pp. 94–106. Springer, Cham (2023). https://doi.org/10.1007/978-3-031-31183-3_8
18. Schneider, J.J., et al.: Influence of the geometry on the agglomeration of a polydisperse binary system of spherical particles. In: ALIFE 2021: The 2021 Conference on Artificial Life (2021). https://doi.org/10.1162/isal_a_00392
19. Schneider, J.J., et al.: Paths in a network of polydisperse spherical droplets. In: ALIFE 2022: The 2022 Conference on Artificial Life (2022). https://doi.org/10.1162/isal_a_00502
20. Schneider, J.J., et al.: Artificial chemistry performed in agglomeration of droplets with restricted molecule transfer. In: De Stefano, C., Fontanella, F., Vanneschi, L. (eds.) WIVACE 2022. CCIS, vol. 1780, pp. 107–118. Springer, Cham (2023). https://doi.org/10.1007/978-3-031-31183-3_9
21. Xavier, J.C., Kauffman, S.A.: Small-molecule autocatalytic networks are universal metabolic fossils. Phil. Trans. R. Soc. A **380**, 20210244 (2022)

# Percolation Breakdown in Binary and Ternary Monodisperse and Polydisperse Systems of Spherical Particles

Johannes Josef Schneider[1,2(✉)], Alessia Faggian[3], Mathias Sebastian Weyland[1], William David Jamieson[4], Jin Li[2], Hans-Georg Matuttis[5], Silvia Holler[3], Federica Casiraghi[3], Aitor Patiño Diaz[3], Lorena Cebolla Sanahuja[3], Martin Michael Hanczyc[3,6], Dandolo Flumini[1], Peter Eggenberger Hotz[1], David Anthony Barrow[2], Pantelitsa Dimitriou[2], Oliver Castell[4], and Rudolf Marcel Füchslin[1,7]

[1] Institute of Applied Mathematics and Physics, School of Engineering, Zurich University of Applied Sciences, Technikumstr. 9, 8401 Winterthur, Switzerland
johannesjosefschneider@googlemail.com, {scnj,weyl,flum,eggg,furu}@zhaw.ch

[2] School of Engineering, Cardiff University, Queen's Buildings, 14-17 The Parade, Cardiff CF24 3AA, Wales, UK
{LiJ40,Barrow,dimitrioup}@cardiff.ac.uk

[3] Laboratory for Artificial Biology, Department of Cellular, Computational and Integrative Biology (CIBIO), University of Trento, 38123 Trento, Italy
{alessia.faggian,silvia.holler,federica.casiraghi,aitor.patino, lorena.cebolla,martin.hanczyc}@unitn.it

[4] Welsh School of Pharmacy and Pharmaceutical Science, Cardiff University, Redwood Building, King Edward VII Avenue, Cardiff CF10 3NB, Wales, UK
{jamiesonw,CastellO}@cardiff.ac.uk

[5] Department of Mechanical Engineering and Intelligent Systems, The University of Electrocommunications, Chofu Chofugaoka 1-5-1, Tokyo 182-8585, Japan
hg@mce.uec.ac.jp

[6] Chemical and Biological Engineering, University of New Mexico, MSC01 1120, Albuquerque, NM 87131-0001, USA

[7] European Centre for Living Technology, Ca' Bottacin, Dorsoduro 3911, 30123 Venice, Italy
https://www.zhaw.ch/en/about-us/person/scnj/

**Abstract.** We perform computer simulations of an agglomeration process for monodisperse and polydisperse systems of spherical particles in a cylindrical container, using a simplified stochastic-hydrodynamic model. We consider a ternary system with three particle types $A$, $B$, and $C$, in which only connections of the type $A - B$ can be forged, while any other connections with particles of the same type or with $C$-particles are forbidden, and for comparison a binary system with two particle types

This work has been partially financially supported by the European Horizon 2020 project *ACDC – Artificial Cells with Distributed Cores to Decipher Protein Function* under project number 824060.


M. Villani et al. (Eds.): WIVACE 2023, CCIS 1977, pp. 161–174, 2024.
https://doi.org/10.1007/978-3-031-57430-6_13

$A$ and $C$, in which only connections of the type $A - A$ can be formed. We study the breakdown of the percolation in the agglomeration at the bottom of the cylinder with an increasing fraction of $C$-particles.

**Keywords:** percolation · polydisperse · binary system · ternary system

# 1 Introduction

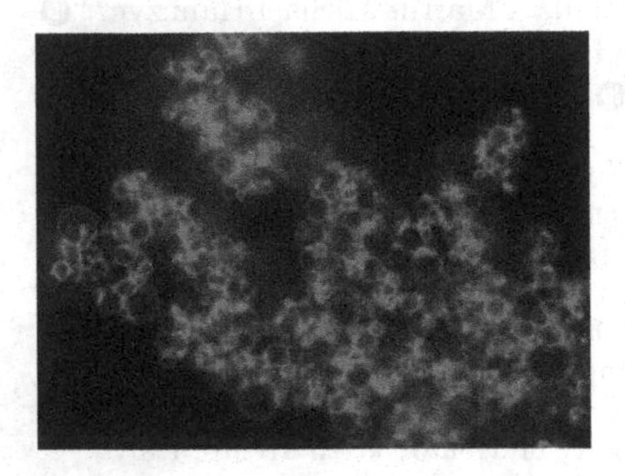

**Fig. 1.** Left: Snapshot of an agglomeration of droplets recorded from an experiment. Right: Agglomeration of a polydisperse system of 2,000 spherical particles with types $A$ and $B$, depicted as red and green, at the bottom of a cylindrical container, obtained in a computer simulation. (Color figure online)

We intend to develop a probabilistic compiler [3,22] to aid the three-dimensional agglomeration of droplets filled with various chemicals (see Fig. 1) in a specific way in order to e.g. allow the creation of desired macromolecules via a successive reaction scheme [12,13,18,19]. Neighboring droplets can form connections, either by forming bilayers [7] or by getting glued to each other by matching pairs of single-stranded DNA [4], as sketched in Fig. 2. Chemicals contained within the droplets can move to neighboring droplets either directly, as hydrophobic compounds can be exchanged between adjacent oil droplets at the contact face, or, if the oil droplets are contained in a hull comprised of amphiphilic molecules like phospholipids, through pores within these bilayers. An example for such a pore is shown in Fig. 3. Thus, a complex bilayer network is created [16], with the droplets being the nodes of this graph and the existing connections being the edges between the corresponding droplets. In such bilayer networks, a controlled successive reaction scheme can be effectuated to produce the intended macromolecules. As already demonstrated for a toy example, a gradual reaction network with three educts, two reaction steps, a desired product, and an undesired side product, can achieve a higher yield and a smaller amount of undesired

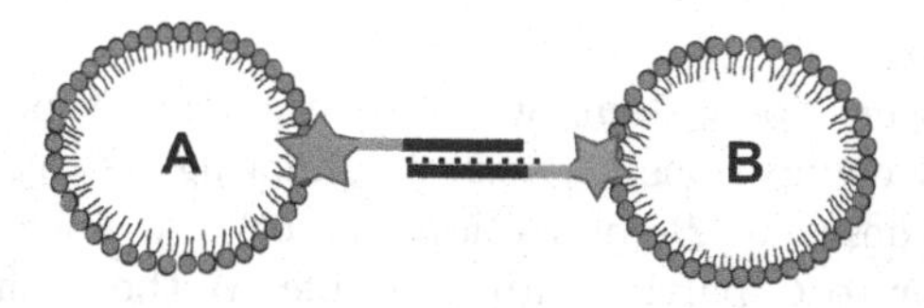


**Fig. 2.** Sketch of a pair of oil-filled droplets in water, to which complementary strands of ssDNA oligonucleotides are attached: The surfaces of the droplets are composed by single-tail surfactant molecules like lipids with a hydrophilic head on the outside and a hydrophobic tail on the inside, thus forming a boundary for the oil-in-water droplet. By adding some single-strand DNA to the surface of a droplet, it can be ensured that only desired connections to specific other droplets with just the complementary single-strand DNA can be formed. Please note that the connection of the droplets in this picture is overenlarged in relation to the size of the droplets. In reality, the droplets have a radius of 1–50 μm, whereas a base pair of a nucleic acid is roughly 0.34 nm in length [1], such that the sticks of connecting DNA strands are roughly 5 nm long.

**Fig. 3.** Alpha hemolysin (aHL) pore: The left picture reveals how the aHL, which is comprised of seven macromolecules, sticks its trunk through the bilayer between two droplets. The right picture presents in detail a cut through the pore formed, revealing the channel through which molecules can move between the adjacent droplets.

side products in an agglomeration of droplets with restricted molecule transfer than in a scenario in which all educts would be put in one well-stirred pot only [17].

For some applications, it is necessary to thin out the network, i.e., to reduce the number of edges connecting nodes in the network. This leads as we will show later to smaller numbers of nodes a node is attached to on average, to smaller cluster sizes and in turn to larger numbers of clusters which are isolated of each other, just as we need them for these applications, in which we either need to better govern the gradual chemical reaction process or in which we need to strongly reduce the maximum number of steps within such a gradual process. In order to achieve this thinning-out, particles which do not connect to any other particle can be added to the system. Within the scope of this paper, we study the effects of these auxiliary particles on the properties of the overall network for two basic scenarios:

- "Ternary" scenario:
  In the ternary scenario, systems with three particle types $A$, $B$, and $C$ are considered. Only connections between neighboring $A$- and $B$-particles can be forged. Besides these $A-B$-connections, no other connections to particles of the same type or to $C$-particles are possible. Without the $C$-particles, this system would form a so-called bipartite network. In our simulations, we only consider the special case that the fraction $f_A$ of $A$-particles equals the fraction $f_B$ of $B$-particles. We will study the changes of the properties of the network for an increasing fraction $f_C$ of $C$-particles, with $f_C = 1 - f_A - f_B = 1 - 2f_A$.
- "Binary" scenario:
  For comparison, we also study a binary scenario with two particle types $A$ and $C$, in which only pairs of neighboring $A$-particles can form connections. Besides these connections of the type $A - A$, there are no other connections, such that also here the $C$-particles serve as auxiliary particles for thinning out the network. Again we want to study the effects of an increasing fraction $f_C$ of $C$-particles on the properties of the network. This scenario can also be considered as a site percolation problem [20], in which the locations of the $A$-particles represent the occupied sites and the locations of the $C$-particles represent the empty sites. The probability $p$ for an occupied site is simply given by $p = f_A = 1 - f_C$.

In our computer simulations, we study both polydisperse systems, in which the radii of the particles differ from each other, and monodisperse systems, in which all particles share the same radius value, in order to mimick experiments of various kinds: The production of droplets using a microfluidic approach, in which an inner stream of fluid within an outer stream of another fluid is broken up in droplets in e.g. a t-junction under specific pressure conditions [7], leads to a rather monodisperse system of droplets. Contrarily, in other experiments, we are repeatedly rubbing a phial filled with water and one drop consisting of oil molecules and amphiphilic molecules over a rough surface, thus sending excitations into the system, which lead to a breakup of the large drop into many small droplets of varying sizes, resulting in a polydisperse system [4].

Within the scope of this paper, we present computational results for simulations based on a simplified stochastic-hydrodynamic model of an agglomeration process of a system of droplets, mimicking experiments. Here we want to focus on the influence of the fraction $f_C$ of auxiliary $C$-droplets on some specific properties of the networks created, which are of crucial importance for the gradual reaction scheme intended. We are especially interested in the question whether there is a percolation transition at some critical value of $f_C$: For an infinitely large system, one expects to get a sharp transition between two phases, with an infinitely large cluster for $f_C$ below the critical value and no such infinitely large cluster for $f_C$ above that value. For finite systems, one gets a smooth transition between a regime with a large cluster dominating the system for small $f_C$ and a regime with no such large cluster but many small clusters for large $f_C$. In order to focus on these questions and to exclude effects from other experimental properties, we simulate the droplets as hard spheres and ignore details of the surface structure of the particles, attractive forces as well as adhesion effects. As the extension of the bilayers is very small and as due to their small radii [2], the droplets keep their spherical shape during the experiments, as shown in Fig. 1, such that this simplified approach is justified.

This paper is organized as follows: In the next section, we sketch briefly how we simulate the agglomeration of droplets in a container. Then we give a short introduction to network analysis and percolation theory, focusing on those network properties for which we will present computational results. As we are mainly interested in the description of the percolation transition with an increasing fraction $f_C$ of auxiliary $C$-droplets, we present results depending on $f_C$ for the decreasing maximum and average number of nodes a node is attached to, for which we find power laws depending on $1 - f_C$, for the increasing number of clusters, for which we get a linear behavior, and for the decreasing size of the largest cluster, which clearly exhibits a percolation transition. Finally, we provide a summary and give an outlook.

## 2 Simulation Details

At the beginning of the simulations, we place $N$ spherical particles at randomly selected positions in a cylindrical container with radius 1 mm and height 4 mm in a way that they do not overlap with each other and that they do not overlap with the walls of the cylinder. For the polydisperse system, we randomly choose the particle radii $r_i$ uniformly from the interval [10–50] µm, whereas we set all radii $r_i \equiv 30\,\mu\text{m}$ for the monodisperse system.

After this initialization, we perform the main simulation which is comprised of $10^7$ time steps of a duration of $\delta t = 10^{-5}$ s. In each time step, the particles are subjected to various forces:

- They sink in water due to gravity $\vec{F}_G$ reduced by the buoyant force $\vec{F}_b$:

$$\vec{F}_G(i) - \vec{F}_b(i) = \frac{4\pi}{3} r_i^3 \left(\varrho_{\text{oil}} - \varrho_{\text{water}}\right) g \tag{1}$$

  For the oil density, we use the value $\varrho_{\text{oil}} = 1.23\,\text{kg/l}$, which is just the density of the oil used in some experiments.
- Secondly, the spatial components $v_{x,y,z}(i)$ of the velocity vectors $\vec{v}(i)$ are subjected to random velocity changes: They are randomly altered by up to ±5% of their absolute values in order to take at least in this small random way into account that the containers are moved by the experimentalists during the agglomeration process.
- The particles are also subjected to the Stokes friction force $\vec{F}_S$:

$$\vec{F}_S(i) = -6\pi\eta r_i \vec{v}(i) \tag{2}$$

  The viscosity of water at 25 °C is $\eta = 0.891$ mPas.

- As in classic hydrodynamics, the concept of added mass [21] is used. When applying Newton's second law, we have to consider an effective mass of the particle, i.e., $\vec{F}(i) = m_{\text{eff}}(i)\vec{a}(i)$. This effective mass is composed of the mass $m(i)$ of particle $i$ and of the added mass $m_{\text{added}}(i)$. This added mass is caused by the inertia of the surrounding fluid, which needs to be deflected or attracted if the particle itself is accelerated or decelerated in the water, and can be determined to being half of the mass of the water displaced by oil particle $i$.

When working with such a set of second order differential equations governing the laws of motion for the particles, the question arises as to which integrator to use. Due to the stochastic nature of random velocity changes, only an Euler scheme with very small time intervals is suitable for the determination of new velocities and positions [6]. In the case of collisions between pairs of particles or between particles and walls, a mostly elastic collision dynamics with 90% elasticity and 10% plasticity is imposed. Overlaps occurring at the end of each time step are resolved as in [9,14].

## 3 Network Analysis and Percolation Theory

For network analysis, we first of all have to define a network related to the problem we intend to study. As mentioned above, we are interested in generating gradual chemical reaction schemes performed in networks of droplets, with neighboring droplets being able to exchange molecules if pores within their bilayers exist or, more theoretically speaking, if a connection between the particle types of the two adjacent droplets is allowed. Then we can define an edge matrix $\eta$ with

$$\eta(i,j) = \begin{cases} 1 & \text{if droplets } i \text{ and } j \text{ are neighbors of each other} \\ & \text{and a connection between them exists} \\ 0 & \text{otherwise.} \end{cases} \tag{3}$$

Two droplets $i$ and $j$ with their midpoints $(x_i, y_i, z_i)$ and $(x_j, y_j, z_j)$ and their radii $r_i$ and $r_j$ are neighboring each other if the condition

$$\sqrt{(x_i - x_j)^2 + (y_i - y_j)^2 + (z_i - z_j)^2} \leq r_i + r_j + 0.1\,\mu\text{m} \tag{4}$$

is fulfilled, i.e., if the distance between their midpoints is smaller or equal to the sum of their radii plus some small offset which we need to introduce because of finite numerical precision. One usually sets $\eta(i,i) \equiv 0$ for all nodes $i$. Such a matrix $\eta$ contains all the information about the network. For this paper, we study both a binary scenario and a ternary scenario, for which we can generate two different edge matrices, considering the different conditions for the existence of a connection.

When analyzing a network, one mostly takes either an atomistic view, looking at the various nodes and determining their network related properties, or a global view, determining clusters of nodes. Clusters are defined as maximum subgroups of nodes in which each node within this cluster can be reached from any other node in the cluster by gradually traversing edges, thus walking along a path from this node perhaps via other nodes in the subgroup to the destined node. More seldomly, networks are considered at an intermediate level, e.g., for the detection of the maximum clique [8], or one asks for the importance of specific nodes for the overall network in a local-global view, see e.g. [11].

When looking at a network from a global point of view, one of the most important questions arising is whether the network is percolating. For an infinitely large network, this means that one has to ask whether there is an infinitely large cluster in the network [20]. In the finite networks resulting from computer simulations, one thus asks whether a dominating cluster exists in the network.

Mostly, the so-called site percolation is studied in which sites on a regular lattice are either occupied or empty and in which each site is connected to all neighboring sites. Theoretically, one finds for infinitely large systems that there is a critical probability $p_{\text{crit}}$ of occupied sites above which an infinitely large cluster exists in the system and below which there is no percolation anymore. Alternatively, also the so-called bond percolation is considered in which all lattice sites are occupied but only a fraction $p$ of the edges exist. Also here one finds such a critical probability $p_{\text{crit}}$ dividing two such regimes. For some scenarios, this critical probability can be calculated exactly, but mostly, one has to make use of computer simulations with increasing system size, to determine the various clusters in the system and the size of the largest cluster, and finally to carefully determine $p_{\text{crit}}$ numerically [20].

## 4 Computational Results

The results presented in Figs. 4, 5, 6, 7 and 8 are averaged over the properties of the final configurations of 100 independently performed simulation runs.

The first observable we have a look at is the number $e$ of edges, which can be derived from the edge matrix $\eta$ with

$$e = \sum_{i<j} \eta(i,j). \tag{5}$$

Figure 4 displays the results for $e$ for the binary and the ternary scenario in simulations of monodisperse systems of 2,000 droplets and polydisperse systems of 2,000 droplets. We generally find that there are more edges in the binary scenario and that the number of edges decreases with an increasing fraction $f_C$ of $C$-particles, which do not connect with each other and with other particles. On average, there are slightly more edges in the monodisperse systems than in the polydisperse systems, both for the binary and for the ternary scenarios.

When taking a local perspective, one of the most important observables for a specific node $i$ is its degree $d(i)$, which can be calculated as

$$d(i) = \sum_{j=1}^{N} \eta(i,j). \tag{6}$$

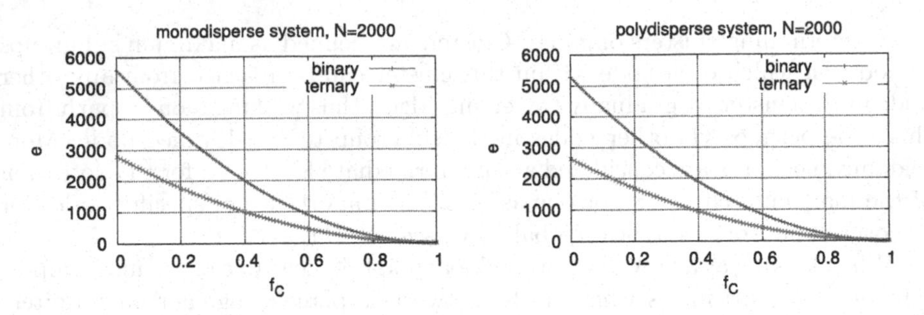


**Fig. 4.** Decrease of the number $e$ of edges with an increasing fraction $f_C$ of $C$-particles for the binary and ternary systems as described in the text: Results are presented for monodisperse systems consisting of 2,000 particles (left) and polydisperse systems consisting of 2,000 particles (right).

Thus, the degree $d(i)$ counts to how many other droplets the droplet $i$ is connected. We are mainly interested in the maximum degree

$$d_{\max} = \max_i d(i) \tag{7}$$

of all nodes. Of course, also $d_{\max}$ decreases with increasing $f_C$, but in a specific way, such that we plot $d_{\max}$ vs. $1 - f_C$ in a double-logarithmic way in Fig. 5. The graphics reveal the existence of a power law for $d_{\max}$. For the monodisperse systems comprised of 2,000 particles, we find a power law of the type

$$d_{\max} = a(1 - f_C)^{1/2}, \tag{8}$$

both for the binary and the ternary scenarios, whereas we get a power law of the type

$$d_{\max} = a(1 - f_C)^{2/3} \tag{9}$$

for the polydisperse systems comprised of 2,000 particles, both for the binary and for the ternary scenarios. The values for the various prefactors $a$ are provided in Table 1.

In the next step, we have a look at the mean value $\langle d \rangle$ of the degrees, which can be calculated as

$$\langle d \rangle = \frac{1}{N} \sum_{i=1}^{N} d(i). \tag{10}$$

$\langle d \rangle$ is related to the overall number $e$ of edges via

$$N \times \langle d \rangle = 2e. \tag{11}$$

Also for $\langle d \rangle$, we find a complex power law behavior depending on $1 - f_C$, such that we plot $\langle d \rangle$ vs. $1 - f_C$ in a double-logarithmic way in Fig. 6. We get both for the binary and for the ternary scenario, both for the monodisperse and for the polydisperse systems the power law

$$\langle d \rangle = a(1 - f_C)^2. \tag{12}$$

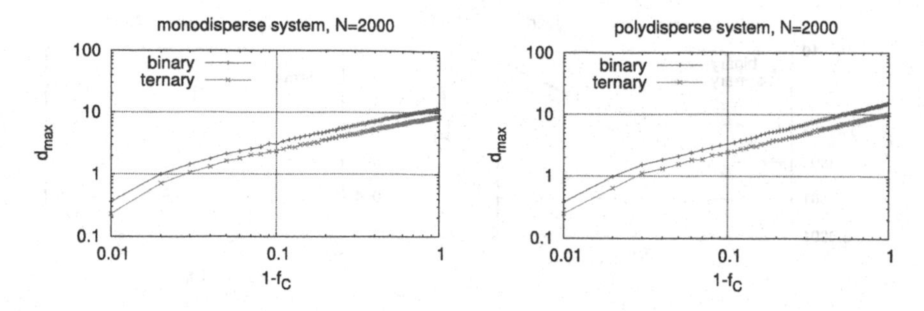


**Fig. 5.** Increase of the maximum degree $d_{\max}$ vs. the remaining fraction $1 - f_C$ of particles not being $C$-particles for the binary and ternary systems as described in the text: Results are presented for a monodisperse system consisting of 2,000 particles (left) and a polydisperse system consisting of 2,000 particles (right).

**Table 1.** Prefactors $a$ found for the power laws as described in the text.

| degree | scenario | system | $a$ |
|---|---|---|---|
| $d_{\max}$ | binary | 2000, mono | 11.5 |
| $d_{\max}$ | binary | 2000, poly | 15.3 |
| $d_{\max}$ | ternary | 2000, mono | 8.6 |
| $d_{\max}$ | ternary | 2000, poly | 10.2 |
| $\langle d \rangle$ | binary | 2000, mono | 5.6 |
| $\langle d \rangle$ | binary | 2000, poly | 5.3 |
| $\langle d \rangle$ | ternary | 2000, mono | 2.8 |
| $\langle d \rangle$ | ternary | 2000, poly | 2.6 |

The prefactors can again be found in Table 1.

Please note that both the maximum degrees and thus also the mean degrees are restricted in size. For the monodisperse system, $d_{\max}$ cannot exceed the value of the so-called kissing number $k$ [15]. The kissing number problem is stated as follows: How many spheres of equal size can be placed around a sphere in their midst touching it without any overlaps? This kissing number equals 12 in three dimensions, as already stated by Newton and proved in the 1950s. For the polydisperse system, there is a related restriction: Here $d_{\max}$ cannot exceed a value of $k$, which depends on the ratio between the radii of the smallest and largest spheres. As the radii of the spherical particles are randomly chosen from the interval [10–50] µm, this ratio could be up to $1:5$, for which we obtained a bidisperse kissing number of 120 [15].

Furthermore, we would like to compare our results for $\langle d \rangle$ for monodisperse systems with results obtained for other configurations of spheres. First of all, let us consider the densest packing of spheres of the same size: The densest packings can be achieved both in a face centered cubic (fcc) lattice and a hexagonal close packing (hcp). For these densest packings, one gets $\langle d \rangle = d_{\max} = 12$, in the

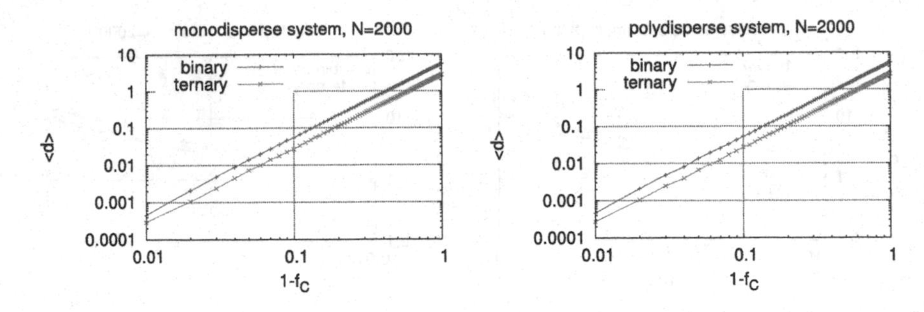


**Fig. 6.** Increase of the average degree $\langle d \rangle$ vs. the remaining fraction $1 - f_C$ of particles not being $C$-particles for the binary and ternary systems as described in the text: Results are presented for monodisperse systems consisting of 2,000 particles (left) and polydisperse systems consisting of 2,000 particles (right).

case of infinitely extended lattices or lattices with periodic boundary conditions. Contrarily, if studying entirely random packings of spheres of the same size, one obtains $\langle d \rangle = 6$ [10], a value which is almost in agreement with our values for the binary scenario at $f_C = 0$ for the monodisperse system. The deviation is due to the finite extension of our agglomerations.

Now we turn to the global view in network analysis and have a look at the number $n$ of clusters in the system, which is plotted vs. the fraction $f_C$ in Fig. 7. For $f_C = 0$, we trivially have only a very small number of clusters in the binary scenario, whereas there is already a significant number of clusters of roughly $n \approx 0.05N - 0.07N$ in the ternary scenario. For small $f_C$, $n$ increases linearly with $f_C$, until it approaches sigmoidally the value $n = N$ in the limit $f_C \to 1$. For the ternary systems, there seems to be a little bending in the curves at $f_C \approx 0.5$.

Finally, we end up at the most important point of our investigation. We consider the size $s_{\text{max}}$ of the largest cluster in the system, which is plotted vs. $f_C$ in Fig. 8. Generally, we get a linear decrease of $s_{\text{max}}$ with increasing $f_C$ for small $f_C$, before a transition takes place, in which the percolation breaks down: For the binary scenario, we find a critical value of $f_C$ of roughly $0.55 \pm 0.05$ both for the monodisperse and the polydisperse system. For the ternary scenario, we get $0.3 \pm 0.05$ both for the monodisperse and for the polydisperse system.

Here we again would like to compare these results with other results obtained for spheres. As already mentioned, the densest packings of spheres of the same size can be achieved in a fcc and a hcp lattice. The critical probability for an infinitely large fcc lattice with periodic boundary conditions has been determined to be $p_{\text{crit}} = 0.198$ [20]. But also hcp lattices on a slab with open boundary conditions and infinite extensions in two dimensions have been studied. The threshold depends on the thickness $h$ (i.e., the number of layers on which the midpoints of the spherical particles are located) of the slab, one gets $p_{\text{crit}} = 0.2828$ for $h = 2$ and $p_{\text{crit}} = 0.2086$ for $h = 16$ in the limit of infinite extension in the other two dimensions [5]. For randomly packed spheres, a threshold of

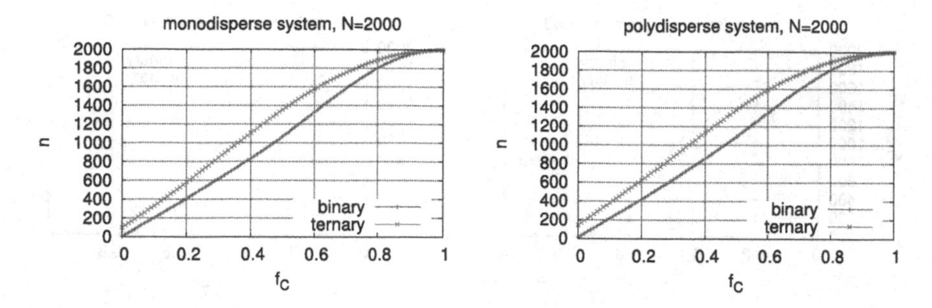


**Fig. 7.** Increase of the number $n$ of clusters with an increasing fraction $f_C$ of $C$-particles for the binary and ternary systems as described in the text: Results are presented for monodisperse systems consisting of 2,000 particles (left) and polydisperse systems consisting of 2,000 particles (right).

$p_{\text{crit}} = 0.31$ [10] was obtained. Thus, we see that the critical values strongly depend on the systems under study. In our case, the spherical particles are neither located on the sites of a regular grid nor placed entirely randomly.

## 5 Summary and Outlook

In this paper, we presented results of simulations for the agglomeration of polydisperse and monodisperse systems of droplets. We were mainly interested in the effects the addition of auxiliary particles, which do not connect to any other particles, has on the networks and their properties. We found a power law behavior for the maximum degrees and mean degrees of the particles depending on the fraction of the auxiliary particles in the system. Furthermore, we detected a percolation breakdown if this fraction exceeds some critical value.

We will continue this study also with other connection scenarios, in which e.g. $A$-particles can connect to other $A$-particles and to $B$- and $C$-particles, while $B$-particles and $C$-particles cannot form connections. This scenario can be easily realized in experiments by only placing the constituents for pore macromolecules exclusively in the $A$-particles. Furthermore, we will extend our study to further system sizes in order to get better estimates for the critical values and also to find out in which way the prefactors $a$ in the power laws found for the maximum and the mean degree depend on the system size $N$.

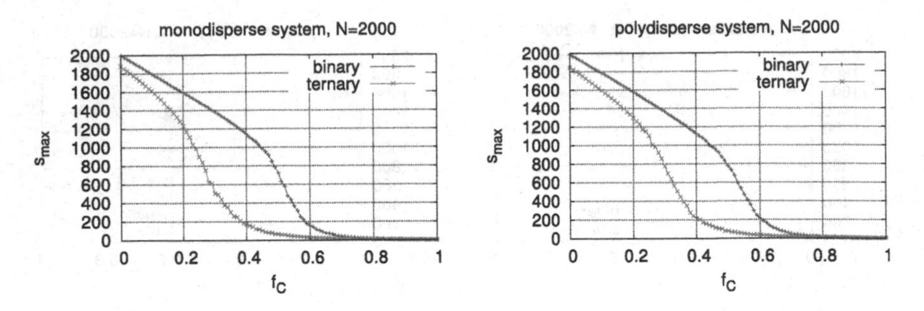


**Fig. 8.** Decrease of the size $s_{\max}$ of the largest cluster with an increasing fraction $f_C$ of $C$-particles for the binary and ternary systems as described in the text: Results are presented for monodisperse systems consisting of 2,000 particles (left) and polydisperse systems consisting of 2,000 particles (right).

**Acknowledgements.** JJS would like to thank Uwe Krey and Ingo Morgenstern for discussions about percolation theory at the University of Regensburg, Germany in the years 1991–1996.

## References

1. Annunziato, A.T.: DNA packaging: nucleosomes and chromatin. Nat. Educ. **1**, 26 (2008)
2. Aprin, L., Heymes, F., Laureta, P., Slangena, P., Le Floch, S.: Experimental characterization of the influence of dispersant addition on rising oil droplets in water column. Chem. Eng. Trans. **43**, 2287–2292 (2015)
3. Flumini, D., Weyland, M.S., Schneider, J.J., Fellermann, H., Füchslin, R.M.: Towards programmable chemistries. In: Cicirelli, F., Guerrieri, A., Pizzuti, C., Socievole, A., Spezzano, G., Vinci, A. (eds.) WIVACE 2019. CCIS, vol. 1200, pp. 145–157. Springer, Cham (2020). https://doi.org/10.1007/978-3-030-45016-8_15
4. Hadorn, M., Boenzli, E., Sørensen, K.T., Fellermann, H., Eggenberger Hotz, P., Hanczyc, M.M.: Specific and reversible DNA-directed self-assembly of oil-in-water emulsion droplets. Proc. Natl. Acad. Sci. **109**(50), 20320–20325 (2012)
5. Horton, M.K., Moram, M.A.: Alloy composition fluctuations and percolation in semiconductor alloy quantum wells. Appl. Phys. Lett. **110**, 162103 (2017)
6. Kloeden, P., Platen, E.: Numerical Solution of Stochastic Differential Equations. Stochastic Modelling and Applied Probability, Springer, Heidelberg (2013). https://books.google.co.jp/books?id=r9r6CAAAQBAJ
7. Li, J., Barrow, D.A.: A new droplet-forming fluidic junction for the generation of highly compartmentalised capsules. Lab Chip **17**, 2873–2881 (2017)
8. Marino, R., Kirkpatrick, S.: Revisiting the challenges of maxclique. CoRR abs/1807.09091 (2018). http://arxiv.org/abs/1807.09091
9. Müller, A., Schneider, J.J., Schömer, E.: Packing a multidisperse system of hard disks in a circular environment. Phys. Rev. E **79**, 021102 (2009)
10. Powell, M.J.: Site percolation in randomly packed spheres. Phys. Rev. B **20**, 4194–4198 (1979)

11. Schneider, J.J., Kirkpatrick, S.: Selfish versus unselfish optimization of network creation. J. Stat. Mech.: Theory Experiment **2005**(08), P08007 (2005). https://doi.org/10.1088/1742-5468/2005/08/p08007
12. Schneider, J.J., Weyland, M.S., Flumini, D., Füchslin, R.M.: Investigating three-dimensional arrangements of droplets. In: Cicirelli, F., Guerrieri, A., Pizzuti, C., Socievole, A., Spezzano, G., Vinci, A. (eds.) WIVACE 2019. CCIS, vol. 1200, pp. 171–184. Springer, Cham (2020). https://doi.org/10.1007/978-3-030-45016-8_17
13. Schneider, J.J., Weyland, M.S., Flumini, D., Matuttis, H.-G., Morgenstern, I., Füchslin, R.M.: Studying and simulating the three-dimensional arrangement of droplets. In: Cicirelli, F., Guerrieri, A., Pizzuti, C., Socievole, A., Spezzano, G., Vinci, A. (eds.) WIVACE 2019. CCIS, vol. 1200, pp. 158–170. Springer, Cham (2020). https://doi.org/10.1007/978-3-030-45016-8_16
14. Schneider, J.J., Müller, A., Schömer, E.: Ultrametricity property of energy landscapes of multidisperse packing problems. Phys. Rev. E **79**, 031122 (2009)
15. Schneider, J.J., et al.: Geometric restrictions to the agglomeration of spherical particles. In: Schneider, J.J., Weyland, M.S., Flumini, D., Füchslin, R.M. (eds.) WIVACE 2021. CCIS, vol. 1722, pp. 72–84. Springer, Cham (2022). https://doi.org/10.1007/978-3-031-23929-8_7
16. Schneider, J.J., et al.: Network creation during agglomeration processes of polydisperse and monodisperse systems of droplets. In: De Stefano, C., Fontanella, F., Vanneschi, L. (eds.) WIVACE 2022. CCIS, vol. 1780, pp. 94–106. Springer, Cham (2023). https://doi.org/10.1007/978-3-031-31183-3_8
17. Schneider, J.J., et al.: Artificial chemistry performed in an agglomeration of droplets with restricted molecule transfer. In: De Stefano, C., Fontanella, F., Vanneschi, L. (eds.) WIVACE 2022. CCIS, vol. 1780, pp. 107–118. Springer, Cham (2023). https://doi.org/10.1007/978-3-031-31183-3_9
18. Schneider, J.J., et al.: Obstacles on the pathway towards chemical programmability using agglomerations of droplets. In: Schneider, J.J., Weyland, M.S., Flumini, D., Füchslin, R.M. (eds.) WIVACE 2021. CCIS, vol. 1722, pp. 35–50. Springer, Cham (2022). https://doi.org/10.1007/978-3-031-23929-8_4
19. Schneider, J.J., Weyland, M.S., Flumini, D., Füchslin, R.M.: Exploring the three-dimensional arrangement of droplets. In: Schneider, J.J., Weyland, M.S., Flumini, D., Füchslin, R.M. (eds.) WIVACE 2021. CCIS, vol. 1722, pp. 63–71. Springer, Cham (2022). https://doi.org/10.1007/978-3-031-23929-8_6
20. Stauffer, D., Aharony, A.: Introduction to Percolation Theory, 2nd revised edn. Taylor & Francis, London (1994)
21. Stokes, G.G.: On the effect of the internal friction of fluids on the motion of pendulums. Trans. Cambridge Philos. Soc. **9**, 8–106 (1851)
22. Weyland, M.S., Flumini, D., Schneider, J.J., Füchslin, R.M.: A compiler framework to derive microfluidic platforms for manufacturing hierarchical, compartmentalized structures that maximize yield of chemical reactions. In: Artificial Life Conference Proceedings, vol. 32, pp. 602–604 (2020). https://doi.org/10.1162/isal_a_00303

# Adaptation and Swarms

# Entangled Gondolas. Design of Multi-layer Networks of Quantum-Driven Robotic Swarms

Maria Mannone[1,2,3(✉)], Norbert Marwan[4,5,6], Valeria Seidita[1], Antonio Chella[1,7], Achille Giacometti[2,3], and Peppino Fazio[3,8]

[1] Department of Engineering, University of Palermo, Palermo, Italy
{mariacaterina.mannone,valeria.seidita,antonio.chella}@unipa.it
[2] European Centre for Living Technology (ECLT), Ca' Foscari University of Venice, Venice, Italy
{maria.mannone,achille}@unive.it
[3] Dipartimento di Scienze Molecolari e Nanosistemi (DSMN), Ca' Foscari University of Venice, Venice, Italy
peppino.fazio@unive.it
[4] Potsdam Institute for Climate Research (PIK), Member of the Leibniz Association, Potsdam, Germany
marwan@pik-potsdam.de
[5] Institute of Physics, University of Potsdam, Potsdam, Germany
[6] Institute of Geosciences, University of Potsdam, Potsdam, Germany
[7] ICAR, National Research Council (CNR), Palermo, Italy
[8] VSB, Technical University of Ostrava, Ostrava, Czechia

**Abstract.** Swarms of robots can be thought of as networks, using the tools from telecommunications and network theory. A recent study designed sets of aquatic swarms of robots to clean the canals of Venice, interacting with computers on gondolas. The interaction between gondolas is one level higher in the hierarchy of communication. In other studies, pairwise communications between the robots in robotic swarms have been modeled via quantum computing. Here, we first apply quantum computing to the telecommunication-based model of an aquatic robotic swarm. Then, we use multilayer networks to model interactions within the overall system. Finally, we apply quantum entanglement to formalize the interaction and synchronization between "heads" of the swarms, that is, between gondolas. Our study can foster new strategies for search-and-rescue robotic-swarm missions, strengthening the connection between different areas of research in physics and engineering.

**Keywords:** swarm intelligence · swarm robotics · telecommunication · quantum computing · multilayer networks

## 1 Introduction: The Core Idea

Multiple simple entities, interacting and achieving a complex task: this could be the preamble on a presentation about flocking birds, schooling fish, or foraging


M. Villani et al. (Eds.): WIVACE 2023, CCIS 1977, pp. 177–189, 2024.
https://doi.org/10.1007/978-3-031-57430-6_14

ants. It could also be the opening phrase of a chapter on networked telecommunications, or of a study on distributed, collective intelligence. They are all instances of distributed dynamic networks, that we can investigate under different points of view, from different disciplines. Here, we focus on quantum-driven swarms of robots, interacting between them and with computational units. Let us present here the core definitions of these topics, and our proposal to join them into an engineering project.

A **swarm of robots** is a set of simple, interacting, decentralized robots. A swarm is robust, that is, the loss of a unit does not affect the behavior of the whole system [10]. A swarm is also scalable, that is, the same behavior is observed even if the size (the number of units) of the swarm changes. **Quantum computing** is a branch of computer science derived by some basic principles of quantum mechanics in Physics [23]. It starts with the definition of *qubit*, the "quantum bit," that, according to the state superposition in quantum mechanics, can assume all the intermediate values between the classical 0 and 1. A **quantum circuit** is a model where initial configurations of qubits undergo a set of transformations, corresponding to reversible logic gates. The qubits are finally measured through a "destructive measure" as in quantum mechanics, and the result is stored in classical variables. We can cite some pioneering applications of quantum computing to artificial intelligence [13,25], robotics [6,14], and swarm robotics [1,12,26]. The application of quantum computing to robotics is motivated by an increased computational efficiency, particularly evident in tasks such as information retrieval. Quantum computing can enhance robots' visual perception (via quantum image processing); thinking (via decision-making under uncertainty); dynamics (finding better solutions for path-planning); data mining (to face NP-hard problems) [21]. More in general, quantum computing is a new paradigm which is largely unexplored, and the conversion itself from classical to quantum algorithms, and the subsequent investigation of issues and advantages, are topics of contemporary interest.

In particular, quantum computing has recently been used to model pairwise interactions between the robots of a swarm in a simulated environment [15,16]. In the context of a search-and-rescue mission, each robot communicates with its peers, transmitting them its position and proximity to the target, in terms of probability amplitude to be in a certain space position, and probability amplitude to be close to the target. In particular, as described in [16], we consider values along the $x$, $y$ axes as the peaks of the wavefunction. The values 0 and 1 are assigned to the extremes of the arena where the robots are moving. Thus, concerning the position, the logic 0 indicates closeness to 0 along the x-axis (y-axis). About the target reaching, the logic 1 indicates *success* and 0 *failure*. Let us initially consider an abstract scenario as a 2D arena of side 1, with robots, schematized as points [16], first randomly exploring the space, then exchanging information about position and "degree of success" in target finding, and deciding where to go next according to the output of the quantum circuit. This is quite an abstract scenario with a search mission, which can be instantiated with different kinds of robots. After each step of individual exploration, the other robots receive the corresponding information, letting it enter a quantum circuit, whose output contains an indication of the space points to be approximately reached during the next step of the simulation. In fact, according to the degree of precision in position location and success in target reaching, the robots can decide to follow a particular peer, or to explore elsewhere. In [18], a classic-only version of the search algorithm was considered. The abstract scenario was contextualized as the tract of a canal in Venice. Aquatic (surface) robots were considered, being organized as multiple swarms. In addition, computing centrals were introduced, one for each swarm. Each robot was able to interact with the computing central of its swarm, while the computing centrals were able to interact with all robots of their swarm, and with the other centrals. The centrals were computers put on gondolas, with the functionality of collecting data from the individual exploration of robots and giving them indications of the portions of water to explore next. Each robot was equipped with GPS sensors, distance sensors, and an underwater camera to observe the ground of the canal. The single robots were coded to identify trash on the seabed and send this information jointly with position and time. The computers on gondolas elaborates on this information. In this study, we extended the algorithm of [16] with the hierarchical network of robots and gondolas, including quantum computing. The scope of the mission is now the search for a target (which can be instantiated according to the specific considered problem). When one of the swarms finds it, the corresponding gondola sends the information to the other gondolas, which start following it. We can use the idea of entanglement of states to indicate such a "locking" of the relative positions of the gondolas between them. We describe such a new strategy in Sect. 2.3, in Algorithm 2.

Figure 1 shows the quantum circuit designed in [16], which computes the truth table of Table 1. What is shown in Table 1 and Fig. 1, is the "mechanism" of pairwise interaction between robots. The overall behavior emerges from these pairwise interactions. In particular, the position along the $x$ axis is described by the first qubit, that is, $q_0 = \alpha_1^x|0\rangle + \beta_1^x|1\rangle$, where 0, 1 indicate the extremes of the horizontal side of the arena, and $\alpha_1^x$, $\beta_1^x$ indicate the probability amplitudes for robot 1 to be on one extreme or on the other one. With a little abuse of notation, in [16], we considered as the "position" the actual value of $\beta_1^x$, considering it as the peak of the wavefunction. Similarly, we define the position along $y$ as the second qubit, that is, $q_1 = \alpha_1^y|0\rangle + \beta_1^y|1\rangle$. The position along $x$, $y$ of robot 2 is defined as $q_3 = \alpha_2^x|0\rangle + \beta_2^x|1\rangle$, $q_4 = \alpha_2^y|0\rangle + \beta_2^y|1\rangle$ respectively. The qubit $q_2$ indicates the reward of the first robot, that is defined here as 1 - the distance from the target. We consider here 0 as *failure*, and 1 as *success*: $q_2 = \gamma_1|0\rangle + \delta_1|1\rangle$. The reward of the second robot is not indicated, because it is computed after that it reaches the position suggested by the outcome of the quantum circuit. Then, the cycle starts again.

What we have described so far is a general approach for an abstract swarm in a 2D arena. What we can do next, is connecting this approach with some more realistic scenario, and proceed raising the level of complexity. In a recent study [18], a set of swarms of aquatic robots designed to clean the canals of Venice, communicating with centralized computational units on gondolas (1-level higher in hierarchy), has been modeled using the language and formalism of network theory in telecommunications [5,7,20,24]. Overall, in this research, we started our analysis from a general case of agents in a 2D arena in [16]. The model of gondolas + robots was conceptualized in a context of energy-efficiency telecommunication protocol, and referred to environmental care of Venice [18]. It was structured as master-slave system, where the "masters" are the gondolas, and the "slaves" are the robots of each swarm. Thus, in this new article, we introduce the formalism of multilayer networks. As an additional novelty, we join the quantum computing element, developed in [16], with the application to a multilayer approach, contextualized in environmental care of the city. The idea is the following: we start with a toy-model for robots and quantum computing, discussing it and its advantages [16], then we add complexity to the formalization, adding physical constraints [18], and using a mathematical framework to well describe the study (here). In this article, we (1) refine the quantum-swarm approach, with robots acquiring information through their sensors and not via an omniscient system; (2) we add this quantum-based decision system to the telecommunication model of aquatic swarms and gondolas; (3) we formalize the whole system via the concept of multilayer networks [3,4], and, finally, (4) we propose quantum entanglement [23] to synchronize the decisions of the centralized units on gondolas at certain points of the exploration, and not at all times and between all robots, as proposed in [11]. Thus, as a homage to Venice, we can nickname this research as *entangled gondolas*. The article is structured as follows. In Sect. 2, we present the original contribution of this study. In Sect. 3, we summarize our findings and discuss some further research developments.

**Table 1.** Truth table for two robots $R_i$, $R_j$ on the plane, not reversible because of the indeterminacy on $x$, $y$ in the case of 0 reward, from [16]. The reward is computed as 1− the distance of a robot from the normalized target. $q_0 \dots q_4$ are the qubits. All robots are exchanging information and according to the reward indication

| $q_0$<br>$x$-pos<br>$R_i$ | $q_1$<br>$y$-pos<br>$R_i$ | $q_2$<br>reward<br>$R_i$ | $q_4$<br>$y$-pos<br>$R_j$ | $q_3$<br>$x$-pos<br>$R_j$ | $q_2$<br>reward<br>$R_i$ |
|---|---|---|---|---|---|
| 0 | 0 | 0 | 0/1 | 0/1 | 0 |
| 0 | 0 | 1 | 0 | 0 | 1 |
| 0 | 1 | 0 | 0/1 | 0/1 | 0 |
| 0 | 1 | 1 | 1 | 0 | 1 |
| 1 | 1 | 1 | 1 | 1 | 1 |
| 1 | 0 | 0 | 0/1 | 0/1 | 0 |
| 1 | 1 | 0 | 0/1 | 0/1 | 0 |
| 1 | 0 | 1 | 0 | 1 | 1 |

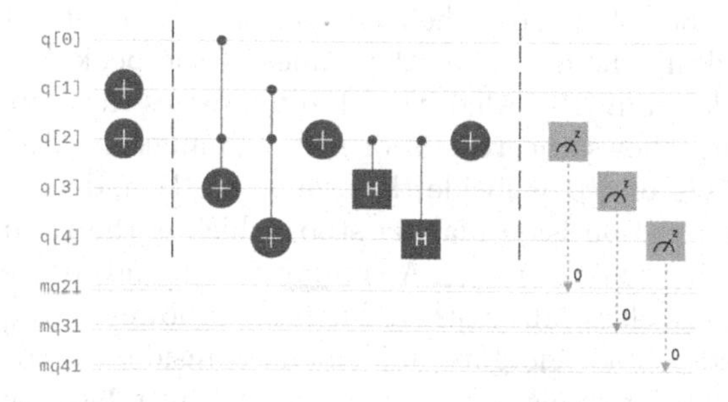


**Fig. 1.** Quantum circuit realizing the truth table of Table 1, from [16]. The circuit is made of NOT, Toffoli, and Hadamard gates. The white "plus" inside a blue circle indicates the NOT gate; the same symbol connected with two other smaller blue circles indicates the Toffoli gate. The symbol containing the red square with the white H indicates the Hadamard gate. The gray boxes with the Z letter characterize the measurement operation. These are standard symbols used in quantum computing, to indicate logic gates. Each line indicates a qubit ($q[0]$, ..., $q[4]$) and a classical bit ($mq21$, $mq31$, $mq41$), where the results of the measurements are stored. At the end of the circuit, there are measurements for each qubit. As an example, the initial configuration with $|q_0\rangle = 0$, $|q_1\rangle = 1$, $|q_2\rangle = 1$ is shown. (Color figure online)

## 2 Novelties: Joining Swarm Robotics, Quantum Computing, and Multilayer Networks

### 2.1 A Quantum-Based Swarm of "Telecommunicating" Robots

In this Subsection, we join the quantum-based decision-making model presented in [16] for a swarm of robots, applying it to a working simulated model for telecommunications. The system presented in [16] had the limitations of an omniscient system, that knew the position of obstacle and target, and where the distance of robots from the target is simply computed as the Euclidean distance between their coordinates. Here, the information on target proximity is acquired through robotic sensors. From [18], we borrow the simulation parameters of aquatic robots and the overall architecture with computational units on gondolas, which will be explicitly used in Subsects. 2.2 and 2.3.

We present now a computational example of a swarm of aquatic robots with telecommunications protocols and quantum computing. The implemented network consists of $N$ aquatic devices, the same as the ones defined in [18] (moving according to a Random WayPoint mobility model — RWP [27]. Each node has a radio coverage radius $R_M$ and a sensing radius (generally optical) $R_S$, able to check if the target is *near* the node. Connections are peer to peer, so they create an Ad-hoc network. Nodes move into a square map.

As aforementioned, robots have visual sensors. The measure of "success" is here called *reward*, and it is measured as 1−(perceived) distance from the target. Possible values of the reward are comprised between 0 and 1. The value 1 is reached when the robots stop their motion on the target. The values of positions and reward are thought of as the values of the peak of the wavefunction, as discussed in detail in [16]. Algorithm 1 summarizes the steps of the proposed code, and Table 2 shows the results of some simulations. In Algorithm 1, we indicate two robots of a pair inside the swarm as R1 and R2. The algorithm is classic, except for the decision-making step, which is the result of the measure from the quantum circuit of Fig. 1. A similar search-and-rescue can also be carried out completely classically with If-Then-Else; however, in [16], the strategy was implemented via a logic gate implemented inside a quantum circuit. The advantages provided by this approach, against up-to-date competition-winner algorithms (CEC 2022 Bound Constrained Single Objective Numerical Optimization benchmark problems; NL-SHADE-RSP with midpoint, [2]), are presented in detail in [16]. In this model, the number of robots can be changed. In this way, we can prove that the quantum protocol can be applied to inter-robot communications in the picture of Fig. 2. In Subsect. 2.2, we describe the overall system in terms of multilayer networks, and in Subsect. 2.3, we apply entanglement to inter-gondolas communications. From Table 2 it can be seen that with a low number of deployed nodes, more time is needed to accomplish the mission (in the table, $t_j - t_i = 120\,\mathrm{s}$). So, the collaborative approach is evident: the information discovered by each node is disseminated, if under coverage, giving the possibility to other robots to reach the target, instead of moving randomly. The code is available from authors upon request.[1]

**Table 2.** Results of ten simulations for a quantum-driven swarm of telecommunicating robots. Each simulation campaign involves the quantum circuit with 1024 shots, that is, 1024 runs of the code for each given configuration of the circuit. The most frequent states obtained in the output are the result of the measure. Map dimensions are 400 × 400 m, each robot has a sensing capacity of $R_S = 20\,\mathrm{m}$ and a coverage radius of $R_M = 50\,\mathrm{m}$. We assume that the robots communicate with the classical Ad-hoc On-demand Distance Vector (AODV) protocol, without any scalability issue, given the low number of involved mobile nodes. Robots, target, and obstacles are initially deployed randomly and when the map bounds are sensed, the robots change their moving direction.

| $N$ | trial | average reward | | | |
|---|---|---|---|---|---|
| | | $t_0$ | $t_1$ | $t_2$ | $t_3$ |
| 2 | 1 | 0 | 0 | 0.002 | 0.08 |
| | 2 | 0 | 0.001 | 0.003 | 0.15 |
| | 3 | 0 | 0 | 0.0015 | 0.22 |
| 10 | 1 | 0 | 0.2 | 0.4 | 0.73 |
| | 2 | 0 | 0.212 | 0.416 | 0.8 |
| | 3 | 0 | 0.34 | 0.436 | 0.828 |

[1] It is the application of the codes in https://github.com/medusamedusa/10_little_ants to the considered case.

**Algorithm 1.** Quantum-driven robots and telecommunications

**Input:** initial position of robots; position of the obstacle (unknown to the robots); position of the target (unknown to the robots)
**while** the swarm barycenter is far from the target **do**
  **while** R1 explores the space **do**
    R1 broadcast its position and reward
    **if** R1 robot finds an obstacle in its visual field **then**
      it changes direction of motion
    **end if**
    **if** R1 robot finds the target in its visual field **then**
      it slows down and starts to approach to it
      its reward slowly increases from 0 to 1 as the robot gets close to the target
      R1 robot remains on
      keep broadcasting its position
      R2 and the other robots get updates on position and reward of R1, and this information enters the **quantum circuit**; its output indicates the positions to be reached next.
    **end if**
    **if** R2's reward is higher than R1's reward **then**
      the coordinates of R2 enters the quantum circuit.
    **end if**
  **end while**
**end while**

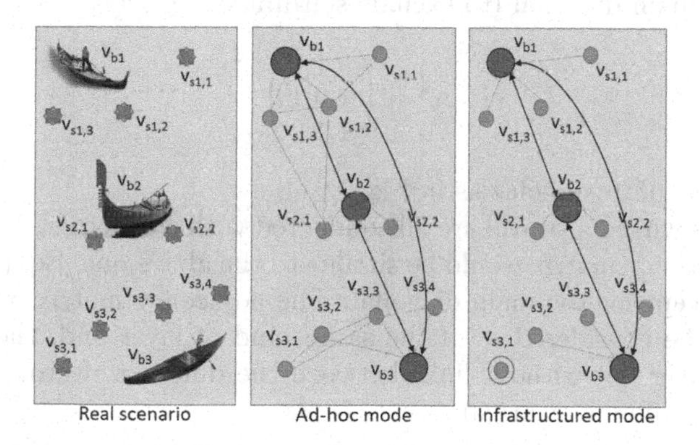


**Fig. 2.** Representation of the overall system with swarms of aquatic robots and computational units on gondolas, from [18].

## 2.2 Multilayer Networks to Model the Interactions Between Robots and Gondolas

The second novelty we introduce here is the hierarchical definition of communicating aquatic swarms and gondolas via the concept of multilayer networks.

Whereas a complex network in general consists of nodes which are connected by links, a *multilayer network* is constituted by a multitude of different kinds of nodes and connections. Therefore, the concept of multilayer networks is very powerful for the description and modeling of systems with multiple types of interactions, subsystems, and multidimensional structures [4]. We considered two hierarchical levels: base stations, responsible for dropping down the robots into water and to collect statistics, and superficial nodes, as illustrated in Fig. 2. Base stations are indicated with $v_{b_i}$, while superficial robots are indicated with $v_{s_{i,j}}$, where $i$, here, is the subscript of the belonging base station. The number of deployed superficial nodes is $N$, while there are $M$ base stations. For more details, see [18]. The most important thing is that $v_{s_{i,j}}$ nodes may communicate with each other directly (Ad-hoc mode) or through their base stations (Infrastructured mode). The links in a (unweighted) complex network can be represented using the adjacency matrix $\mathbf{A}$ with $A_{i,j} = 1$ representing a link between nodes $i$ and $j$ (and 0 for no link). In a multilayer network, we have for each individual layer $\alpha$ a separate $\mathbf{A}^{\alpha}$ (with the greek letters $\alpha$, $\beta$, etc. indicating a separate layer), indicating the intra-layer links. The inter-layer linking, i.e., links between different layers are represented by the adjacency matrix $\mathbf{A}^{\alpha,\beta}$ (with $\alpha$, $\beta$ indicating the corresponding layers $\alpha$ and $\beta$).

For our special case of swarm robots connected to gondolas, we present a multilayer approach taking into account submatrices describing the swarm component, and submatrices describing the gondolas component. The connected gondolas can be represented by one layer, using the long-range communication between them. Since the gondolas are all connected with each other (complete network), the corresponding adjacency matrix consists everywhere of values 1 except at the main diagonal (to exclude self-links):

$$\mathbf{A}^{\mathrm{g}} = \begin{pmatrix} 0 & 1 & 1 \\ 1 & 0 & 1 \\ 1 & 1 & 0 \end{pmatrix} \tag{1}$$

for an example of 3 gondolas as in Fig. 2.

The robots within a swarm are all connected with each other, thus, the corresponding adjacency matrix would be similar to the above one, Eq. (1), consisting of values 1 except for the main diagonal. The adjacency matrix of the different gondolas can be considered as of the same kind of layer and, therefore, forming just one layer but without links between the different swarms (the coloring indicating the different swarms):

$$\mathbf{A}^{\mathrm{s}} = \left(\begin{array}{ccc|cc|cccc} 0&1&1&0&0&0&0&0&0 \\ 1&0&1&0&0&0&0&0&0 \\ 1&1&0&0&0&0&0&0&0 \\ \hline 0&0&0&0&1&0&0&0&0 \\ 0&0&0&1&0&0&0&0&0 \\ \hline 0&0&0&0&0&0&1&1&1 \\ 0&0&0&0&0&1&0&1&1 \\ 0&0&0&0&0&1&1&0&1 \\ 0&0&0&0&0&1&1&1&0 \end{array}\right). \tag{2}$$

The multilayer network is the combination of these adjacency matrices $\mathbf{A}^{\mathrm{g}}$ (here indicated with red) and $\mathbf{A}^{\mathrm{s}}$ with the inter-layer connectivity matrices $\mathbf{A}^{\mathrm{g,s}}$:

$$\mathbf{A} = \begin{pmatrix} \mathbf{A}^{\mathrm{g}} & \mathbf{A}^{\mathrm{s,g}} \\ \mathbf{A}^{\mathrm{g,s}} & \mathbf{A}^{\mathrm{s}} \end{pmatrix} = \left(\begin{array}{ccc|ccccccccc} 0&1&1&1&1&1&0&0&0&0&0&0 \\ 1&0&1&0&0&0&1&1&0&0&0&0 \\ 1&1&0&0&0&0&0&0&1&1&1&1 \\ \hline 1&0&0&0&1&1&0&0&0&0&0&0 \\ 1&0&0&1&0&1&0&0&0&0&0&0 \\ 1&0&0&1&1&0&0&0&0&0&0&0 \\ 0&1&0&0&0&0&0&1&0&0&0&0 \\ 0&1&0&0&0&0&1&0&0&0&0&0 \\ 0&0&1&0&0&0&0&0&0&1&1&1 \\ 0&0&1&0&0&0&0&0&1&0&1&1 \\ 0&0&1&0&0&0&0&0&1&1&0&1 \\ 0&0&1&0&0&0&0&0&1&1&1&0 \end{array}\right). \quad (3)$$

The colored matrices along the diagonal represent the intra-layer communication and the off-diagonal parts (i.e., $\mathbf{A}^{\mathrm{s,g}}$) the inter-layer communications, i.e., the exchange of information between robots in a swarm and their connected centralized gondola.

In an extension, we could also think of exchanging information between the robots of different swarm directly (without the gondola layer). Then the off-diagonal elements in $\mathbf{A}^{\mathrm{s}}$ in Eq. (2) would have values 1, too.

This kind of representation can be used to model the information flow through the network. Studies with diffusive processes have shown that the diffusion can be much faster in multilayer networks than in the separate single layers [8], or that multiplex networks can enhance congestions [22]. Such a model approach can be based on a Markov chain model [9], a simple standardized model [22], or diffusion model [8]. A block-matrix representation for a single swarm was proposed in [15,16]. In that representation, diagonal submatrices contained the information on single robots, while the off-diagonal submatrices represented the pairwise interactions. As one of the next steps of research, we can develop a comprehensive formalism, where we connect the parameters of each single robot with the overall architecture of swarms and gondolas.

### 2.3 Entanglement Between "Gondolas"

The third and last novelty we introduce in this study is the application of entanglement to force gondolas follow each others' change of positions if certain conditions are verified. Thus, we borrow the (theoretical) use of the entanglement, computationally used in quantum circuits, to model interactions between the "heads," that is, the central units on gondolas.

In quantum mechanics, the entanglement is a non-local phenomenon, where two different particles behave as being part of the same physical system, despite possible ample space distance [23]. It means that, if a measure is performed on a particle, forcing it to enter into the eigenstate of the solution (destructive measure in physics), automatically, and instantaneously, the other particle switches its configuration to the same or opposite value, according to the overall state definition. In the case of classical devices as robots, we cannot think of instantenous exchange of information. Nevertheless, the core idea of considering multiple elements as part of the same physical system can be borrowed. And this idea can be considered as central to the very same concept of swarms: multiple units acting as elements of the same object.

---

**Algorithm 2.** Mechanism of entanglement-setup for gondolas

---

**Input:** max reward in each swarm; average reward in each swarm; barycenter position in each swarm
Each gondola computes and broadcasts these information to the other gondolas
**if** There is a swarm whose robots have max absolute individual reward; max average reward; more precise barycenter position **then**
    The *entangled condition* is activated
    The other gondolas switch their indication of position according to the one of the most successful swarm, following the eventual changes of position of the corresponding gondolas
**else if** None of the swarms is successful **then**
    The independent search of each swarm and their gondolas keeps going
**else if** There is a swarm presenting the worse values of the initial inputs **then**
    The *entangled condition* is activated in the opposite sense
    The other gondolas switch their indication of position as the opposite (randomly exploring in opposite regions of space) to the one of the most successful swarm
**end if**

---

In our study, the entanglement can intervene as part of the inter-gondolas communication, when the barycenter (and the majority) of the robots in a swarm reached the target. So, the information is acquired by the computational unit in the corresponding gondola, and the message "success of the search-and-rescue mission of my swarm" is transmitted to the other gondolas, to synchronize their movements. This is, of course, only an instance of modeling. In general, the inter-gondolas synchronization can be activated according to some threshold values of swarm-success or failure, depending on the specific mission assigned to each swarm. The synchronization between gondolas can be expressed via correlated, anticorrelated, or uncorrelated states. From quantum mechanics, we can borrow the idea of Bell states, maximally entangled. For instance, we can consider the Bell state $\frac{1}{\sqrt{2}}(|00\rangle + |11\rangle)$, where the first qubit exemplifies here the (mean) reward of the most successful swarm as transmitted by its leading gondola, and the second qubit, the activation of the entangled condition for the other gondola. If the reward is high, then the other gondola, with its corresponding swarm, follows the successful one. If even the most successful swarm has a reward which is still quite low, and thus it is considered as a logic 0, then the other gondola is unticorrelated to the other one, and explores elsewhere. If the reward of even the most successful swarm is considered as 0.5, than no correlation or unticorrelation is activated, and thus, there is no entanglement. In a quantum circuit, a Bell state is easily realized with a Hadamard gate (H) and a controlled NOT gate (CNOT). Algorithm 2 presents a possible organization of this step of research, which is to be implemented in next applications. It is mainly a "switch" that activates and de-activates the "entangled state" when the IF conditions are no longer verified.

## 3 Discussion and Conclusions

In this article, we proposed a study to join efforts from different disciplines, namely swarm robotics, telecommunications, and network theory. The considered problem, that is, the formalization and computational analysis of a set of swarms of robots interacting with central units, has been investigated in light of diverse languages and formal approaches. In particular, we presented a computational example of a swarm of aquatic robots, interacting via broadcast communications, where the decision-making system is based on quantum computing. Then, we considered the case of multiple swarms of robots, where each swarm is also exchanging messages with a central unit put into a boat, specifically a gondola. Units on gondolas are also able to exchange messages between them. To model the overall system of swarms and gondolas, we can borrow the formalism of multilayer networks. Finally, the synchronization between gondolas, occurring at certain points of the simulation—e.g., when one of the swarms was overall successful in finding the target of the mission—can be modeled using key ideas from entanglement in quantum mechanics.

This study can open new questions, mainly concerning the limits of applicability of the proposed system of communication and decision-making in real devices, taking into account limits of battery, noise disturbing the message exchange, and limits of internet connectivity. In fact, internet is required to remotely access IBM quantum simulators and quantum computers, which we exploited in our working example. In [18], an essential model of aquatic robot had been modeled, with four propellers, cameras for object detection, GPS sensors, and distance sensors. Thus, a first assessment of practical limitations was feasible. However, in that study, the quantum computing for robotic decision-making had not been introduced yet. And that was one of the novelty of the present research. As a possible strategy to overcome limitations of our study, we can choose a model of existing aquatic robot used in swarms, and conduct a sequence of experiments, to refine the models and highlight potential, unexpected problematics. A series of experiments with real devices could also help us more realistically estimate the advantage provided by quantum efficiency, comparing it against the computational resources needed by quantum devices, even if accessed remotely, and the time delay induced in communications. Further developments can include the use of sonification to highlight patterns of behavior in swarms at different hierarchical levels. The new development can draw upon swarm-robotic sonification [17] and soundscape analyses [19].

## References


1. Atchade-Adelomou, P., Alonso-Linaje, P., Albo-Canals, J., Casado-Fauli, D.: qRobot: a quantum computing approach in mobile robot order picking and batching problem solver optimization. Algorithms **14** (2021)
2. Biedrzycki, R., Arabas, J., Warchulski, E.: A version of NL-SHADE-RSP algorithm with midpoint for CEC 2022 single objective bound constrained problems. In: IEEE, Padua, Italy (2022)
3. Boccaletti, S., Bianconi, G., Criado, R., Genio, C.: The structure and dynamics of multilayer networks. Phys. Rep. (2014)
4. De Domenico, M., Granell, C., Porter, M., Arenas, A.: The physics of spreading processes in multilayer networks. Nat. Phys. (2016)
5. De Rango, F., Palmieri, N., Yang, X.S., et al.: Swarm robotics in wireless distributed protocol design for coordinating robots involved in cooperative tasks. Soft. Comput. **7**(4), 4251–4266 (2018)
6. Dong, D., Chen, C., Li, H., Tarn, T.: Quantum reinforcement learning. IEEE Trans. Syst. Man Cybern. Part B (Cybern.) **38**, 1207–1220 (2008)
7. Fazio, P., Mehic, M., Voznak, M.: Effects of sampling frequency on node mobility prediction in dynamic networks: a spectral view. Digit. Commun. Netw. **9**, 1009–1022 (2022)
8. Gómez, S., Díaz-Guilera, A., Gómez-Gardeñes, J., Pérez-Vicente, C., Moreno, Y., Arenas, A.: Diffusion dynamics on multiplex networks. Phys. Rev. Lett. **110**, 028701 (2013)
9. Granell, C., Gómez, S., Arenas, A.: Competing spreading processes on multiplex networks: awareness and epidemics. Phys. Rev. E **90**, 012808 (2014)
10. Hamann, H.: Swarm Robotics: A Formal Approach. Springer, Cham (2018)
11. Ivancevic, V.: Entangled swarm intelligence: quantum computation for swarm robotics. Math. Eng. Sci. Aerosp. **7**, 441–451 (2016)
12. Koukam, A., Abbas-Turki, A., Hilaire, V., Ruichek, Y.: Towards a quantum modeling approach to reactive agents. In: 2021 IEEE International Conference on Quantum Computing and Engineering (QCE) (2021)
13. Kwak, Y., Yun, W., Jung, S., Kim, J., Kim, J.: Introduction to quantum reinforcement learning: theory and PennyLane-based implementation. In: International Conference on Information and Communication Technology Convergence (ICTC) (2021)
14. Lamata, L., et al.: Quantum mechatronics. Electronics **10**, 2483 (2021)
15. Mannone, M., Seidita, V., Chella, A.: Categories, quantum computing, and swarm robotics: a case study. Mathematics **3**(372) (2022)
16. Mannone, M., Seidita, V., Chella, A.: Modeling and designing a robotic swarm: a quantum computing approach. Swarm Evol. Comput. **79**(101297) (2023)
17. Mannone, M., Seidita, V., Chella, A.: The sound of swarm. Auditory description of robotic movements. ACM Trans. Hum.-Robot Interact. **12**(4), 1–27 (in press)
18. Mannone, M., Seidita, V., Chella, A., Giacometti, A., Fazio, P.: Energy and SNR-aware robotic swarm coordination for aquatic cleaning operations. In: 97th IEEE Vehicular Technology Conference (VTC), Florence, Italy, pp. 1–7 (2023, in press)
19. Haselhoff, T., et al.: Complex Networks for Analyzing the Urban Acoustic Environment. EarthArXiV. https://eartharxiv.org/repository/view/5371/
20. Palmieri, N., Yang, X.S., De Rango, F., et al.: Comparison of bio-inspired algorithms applied to the coordination of mobile robots considering the energy consumption. Neural Comput. Appl. **31**, 263–286 (2019)

21. Srivastava, S.: Quantum Robotics: Applications of Quantum Computing in Robotic Science. Analytics Insight (2020). https://www.analyticsinsight.net/quantum-robotics-applications-quantum-computing-robotic-science/
22. Solé-Ribalta, A., Gómez, S., Arenas, A.: Congestion induced by the structure of multiplex networks. Phys. Rev. Lett. **116**, 108701 (2016)
23. Stolze, J., Suter, D.: Quantum Computing: A Short Course from Theory to Experiment. Wiley, Hoboken (2004)
24. Tropea, M., Palmieri, N., De Rango, F.: Modeling the coordination of a multiple robots using nature inspired approaches. In: Cicirelli, F., Guerrieri, A., Pizzuti, C., Socievole, A., Spezzano, G., Vinci, A. (eds.) WIVACE 2019. CCIS, vol. 1200, pp. 124–133. Springer, Cham (2020). https://doi.org/10.1007/978-3-030-45016-8_13
25. Wichert, A.: Principles of Quantum Artificial Intelligence. World Scientific (2020)
26. Zhu, K., Jiang, M.: Quantum artificial fish swarm algorithm. In: Proceedings of the 8th World Congress on Intelligent Control and Automation (2010)
27. Bahadori, F., et al.: The sum secrecy rate of NOMA-enabled VLC network with the random-way point mobility model. In: Proceedings of the Third South American Colloquium on Visible Light Communications (SACVLC) (2021)

# Generalizations of Evolved Decision-Making Mechanisms in Swarm Collective Perception

Dari Trendafilov(✉), Ahmed Almansoori, Timoteo Carletti, and Elio Tuci

Namur Institute for Complex Systems, University of Namur, Namur, Belgium
{dari-borisov.trendafilov,ahmed.almansoori,timoteo.carletti,
elio.tuci}@unamur.be

**Abstract.** This paper presents a study investigating the generalization characteristics of two neuro-controllers underpinning decision-making mechanisms in a swarm of robots engaged in a collective perception task. The neuro-controllers are both designed—using evolutionary computation—to operate in a randomly distributed cues environment, but under different conditions for what concerns the length of the communication range characterising the robots interactions. For one neuro-controller, the communication range during the design phase is 30 cm, while for the other is 50 cm. The aim of the study is to explore the robustness and the limitations of the two distinct neuro-controllers across a range of different conditions and to establish the optimal bounds on the swarm communication range for this collective perception task. To examine the performance of the two neuro-controllers we conduct a series of post-evaluations in 45 distinct environments, given by nine different distributions of the perceptual cues, and five different communication ranges (i.e., 10, 20, 30, 40, and 50 cm). The results demonstrate that the neuro-controller evolved with a swarm communication range of 30 cm generalizes better and exceeds the performance of the other neuro-controller evolved with 50 cm communication range in a vast majority of the post-evaluation conditions. However, the swarm performance degrades in conditions with patchily distributed perceptual cues and/or very short communication range.

**Keywords:** Swarm robotics · Collective perception · Evolutionary robotics

## 1 Introduction

Swarm robotics studies multi-robot systems in which each robot has its own controller, perception is local and communication is based on spatial proximity [10]. The group-level response emerges from a self-organisation process [7], based on the interaction between the robots and their physical environment. However, the autonomous nature of this process poses a challenge for designers, since it is notoriously difficult to infer which set of individual actions leads to the emergence of a desired collective response. Moreover, traditional design methods lack


M. Villani et al. (Eds.): WIVACE 2023, CCIS 1977, pp. 190–201, 2024.
https://doi.org/10.1007/978-3-031-57430-6_15

the ability to tackle problems and swarms of increasing complexity in uncertain and unpredictable environments. This further intensifies the need for fundamental and generic automated methodologies for modulating collective behaviour, with the potential to circumvent tedious trial-and-error model tuning.

One type of widely studied paradigms in swarm robotics is the "best-of-n" problem set [18,21], which requires the swarm to reach a consensus on the best among a number of available options. Consensus achievement is a process in which swarm members exchange their opinions with each other and eventually converge to a unique opinion. These studies instigate the search for controllers that perform robustly in different conditions, while at the same time optimize the utilization of critical device-operating resources. For example, emitting longer-range signals for swarm communication inevitably contributes to higher energy consumption and negatively impacts the autonomy of the robots. Since swarm communication is inherently local, it is important to establish the optimal bounds for the maximal communication range (and therefore constrain the energy consumption) and the exact trade-offs with respect to swarm performance in a particular task.

In this paper, we investigate the ability of two particular neural models [1] to generalize the opinion selection in a swarm of robots, engaged in the collective perceptual discrimination task, across a range of qualitatively and quantitatively different conditions, while preserving their effectiveness. We evaluate one previously developed neural model [1–3] and one recently evolved controller over a set of test conditions in the collective perceptual discrimination task, while varying the environmental patterns with respect to the spatial distribution of the options and the communication range. We use the nine benchmark environments proposed in [4] in which the options are more patchily distributed than the environment experienced by the swarm during the control system design phase. To our knowledge, this is the first study exploring swarm performance across environmental patterns and communication ranges simultaneously. The results of this study highlight important time vs. accuracy trade-offs, elucidate the interplay between swarm communication range and generalization capabilities and contribute to providing better awareness about the robustness of evolved controllers for swarm robotics. In the next section we present the related previous work, followed by the methodology of our study, the presentation of results and the discussion of salient observations, and finally we close with conclusions and directions for future work.

## 2 Related Work

For designing large groups of robots, which coordinate and cooperatively perform a task, swarm robotics takes inspiration from natural self-organizing systems and attempts to recreate the emergence of collective behaviour from simple local interaction rules [12,22]. Through the design of individual robot behaviour, swarm robotics aims to achieve locally coordinated interactions that results in a self-organized collective behaviour [6,9,11]. Collective decision-making mechanisms, designed by behaviour-based modular control systems, have demonstrated their effectiveness in a variety of scenarios [15,19,20]. However, the adaptability of these swarms to unexpected and unpredictable circumstances tends to be limited by the designer imposed bias. Further research is required to design collective decision-making mechanisms that allow swarms of robots to mimic natural swarms with respect to robustness, scalability, and flexibility [10].

The collective perceptual discrimination task for swarm of robots has been originally introduced by [14], who used a binary version of this scenario to design and evaluate individual mechanisms underpinning the collective decision-making process. In this task, the swarm explored a close arena patched with tiles, randomly painted in black and white, with the aim to collectively decide which colour is dominant. The two colours are the options or features, and the proportion with which each colour covers the arena floor corresponds to the option/feature quality. The goal is to design individual opinion selection mechanisms that allow the swarm to converge on the desired consensus state (i.e., all robots sharing the correct opinion about the arena colour dominance). Various individual mechanisms for opinion selection have been developed, from the classical hand-crafted solutions, based on the Voter model, the Majority rule, and their variants [21], to more recent ones, based on the synthesis of artificial neural networks [1]. More recently, an opinion selection mechanism based on artificially synthesised neural network using evolutionary algorithms [2] have proved effective for the collective perceptual discrimination task [19]. Further studies with evolved controllers [3] demonstrated that the neural network based opinion selection mechanism is more effective and scalable than the Voter model [19] in a set of environmental patterns, however, with a fixed swarm communication range. The perceptual discrimination task has been used by [19] to investigate the performance of various decision-making strategies for swarm of robots while varying the options quality (i.e., the features ratio) for controlling task difficulty. [16] explored further variations of this task, characterised by the presence of byzantine robots, i.e., robots that communicate deceptive messages with the intent to entice the swarm to converge on a consensus to a non-optimal choice. [8] investigated scenarios with more than two options/features. Arguing that the key determinant of the difficulty of the perceptual discrimination task for swarms of robots required to choose the best option is the features' distribution, [4] proposed a set of nine structurally different variations in the environmental topology of the patterns (Fig. 1) and a set of measures for their characterization, without considering the communication range as an important system variable. Their work was further expanded by [17], who proposed a universal and generic

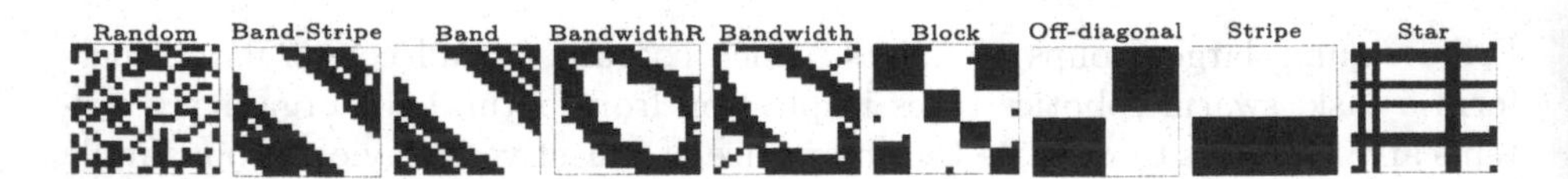


**Fig. 1.** The benchmark patterns used in our study, proposed by [4] and employed in other perceptual discrimination tasks [1] to assess the swarm performance.

measure of task difficulty, which takes into account not only the environmental complexity, but also the agent's capabilities. More recently, [1] used this set of patterns to evaluate the effectiveness of neural network-based decision-making mechanisms with a fixed communication range. This set of spatial distributions of the perceptual cues have been evaluated in [5] with a decision-making mechanism tackling spatial correlations in unknown environments statistically.

## 3 Methods

This study is based on the collective perceptual discrimination task as described in [1,4], and is conducted in a simulation environment represented by a square arena of $2 \times 2$ m, whose floor is covered with black and white tiles (see Fig. 2a), $10 \times 10$ cm each, distributed according to the patterns shown in Fig. 1. The dominant colour (either black or white) covers 55% of the arena floor and corresponds to the best quality option/feature, while the other colour covers the remaining 45%. In this scenario, a swarm of 20 robots navigate the arena with the task to reach a consensus on the type of environment on which they are placed. Consensus refers to a state in which all robots share the same opinion on the arena colour dominance.

Our robot model simulates the widely popular in the research community e-puck2 robot [13], which is equipped with eight proximity infra-red sensors, a binary floor colour sensor, and a range&bearing board for local communication. Swarm communication consists of emitting a binary signal, which represents the robot's current opinion about the arena colour dominance. To compensate for the simulation–reality gap, 10% uniform noise is added to all sensor readings, motor outputs and robot position.

Initially, a homogeneous swarm of 20 robots is distributed randomly in the arena. During the evaluation the robots explore the arena with a random walk

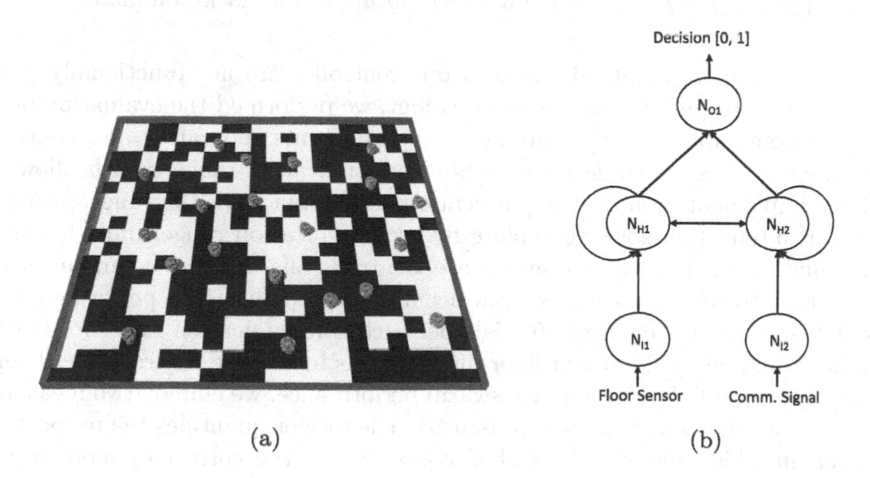


**Fig. 2.** (a) Simulated arena of the collective perceptual discrimination task. (b) Multi-layer CTRNN underpinning the opinion selection of a single agent.

with a fixed step length (5 s, at 20 cm/s), and turning angles chosen from a wrapped Cauchy distribution, while avoiding obstacles (arena walls and neighbours) for up to 1000 s. On every iteration, the robots sample the arena under their body and communicate their opinion on the dominant colour to spatially proximal robots. The objective of the swarm is to reach a consensus (i.e., all robots sharing the same opinion) on the correct colour dominance. The process underpinning the development of the individual opinion is regulated by a continuous time recurrent neural network (Fig. 2b), synthesised using evolutionary algorithms. In this study, we have evaluated and compared the performance of one previously developed controller [1] designed to operate with a communication range of 50 cm, and one recently evolved neuro-controller designed to operate with a shorter 30 cm communication range.

A simple evolutionary algorithm with linear ranking is used with a population of 100 genotypes. Generations are produced by a combination of selection with elitism, and mutation. For each new generation, the highest scoring individuals are retained unchanged. The remainder of the new population is generated by binary tournament selection from the 70 best individuals of the old population. Each genotype is a vector comprising 15 real values with 10 connections, 2 decay constants and 3 bias terms. At the beginning of each evaluation trial, each genotype is decoded into a neuro-controller, which is cloned on each robot. Robots are randomly placed in the arena with a random orientation. The fitness of a genotype is its average swarm evaluation score after it has been assessed twice for both, black-dominant and white-dominant environments. In each trial $e$, the opinion of robot $r$ is evaluated at every time $t$ (i.e., $O_t^r$) for every robot in the swarm, and the swarm is rewarded by the following function:

$$F_e = \begin{cases} \frac{2}{T}\sum_{t=T/2}^{T}\sum_{r=1}^{R} o_t^r \text{ if swarm located in a white-dominant env.} \\ \\ \frac{2}{T}\sum_{t=T/2}^{T}\sum_{r=1}^{R}(1 - o_t^r) \text{ if swarm located in a black-dominant env.} \end{cases}$$

Since the operational principles of our controller are not functionally symmetrical with respect to the dominant colour, we performed the evaluation both in black-dominant and white-dominant environments separately and contrast the outcome. The maximal robot–robot communication range, which allows a reliable implementation on the physical e-puck2 robot with the range&bearing board, is 50 cm, therefore we explore five communication range limits between 10 cm and 50 cm. In order to investigate the trade-offs between communication range, time-to-consensus and swarm accuracy, we analyze the performance of simulated robot swarms over 50 trials per each communication range for black-dominant and white-dominant floor distributions in all nine environmental conditions (patterns). To characterize swarm performance, we employ two measures – decision accuracy and time-to-consensus. The former quantifies the proportion of trials in which the swarm reached consensus on the correct opinion/option and the latter characterizes each successful trial.

## 4 Results

In order to explore the relationship between the time the swarm spent on the task, as a key performance indicator, and the swarm communication range, we averaged the execution time of all 50 trials per condition, including the time-to-consensus of successful and the time limit (1000 s) of unsuccessful trials. In the following, for brevity, we will denote the controllers evolved with communication range of 50 cm and 30 cm with NN50 and NN30, respectively. The results reveal decreasing trends (Fig. 3) with the increase of communication range for all environmental patterns, as expected, however, with significant variability in the shapes and slopes. At one extreme, for the homogeneous Stripe pattern, the swarm achieved negligible success across conditions, revealing the limitations of the controllers. At the other extreme, the curves for the Random pattern show consistent monotonic trends of execution time as the communication range increases, outperforming the rest with minor exceptions. This is unsurprising, as the neuro-controllers are evolved in the Random pattern environment. The generalization capabilities of the controllers are best evidenced for the Star pattern, which approaches the performance levels of the Random pattern as the communication range increases and surprisingly exceeds those in the largest range.

Interestingly, NN30 demonstrates faster average execution times than NN50 overall, in particular with respect to more challenging environmental patterns. Strikingly, in the communication range of 50 cm NN30 is superior to NN50 for most patterns, which is surprising, provided that NN50 was evolved for the communication range of 50 cm in particular. The results reveal that NN30 is narrowing the performance gap between the Random and the Star patterns and the rest of the pack, which is large in the case of NN50.

Figures 4 and 5 present the accuracy and the time-to-consensus distributions respectively, for all conditions (pattern, communication range, dominant colour), which provide further insights about the generalization abilities of the controllers. Both controllers generalize rather well in the Star and the Random patterns across the range of 20 cm to 50 cm.

Figure 4 reveals the superior overall accuracy of NN30 over NN50 across conditions. NN30 performs markedly better than NN50 in the shorter communication ranges of 20–30 cm, however, it excels in the larger ranges as well. NN30 generalizes better than NN50 across environmental patterns in the communication ranges of 30 cm to 50 cm, and has a significantly stronger performance also in the ranges of 10–20 cm. Furthermore, the evolutionary bias between black and white dominant environments, strongly evident in the performance of NN50, mostly vanishes in the case of NN30.

The time-to-consensus distributions (Fig. 5) exhibit large variability across patterns and communication ranges. The exception is confined to the Random pattern and to some extent to the Star pattern, with more compact time distributions. The results reveal a quasi-monotonic relationship between execution time and communication range for most patterns, as increasing the range increases both the average time and its variance. Increasing the time limit to 1000 s appears to allow to a certain degree the convergence to consensus for certain patterns as the communication range decreases. The time distributions for NN30 reveal that while 500 s are sufficient for convergence in the largest communication range for the majority of the conditions, this time limit is insufficient for shorter communication ranges, which require 600–800 s for some patterns.

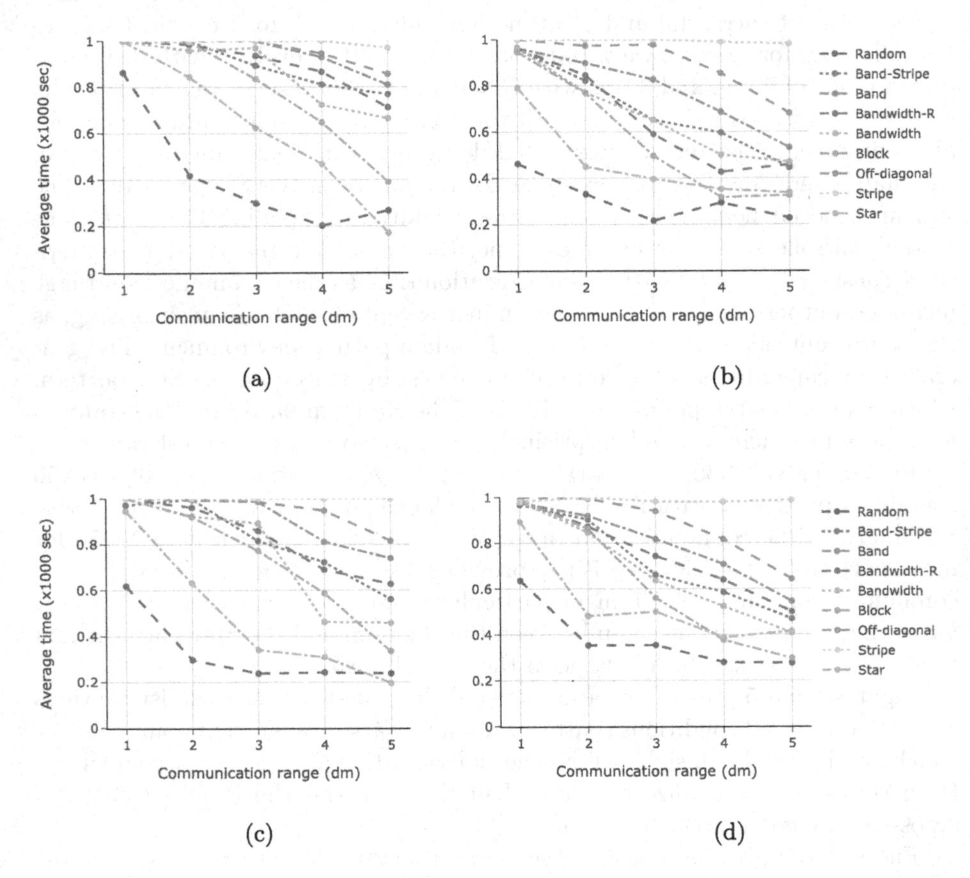


**Fig. 3.** Average execution time over 50 trials per communication range (on the x axis in decimeters) in all nine environmental conditions for black-dominant (a) NN50 and (b) NN30, and for white-dominant environments (c) NN50 and (d) NN30. For one type of environment—Stripe—the swarm achieved only negligible success across-the-board. For the Random environment, the graphs show a steady decrease of execution time as communication range increases, with minor exceptions. The trends for NN30 reveal better performance than NN50, overall.

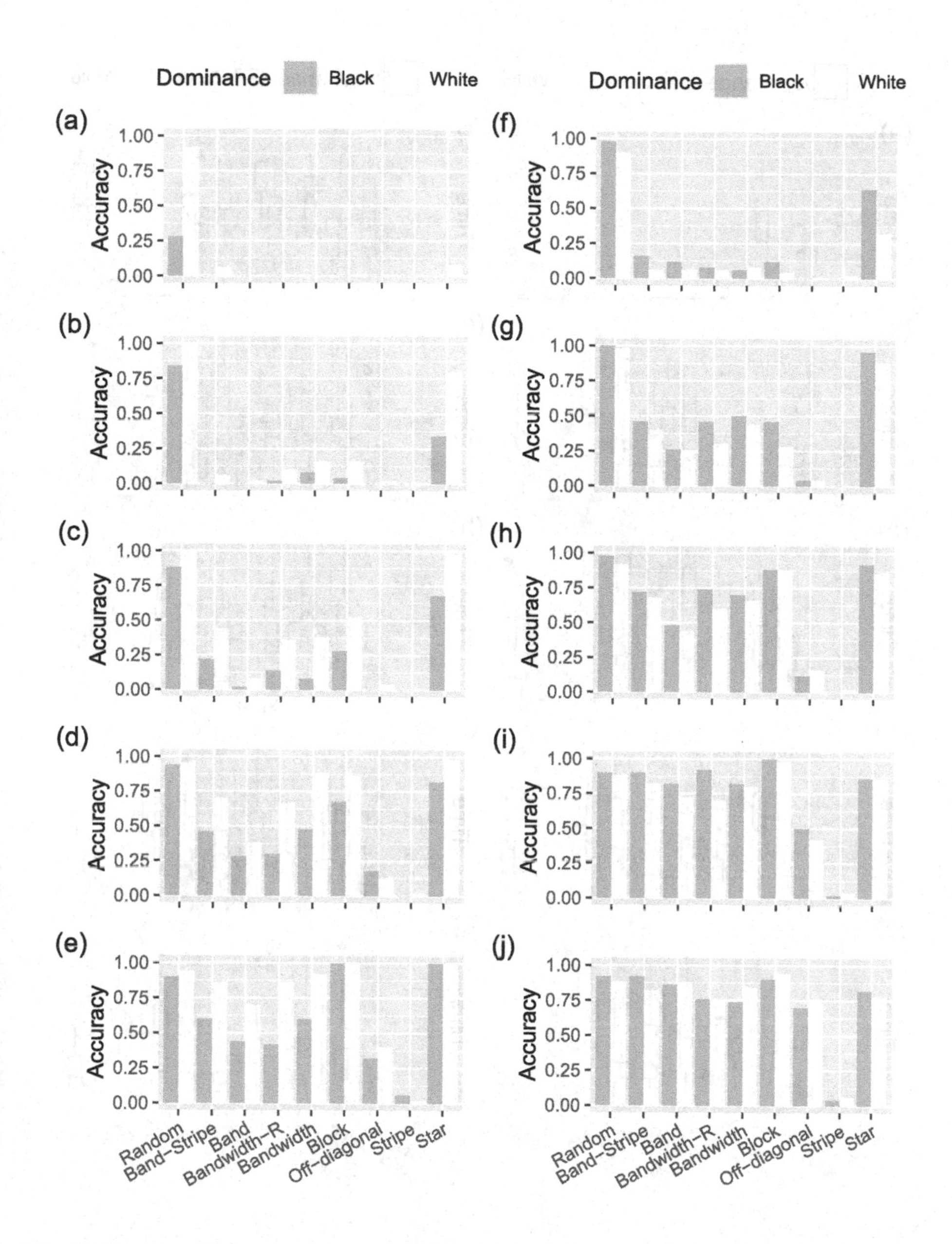


**Fig. 4.** Swarm consensus accuracy of successful trials, recorded in 50 simulation trials per condition in the nine environmental patterns and five communication ranges: for NN50 (a) 10 cm, (b) 20 cm, (c) 30 cm, (d) 40 cm, and (e) 50 cm, and for NN30 (f) 10 cm, (g) 20 cm, (h) 30 cm, (i) 40 cm, and (j) 50 cm. NN30 achieved higher accuracy than NN50 overall across conditions. The superior performance of NN30 over NN50 is most pronounced in the communication ranges of 20–30 cm, and extends to the larger ranges as well.

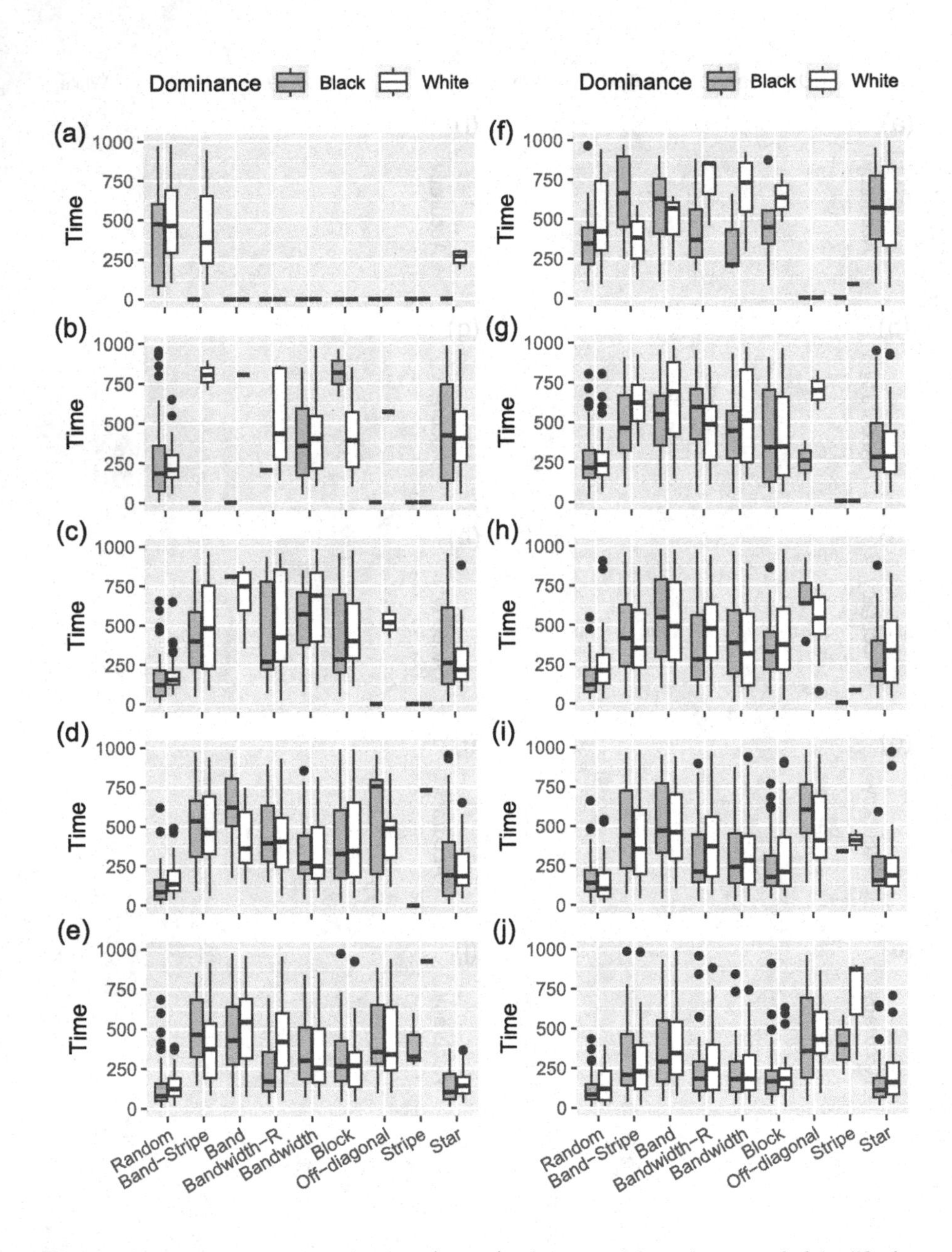


**Fig. 5.** Consensus time distribution (in sec) of successful trials, recorded in 50 simulation trials per condition in the nine environmental patterns and five communication ranges: for NN50 (a) 10 cm, (b) 20 cm, (c) 30 cm, (d) 40 cm, and (e) 50 cm, and for NN30 (f) 10 cm, (g) 20 cm, (h) 30 cm, (i) 40 cm, and (j) 50 cm. NN30 achieved faster consensus and more compact time distributions than NN50 overall. The effect that decreasing communication range has on time is consistent for both controllers, as the consensus time and its variance both increase.

## 5 Discussion

The results of this study complement the findings reported in [1] by extending the scope of the test conditions. In this work, we have evolved a new neurocontroller using the same mechanism as in [1], however, restricting the communication range to 30 cm instead of 50 cm. We have evaluated both controllers, the new one (NN30) and the previously reported one (NN50) in [1], in the collective perceptual discrimination task on the nine environmental patterns using five communication settings, ranging from 10 cm to 50 cm, with the aim to explore the limits of generalizability of the controllers. Anticipating slower than typical convergence to consensus, in this study, we have increased the cut-off time from 400 s to 1000 s, which proved beneficial overall. Certain patterns required longer time for convergence (600 800 s), especially with decreasing communication ranges. This insight is an important empirical benchmark, given that the typical cut-off time used by the research community in this type of studies is 400 s.

The results demonstrate the superior generalization capabilities of NN30 over NN50 across conditions. The NN30 controller, surprisingly, outperforms NN50 in the largest communication range, for which NN50 was evolved in particular. The effect of the evolutionary bias between black-dominant and white-dominant environments is also significantly reduced by NN30, which is a major step forward. As expected, performance drops with decreasing communication range, gradually for NN30, and more dramatically for NN50. More clustered pattern types, e.g., Stripe, proved difficult for both controllers, achieving very low accuracy in all conditions.

The results highlight that the overall performance drops with decreasing communication range across all environmental patterns, as expected, however at a different rate for different patterns. Time-to-consensus increased gradually for all pattern types as the communication range decreased, as expected, however, the increase for the Random pattern was negligible and for the Star pattern moderately steeper, which highlights the robustness of the controllers. Interestingly, the Star and the Block patterns outperformed the Random pattern in some of the conditions. Furthermore, in the white-dominant environment the NN50 performance for the Random pattern was unexpectedly higher in the ranges of 20 cm and 30 cm, compared to the ranges of 40 cm and 50 cm, an artefact of the evolutionary bias effect. Similarly, NN30 achieved higher accuracy in the range of 20 cm than 30 cm for the Random pattern.

## 6 Conclusion

This paper presents an investigation into the performance of two neural network controllers recently evolved for the collective perceptual discrimination task with swarm communication ranges of 50 cm and 30 cm respectively. The controllers were evaluated in a simulated homogeneous swarm of 20 robots in a set of conditions, varying the structural distribution of the arena floor pattern and the swarm communication range. The results indicate the ability of this type of neural models to generalize their decision-making behaviour towards swarm consensus in a range of test conditions, and highlight their limitations. This work demonstrates the potential of the evolutionary approach to automatically design decision-making mechanisms for swarm robotics and represents an important milestone towards the development of robust controllers that adapt successfully to unexpected conditions while retaining their performance. Future research will focus on the evolution of controllers for various conditions in order to identify the optimal configurations and establish the precise performance bounds and trade-offs in swarm collective perception.

**Acknowledgements.** This project has received funding from the European Union's Horizon 2020 research and innovation programme under the Marie Skło-dowska-Curie grant agreement No 101034383 and the CERUNA doctoral fellowship by the University of Namur. Computational resources have been provided by the Consortium des Équipements de Calcul Intensif (CÉCI), funded by the Fonds de la Recherche Scientifique de Belgique (F.R.S.-FNRS) under Grant No. 2.5020.11 and by the Walloon Region.

## References

1. Almansoori, A., Alkilabi, M., Tuci, E.: Further investigations on the characteristics of neural network based opinion selection mechanisms for robotics swarms. In: Correia, J., Smith, S., Qaddoura, R. (eds.) EvoApplications 2023. LNCS, vol. 13989, pp. 737–750. Springer, Cham (2023). https://doi.org/10.1007/978-3-031-30229-9_47
2. Almansoori, A., Alkilabi, M., Colin, J.N., Tuci, E.: On the evolution of mechanisms for collective decision making in a swarm of robots. In: Schneider, J.J., Weyland, M.S., Flumini, D., Füchslin, R.M. (eds.) WIVACE 2021. CCIS, vol. 1722, pp. 109–120. Springer, Cham (2022). https://doi.org/10.1007/978-3-031-23929-8_11
3. Almansoori, A., Alkilabi, M., Tuci, E.: A comparative study on decision making mechanisms in a simulated swarm of robots. In: 2022 IEEE Congress on Evolutionary Computation (CEC), pp. 1–8. IEEE (2022)
4. Bartashevich, P., Mostaghim, S.: Benchmarking collective perception: new task difficulty metrics for collective decision-making. In: Moura Oliveira, P., Novais, P., Reis, L. (eds.) EPIA 2019. LNCS, vol. 11804, pp. 699–711. Springer, Cham (2019). https://doi.org/10.1007/978-3-030-30241-2_58
5. Bartashevich, P., Mostaghim, S.: Multi-featured collective perception with evidence theory: tackling spatial correlations. Swarm Intell. **15**(1), 83–110 (2021)
6. Boudet, J.F., et al.: From collections of independent, mindless robots to flexible, mobile, and directional superstructures. Sci. Robot. **6**(56) (2021). https://doi.org/10.1126/scirobotics.abd0272
7. Camazine, S., Deneubourg, J., Franks, N., Sneyd, J., Theraulaz, G., Bonabeau, E.: Self-Organization in Biological Systems. Princeton University Press, Princeton (2001)
8. Ebert, J., Gauci, M., Nagpal, R.: Multi-feature collective decision making in robot swarms. In: 17th International Conference on Autonomous Agents and Multiagent Systems, AAMAS 2018, pp. 1711–1719 (2018)

9. Ferrer, E.C., Hardjono, T., Pentland, A., Dorigo, M.: Secure and secret cooperation in robot swarms. Sci. Robot. **6**, eabf1538 (2021)
10. Hamann, H.: Swarm Robotics: A Formal Approach. Springer, Cham (2018). https://doi.org/10.1007/978-3-319-74528-2
11. Hasselmann, K., Ligot, A., Ruddick, J., Birattari, M.: Empirical assessment and comparison of neuro-evolutionary methods for the automatic off-line design of robot swarms. Nat. Commun. **12**(4345), 1–11 (2021)
12. Kube, C.R., Zhang, H.: Collective robotics: from social insects to robots. Adapt. Behav. **2**(2), 189–218 (1993). https://doi.org/10.1177/105971239300200204
13. Mondada, F., et al.: The e-puck, a robot designed for education in engineering. In: Proceedings of the 9th International Conference on Autonomous Robot Systems and Competitions, vol. 1, pp. 59–65 (2009)
14. Morlino, G., Trianni, V., Tuci, E.: Collective perception in a swarm of autonomous robots. In: Proceedings of the International Joint Conference on Computational Intelligence, vol. 1, pp. 51–59 (2010)
15. Scheidler, A., Brutschy, A., Ferrante, E., Dorigo, M.: The k-unanimity rule for self-organized decision-making in swarms of robots. IEEE Trans. Cybern. **46**, 1175–1188 (2016)
16. Strobel, V., Ferrer, E., Dorigo, M.: Managing byzantine robots via blockchain technology in a swarm robotics collective decision making scenario. In: Proceedings of the 17th International Conference on Autonomous Agents and Multi-Agent Systems, pp. 541–549 (2018)
17. Trendafilov, D., Almansoori, A., Carletti, T., Tuci, E.: The role of the environment in collective perception: a generic complexity measure. In: ALIFE 2023: Ghost in the Machine: Proceedings of the 2023 Artificial Life Conference, p. 13 (2023)
18. Trianni, V., Tuci, E., Ampatzis, C., Dorigo, M.: Evolutionary swarm robotics: a theoretical and methodological itinerary from individual neuro-controllers to collective behaviours. In: Vargas, P.A., Paolo, E.D., Harvey, I., Husbands, P. (eds.) The Horizons of Evolutionary Robotics, pp. 153–178. MIT Press (2014)
19. Valentini, G., Brambilla, M., Hamann, H., Dorigo, M.: Collective perception of environmental features in a robot swarm. In: Dorigo, M., et al. (eds.) ANTS 2016. LNCS, vol. 9882, pp. 65–76. Springer, Cham (2016). https://doi.org/10.1007/978-3-319-44427-7_6
20. Valentini, G., Hamann, H., Dorigo, M.: Efficient decision-making in a self-organizing robot swarm: on the speed versus accuracy trade-off. In: Proceedings of the 2015 International Conference on Autonomous Agents and Multiagent Systems (AAMAS), pp. 1305—1314. International Foundation for Autonomous Agents and Multiagent Systems (2015)
21. Valentini, G., Ferrante, E., Dorigo, M.: The best-of-n problem in robot swarms: formalization, state of the art, and novel perspectives. Front. Robot. AI **4**, 9 (2017)
22. Werfel, J., Petersen, K., Nagpal, R.: Designing collective behavior in a termite-inspired robot construction team. Science **343**(6172), 754–758 (2014)

# An Investigation of Graceful Degradation in Boolean Network Robots Subject to Online Adaptation

Michele Braccini[1(✉)], Paolo Baldini[1], and Andrea Roli[1,2]

[1] Department of Computer Science and Engineering (DISI), Campus of Cesena, Alma Mater Studiorum Università di Bologna, Cesena, Italy
{m.braccini,p.baldini,andrea.roli}@unibo.it

[2] European Centre for Living Technology, Venice, Italy

**Abstract.** The ability to resist to faults is a desired property in robotic systems. However, it is also hard to obtain, having to modify the behavior to face breakdowns. In this work we investigate the robustness against sensor faults in robots controlled by Boolean networks. These robots are subject to online adaptation—i.e., they adapt some structural properties while they actually act—for improving their performance at a simple task, namely phototaxis. We study their performance variation according to the number of faulty light sensors. The outcome is that Boolean network robots exhibit graceful degradation, as the performance decreases gently with the number of faulty sensors. We also observed that a moderate number of faulty sensors—especially if located contiguously—not only produces a negligible performance decrease, but it can be sometimes even beneficial. We argue that online adaptation is a key concept to achieve fault tolerance, allowing the robot to exploit the redundancy of sensor signals and properly tune the dynamics of the same Boolean network depending on the specific working sensor configuration.

**Keywords:** Boolean networks · Robot · Online adaptation · Graceful degradation · Fault-tolerance

## 1 Introduction

Research in robotics has often been influenced by biology. Sometimes the goal is to explore natural behaviors in a controlled setting, some others to bring the effectiveness of natural strategies to artificial applications [5,9,11,17,19]. Nevertheless, one often neglected property is the natural ability to survive to faults. The motivations are many, and rooted in the differences between artificial and biological bodies. For instance, organisms are able to recover from limited damages regenerating tissues, while machines are not. Even the perception capabilities differ, with biological bodies having many more redundant and degenerate sensors. This directly enhances the survival of an organism, since in case of faults essential information can be retrieved—and reconstructed—from the remaining


M. Villani et al. (Eds.): WIVACE 2023, CCIS 1977, pp. 202–213, 2024.
https://doi.org/10.1007/978-3-031-57430-6_16

signals. From this point of view, the gap between artificial and biological agents is still huge and far from being filled.

Although we cannot expect to achieve the same resilience of living organisms soon, we can still aim at improving the current state of robotics. Many robots work in a fragile equilibrium, where a single breakdown can lead to the catastrophic fault of the entire system. This problem is too often ignored, as it is frequently possible to reach the robot and recover the damage. Nevertheless, this binds the robot existence to us and prevents the creation of really autonomous robots, i.e., robots able to operate without the need of human intervention [18]. Autonomy is important especially in situations in which the robot have to act in an isolated environment, for example during rescue operations in hostile environments. We believe that the first step to obtain autonomous robots consists in making them less dependent on human support. This can be achieved by improving their resilience to faults, i.e., by making them able to continue their duties when subject to breakdowns. Obviously, the limited amount of sensors and actuators in a robot still makes it sensible to faults. Nevertheless, we aim at a graceful degradation of performance according to the amount of damage. This property alone is already an important improvement over the current state, where the lack of a single cue can lead to a critical decrease in performance.

Previous attempts to tackle the problem consisted in defining an internal representation of the robot structure, allowing to adapt the behavior as soon as some part of the body were detected to be damaged [4]. The robot synthesizes a new behavior simulating its self-model and finding the most suitable solution to the fault. This technique is very powerful, but also highly computationally demanding. This makes the approach unsuitable for computationally limited robots. In this work we build upon previous works of online adaptation in order to provide an alternative approach for minimally cognitive agents [1,5,6]. The idea is to continuously adapt the behavior with a simple strategy in order to maximize a performance metric. We expect this approach to lead to a graceful degradation of performance, as the reaction of the robot adapts to exploit the best information available. However, as the adaptation process is not instantaneous, we expect the performance to briefly drop before each recovery. The decrease directly correlates with the importance of the faulted signal in the reaction of the robot. Therefore, it is expected that different faults will recover in different amounts of time. In order to overcome this problem, we propose to analyze the performance of a robot that does not have any innate behavior or any sort of reactive mechanism. This allows us to assess the average time needed to recover, regardless of any previous known behavior.

Our investigation consists in evaluating the graceful degradation of performance in robots controlled by Boolean networks (BN). During the adaptation, only the couplings between robot sensors and BN nodes undergo changes, leaving the BN itself unchanged. The performance is evaluated in a phototaxis task, in which the robot has to reach a light source. The results aim at providing a first assessment as to whether online adaptation induces a graceful degradation of performance in BN-robots that perform phototaxis. Finally, we also try to understand to what extent the kinds of damage affect the performance.

This paper is structured as follows. In Sect. 2 we give an introductory description of BNs and their working principle. In Sect. 3 we describe the adaptive approach used in the experiments. Section 4 delves into experimental details. Following, Sect. 5 presents and discusses the results of the study. Finally, in the conclusion section we summarize the work done and we illustrate future explorations on the topic.

## 2 Boolean Networks

The core of the robot controller used in this work is a Boolean network (BN). BNs can be represented as a discrete set of Boolean variables and Boolean functions. The state of a variable depends on the state of the neighbors and on the transition function that controls it. Since their introduction as an abstract model of gene regulatory networks [14], BNs have been the subject of many works investigating their computational and dynamical properties. Notably, they effectively capture significant biological phenomena, such as cell differentiation [7,8,12,13,16,23,24]. One interesting application is in the context of robotics, where they have been used to develop adapting and evolving[1] robots [5,22].

A specific class of BNs is Random Boolean Networks. Those are created randomly according to some rules, such as the (possibly average) number of neighbors $K$ of each node. Also the transition functions are randomly generated, usually creating a mapping from the state of the neighbors to the resulting state of the variable. This can be done by creating a table enumerating the possible input states and by randomly setting the corresponding output. The setup can be influenced by a bias $p$ representing the probability of an output to be set to 1. Different combinations of values for $p$ and $K$ change the dynamic of the network towards different dynamic regimes [15]. BNs possess an interesting dynamics that can be divided in two regimes: order and chaotic. The ordered region is characterized by a short propagation of the signals, and by the tendency of the system to return to a stable state. The chaotic regime enhances and favors the perpetuation of perturbations that may permanently change the state evolution of the system. The result is that the state of a chaotic network is never stable. A critical region lies at the boundary between these two regimes, separating them. In this condition the perturbations propagate more than in the ordered region, but fade away faster than in the chaotic one. Many works suggest critical BNs favors computation, and indeed have been successfully used to complete different tasks, such as classification, filtering and control [21]. Even BNs evolved to solve specific tasks were often found to be critical [2].

[1] This mostly thanks to their simple encoding and mutation.

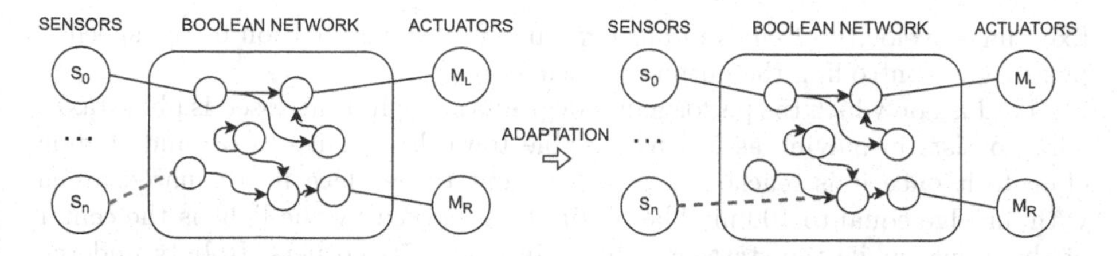


**Fig. 1.** Example of an adaptation step. The subset of sensor-node couplings to be reconnected is indicated by a red dashed line on the left. The reconnected couplings are represented by a blue dashed line on the right. As visible, the adaptation affects only the node to which a sensor connects, and leaves unchanged the topology of the BN. (Color figure online)

## 3 Adaptation

We claim that the ability to adapt is a major facilitator to achieve a graceful degradation of performance. In this work, we use a previously proposed adaptive mechanism for the adaptation of the robot behavior [6]. This is designed to work on network based control systems, and has been already successfully used for the adaptation of robots controlled by random BNs.

In this schema, the BN mediates the signals from the sensors in order to produce an adequate output to control the actuators. The BN itself is random, and never changes during the experiment. Indeed, the adaptive process does not directly affect the BN. The only way to modify the behavior of the robot is thus by modifying the couplings between network nodes and sensors/actuators (see Fig. 1). Specifically, in this work we only change the sensory couplings, i.e., those from the sensors to the BN. The idea is that it is possible to perturb the internal state of the BN in order to generate a desired response, i.e., a desired output which is used to properly control the actuators. Differently, we can state that the adaptation modifies the point of view of the robot, increasing or decreasing the focus on (i.e., effect of) some signals. When the robot behavior adapts, a subset of couplings between the robot light sensors and a BN controller is chosen and modified. The new couplings must connect to different nodes of the BN. Moreover, no more than one sensor can connect to a single node. Previous works demonstrated that this simple adaptive strategy is powerful enough to produce complex behaviors [5].

## 4 Experimental Setting

The robot used in the experiment is a *foot-bot*, simulated in ARGoS [3,20]. The foot-bots are equipped with 24 light sensors placed in a circular ring around the chassis of the robot, and can move by acting on the motors of two wheels. Before being passed to the BN, the values of the sensors are encoded in binary form according to a threshold $t = 0.2$. The outputs provided to the two wheels are either 0 or 1, converted either into a stopped wheel or a wheel that provides a fixed linear velocity (2 cm/s in our experiments). A modulation of the speed is possible by controlling the output at each step.

The task on which the performance degradation will be assessed is phototaxis. This consists in moving as fast as possible towards a light source, and staying close to it once it is reached. The experiment takes place in a squared arena with an edge equal to 100 m. The destination marked by the light is the center of the arena, while the starting point is in one of its corners. Robots undergo an adaptation process composed of 1200 epochs, each lasting one minute in simulation time. At the start of each epoch the robot adapts its controller, that is then evaluated at the end the epoch. At the end of each epoch the robot evaluates its performance as the difference between initial and final distance from the light. For ease of the experiments we provide this information directly to the robot. However, in principle robots are situated in their environment and should use only the information they can get from their sensors. For example, in this case we could use the variation of light intensity read by the light sensors. If the performance is not worse than the best one achieved so far, the incumbent mapping becomes the starting point for subsequent adaptations. Otherwise, the previous mapping is restored and adapted once more. In this experiment the adaptation can affect up to 6 connections from sensors to BN nodes.[2]

As we are in an online setting, besides the final distance to the light, we are also interested in the life performance of the robot, i.e., in the performance they achieve in the whole duration of a single experiment. For this reason, we assess the performance according to three different measures: *(a)* the distance from the light along the epochs; *(b)* the *run length distribution* (RLD), which represents the fraction of successful runs at each epoch;[3] *(c)* the final distance from the light.

Since a real robot may undergo many types of fault, we analyze how the performance changes in different situations. (i) The first case considers the event in which the sensors simply stop to produce any output. For instance, this may be due to the disconnection of a cable. Although the BN node to which the broken sensor connects can be changed, its signal will never affect the internal state of the BN controller. (ii) The second case consists in sensors producing random outputs. Differently from before, those still perturb the BN internal state. However, their signal can be considered noise, and is therefore expected to be detrimental. (iii) The third scenario considers the output of the faulted sensors to block in a fixed (random) state. This simulates the case of a short circuit in which the output connection touches the source voltage or the ground.

All these type of damages can affect randomly picked sensors (case $R$) or a contiguous set of sensors (case $A$). The aim is to represent both random faults and breakdowns due to external events. For instance, a robot passing near a strong heat source or crashing against an obstacle may be subject to localized damages. In this case, the affected sensors would likely be physically contiguous.

[2] The number of changes is randomly chosen in 1–6.

[3] We consider a run successful if the robot reaches a point in the arena at a distance less than or equal to a given threshold value $d_\theta$.

We tested the working conditions with 0, 3, 6, 9, ..., 24 faulty sensors. The faulty sensors are chosen at random at the beginning of each replica of the experiment. We collected results on 300 replicas for each type of damage.

## 5 Results

We expect to identify a graceful degradation of performance, where it decreases according to the number of faulted sensors. Overall, the results support this hypothesis, as we can observe from the barplots depicting the average final distance from light reached at the end of the 1200 epochs (see Fig. 2). This function estimates the capability of the robot to adapt to faults in the long run. Apart from some fluctuations due to variance in the experiments, the higher the number of faulty sensors, the higher the final distance from light. Notably, the performance decreases sensibly when the number of faulty sensors is fairly high.

The RLD is computed at a given target distance $d_\theta$ from the light and provides an estimation of the efficiency of the adaptive process to exploit the (limited) resources the robots can use: the steeper and the higher the curve, the better the performance at reaching for the first time a distance less than or equal to $d_\theta$ from the light. We computed the RLD with $d_\theta = 1\,\text{m}$ and $d_\theta = 5\,\text{cm}$. The plots are depicted in Fig. 3 and Fig. 4, respectively. Also in this case we observe a gradual decline in performance, as the number of defective sensors increases. Nevertheless, here a striking difference emerges between the cases $R$ and $A$: when contiguous sensors are damaged, the RLD does not decrease until a large fraction of sensors is touched. In some cases (*i* and *ii*) it seems even better to have few sensors out of order. A possible explanation of this phenomenon is that when contiguous sensors are damaged the remaining sensors are sufficient to provide the robot the necessary information to navigate the arena correctly. Less inputs also imply a search among a lower number of working configurations, therefore the adaptation process can proceed faster.

Finally, the distance from light at each epoch provides a complementary view of robot performance, as it captures the performance in time, taking also into account the performance achieved in the adaptation attempts (i.e. the adaptation epochs). The distance from light—averaged across the replicas for each epoch—is plotted in Fig. 5. We can observe a qualitative behavior analogous to the previous cases. It is remarkable that, also under this evaluation, the performance with few damaged sensors is still equivalent to—or even better than—the one in which all sensors are working. This observation reinforces our previous hypothesis, as it provides an online view of the impact of having less information. We conclude by observing that when all the 24 sensors do not work and just provide noise or fixed random values, the distance from light decreases anyway, as some lucky combinations can drive the robot towards the light even if it does not perceive it.

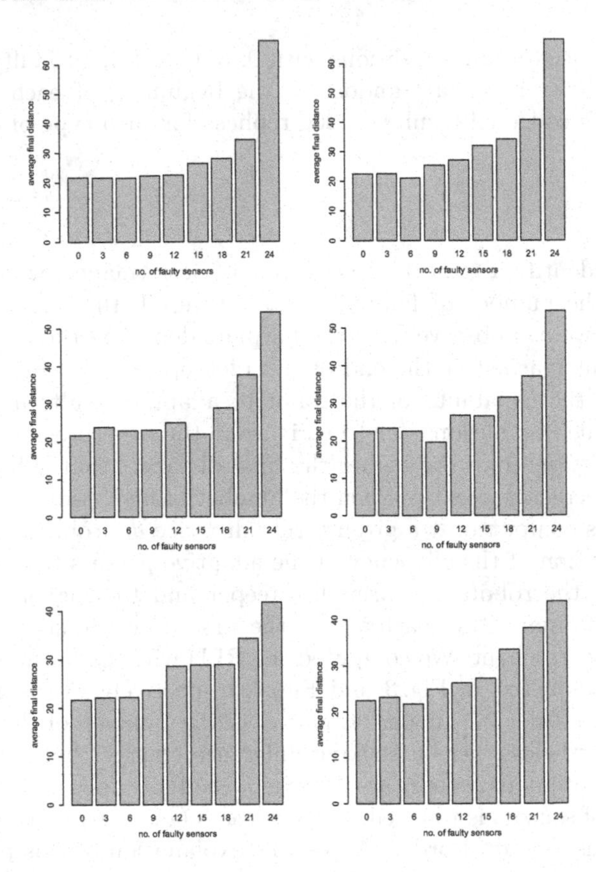


**Fig. 2.** Barplots of final distance from the light, averaged across 300 replicas. Columns: left, case $R$ – random picked sensors; right, case $A$ – contiguous faulty sensors. Rows corresponds to kinds of damage (from the top, $i$: detached sensor, $ii$: noisy sensor, $iii$: random fixed value).

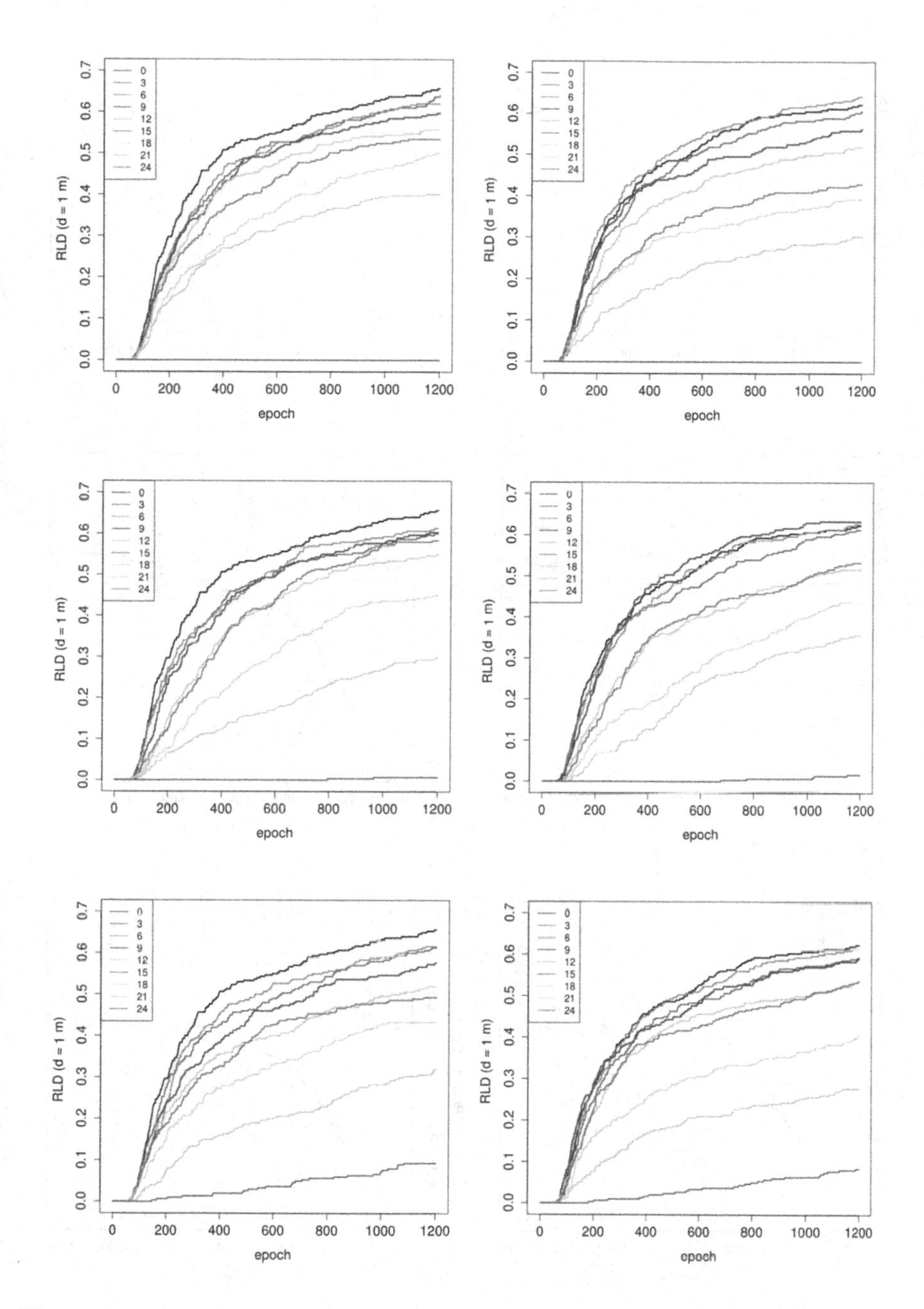


**Fig. 3.** Run length distribution for target distance equal to **1 m**. A point $(x, y)$ in the plot represents the fraction of replicas $(y)$ that achieved a distance less than or equal to 1 m at epoch $x$. Columns: left, case $R$ – random picked sensors; right, case $A$ – contiguous faulty sensors. Rows corresponds to kinds of damage (from the top, $i$: detached sensor, $ii$: noisy sensor, $iii$: random fixed value).

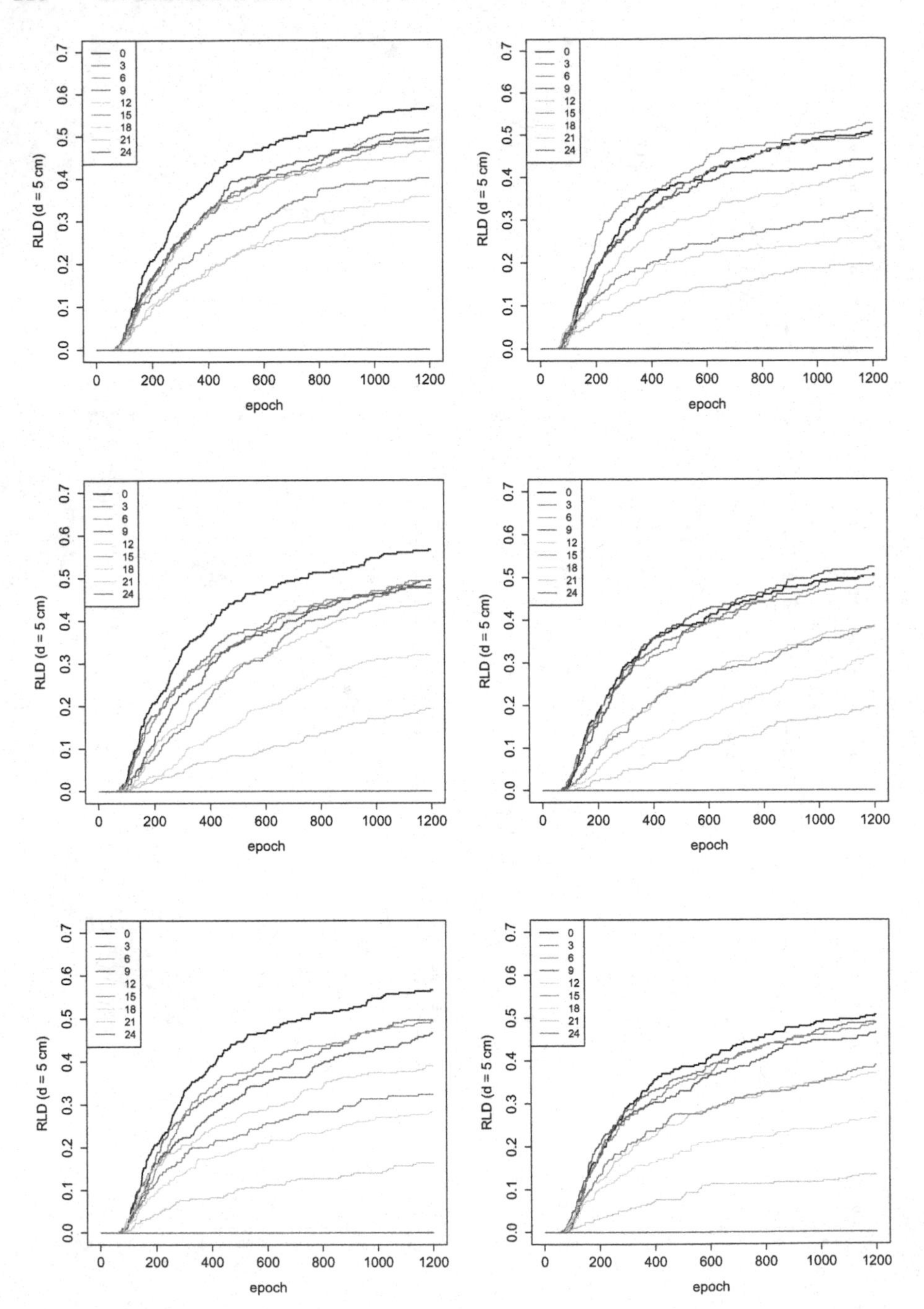


**Fig. 4.** Run length distribution for target distance equal to **5 cm**. A point $(x, y)$ in the plot represents the fraction of replicas $(y)$ that achieved a distance less than or equal to 5 cm at epoch $x$. Columns: left, case $R$ – random picked sensors; right, case $A$ – contiguous faulty sensors. Rows corresponds to kinds of damage (from the top, $i$: detached sensor, $ii$: noisy sensor, $iii$: random fixed value).

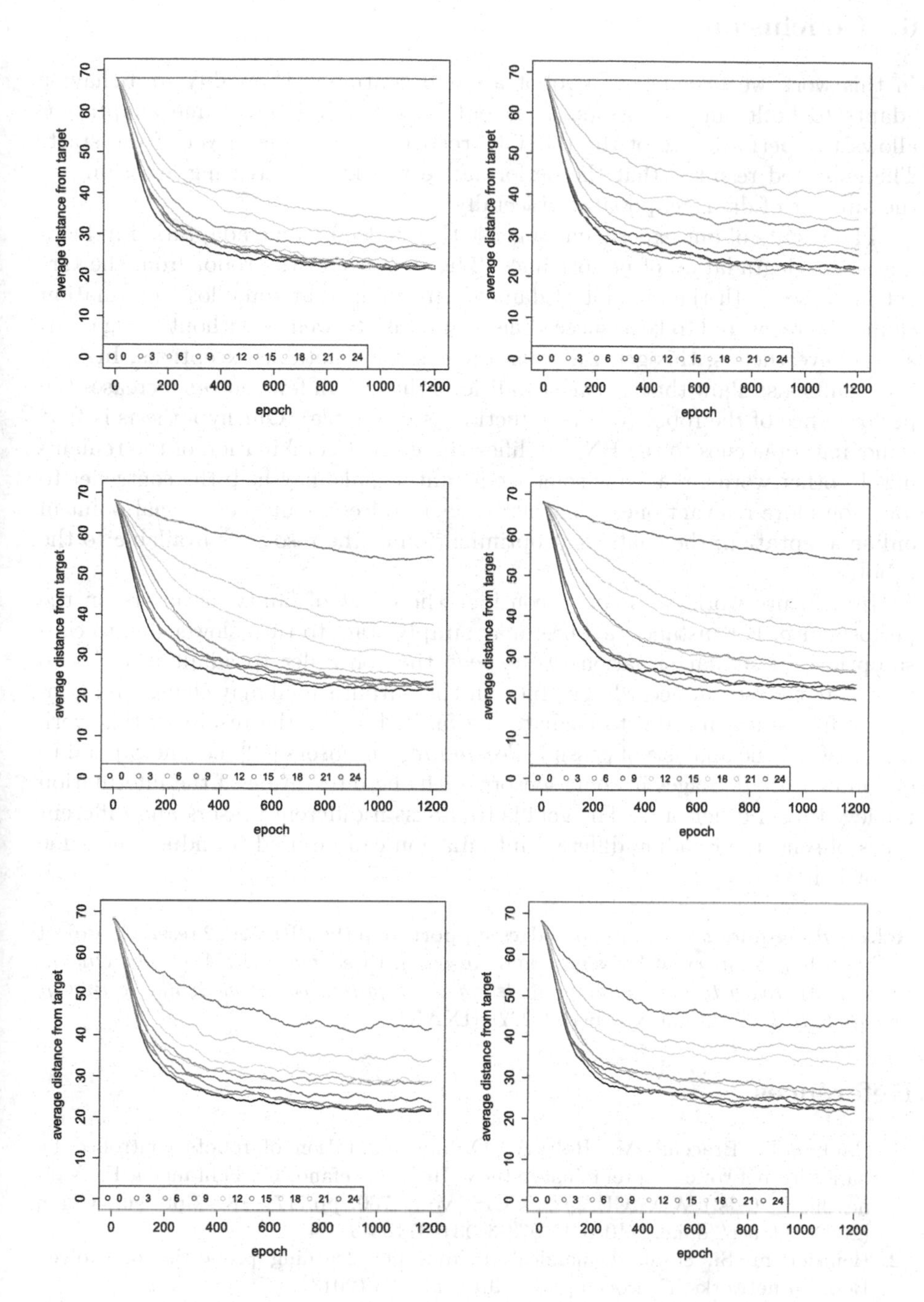


**Fig. 5.** Distance from the light along adaptation epochs, averaged across 300 replicas. Columns: left, case $R$ – random picked sensors; right, case $A$ – contiguous faulty sensors. Rows corresponds to kinds of damage (from the top, $i$: detached sensor, $ii$: noisy sensor, $iii$: random fixed value).

## 6 Conclusion

In this work we tested how a robot able to continuously modify its behavior adapts to faults in its sensors. The goal is to verify if an online adaptation allows the performance of the robot to recover to a similar level of the start. The expected result is that the performance will keep decreasing according to the amount of damage, possibly gracefully.

The results of our experiment suggest that indeed online adaptation induces a graceful degradation of performance. The distance of the robot from the target increases with the amount of damage, meaning that some lost information cannot be recovered to keep succeeding in the task as well as without being damaged. Nevertheless, this general situation has some notable exceptions. In fact, the results also show that in some conditions the lack of few sensors increases the performance of the robot (or the reduction is negligible). Our hypothesis is that removing some cues to the BN simplifies the correct combination of the remaining. In other words, removing some irrelevant signals may help the controller to focus on more relevant ones. This last aspect addresses another crucial point in online adaptation: the contingent optimization of the resources available to the robot.

In a future work, we plan to consider the effect of faulty actuators on the performance. For instance, a wheel may simply start to turn slower due to consumption. In similar situations we expect the controller to adapt in order to modulate the output accordingly. Indeed, the output signal may change in intensity of frequency in order to mediate the fault. Finally, the results of this work push toward the analysis of possible *degeneracy* of sensors [10], i.e. the capability of compensate damages in some sensors with the integration of the information retrieved from other ones. The goal is to assess if different sensors and different *types* of sensors providing different information can be used to induce the same robot behavior.

**Acknowledgements.** AR acknowledges support from the PRIN 2022 research project of the Italian Ministry of University and Research titled *Org(SB-EAI) – An Organizational Approach to the Synthetic Modeling of Cognition based on Synthetic Biology and Embodied AI* (Grant Number: 20222HHXAX).

## References

1. Baldini, P., Braccini, M., Roli, A.: Online adaptation of robots controlled by nanowire networks: a preliminary study. In: De Stefano, C., Fontanella, F., Vanneschi, L. (eds.) WIVACE 2022. CCIS, vol. 1780, pp. 171–182. Springer, Cham (2022). https://doi.org/10.1007/978-3-031-31183-3_14
2. Benedettini, S., et al.: Dynamical regimes and learning properties of evolved Boolean networks. Neurocomputing **99**, 111–123 (2013)
3. Bonani, M., et al.: The marXbot, a miniature mobile robot opening new perspectives for the collective-robotic research. In: 2010 IEEE/RSJ International Conference on Intelligent Robots and Systems, Taipei, Taiwan, 18–22 October 2010. IEEE (2010)

4. Bongard, J., Zykov, V., Lipson, H.: Resilient machines through continuous self-modeling. Science **314**(5802), 1118–1121 (2006)
5. Braccini, M., Roli, A., Barbieri, E., Kauffman, S.: On the criticality of adaptive Boolean network robots. Entropy **24**, 1368 (2022)
6. Braccini, M., Roli, A., Kauffman, S.: A novel online adaptation mechanism in artificial systems provides phenotypic plasticity. In: Schneider, J.J., Weyland, M.S., Flumini, D., Füchslin, R.M. (eds.) WIVACE 2021. CCIS, vol. 1722, pp. 121–132. Springer, Cham (2021). https://doi.org/10.1007/978-3-031-23929-8_12
7. Braccini, M., Roli, A., Villani, M., Montagna, S., Serra, R.: A simplified model of chromatin dynamics drives differentiation process in Boolean models of GRN. In: Artificial Life Conference Proceedings. MIT Press, Cambridge (2019)
8. Braccini, M., Roli, A., Villani, M., Serra, R.: Dynamical properties and path dependence in a gene-network model of cell differentiation. Soft. Comput. **25**(9), 6775–6787 (2021)
9. Dorigo, M., et al.: Evolving self-organizing behaviors for a swarm-bot. Auton. Robots **17**(2), 223–245 (2004)
10. Edelman, G., Gally, J.: Degeneracy and complexity in biological systems. PNAS **98**(24), 13763–13768 (2001)
11. Floreano, D., Mattiussi, C.: Bio-Inspired Artificial Intelligence: Theories, Methods, and Technologies. MIT Press, Cambridge (2008)
12. Huang, S., Eichler, G., Bar-Yam, Y., Ingber, D.: Cell fates as high-dimensional attractor states of a complex gene regulatory network. Phys. Rev. Lett. **94**(12), 128701 (2005)
13. Huang, S., Ernberg, I., Kauffman, S.: Cancer attractors: a systems view of tumors from a gene network dynamics and developmental perspective. Semin. Cell Dev. Biol. **20**(7), 869–876 (2009)
14. Kauffman, S.A.: Metabolic stability and epigenesis in randomly constructed genetic nets. J. Theor. Biol. **22**(3), 437–467 (1969)
15. Luque, B., Solé, R.: Phase transitions in random networks: simple analytic determination of critical points. Phys. Rev. E **55**(1), 257 (1997)
16. Montagna, S., Braccini, M., Roli, A.: The impact of self-loops on Boolean networks attractor landscape and implications for cell differentiation modelling. IEEE/ACM Trans. Comput. Biol. Bioinform. **18**(6), 2702–2713 (2020)
17. Nolfi, S., Floreano, D.: Evolutionary Robotics: The Biology, Intelligence, and Technology of Self-Organizing Machines. MIT Press, Cambridge (2000)
18. Pfeifer, R., Scheier, C.: Understanding Intelligence. The MIT Press, Cambridge (2001)
19. Pfeifer, R., Lungarella, M., Iida, F.: Self-organization, embodiment, and biologically inspired robotics. Science **318**(5853), 1088–1093 (2007)
20. Pinciroli, C., et al.: ARGoS: a modular, multi-engine simulator for heterogeneous swarm robotics. Swarm Intelligence (2012)
21. Roli, A., Villani, M., Filisetti, A., Serra, R.: Dynamical criticality: overview and open questions. J. Syst. Sci. Complex. **31**(3), 647–663 (2018)
22. Roli, A., Manfroni, M., Pinciroli, C., Birattari, M.: On the design of Boolean network robots. In: Di Chio, C., et al. (eds.) EvoApplications 2011. LNCS, vol. 6624, pp. 43–52. Springer, Cham (2011). https://doi.org/10.1007/978-3-642-20525-5_5
23. Serra, R., Villani, M., Graudenzi, A., Kauffman, S.: Why a simple model of genetic regulatory networks describes the distribution of avalanches in gene expression data. J. Theor. Biol. **246**(3), 449–460 (2007)
24. Villani, M., Barbieri, A., Serra, R.: A dynamical model of genetic networks for cell differentiation. PLoS ONE **6**(3), e17703 (2011)

# Blockchain-Empowered PSO for Scalable Swarm Robotics

Franco Cicirelli, Emilio Greco, Antonio Guerrieri, Antonio Francesco Gentile, Giandomenico Spezzano(✉), and Andrea Vinci

ICAR-CNR, Via P. Bucci 8/9C, 87036 Rende, Italy
{franco.cicirelli,emilio.greco,antonio.guerrieri,antoniofrancesco.gentile, giandomenico.spezzano,andrea.vinci}@icar.cnr.it

**Abstract.** Swarm robotics is an innovative field that utilizes collective behavior principles to design systems where multiple robots coordinate through simple rules and interactions. It faces the challenges of decentralized governance, security, and scalability. Due to its decentralized optimization capabilities, Particle Swarm Optimization (PSO) has shown promise for controlling robot swarms. However, implementing PSO in a distributed manner still poses problems in achieving full scalability and fault-tolerant operation. Blockchain, a decentralized system that securely stores and distributes data, enables transparent and autonomous communication among robots. Integrating blockchain with PSO can potentially revolutionize swarm robotics by providing secure and decentralized coordination through Decentralized applications (Dapps). The work proposed here demonstrates the application of blockchain technology, utilizing adhoc techniques, to manage a swarm of robots in conjunction with particle swarm optimization for solving navigation paths. In particular, the emergent Tendermint platform is exploited as a lean blockchain infrastructure for supporting asynchronous swarm robotics applications by showing its main advantages compared to a more traditional blockchain platform.

**Keywords:** Blockchain · swarm robotics · swarm intelligence · Particle Swarm Optimization (PSO) · Tendermint

## 1 Introduction

Swarm robotics harnesses collective problem-solving capabilities to achieve efficient results. Inspired by natural systems, multiple robots work together through local communication and sensing, resulting in desired collective behavior [1]. Utilizing a cooperative algorithm, this decentralized approach eliminates the need for external centralized control. Autonomous and scalable robots maximize performance, with the ability to add new units easily. The swarm remains robust in the presence of individual robot failure. The decentralized and flexible nature ensures continued coordination.


M. Villani et al. (Eds.): WIVACE 2023, CCIS 1977, pp. 214–227, 2024.
https://doi.org/10.1007/978-3-031-57430-6_17

Swarm robotics finds applications in miniaturization tasks like nanorobotics, distributed sensing, surveillance, and environment monitoring [2]. It helps to investigate environmental parameters, search and rescue, and detect hazards such as spills, pollution, and radioactivity. Swarm robots can cover broad regions, making them suitable for dispersed monitoring and large-scale agricultural applications. They also excel in solving problems in IoT systems. However, transitioning swarm robotics from academia to real-world scenarios poses challenges. Consensus is achieved through local communications without global knowledge, necessitating distributed decision-making algorithms. Furthermore, flawed or malicious agents, called Byzantine robots, pose risks to objectives and security [3]. Spanish engineer Eduardo Castelló [4] combines robotics and blockchain to address these challenges. He aims to enable simple machines to collaborate, exchange information, and contribute to knowledge while ensuring security and consensus. Integrating blockchain technology into swarm robotics enhances data security, enables cross-referencing, and accelerates task execution [5]. Castelló's study explores the feasibility of using blockchain-based swarm intelligence. By leveraging blockchain, swarm robotics participants share a common worldview, facilitating consensus without a centralized authority. Blockchain provides a shared, secure, immutable ledger for transactions, tracking assets, and establishing trust. It eliminates the need for intermediaries. Dapps [6] are distributed applications that run on a blockchain network of computers, establish secure swarm coordination mechanisms, and exclude Byzantine members. The goal is a safe and reliable consensus among all participants.

Our research uses PSO combined with a blockchain network to coordinate multiple robots as a swarm, allowing them to reach a desired solution through a decentralized and self-organizing approach. The paper's original contribution is twofold: (I) it exploits a lightweight blockchain layer developed on top of the emergent Tendermint platform; (ii) it proposes a decentralized implementation of the asynchronous PSO compliant with a blockchain environment. As a beneficial side-effect, by leveraging this decentralized implementation and blockchain, robots can dynamically join or leave the application without affecting the system's evolution, thus opening dynamic scalability. Specifically, in our implementation, each robot is treated as a particle in the PSO algorithm. This allows for local best updates of the velocity and position of each robot without relying on a central authority, and global best values updates are not broadcast but are shared. The exchange of information and the velocity modification is carried out by each particle, based on their own experiences and those of the other particles, by using the blockchain. To validate the whole approach, a PSO scenario is considered in which robots in a swarm have to search for some targets deployed in a bi-dimensional space.

The paper is structured as follows. Section 2 provides an overview of the existing PSO frameworks for swarm robotics, while Sect. 3 offers a description of blockchain technology and introduces the Tendermint platform. Section 4 discusses the implementation of blockchain PSO on Tendermint and presents the experimental setup and results. Finally, the conclusions end the paper by summarizing this study's essential findings and contributions.

## 2 PSO in Swarm Robotics

PSO, developed by Kennedy and Eberhart in 1995 [7], is an optimization technique inspired by bird flocking and fish schooling. From this perspective, in the PSO, movement rules are assigned to agents (particles), leading them to explore the assigned spatial domain to search target points. These points minimize a certain objective function. During the exploration, each agent can also share information with the rest of the swarm. Therefore, given the optimization problem of a function $f$ of $n$ variables on a set or feasible region $X$, the $f : \mathbb{R}^n \to \mathbb{R}$ is called the objective function while each point $x = (x_1, x_2, \ldots, x_n) \in X$ constitutes a feasible solution. The problem therefore consists in determining a point $x^* \in X$, which makes the function $f$ minimal:

$$x^* : f(x^*) = \min_{x \in X} f(x) \tag{1}$$

As an example, let's consider a two-dimensional $XY$ space. For each agent $i$, the position of the agent at a given instant $t$ is given by the coordinates $(x_i(t), y_i(t))$, with $i \in \{1, 2, \ldots, M\}$, where $M$ is the number of agents in the swarm. By indicating with $(x_{target}, y_{target})$ the position to be reached, it is possible to evaluate the current distance of the agent $i$ to the target as:

$$d = f(x_i(t), y_i(t)) = \sqrt{(x_i(t) - x_{\text{target}})^2 + (y_i(t) - y_{\text{target}})^2} \tag{2}$$

To reach the desired point and, therefore, ensure that the distance described above is finally equal to zero, the movement of each agent $i$ can be determined by taking into account two factors:

- the ability to remember the own local current best *personal* position so far achieved, which is indicated with the coordinates $pBest = (pBestX, pBestY)$;
- the current *global* best position, represented with the coordinates $gBest = (gBestX, gBestY)$, evaluated by considering the whole swarm.

The particles' positions are updated based on a weighted sum of vectors determined by equations involving their current positions, the best-found personal positions (*pBest*), and the global best-found position (*gBest*) among all particles. In this paper, the Heppner model [17] is adopted, which considers that each agent always moves on time-step bases, i.e., the time is discrete, and a certain random quantity, i.e., a random perturbation, is considered during movements.

More formally, let $X$ be the feasible region belonging to the continuous $n$-dimensional space in which the particles can move, let $t$ be the current time step of the system, let $i \in \{1, 2, \ldots, M\}$ be the index of an agent in the system, and let $f : \mathbb{R}^n \to \mathbb{R}$ be the objective function to minimize. Three $n$-dimensional vectors can represent the status of each particle :

- $\mathbf{x}_i(t) = (x_{i,1}(t), x_{i,2}(t), \ldots, x_{i,n}(t))$ indicating the position of the agent $i$ at time $t$

- $\mathbf{pBest}_i(t) = (pBest_{i,1}(t), pBest_{i,2}(t), \ldots, pBest_{i,n}(t))$, indicating the local best position of the agent $i$ at time $t$, where:

$$\mathbf{pBest}_i(t) = \arg\min_{x \in E_t^i} f(\mathbf{x})$$

and

$$E_t^i = \{\mathbf{x}_i(0), \mathbf{x}_i(1), \ldots, \mathbf{x}_i(t)\}$$

is the set of all the positions explored by the agent $i$ up to time $t$
- $\mathbf{v}_i(t) = (v_{i,1}(t), v_{i,2}(t), \ldots, v_{i,D}(t))$ indicating the velocity of the agent $i$ at time $t$.

A particle communicating with each other particle in the swarm is influenced by their own experience and, above all, by the swarm best position $gBest$ chosen among all the local best positions in the swarm:

$$\mathbf{gBest}(t) = \arg\min_{x \in \cup_{i=1}^{M} E_t^i} f(\mathbf{x})$$

To define the movement of particles, starting from the laws of dynamics, both **gBest** and $\mathbf{pBest}_i$ have to be used. In other words, the swarm is more than just a collection of particles; a particle by itself does not have enough capability to solve the whole problem, and progress relies on particle interaction.

In this paper, we adopted the following rules to determine the dynamics of a particle $i$:

$$\mathbf{v}_i(t+1) = \mathbf{v}_i(t) + c_1\mathbf{r}_1(t)(\mathbf{pBest}_i(t) - \mathbf{x}_i(t)) + c_2\mathbf{r}_2(t)(\mathbf{gBest}(t) - \mathbf{x}_i(t)) \quad (3)$$

$$\mathbf{x}_i(t+1) = \mathbf{x}_i(t) + \mathbf{v}_i(t)\Delta t = \mathbf{x}_i(t) + \mathbf{v}_i(t) \quad (4)$$

Where:

- $c_1$ and $c_2$ are positive constants, called acceleration coefficients, used for scaling the contribution of the $pBest$ and $gBest$ components, respectively;
- $r_{i,d}(t)$ is a vector of a random number between (0,1) for favoring space exploration.

In the original PSO algorithm, particle updates occur synchronously, with the entire swarm's fitness being evaluated before the particle update process takes place. Conversely, particles can update their velocities and position by asynchronous updates after the gBest is evaluated. This introduces the challenge of carrying out the particle's search with imperfect information. However, asynchronous updates proved advantageous in practical applications such as swarm robotics search problems, where robots can continuously move based on available information without waiting for the entire swarm to complete its current evaluation. In addition, asynchronous updates permit support changes in the number of robots in the system, opening to dynamic system behavior and scalability.

PSO resemblance to physical robotic swarms has led to its proposal as a control strategy for swarm robotic systems. Modified PSO algorithms and other swarm intelligence algorithms like Glowworm Swarm Optimization (GSO), Artificial Bee Colony Optimization (ABCO), Bacterial Foraging Optimization (BFO), the Firefly algorithm, and the Bees algorithm have been explored for obstacle avoidance and target localization tasks [8]. The distributed PSO (dPSO) implementation for robotic search applications presented in [9] stresses distributiveness, simplicity, and scalability concerning the number of robot particles. The concept behind the dPSO algorithm revolves around the possibility of each robot conducting measurements, updating its position and velocity, updating its personal best measurement, and personal best location (if needed). Additionally, if a particle discovers a global best measurement/position, it broadcasts this information to other robots. Consequently, each robot performs all velocity and position updates locally, while global best updates are shared through broadcasting. As a result, the amount of information exchanged between robots is minimized, thus enhancing the algorithm's scalability. The dPSO can behave as a parallel variant of the PSO. However, the primary objective of the dPSO is to reduce communication overhead between particles or robots and yield favorable results with complex search functions.

A parallel asynchronous PSO implementation is presented in the literature and was developed using the multi-agent approach [10]. The multi-agent approach can generally handle a problem by breaking it down into simpler subproblems so that agents have to focus only on sub-tasks of the general problem. The multi-agent approach is used to achieve parallelism and the simultaneous execution of calculations (possibly linked) to speed up the processing of intensive computer problems and perform many operations in a limited time. The presence of a central entity called *broker* can characterize the Multi-Agent Systems (MAS). All the system agents can communicate among themselves using the *broker* as a postman to the other agents of the group. Before interacting with the system, each agent must declare himself to the *broker* and provide him with his characteristics. The advantage of this centralized communication architecture is the ease of adding and removing an agent: for a new agent to integrate, it is sufficient to establish an interaction protocol with the *broker*. The downside is that centralization can slow down exchanges and communication in the system.

Instead of a *broker*, a MAS can use a distributed communication approach. In this case, each agent can interact with the other agents without an intermediary. Each element must maintain a knowledge base that describes the characteristics and addresses of the agents with whom it collaborates. When an agent wants to provide a new service, he must inform everyone else about this new service. The advantage of the distributed approach is that it facilitates exchanges and improves communication. The downside is that the new service provision requires updating each agent's knowledge base.

In [11], the paper presents a blockchain-powered PSO algorithm for swarm robotics. However, this specific implementation does not incorporate asynchronous functionalities. Our work stands out due to the utilization of Tendermint. Tendermint presents a distinctive perspective within the blockchain landscape. According to its vision, every application should operate on a dedicated blockchain, which sets it apart from Ethereum, where numerous applications coexist on a shared blockchain. Moreover, Tendermint employs a proof-of-stake consensus mechanism that promotes interoperability and is well-suited for the Internet of Things (IoT) environment. Consequently, Tendermint can reach thousands of transactions per second, representing a considerable improvement [14].

## 3 Blockchain and the Tendermint System

Blockchain technology enables a distributed database's creation, coordination, and synchronization. It consists of blocks containing transactions between network nodes, forming a chain. Transactions are validated by network nodes, establishing a distributed network of trust. Consensus mechanisms ensure agreement on the network's current state, guaranteeing fault tolerance and security.

In summary, blockchain is a secure, distributed, immutable data storage system shared among network actors. It maintains a public, auditable ledger of transactions and combines various fields such as cryptography, distributed systems, and finance. The technology allows the creation of Dapps that run on the blockchain, enabling user interaction with smart contracts. Dapps encompass a range of applications, including exchanges, loans, games, and payment terminals.

Tendermint [12] is a blockchain system that introduced a slashing variant of proof-of-stake consensus. It replaced Nakamoto Consensus [13], which used proof-of-work [15] or stake. Early concerns about proof-of-stake were addressed with validator deposits that could be slashed for Byzantine behavior. Tendermint's consensus algorithm is the Practical Byzantine Fault Tolerance [16] (PBFT), which provided the first optimal Byzantine fault-tolerant algorithm for practical use, built on top of an efficient gossip layer. It supports open-membership peer-to-peer networking and can be used in public or private settings.

Tendermint is a distributed message-passing system with dynamic nodes. Nodes communicate through encrypted point-to-point channels and rely on gossip for indirect communication. It tolerates benign and Byzantine faults, with a subset of nodes acting as validators. The set of validators is dynamic and known by all nodes. Tendermint assumes partial synchrony and relies on partial synchrony for progress.

Tendermint aims to be fast, flexible, and language-independent for large-scale, geographically distributed environments. It enables the development of open, public, and general-purpose blockchain applications. Tendermint is a significant component of the Cosmos Network project, facilitating communication between blockchains through sidechains. Tendermint's goals include deployment flexibility, scalability, Byzantine fault tolerance, language independence, and support for light clients. Tendermint proposes an application-based blockchain by separating the application layer from the consensus and networking layers. With Ethereum, the programmer usually codes a smart contract deployed and executed by the Ethereum Virtual Machine. With Tendermint, the programmer codes an application compiled into a process. Its architecture includes peer-to-peer communication, *mempool*, consensus, and application interface modules.

## 4 Blockchain-Based PSO for Swarm Robotics

Swarm robotic systems can utilize particle swarm optimization (PSO) as a control strategy. Each robot acts as a particle in PSO, updating its position, velocity, and personal best measurements. Swarm intelligence, facilitated by PSO, enables decentralized behavior and collective decision-making in robotic swarms.

Combining blockchain with robotic swarm systems offers improved security, autonomy, flexibility, and profitability. Blockchain allows agents to reach agreements and record them without a central authority. Swarm members can use blockchain to achieve consensus through voting, associating addresses with voting options. This ensures a quick, safe, and verifiable agreement among robots. The following section presents a blockchain-based PSO, bPSO, implemented as a Dapp using the Tendermint framework.

### 4.1 A Tendermint PSO Implementation

Our implementation of the PSO algorithm on the blockchain network is similar to the broker-based approach in the MAS system. The blockchain replaces the work of the broker. A new node that wants to be added to the chain must declare itself and use the formalism of the chain to which it subscribes; each node does not know the other nodes' number and locations and interacts only with the chain.

You can choose between asynchronous or synchronous versions when implementing the PSO algorithm. In the synchronous approach, after each optimization operation, each element in the swarm computes its local best, sends such a value to the blockchain, and then waits for all the other particles to execute the same operation on a synchronization barrier. After the synchronization barrier is satisfied, the particles ask the blockchain for the global best value and continue their work.

The asynchronous version, depicted in Fig. 1, aligns well with the utilization of a blockchain since the only information exchanged with the swarm is the value of the *gBest* calculated so far. To maintain a fully asynchronous behavior, each node, having improved its *gBest* value, communicates its *gBest* to the blockchain and continues its processing by utilizing the *gBest* stored in it. Further details of the asynchronous mode will be provided in the following.

### 4.2 Asynchronous bPSO Implementation in Tendermint

The most straightforward approach to implement the asynchronously distributed PSO in Tendermint is creating an external Dapp that uses the functions and data that the blockchain provides through the Application BlockChain Interface (ABCI) application, which is the way Tendermint provides to implement

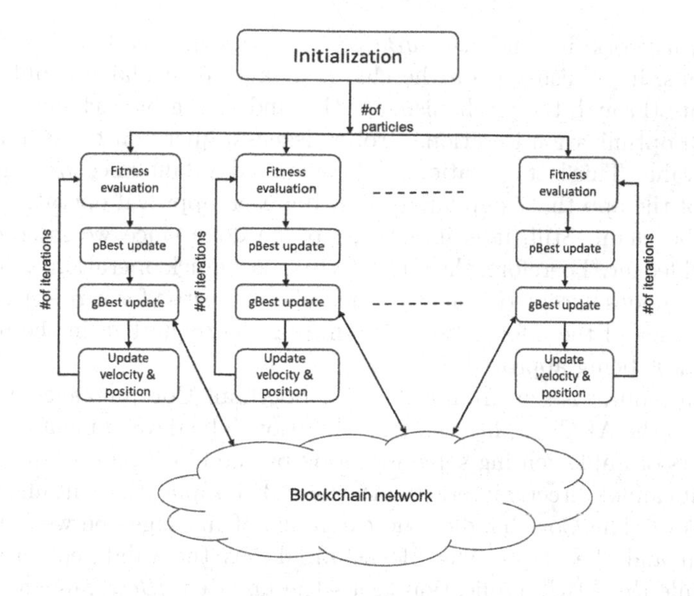


Fig. 1. Asynchronous PSO with blockchain.

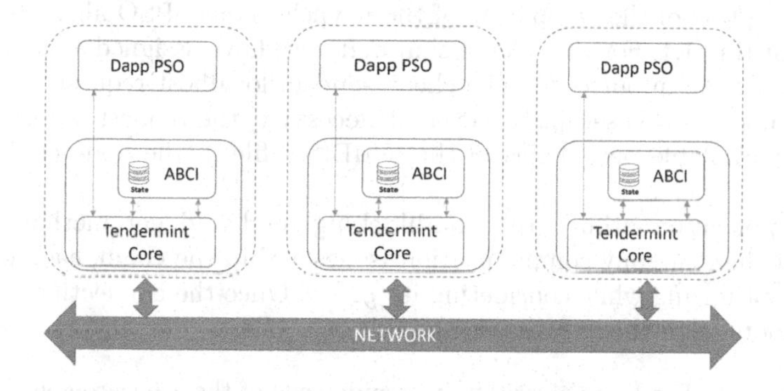


Fig. 2. Software Architecture.

custom applications. The Dapp software runs on each node in a decentralized architecture, as shown in Fig. 2. The Dapp that executes the PSO Java code will then communicate with the blockchain using the RPC (Remote Procedure Call) protocol on port 26657. At the same time, the ABCI will use the JTendermint library and then communicate with the Tendermint Core through port 26658 on four sockets (*Consensus connection*, *Mempool connection*, *Info connection*, *Snapshot connection*) predefined by the ABCI protocol.

The ad hoc blockchain architecture (Tendermint Core + ABCI) to receive and manage the requests of the Dapp PSO (for the execution of the asynchronous PSO algorithm) is shown in Fig. 2. The PSO Dapp will have to manage several robots that optimize the entire solution space. Each robot performs the search for the optimal solution in parallel. Whenever a robot residing on a node of the blockchain network finds a better *gBest* value than the one found so far by the other groups, it publishes it on the chain, making it available (public) for the other robots through the mechanisms of the underlying blockchain.

At each optimization iteration, a robot issues a query on the chain to verify the *gBest* value. This last operation might seem redundant since *gBest* represents the state of the app that is updated following each approval operation of a new block in the chain. Still, it is important to be done since we interact with a distributed ledger. Therefore, the query (which is a quick operation as it does not affect other nodes except the one to which the request is forwarded) guarantees the consistency of the information, suspending the request if, in the meantime, a new block is being approved.

When a request is sent from a node, Tendermint Core manages it by transmitting it to the ABCI application for validation. The development of the ABCI app consists of implementing some methods by the blockchain protocol.

A client cannot directly interact with the ABCI application but must use Tendermint Core. The Core handles the exchange of messages between the ABCI application and the Dapp. The blockchain keeps the validated data sent by Dapps, while the ABCI application keeps the updated *gBest* value at each iteration.

A description of the Dapp code of the asynchronous bPSO algorithm is provided in Listing 1.1. For the sake of simplicity, we have assigned a client to each Tendermint node in our code. The client sends a localhost request to its corresponding node in this setup. However, if necessary, the request is forwarded to one of the available peer nodes in the addBlock file in the node configuration folder.

Our application utilizes the non-blocking send and get methods. Consequently, if there are any communication issues within the chain, each robot will continue optimizing while considering its *gBest*. Once the connection is reestablished, robots align themselves with the chain's *gBest*.

**Listing 1.1.** The Dapp algorithm in pseudo-code of the asynchromous bPSO.

```
 Vector position; Vector velocity; // Current position and velocity. A random starting position is
     considered.
Vector pBestPos; Double pBestVal = MaxDouble; // Personal best solution.
Vector gBestPos; Double gBestVal = MaxDouble; // Global best solution.
while (target not found or max iterations not reached)
    value = evaluate_fitness(position); // evaluate the distance, using Eq. (2)
    if (value < pBestVal) // Update Personal best solution value and location if necessary
        pbestVal = value;
        pbestPos = position;
        if (pBestVal < gBestVal) // Update Global best solution value, if necessary
            sendgbest(pbestVal,pbestPos); // Send the new value of gBest into the blockchain
        end
    end
    result = null;
    result = get_gBestWithTimeout(); // Returns the gBest value and position calculated in the chain
          through a non-blocking call
    // Update gbest value and position based on the result obtained
    if (result == null) // Robot is offline
        gbestVal = pbestVal;
        gBestPos = pBestPos;
    else // Robot is online
        gbestVal = result.gbestVal;
```

```
        pBestPos = result.gBestPos;
    end
    // Move robot based on PSO rules
    velocity = updateParticleVelocity(); // update velocity, using Eq. (3)
    position = updateParticlePosition(); // update position, using Eq. (4)
End while
```

A description of the ABCI application is instead provided in Listing 1.2. The ABCI application performs data validation of each transaction through the CheckTx method. In this phase, the transition does not cause any state change (nothing is written in the database managed by the application). If the preliminary application determination is valid, Tendermint Core broadcasts and synchronizes the transaction to all network nodes configured for mining. At fixed time intervals, the network creates a new block with all the transactions validated during this time frame. Once a new block has been added to the blockchain, each node will locally execute all transactions included in the block, using the node's local ABCI application for processing. At this point, the ABCI application can update its database to store changes in the application state. A new node that you wanted to add to the network, carrying out all the transitions starting from the genesis block, would synchronize the data and state of the blockchain.

**Listing 1.2.** The ABCI algorithm in pseudo-code for the asynchronous bPSO.

```
// -------------- These methods are overridden and triggered by the Tendermint Core
    --------------
// When a block is committed in the blockchain, Tendermint makes a list of DeliverTx requests (one
    for each transaction),
//This is to check if a malicious proposer eventually put a wrong value in the transaction.
// At this stage, we also update the status (stgBestVal, stgBestPos) of our application.
@Override requestDeliverTx(pBestVal, pBestPos)
    //localMemory.stgBestVal is the stored best value obtained by each robot in the swarm.
    //localMemory.stgBestPos is the stored best position obtained by each robot in the swarm.
    code = validate(pBestVal, pBestPos); //check correctness of the received parameters
    if (code == 0) //if parameters are OK
        if(pBestVal < localMemory.stgBestVal) // update gBest value and position
            localMemory.stgBestVal = pBestVal;
            localMemory.stgBestPos = pBestPos;
    end
end
// -------------- These methods are triggered by Dapp client --------------
// requestCheckTx is triggered when a transaction is received by the Dapp.
// It receives the value and position of pBest sent by the Dapp using the sendgbest(pBestVal,
    pBestPos) method.
@Override requestCheckTx(pBestVal, pBestPos);
    code = validate(pBestVal, pBestPos); //check correctness of the received parameters
    if (code == 0)
        return OK;
    else
        return NOTOK;
    end
end
// It is triggered by the Dapp when calling the get_gBestWithTimeout() method without engaging
    consensus.
// It returns the value and position of gBest.
@Override Query(msg);
    if(msg == "request of gBest")
        return (localMemory.stgBestVal, localMemory.stgBestPos);
    end
end
```

### 4.3 Results and Evaluations

For validation purposes, the approach was applied to a simple robot-search simulation scenario. The goal is to describe the architecture setup on which the system is executed, along with some execution paths showing that the system properly evolves. Specifically, it was considered a deployment scenario composed of four Raspberry Pi4 connected through an Ethernet switch. Docker-based containerization [18] on Raspberry Pi devices with limited resources was used to deploy the application across the different nodes in the network. Each device was used for executing the code simulating the behavior of a robot and for implementing a blockchain network node. On each container, we have configured the Ubuntu operating system. The working environment was obtained by installing the following libraries: jtendermint0.32.3 [19] and openjdk-17-jdk [20]. In the container, we deployed the developed ABCI and Dapp applications. The containers communicate towards the outside through the communication ports 26657–26666 of the Tendermint Core protocol. In the experiment, we set the *httpURLConnectionTimeout* to 1 s for the non-blocking ABCI-Dapp communications. The constants $c_1$ and $c_2$ of Eq. 3 have been set to 2, while the number of iterations was set to 300. The simulation scenario refers to a 30 m × 30 m room with multiple target points. Ten different target points were chosen, as depicted in Fig. 3. These target points were strategically positioned throughout the room, some near the center and others near the edges. A variable number of robots, ranging from 2 to 10, was considered. During the simulation, the robots must reach the target point (one per simulation) assigned within the search space. The function of Eq. 2 is employed as the objective function, where $(x_i(t), y_i(t))$ is the position of the robot and $(x_{target}, y_{target})$ is the position of the target.

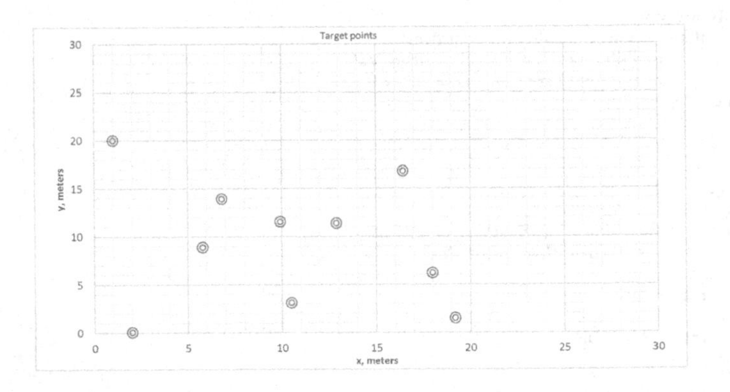


**Fig. 3.** Map of the room showing the ten different target points.

At $t = 0$, the robots were randomly deployed across the search space. Figure 4 and Fig. 5 showcase sample plots with three and four robots, respectively, illustrating the search progress of the *gBest* of the robots using bPSO to locate the target point (represented by a circle symbol). In both cases, a robot is positioned

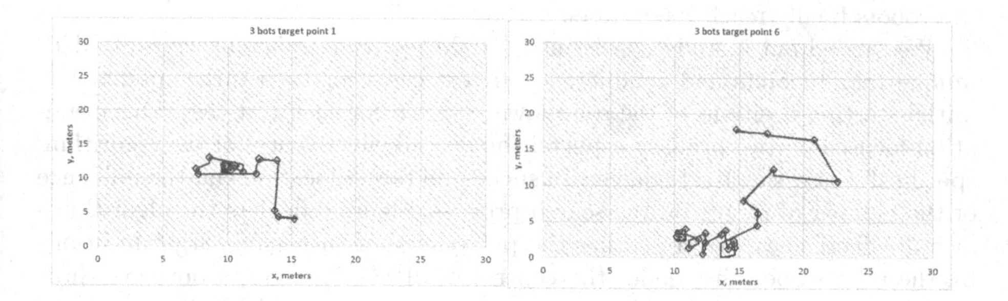


**Fig. 4.** Plot of the gBest of three robots using the bPSO algorithm.

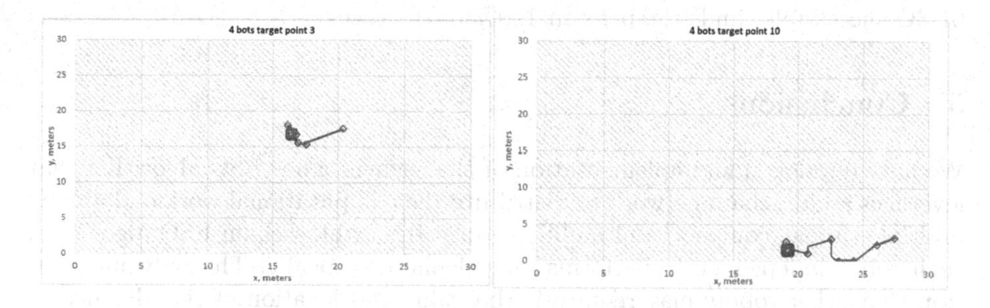


**Fig. 5.** Plot of the gBest of four robots using the bPSO algorithm.

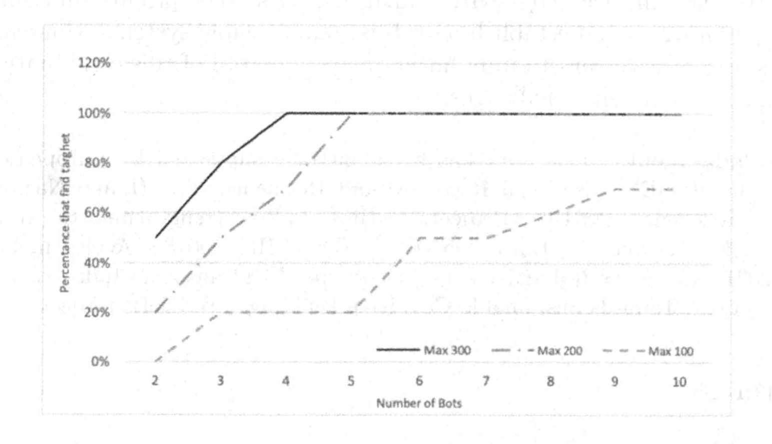


**Fig. 6.** Percentage of successful searches (out of 1000) for different numbers of robots.

at the center and another at the edge, serving as target points. In all the cases, the robots finally reach the targets.

We carried out a study consisting of 1000 test runs, wherein we utilized 10 unique target points and conducted 100 test cases for each target point. The initial starting positions of the robots were varied for each test case. Our evaluation focused on two primary aspects: the overall effectiveness of the algorithm, specifically assessing the frequency of successful target location and the influence of the number of robots on the search process. Figure 6 depicts the algorithm's overall effectiveness by presenting the percentage of successful target locations by the robots operating under the control of bPSO. The graph portrays three distinct lines: a gray dotted line representing a maximum of 100 iterations per robot, a red dash-dot line indicating the number of bots reaching the target within 200 iterations, and a solid black line representing the number of robots reaching the target within 300 iterations. The graph demonstrates that increasing the number of robots yields improved results. With five or more robots, the bPSO successfully finds the peak in 100% of the cases.

## 5 Conclusions

We have developed an implementation of the asynchronous PSO algorithm that leverages a blockchain network to distribute the computational workload among multiple simple, compact, and mobile robots. In our blockchain PSO algorithm, each robot independently performs all calculations locally. The only information that other robots may require is the value and location of the global best solution, referred to as *gBest*. Our algorithm is designed such that *gBest* is transmitted only when a robot discovers a value in the search space that surpasses the current *gBest*. As a result, communication between robots is minimal and occurs only when a significant point is found during the search process. Integrating blockchain technology with swarm robotics holds promise in addressing existing limitations and establishing robust and scalable systems. However, further research is necessary to fully harness the potential of this combination and overcome the remaining challenges.

**Acknowledgements.** This work has been partially supported by European Union - NextGenerationEU - National Recovery and Resilience Plan (Piano Nazionale di Ripresa e Resilienza, PNRR) - Project: "SoBigData.it - Strengthening the Italian RI for Social Mining and Big Data Analytics" - Prot. IR0000013 - Avviso n. 3264 del 28/12/2021 and by the Italy-CNR, "Le Scienze per le TRansizioni Industriale, Verde ed Energetica": Towards Sustainable Cognitive Buildings (ToSCoB) project.

## References

1. Khaldi, B., Cherif, F.: An overview of swarm robotics: swarm intelligence applied to multi-robotics. Int. J. Comput. Appl. 3–10 (2015)
2. Spezzano, G. (ed.): Swarm Robotics, Applied Sciences. MPDI Books, Basel (2019)

3. Castro, M., Liskov, B.: Practical Byzantine fault tolerance and proactive recovery. ACM Trans. Comput. Syst. (TOCS) **20**(4), 398–461 (2002)
4. Castelló, F.E.: The blockchain: a new framework for robotic swarm systems. arXiv (2016). http://arxiv.org/abs/1608.00695
5. Lopes, V., Alexandre, L.: An overview of blockchain integration with robotics and artificial intelligence. Ledger J. **4**, 1–6 (2019)
6. Cai, W., Wang, Z., Ernst, J.B., Hong, Z., Feng, C., Leung, V.C.M.: Decentralized applications: the blockchain-empowered software system. IEEE Access **6**, 53019–53033 (2018). https://doi.org/10.1109/ACCESS.2018.2870644
7. Kennedy, J., Eberhart, R.: Particle swarm optimization. In: Proceedings of ICNN 1995 - International Conference on Neural Networks, vol. 4, pp. 1942–1948 (1995). https://doi.org/10.1109/ICNN.1995.488968
8. Senanayake, M., Senthooran, I., Barca, J.C., Chung, H., Kamruzzaman, J., Murshed, M.: Search and tracking algorithms for swarms of robots: a survey. Robot. Auton. Syst. **75**, 422–434 (2016)
9. Hereford, J.M., Siebold, M., Nichols, S.: Using the particle swarm optimization algorithm for robotic search applications. In: 2007 IEEE Swarm Intelligence Symposium, pp. 53–59 (2007). https://doi.org/10.1109/SIS.2007.368026
10. Gazi, V., Ordonez, R.: Particle swarm optimization based distributed agreement in multi-agent dynamic systems. In: IEEE Symposium on Swarm Intelligence, Orlando, FL, USA, pp. 1–7 (2014). https://doi.org/10.1109/SIS.2014.7011792
11. Adeoti, T., Olasoji, Y.: Control and path planning of mobile swarm robots using blockchain technology with particle swarm optimization. Eur. J. Electr. Eng. Comput. Sci. **6**, 51–60 (2022). https://doi.org/10.24018/ejece.2022.6.4.456
12. Braithwaite, S., et al.: A tendermint light client. arXiv preprint arXiv:2010.07031 (2020)
13. Nakamoto, S.: Bitcoin: a peer-to-peer electronic cash system. Cryptography Mailing (2009). https://metzdowd.com
14. Ethan Buchman: Tendermint: Byzantine fault tolerance in the age of blockchains. Ph.D. thesis (2016). https://atrium.lib.uoguelph.ca/xmlui/handle/10214/9769
15. Vranken, H.: Sustainability of Bitcoin and blockchains. Curr. Opin. Environ. Sustain. **28**, 1–9 (2017)
16. Castro, M., Liskov, B., et al.: Practical Byzantine fault tolerance. In: Proceedings of the Third Symposium on Operating Systems Design and Implementation (1999)
17. Heppner, F.: A stochastic nonlinear model for coordinated bird flocks. The ubiquity of chaos (1990)
18. Docker. https://www.docker.io
19. Tendermint Core 0.32.3. https://github.com/zlyzol/tendermint-0.32.3.git
20. Java OpenJDK. https://jdk.java.net/archive/

# Hybrid GP/PSO Representation of 1-D Signals in an Autoencoder Fashion

Giulia Magnani, Monica Mordonini, and Stefano Cagnoni(✉)

Department of Engineering and Architecture, University of Parma, Parma, Italy
{giulia.magnani1,monica.mordonini,stefano.cagnoni}@unipr.it

**Abstract.** A popular low-dimensional latent representation that retains as much information as possible about a data set is the one represented by the output of a hidden layer located within a neural net having homomorphic input and output layers, termed autoencoder, trained to produce a copy of the input data as its output.

This exploratory paper suggests that reformulating the problem as a GP-based symbolic regression can achieve the same goal. The latent representation, in this case, is obtained as a byproduct of the solution to the problem of finding a parametric equation that represents a model of a family of signals (functions) that share the same equation, differing only for the values of a set of free parameters that appear in their definition.

This hypothesis is supported by a simple proof of concept based on the results of symbolic regression of a set of Gaussian functions. A discussion of possible issues that might need to be tackled when the method is applied to more complex real-world data and of the corresponding possible countermeasures concludes the paper.

**Keywords:** Latent Representations · Genetic Programming · Symbolic Regression · Machine Learning

## 1 Introduction

Deep Learning (DL) is undoubtedly one of the driving forces of the current Artificial Intelligence (AI) revolution. As soon as massive and often open data sets, along with moderately priced high-performance computing resources, were made widely available soon after the turn of the century, ideas and computation models that had been lingering in scientists' drawers for some decades could, at last, become concrete and effective Machine Learning (ML) applications. In most cases, they could outperform any previously developed approach by far and soon proposed themselves as "The solution" to many complex ML and Pattern Recognition (PR) problems. The general applicability of such methods and the possibility of quickly setting up scalable architectures that support them make it possible to reach limits that make the curse of dimensionality less scary.

At the same time, however, the velocity at which this process has progressed, along with the millions of users to whom such technologies have been made avail-


M. Villani et al. (Eds.): WIVACE 2023, CCIS 1977, pp. 228–238, 2024.
https://doi.org/10.1007/978-3-031-57430-6_18

able, has led to questioning the principle by which what limited many theoretically possible achievements in ML and AI was their intrinsic practical infeasibility, giving for granted that what was feasible was also sustainable. Especially after the development and wide availability of Large Language Models (LLMs), the present forecast on the diffusion of DL applications, and the near-future estimate of the consequent electric and computing power requirements urge us to consider sustainability a most relevant issue. However, even if we supposed all imaginable power to be available, data requirements would raise sustainability concerns. The applications for which DL has been most beneficial are certainly those related to multimedia technology: Image Analysis (IA) and Natural Language Processing (NLP). This does not come as a surprise if one thinks about the contemporary flourishing of the user-fed image and text databases on which social networks rely. This consideration raises a further issue since the requirement for vast amounts of data practically rules out a whole set of problems, of which medical applications are possibly the most frequent examples. For these problems, on the one hand, insufficient data are available to build appropriate models using DL. On the other hand, data augmentation methods are at high risk of generating unreal cases.

Finally, one of the main and most often cited drawbacks of DL is the lack of explainability and interpretability of the generated solutions, often characterized by millions (when not billions) of parameters.

The above considerations show that DL cannot be considered a universal solution despite its superior performance and general applicability. This calls for the study and development of alternative options that can trade a reasonable decline in performance for computationally lighter, less data-demanding, and more interpretable solutions.

Among Evolutionary Computation (EC) methods [4], Genetic Programming [12] has been proposed as a possible candidate for substituting DL effectively. Considering the typical structure of a deep network, represented by a set of cascaded layers, it is possible to group layers such that each group represents one of a set of cascaded functions, each playing a specific role. This is, for example, the classical case of image classifiers based on Convolutional Neural Networks (CNNs), in which a first set of convolutional and pooling layers acts as a feature extractor, i.e., a function having an image as input and a set of features as output, followed by a second set of dense layers that act as a classification function, i.e., a function having a set of features as input and a label or a probability distribution as output. Since GP's primary goal is evolving symbolic representations of functions, the above schematization justifies the choice of GP as a substitute for DL. The following considerations further support this choice:

- It has often been demonstrated that GP solutions can be evolved based on a limited set of "training" examples.
- A symbolic representation of a function composed of a set of basic high-level operators is more informative and interpretable than the sparse representation of deep neural nets, comprising thousands of instances of the same elementary, threshold-like operator and millions of associated parameters. This observa-

tion could be summarized as follows: "While GP-based solutions are encoded by the structure of the function that represents them, DL-based solutions are encoded by the function parameters." Such a distinction, therefore, indicates that GP-based solutions are more accessible to interpret than DL-based ones.

GP's capacity to evolve functions at their symbolic representation level fits very well the current request for explainable Artificial Intelligence (XAI) methods [2,9], primarily when the decision they take affects human beings, as stated by the EU in the recently proposed AI Act [1].

### 1.1 Genetic Programming and Latent Data Representations

Therefore, aiming to substitute DL feature extraction layers, GP has been extensively used for finding low-dimensional non-linear representations of images or high-dimensional data, as recently described in [5]. As mentioned above, in its simplest and most general structure, a classification architecture based either on classical pattern recognition methods or DL can be modeled as a pipeline composed of a feature extractor that computes a new representation of the input and a cascaded classifier whose performance is optimized by having such a representation as its input.

Autoencoders [3] are a common and powerful way to obtain such a latent intermediate representation in a self-supervised way, using neural architectures having a single hidden layer up to very deep neural networks. GP has been used to develop autoencoders or autoencoder-like structures, as described, for example, in [7] or in [13], where a GP-evolved multi-tree function representation substitutes the encoder.

This paper proposes a method of generating latent representations of time-dependent signals, somehow inspired by autoencoders, based on the concurrent use of GP-based symbolic regression and Particle Swarm Optimization (PSO) [11] of the free parameters of the symbolic function thus obtained.

In the following chapters, we will first describe the proposed approach, starting from its inspiring principles taken from symbolic regression. We will then show that modeling families of signals by functions characterized by a sufficient number of free parameters can help reconstruct possible instances of the signal with high precision, using the parameter values that optimize their reconstruction as their latent representation. The paper will then describe some preliminary results obtained, as a proof of concept, on a set of simple, noiseless Gaussian functions, suggesting possible future applications focused on medical signals.

### 1.2 Autoencoders

In their most typical realization, autoencoders are deep or shallow feedforward neural nets having equally-sized input and output layers (I, O) and at least one lower-dimensional (with respect to the input/output layers) hidden layer H. They are trained to reproduce, as their output, the same patterns they receive in input. As a result, the output of H can be seen as a lower-dimensional encoding

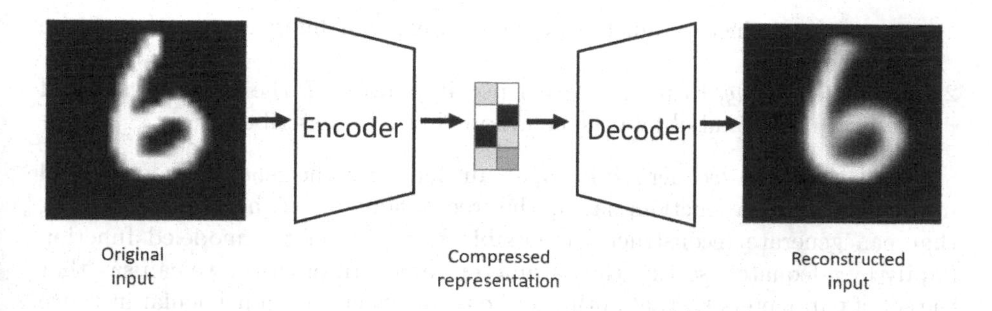


**Fig. 1.** Autoencoder for image representation (from [3])

of the input, from which the input pattern can be recovered by the section of the net comprised between H and O ([H:O]). Because of this, the network section [I:H] is termed *encoder*, while the section [H:O] is termed *decoder*.

In pattern recognition, autoencoders can synthesize large-dimensional inputs without losing information content, generating easier-to-handle compact representations of the input data that often lead to better performances when training classifiers.

Also, autoencoders have a crucial role in generative approaches since the mapping, produced by the decoder, from the latent representation to the data domain is such that appropriately feeding it with random tuples generates a process that samples the same statistical distribution of the data by which it was trained. This means it creates new patterns belonging to the same *family* as the input data, finding obvious applications in data augmentation and knowledge transfer (Fig. 1).

## 2 GP2SO: Symbolic Regression-Based Representation of Time-Dependent Signals

Unlike the sub-symbolic approach of Neural Network (NN)-based regression, symbolic regression consists of finding a symbolic representation of an input/output mapping, i.e., finding the expression of a function that best approximates a sampling of the unknown mapping.

In most cases, symbolic regression is applied to a single mapping, e.g., a law expressing a causal relationship between the measures of two physical quantities based on a set of observations, as is typical of the experimental method or inductive learning.

Suppose, instead, that one wants to model different instances of the same phenomenon (e.g., several Gaussian functions) that share a common model and expression but can be differentiated as the shift/scaling/deformation of that basic pattern. In that case, we can divide the process into two steps:

1. *Symbolic regression*, to find the expression of a common model $\hat{f}(t, \mathbf{K})$, that represents the common dependency of all instances on time $t$, but also includes

some undetermined parameters $\mathbf{K}$ whose values can differentiate each instance from the others.
2. *Parameter fitting*, to find, for each possible instance, the optimal values of the parameters and obtain the equation that fully describes it.

Similarly to the decoder section of an autoencoder, the general model evolved in the first step is a function, having the free parameters of the model as inputs, that can generate/reconstruct all possible instances of the modeled function family by adequately setting the parameter values. In practice, we can say that the set of parameters $\mathbf{K}_i$ that optimize the reconstruction of a particular instance $I_i$ is a latent, lower-dimensional representation of $I_i$, i.e., $I_i = \hat{f}(t, \mathbf{K}_i)$.

The expected advantages of using an evolved expression instead of an autoencoder are the following:

1. In the ideal case of noiseless signals, a single instance of the function family could be a sufficient training set for modeling any other instance. Suppose the unknown functions we are modeling are Gaussian; if the model evolved by GP is correct (an exponential function with two free parameters: mean and standard deviation), then one only needs to determine the values of the two parameters to characterize all other possible Gaussian functions.
2. GP evolves directly interpretable expressions, provided the model is not too complex.
3. The concurrent use of PSO for parameter optimization compensates for the problems that GP often finds when it has to evolve functions containing constants that need to be precisely set, as discussed in [14].

In Sect. 4, we describe how we used our approach (GP2SO) to evolve, as a proof of concept, a model representing all possible Gaussian functions using only the sampling of three specific Gaussians to derive it.

## 3 Implementation

GP2SO has been implemented using the Python package DEAP (Distributed Evolutionary Algorithms in Python) [6], a popular tool that provides classes and easily customizable code for both GP and PSO.

### 3.1 Function and Terminal Set

We have used traditional tree-based GP, with an essential function set $\mathcal{F}$, including the four arithmetic operators (with protected division), protected exponential, sine, and cosine.

The terminal set $\mathcal{T}$ is peculiar to our method. Besides the independent variable (e.g., time, if the function represents a signal) and some ephemeral random constants, as usual, it also includes the free parameters that PSO is then expected to set for fitting the different instances, both during training and in computing the embeddings when the model is applied to new data.

### 3.2 Fitness Function

The fitness function of the GP optimizer is applied to a GP tree $T$ representing the model $\hat{f}(t, \mathbf{K})$ and does nothing but pass $\hat{f}$ to the PSO, whose duty is to find the optimal values for the free parameters $\mathbf{K}$, i.e., the values that minimize the sum of squared differences between $\hat{f}$ and the target function to be approximated. Such a value will be returned to the GP fitness function and assigned to $T$. Alternative fitness functions, such as the cosine distance, could also be considered.

## 4 Proof of Concept

We generated a training set including the sampling of three Gaussian functions $G(x) = \frac{1}{\sqrt{2\pi}\sigma} \cdot e^{-\frac{(x-\mu)^2}{2\sigma^2}}$ having $\mu = \{0, -3, 4\}$ and $\sigma = \{4, 1, 2\}$, respectively, sampled in the interval $x = [-10, 10)$ with a step of 0.01, obtaining 200 samples per function, thus a total of 600 samples in the whole training set.

GP was run with a population of 300 individuals, crossover probability of 0.6, mutation probability of 0.2, and tournament selection, with tournament size equal to 7. PSO relied on a swarm of 50 particles, run for 150 iterations for each fitness evaluation of a GP tree, with the inertia coefficient $w$ set to 0.7 and local and global attraction coefficients $c_1$ and $c_2$ set to 1.7.

Although the exponential function was included in $\mathcal{F}$, and the direct expression of a Gaussian function was within the GP search domain, the rather complex structure of the exponent and the normalization factor of the Gaussian made it simpler for GP to evolve the following model:

```
pDiv(k2, add(k1, mul(add(x, k4), mul(add(x, k4), add(k3, mul(add(x, k4),
mul(add(x, k4), add(k5, mul(add(x, k4), mul(add(x, k4), mul(add(x, k4),
mul(add(x, k4), k6))))))))))))
```

that represents the polynomial model:

$$\hat{G}(x, k_i, i = 1 \ldots 6) = \frac{k_2}{(k_1 + k_3(x + k_4)^2 + k_5(x + k_4)^4 + k_6(x + k_4)^8)}$$

To verify the generality of the evolved model, we tested its capacity to reconstruct unseen instances of Gaussian functions generating 300 random pairs $(\mu, \sigma)$, $\mu \in [-7, 7], \sigma \in [0.5, 4]$, corresponding to as many test instances. All of them were sampled within the same domain $[-10, 10)$ as the training instances.

We then fit the model to each of the test instances by computing their embedding, i.e., by invoking PSO to find the optimal values of $\mathbf{K} = \{k_1, k_2, ..., k_6\}$ that minimize the mean squared error of the reconstruction. To ensure the best possible parameter set could be found, exploiting the fast convergence properties of the algorithm, PSO was run five times for each test function, and the best of the five parameter sets was finally used.

Results demonstrated that the model we found was virtually equivalent to the original Gaussian function, the maximum total squared error over the 200 samples being lower than $10^{-4}$ in 286 cases out of 300 (95.34%), the worst results being 0.101 and 0.0091.

After evaluating regression quality, we tried to analyze how consistent each parameter set used to embed a Gaussian function was with its natural characterization as a (mean, standard deviation) pair. We first projected the 6-dimensional embedding obtained for each function in two dimensions using tSNE [8]. We then used the two representations to create two 2-D plots in each of which each point represents a function; the x and y coordinates are from one of the two representations (tSNE parameters or mean/standard deviation), while the other representation is used to encode color. In particular, in Fig. 2, in the upper graph, the mean was encoded as the Red color component, and the standard deviation was encoded as the Blue component, while, in the lower graph, the Red and Blue color components encode the two tSNE coordinates.

In the two plots, one can observe a strong correlation between color and position (or, equivalently, between color difference and distance), which suggests that the embedding obtained through the regression can preserve the topological properties of the original representation, i.e., that the representations of two function instances that are similar in the mean-standard deviation space are also similar in the embedding space.

## 5 Possible Applications, Preliminary Tests, and Open Problems

Our very preliminary tests show that the hybrid GP/PSO approach can find a model that parametrically represents a family of curves/functions, thanks to GP's symbolic regression capabilities. The application of PSO to the model thus derived can transform each curve instance into a vectorial embedding, represented by the set of parameter values that minimize the error in recovering the original signal based on the GP model.

This encoding method is functionally and semantically similar to using neural network-based autoencoders since it allows one to evolve an encoder/decoder pair in an unsupervised way. The encoder is obtained by GP-based symbolic regression of only a few instances of the functions of interest. The decoder is obtained by applying PSO to the free parameters of the GP-evolved general model. Therefore, we are considering an "asymmetric" hybrid encoding/decoding scheme with the same applications as neural network-based autoencoders. Among these, we can highlight:

- Finding a low-dimensional representation of a family of functions (as the Gaussians used in the example) or finding the model underlying a physical phenomenon whose structure is represented as a symbolic equation, while the peculiarities that allow one to distinguish one instance of the phenomenon from another are encoded as a vector of real numbers.
- Using the evolved model to implement generative approaches where a random generation of parameter vectors from a proper distribution can produce new

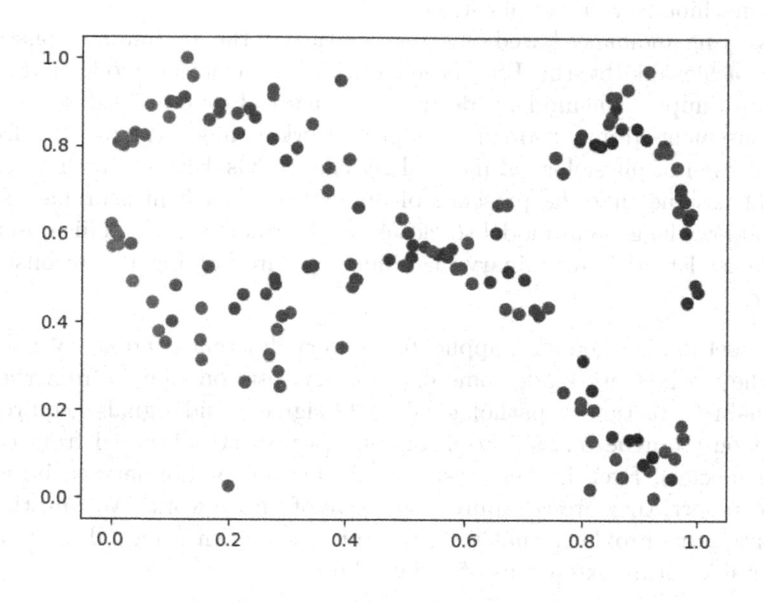


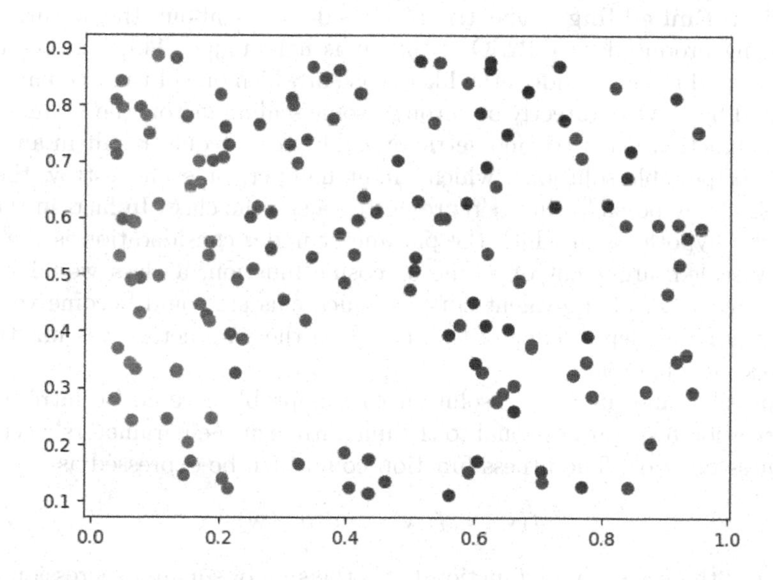


**Fig. 2.** Above: 2-D projection of the six-dimensional embedding of 200 Gaussian functions using 2-component tSNE (color encoding: Red component = mean; Blue component = standard deviation). Below: mean (x-axis) vs. standard deviation (y-axis) representation of 200 Gaussians (color encoding: Red component = first tSNE coordinate; Blue component = second tSNE coordinate)

synthetic data that could be used, for instance, to augment a training dataset in a machine learning application.
- Classifying anomalies based on the magnitude of the minimum reconstruction error achievable by the PSO-based optimization of the model parameters' vector. Suppose the model is derived from a set of time series describing, e.g., measurements from sensors in a properly working industrial plant or from the acquisition of physiological medical signals (ECGs/EEGs). In that case, one might assume that the presence of anomalies in such measurements would not follow the general model so closely as "normal" signals. This observation could lead to defining binary classifiers by thresholding the reconstruction error.

The method is, therefore, applicable to very different contexts. While developing the method, we made some exploratory tests on signals from the medical domain (detection of pathological ECG signals) and signals acquired from an electronic tongue (classification of tomato preserve obtained from different tomato species). Even if these tests are by far too preliminary to be worth a detailed report, they are certainly suggestive of future work. We can, therefore, anticipate some problems that need to be tackled when using GP2SO and suggest possible future extensions of the method.

**Function Embedding.** One trivial consideration about the nature of the embedding produced by GP2SO is that it is not unique. This can be trivially demonstrated if one considers the likely case in which one of the free parameters optimized by PSO is, directly or through some scaling factor, the argument of a periodic function like a trigonometric one. That case would result in an infinite number of possible solutions, which might hamper, or even destroy, the compactness of any possible cluster representing a specific class. In fact, in the most optimistic hypothesis in which the parameter under consideration is the single, possibly scaled, argument of a sine or cosine function, a class would be fragmented into a set of equivalent clusters. Such clusters could become very hard to identify if the dependency of the model on the parameters was not trivially expressed or separable.

A possible and immediate solution to this problem could be introducing a regularization term proportional to the magnitude of the parameters' vector into the fitness function. The fitness function could then be expressed as

$$\hat{f}(\mathbf{w}) = f(\mathbf{w}) + norm_x(\mathbf{w}),$$

where $f$ is the basic fitness function (e.g., the sum of squared regression errors) and $norm_x(\mathbf{w})$ the $x$-degree norm of the parameter vector $\mathbf{w}$. This way, we would introduce a bias in the fitness function that, in case of equivalent solutions, would make the one characterized by the smallest values of $\mathbf{w}$ preferable to the others.

**Reconstruction-Oriented Representations.** Many popular data representations designed for maximizing the preservation of the original information are often also used for classification, as happens, for instance, with Principal Component Analysis. Unfortunately, there is no guarantee that such a representation will also be optimal for classification. This is often the case with biomedical signals, in which the components that characterize anomalies from physiological signals have very low amplitude with respect to the signal's total energy, causing the encoding to filter them out. The use of different, scale independent distance measures as fitness functions could limit this partial reconstruction problem. When using this method to solve classification problems, a possible countermeasure could also be adding the residuals of the signal reconstruction to the embedded representation, followed by further feature selection to identify the most relevant components for classification.

**Implementation Efficiency.** While the low requirements in terms of the amount of training data can undoubtedly be a plus also from the viewpoint of computation load, GP2SO is certainly not intrinsically efficient, requiring an entire PSO run for each GP fitness case evaluation during training, and another PSO run for encoding each new unseen data instance after the model has been evolved. However, the choice of PSO as an algorithm in the second-level optimization (i.e., the optimization model's free parameters) is justified by PSO search efficiency and, even more, by its algorithmic structure that lends itself very naturally to a GPU-based implementation [10] with speed-ups easily reaching two orders of magnitude using a single-thread implementation as a reference.

## 6 Conclusions

This position paper has argued and demonstrated, solving a straightforward toy problem, that Genetic Programming-based parametric symbolic regression can replace Deep Learning in developing latent data representations following an autoencoder-like approach. The models developed by GP2SO, an algorithm that implements such an approach, have several advantages over the corresponding representations obtained by DL, the most relevant of which are related to their lower requirements of training data and higher interpretability.

The paper has anticipated possible practical applications of GP2SO to real-world data and problems, mainly focusing on representing and classifying one-dimensional signals. In doing so, it has highlighted some issues likely to occur in GP2SO practical applications and suggested possible solutions to circumvent them. Future work will thoroughly evaluate GP2SO's potential in applications like classification and anomaly detection and increase its computation efficiency by exploiting PSO's intrinsic parallelism.

**Acknowledgments.** Giulia Magnani is supported by the METROFOOD-IT project, funded by the PNRR - Mission 4 "Education and Research" Component 2: from research to business, Investment 3.1: Fund for the realization of an integrated system of research and innovation infrastructures-IR0000033 (D.M. (Ministerial Decree) Prot. n. 120 of 21 June 2022).

## References


1. The Artificial Intelligence Act. https://artificialintelligenceact.eu/. Accessed 13 Jan 2024
2. Bacardit, J., Brownlee, A.E.I., Cagnoni, S., Iacca, G., McCall, J., Walker, D.: The intersection of evolutionary computation and explainable AI. In: Proceedings of the Genetic and Evolutionary Computation Conference Companion, pp. 1757–1762. GECCO 2022, Association for Computing Machinery, New York, NY, USA (2022)
3. Bank, D., Koenigstein, N., Giryes, R.: Autoencoders. arXiv preprint arXiv:2003.05991 (2020)
4. Eiben, A.E., Smith, J.E.: Introduction to Evolutionary Computing. Springer, Heidelberg (2015). https://doi.org/10.1007/978-3-662-44874-8
5. Fan, Q., Bi, Y., Xue, B., Zhang, M.: Genetic Programming for image classification: a new program representation with flexible feature reuse. IEEE Trans. Evol. Comput. **27**(3), 460–474 (2023)
6. Fortin, F.A., De Rainville, F.M., Gardner, M.A., Parizeau, M., Gagné, C.: DEAP: evolutionary algorithms made easy. J. Mach. Learn. Res. **13**, 2171–2175 (2012)
7. Lensen, A., Xue, B., Zhang, M.: Genetic Programming for manifold learning: preserving local topology. IEEE Trans. Evol. Comput. **26**(4), 661–675 (2022)
8. van der Maaten, L., Hinton, G.: Viualizing data using t-SNE. J. Mach. Learn. Res. **9**, 2579–2605 (2008)
9. Mei, Y., Chen, Q., Lensen, A., Xue, B., Zhang, M.: Explainable artificial intelligence by genetic programming: a survey. IEEE Trans. Evol. Comput. **27**, 621–641 (2022)
10. Mussi, L., Daolio, F., Cagnoni, S.: Evaluation of parallel particle swarm optimization algorithms within the CUDA$^{TM}$architecture. Inf. Sci. **181**(20), 4642–4657 (2011)
11. Poli, R., Kennedy, J., Blackwell, T.: Particle swarm optimization. Swarm Intell. **1**(1), 33–57 (2007)
12. Poli, R., Langdon, W.B., McPhee, N.F.: A field guide to Genetic Programming (2008). https://www.lulu.com/
13. Schofield, F., Slyfield, L., Lensen, A.: A genetic programming encoder for increasing autoencoder interpretability. In: Pappa, G., Giacobini, M., Vasicek, Z. (eds.) EuroGP 2023. LNCS, vol. 13986, pp. 19–35. Springer, Cham (2023). https://doi.org/10.1007/978-3-031-29573-7_2
14. Vanneschi, L., Mauri, G., Valsecchi, A., Cagnoni, S.: Heterogeneous cooperative coevolution: strategies of integration between GP and GA. In: Proceedings of the 8th Annual Conference on Genetic and Evolutionary Computation (GECCO 2006), pp. 361–368 (2006)

# Learning

# Local Delay Plasticity Supports Generalized Learning in Spiking Neural Networks

Jørgen Jensen Farner[1], Ola Huse Ramstad[1,2], Stefano Nichele[1,3], and Kristine Heiney[1,4](✉)

[1] Department of Computer Science, Oslo Metropolitan University, Oslo, Norway
kristine.heiney@gmail.com

[2] Department of Neuromedicine and Movement Science, Norwegian University of Science and Technology, Trondheim, Norway

[3] Department of Computer Science and Communication, Østfold University College, Halden, Norway
stefano.nichele@hiof.no

[4] Department of Computer Science, Norwegian University of Science and Technology, Trondheim, Norway

**Abstract.** We propose a novel local learning rule for spiking neural networks in which spike propagation times undergo activity-dependent plasticity. Our plasticity rule aligns pre-synaptic spike times to produce a stronger and more rapid response. Inputs are encoded by latency coding and outputs decoded by matching similar patterns of output spiking activity. We demonstrate the use of this method in a three-layer feed-foward network with inputs from a database of handwritten digits. Networks consistently showed improved classification accuracy after training, and training with this method also allowed networks to generalize to an input class unseen during training. Our proposed method takes advantage of the ability of spiking neurons to support many different time-locked sequences of spikes, each of which can be activated by different input activations. The proof-of-concept shown here demonstrates the great potential for local delay learning to expand the memory capacity and generalizability of spiking neural networks and offers new perspectives on how to configure neuromorphic hardware.

**Keywords:** delay plasticity · local learning · spiking neural networks · Izhikevich neuron · generalized learning

## 1 Introduction

The firing rate has long been considered the primary mode by which neurons convey information. This framework has been translated into the way in which artificial neural networks (ANNs) in artificial intelligence and machine learning applications encode information: in classic ANNs, downstream neurons receive

S. Nichele and K. Heiney—These authors are co-senior authors.


M. Villani et al. (Eds.): WIVACE 2023, CCIS 1977, pp. 241–255, 2024.
https://doi.org/10.1007/978-3-031-57430-6_19

inputs as a weighted sum of the activity from upstream neurons and output a continuous value representing their activation level.

However, the rise of spiking neural networks (SNNs) in artificial intelligence [1] has brought into focus the importance of spike times and their use in encoding information [2]. Information can be encoded in spike times with much greater efficiency than in firing rates and endows networks with a greater computational capacity than coding with firing rates alone [2–4]. A toy example of a network reliant on a temporal code is shown in Fig. 1 [5]. As shown in Fig. 1(a), each connection between a pair of neurons is defined by a transmission delay, and the delays are not uniform in the network. The effect of this is that the pre-synaptic neurons must fire in a precise time-locked sequence in order for either of the post-synaptic neurons to activate (Fig. 1(b)).

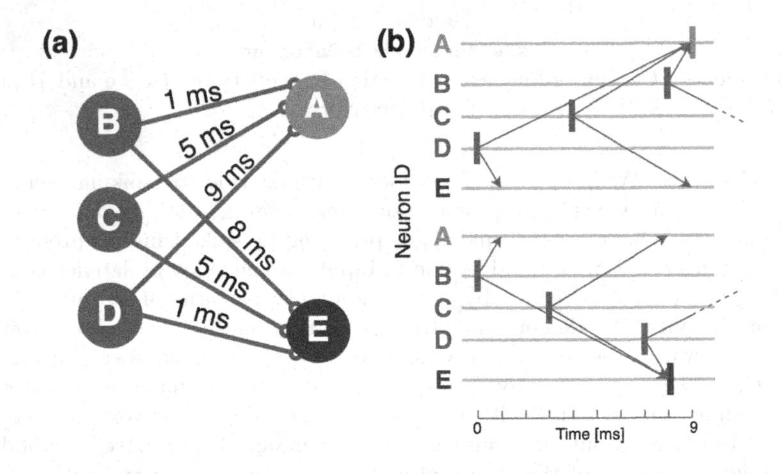


**Fig. 1.** Exmaple of polychrony in a toy feedforward spiking network. Pre-synaptic neurons activate a different post-synaptic neuron depending on the pre-synaptic activation order. (a) Pre-synaptic neurons B, C, and D are connected to post-synaptic neurons A and E with spike transmission delays as defined in the schematic. (b) Only certain time-locked sequences of neurons B, C, and D firing elicit spikes from neuron A or E. Figure adapted from Izhikevich [5].

Expanding this effect of temporal spike alignment to a larger network of neurons gives rise to the phenomenon of polychrony. As defined by Izhikevich [5], a polychronous group is an ensemble of neurons that can produce multiple "reproducible time-locked but not synchronous firing patterns" depending on how they are activated. In our simple toy example, the two patterns of activity shown in Fig. 1(b) are two such time-locked patterns; hereafter, we refer to these patterns as polychronous group patterns (PGPs). These patterns may be associated with memories stored in the neural network: a particular input elicits a particular PGP, and the greater the number of PGPs to be supported by a network, the larger the set of inputs the network responds to. However, it is not trivial to select transmission delays in a network to optimize the patterns the network can support, and so a method of training the delays in an SNN would promote their efficient use in artificial intelligence applications.

Classically, the most widely discussed mechanism associated with learning in the brain is Hebbian plasticity [6,7]. This theory on neural learning states that when one neuron causes repeated excitation of another, the efficiency with which the first cell excites the second is increased. In other words, the weights of connections change in a local, activity-dependent manner. Local plasticity rules, such as spike-timing-dependent plasticity (STDP) [8], that change synaptic weights in an activity-dependent manner are of great interest in the context of unsupervised deep learning in deep spiking neural networks (SNNs) [9]. But why should plasticity in SNNs be confined to synaptic weights, when we are aware of a much richer repertoire of plastic changes that occur in the brain [10,11]? In particular, there is evidence that neurons may change the speed of spike transmission in an activity-dependent manner [12,13]. This type of delay plasticity would allow networks to encode information and learn using spike times, and a similar type of learning could be translated to neuromorphic event-based hardware [14,15].

Delay plasticity has been explored in SNNs, but the majority of studies have used supervised methods, with most combining weight and delay training [16–21]. Since Taherkhani et al. [18] introduced the idea of training delays to reduce the need for weight adjustments, a number of other studies have similarly shown improved performance, greater sparsity, and reduced burden on weight training when delay training is introduced into the network [16–19]. One study has shown that training delays using backpropagation in a network with fixed weights can yield performance comparable to conventional weight training [22].

Supervised methods come with many drawbacks, including high requirements for memory and less flexibility for real-time applications. The development of a local delay learning rule that uses time-based coding would allow the advancement of more robust and flexible neuromorphic computing devices. In one study, an unsupervised method was used to train spike transmission delays in only the readout layer of a reservoir [23]. Another particularly relevant and compelling recent study has considered local, unsupervised delay training in a weightless SNN, combined with encoding inputs using relative latencies and decoding using a time-to-first-spike (TTFS) scheme [24]. In this study, the mode of training was local and activity-dependent, and weights were not involved in the training; this is very similar to the approach we used in the current study. The main difference between the present study and the work by Hazan et al. [24] is the method of decoding outputs; in their TTFS scheme, each output neuron corresponds to a single input class and the first neuron to spike prevents firing in the remainder of the output layer, in a winner-take-all fashion. In contrast, we focus on the overall pattern of activity produced by the output neurons, which has implications for generalized learning, as will be discussed later.

In this paper, we present a local activity-dependent delay plasticity algorithm for unsupervised learning with spike times [25]. In this learning rule, the relative timing of pre- and post-synaptic spikes locally alters the delay of the connection, causing any subsequent spike transmission between a pair of neurons to occur at a different speed; the weights remain fixed at constant uniform values. The mechanism of our method is to better align all pre-synaptic spikes causally related to a post-synaptic spike, with the purpose of producing a faster and stronger response in the post-synaptic neuron, in order to produce PGPs that reflect the structure of the input data. We applied our developed delay learning method to the classification of handwritten digits [26] in a proof-of-concept and demonstrated that training delays in a feedforward SNN is an effective method for information processing and classification. Our networks consistently outperformed their untrained counterparts and were able to generalize their training to a digit class unseen during training.

It should be noted that the purpose of this study was not to benchmark the developed method. Rather, we aim here to explore the behavior that arises with the proposed delay learning framework and understand how delays can be used to develop efficient coding schemes and advance generalized learning. We hope to further advance the framework presented here to take advantage of the richness of activity that we see emerge in response to our delay plasticity algorithm.

## 2 Delay Learning in Spiking Neural Networks

This section presents the delay learning framework developed in this study [25][1]. Section 2.1 describes the proposed activity-dependent delay plasticity algorithm we developed. The aim of this algorithm is to cause any pre-synaptic spikes that elicit post-synaptic activity to better align in their arrival at the post-synaptic neuron, by adjusting the transmission delays between the pre- and post-synaptic neurons. Section 2.2 presents the encoding and decoding approaches of latency coding (LC) and polychronous group pattern (PGP) clustering used in our delay learning framework. In LC, input values are encoded as relative latencies between the spike times of input neurons. PGP clustering assigns different output spiking activity to output labels by comparing the similarity of the activity patterns.

### 2.1 Activity-Dependent Delay Plasticity

The goal of our proposed learning method is to consolidate the network activity associated with similar inputs that constitute a distinct input class, so that the network will produce similar patterns of activity to be read out. With this aim in mind, the delays of pre-synaptic neurons that together produce activity in a post-synaptic neuron are adjusted to better align the arrival of their spikes at the post-synaptic neuron. Our framework was developed using Izhikevich regular spiking (RS) neurons [27].

Analogous to how STDP potentiates connections between causally related neurons to enhance the post-synaptic response, our delay plasticity mechanism

[1] Code available at https://github.com/DelayLearninginSNN/DelayLearninginSNN.

increases the post-synaptic response by better aligning causally related pre-synaptic spikes. This alignment process is illustrated in Fig. 2 for the case of four pre-synaptic neurons connected to one post-synaptic neuron. As shown in this figure, the pre-synaptic spikes (purple lines) that arrive (green lines) before the post-synaptic spike (blue line) are pushed towards their average arrival time (yellow line).

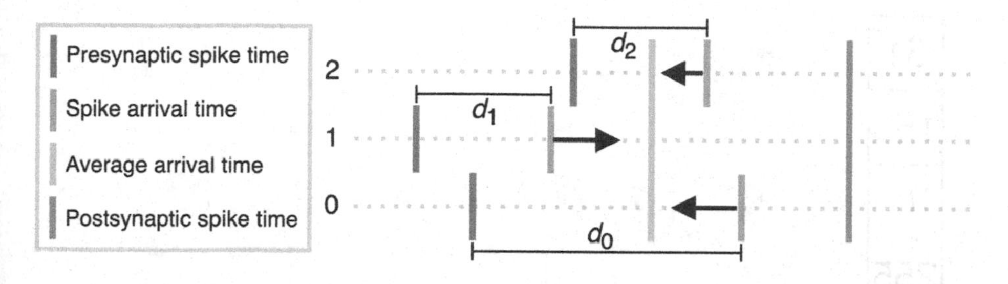


**Fig. 2.** Schematic overview of the delay learning mechanism. Purple vertical lines indicate presynaptic spike initiation times, green lines indicate presynaptic spike arrival times according to their delays $d_i$, and the blue line indicates the post-synaptic spike time. The learning mechanism works by pushing pre-synaptic spikes that arrive before the post-synaptic spike towards their average arrival time, indicated by the yellow line. (Color figure online)

The delay $d_{i,j}$ between pre-synaptic neuron $i$ and post-synaptic neuron $j$ changes according to the following equation:

$$\Delta d_{i,j} = -3\tanh\left(\frac{t_i + d_{i,j} - \bar{t}_{\text{pre}}}{3}\right), \quad 0 \leq \Delta t_{\text{lag}} < 10 \text{ ms}, \tag{1}$$

where $t_i$ is the spike time of neuron $i$, $\bar{t}_{\text{pre}}$ is the average pre-synaptic arrival time across all neurons with spikes arriving within 10 ms before the post-synaptic spike, and $\Delta t_{\text{lag}} = t_j - t_i + d_{i,j}$ is the time lag between when the pre-synaptic spike arrives at the post-synaptic neuron and when the post-synaptic neuron fires. The time window of 10 ms was selected because this is the window in which a pre-synaptic spike elicits a post-synaptic response.

### 2.2 Encoding and Decoding with Spike Times

To evaluate our learning rule in isolation, we consider here connections with static homogeneous weights; only the delay of the connections is allowed to change. Thus, encoding and decoding strategies must take advantage of spike timing. Here we describe the approaches of latency coding (LC) and polychronous group pattern (PGP) clustering used in our framework (Fig. 3).

In LC, inputs are encoded in the relative spike times of the input neurons. That is, input channels with a value of 0 will fire first, followed by other channels in order of increasing input value. Through experimentation, we determined that rescaling the dynamic range to relative latencies of [0, 40 ms] produced good results; details on this experimentation can be found in the thesis that serves as the basis for this study [25].

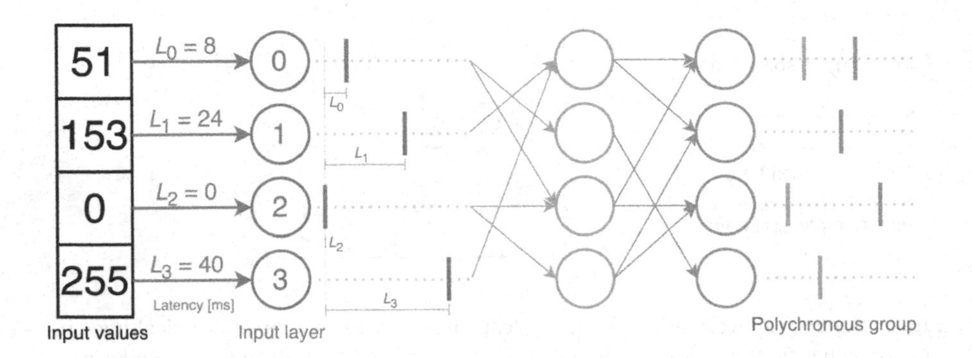


**Fig. 3.** Illustration of the encoding and decoding methods. Input values are encoded as spike latencies, as shown on the left side of the network. PGPs are defined as sets of sequential activity in the output layers triggered by inputs, and they are clustered in a hierarchical manner by checking the ratio of matching spikes with other PGPs.

We confirmed that LC is an appropriate method of encoding to accompany our delay learning algorithm by evaluating the behavior of small networks in response to different inputs. We considered networks with two or three input neurons and one output neuron and classified the trajectories of the connection delays as they changed over time according to the delay learning algorithm. The connections between the neurons were initialized with delays ranging from 18–22 ms. Under inputs encoded as firing rates, the majority of networks showed diverging delay trajectories. Convergence was only achieved when the firing rates of the input neurons were equal or multiples of each other, and cyclic delays emerged when the initialized delays were aligned with the offsets produced by the different input rates. Under alternating sets of LC inputs, however, the network delays generally either converged or followed cyclic patterns along with the alternating inputs, indicating a more stable response to this type of input.

Our decoding approach of PGP clustering is based on the concept of polychronization, where an ensemble of neurons can produce many time-locked patterns of activity, as described in the introduction [5]. This decoding concept is compatible with encoding by LC, as PGPs can be considered analogous to the patterns of spikes produced when encoding an input by LC. Thus, a PGP is one such time-locked pattern in the output layers of a network, consistently produced in response to the same input.

Because different inputs from the same class do not activate precisely the same input neurons, we also introduced an unsupervised method of clustering PGPs into output classes. A given PGP is described as a nested set in our framework. Each neuron that fires in a given layer is appended to the corresponding level of the set. Presynaptic spikes from an earlier layer are connected to post-synaptic spikes they participate in eliciting. The resulting PGP describes how activity flows through the network, retaining the ordinal relationships of the spikes without explicitly including time.

We cluster these output PGPs using hierarchical clustering. First, a pair of PGPs is given the same label if the number of intersecting elements in the set is greater than 95% of the mean number of spikes in the two PGPs. Each cluster obtained by this pairwise comparison is then described by the pattern averaged over all PGPs in the cluster. Thus, the number of labels does not necessarily match the number of classes, and may range from 1 to the total number of training input instances. This is repeated while iteratively dropping the matching threshold by 5% until the target threshold $\theta$ is reached. In our proof-of-concept, we considered $\theta = 80\%$ and 90%.

## 3 Proof-of-concept: Classification of Handwritten Digits

To demonstrate the utility of our proposed delay learning method, we applied it to the classification of handwritten digits from the Modified National Institute of Standards and Technology (MNIST) database [26]. In this preliminary proof-of-concept, we trained networks on only two digit classes and then tested its performance on classifying a third digit class in addition to the two trained classes to evaluate the networks' ability to generalize their training.

### 3.1 Experimental Setup

The MNIST dataset consists of images of $28 \times 28$ pixels [26]. We first scaled these images down to a size of $10 \times 10$ and assigned an input neuron to each pixel. The details of our experimental setup are given in Table 1. We used feedforward networks with three layers, including the input layer, and fixed homogeneous connection weights. Each neuron is connected with a neuron in the next layer with a probability of 0.1.

**Table 1.** Network architecture and experimental parameters

| Layer size | Number of layers | Connection probability | Weight | Digits (unseen) | Train instances | Test instances | PGP match threshold |
|---|---|---|---|---|---|---|---|
| 100 | 3 | 0.1 | 6 | 0, 1, (2) | 20 | 25 | 80%, 90% |

In each iteration of the experiment, a feedforward network was generated with connectivity between layers according to the connection probability, and each connection was assigned an initial delay randomly drawn from the set of integers between 0 and 40 ms. We then provided inputs from the selected digit classes to this untrained network with local plasticity switched off to give a performance baseline for random delays. In the training phase, different inputs of the same digit classes were fed into the network with local delay plasticity switched on. Following training, we again switched off local plasticity and provided the same set of inputs as given in the baseline test phase to assess the performance of the trained network. One digit class was selected as an "unseen" class, i.e., a class presented during testing but not training, to evaluate the network's ability to generalize.

### 3.2 Delay Training Improves Classification Accuracy

As described in Sect. 2.2, outputs are decoded in our framework by clustering the PGPs produced by the network. In our proof-of-concept, this means taking the firing patterns from the final two layers (i.e., excluding the input layer) in response to each of the inputs and assigning each a label based on how similar it is to the other output patterns. Because the clustering is dependent on a similarity threshold, the number of labels is not predetermined. To evaluate the accuracy of the network, the correct label is considered to be the label most commonly applied to outputs from a given input class, and the accuracy is then the ratio of the number of instances assigned that label to the total number of instances presented to the network. In cases where all outputs have the same label regardless of input class, the accuracy is 0% because the network has failed to separate the input classes.

Figure 4 shows the accuracy after training plotted against the accuracy before training. Each data point represents a network, and points above the dashed line are networks whose performance was improved by training. Figure 4(a) shows the test results for the two training digit classes (0 and 1), which were presented to the network during the training phase. In nearly all cases where the network could separate the digit classes (accuracy $> 0\%$), the trained network performed better than the corresponding untrained network. However, some networks were unable to separate the classes (2.4% and 45% of networks for PGP thresholds $\theta = 90\%$ and 80%, respectively).

### 3.3 Networks with Plastic Delays Generalized Training to an Unseen Input Class

As stated previously, the test phases before and after training consisted of presenting the network with three digit classes (0, 1, and 2) despite only training with two of these classes (0 and 1). Figure 4(b) shows the test results for the third "unseen" digit class (2). These results demonstrate that some networks were able to generalize their learning to a digit class not input to the network during training. Here, the accuracy remained low for the more stringent matching threshold of $\theta = 90\%$ but reached up to 64% for $\theta = 80\%$ (mean accuracy 32% in 38 networks able to separate the unseen class). Flexibility with the PGP threshold can thus allow a network to generalize its training to unseen classes while maintaining good performance on trained classes.

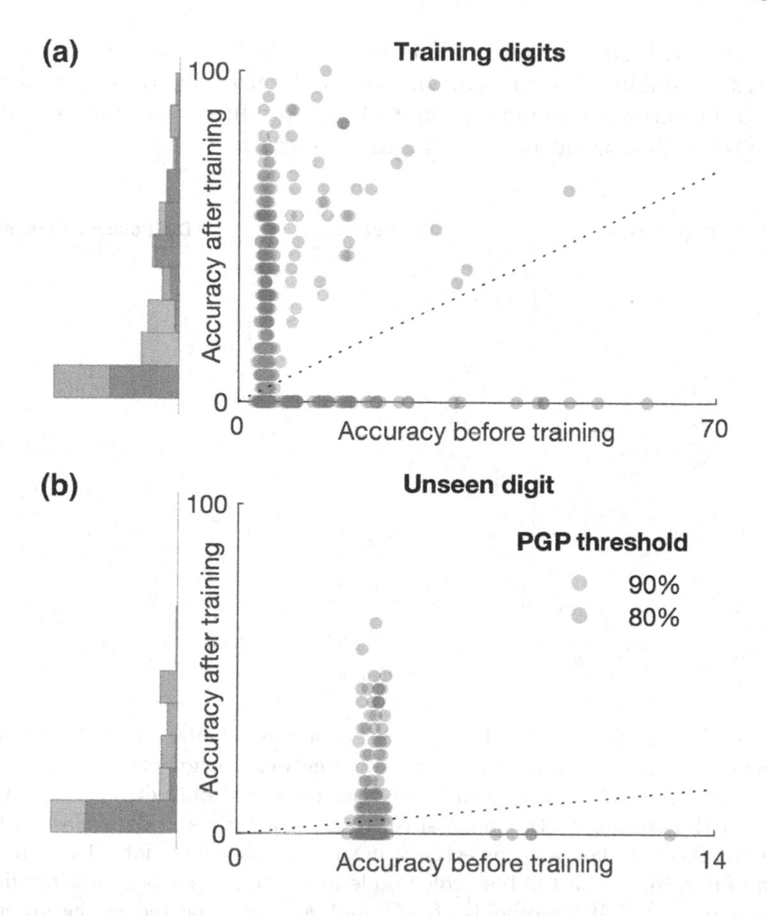


**Fig. 4.** Accuracy of classifying handwritten digits before and after training using delay learning. (a) Two training digit classes (0,1), $N = 500$ networks. (b) One unseen digit class (2), $N = 100$ networks. Results are plotted with jitter for the sake of visualization. Histograms show the accuracy distribution after training. Accuracy of 0 indicates non-separable classes.

## 3.4 Output Activity Patterns Before and After Training

Examples of the activity in the output layers before and after training are shown in Fig. 5, with correct trials colored different shades of blue according to trial number and incorrect trials colored orange (here, labels were determined with a PGP matching threshold of $\theta = 80\%$). The neurons are ordered according to their mean spike time across all trials with inputs in digit class 0; note that the top and bottom rows have different neuron orders.

These raster plots demonstrate the way the delay learning pushes the network to produce similar patterns (PGPs) when presented with inputs from the same class, as evidenced by the greater overlap of activity patterns after training. Prior to training, the network activity is less structured overall and sparser in the final layer (neurons 101–200), whereas after training, the final layer is more active, and consistent spiking patterns can be observed across many inputs from the same class. In particular, inputs in digit class 1 produce very similar patterns, with very few spikes deviating from the main pattern.

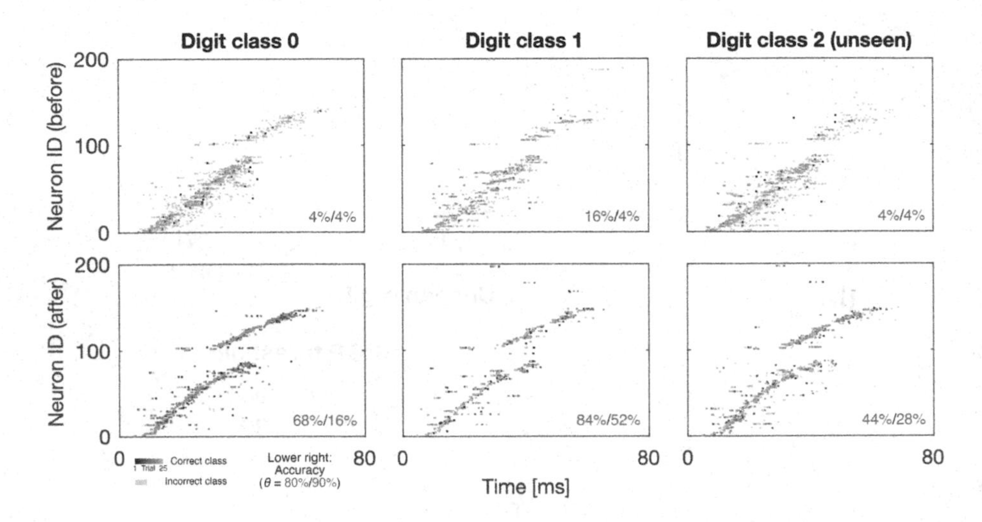


**Fig. 5.** Raster plots of activity in layers 2 and 3 (neurons 1–100 and 101–200, respectively) before and after training for an example network. Digit classes 0 and 1 were used for training, and 2 is an unseen third class presented only during testing. Neurons are sorted according to the mean spike time for all trials in digit class 0. Colors represent whether the class was correctly or incorrectly identified, for a PGP matching threshold of $\theta = 80\%$, with the blue color scale for the correct label showing different trials. Accuracies at PGP thresholds of 80% and 90% are reported in the lower right corner of each plot.

To more clearly demonstrate the distinct representations that emerge in the network as a result of training, Fig. 6 shows the mean activity patterns during correctly labeled trials, sorted according to the mean spike time of the neurons during each digit presentation. In these plots, only a single time is plotted for each active neuron, with this time representing the average time over all of its spikes during correct trials. These figures highlight the effect of the training: more neurons are active, especially in the second layer, and a representational structure in the activity forms.

Although our method yields networks with fair classification performance, the spiking patterns shown here indicate one drawback that will be addressed in future work: the representations of each digit class are quite similar, which can cause erroneous class assignment. This may be solved by introducing competition in the network to encourage more diverse representations and thus greater separability among output patterns.

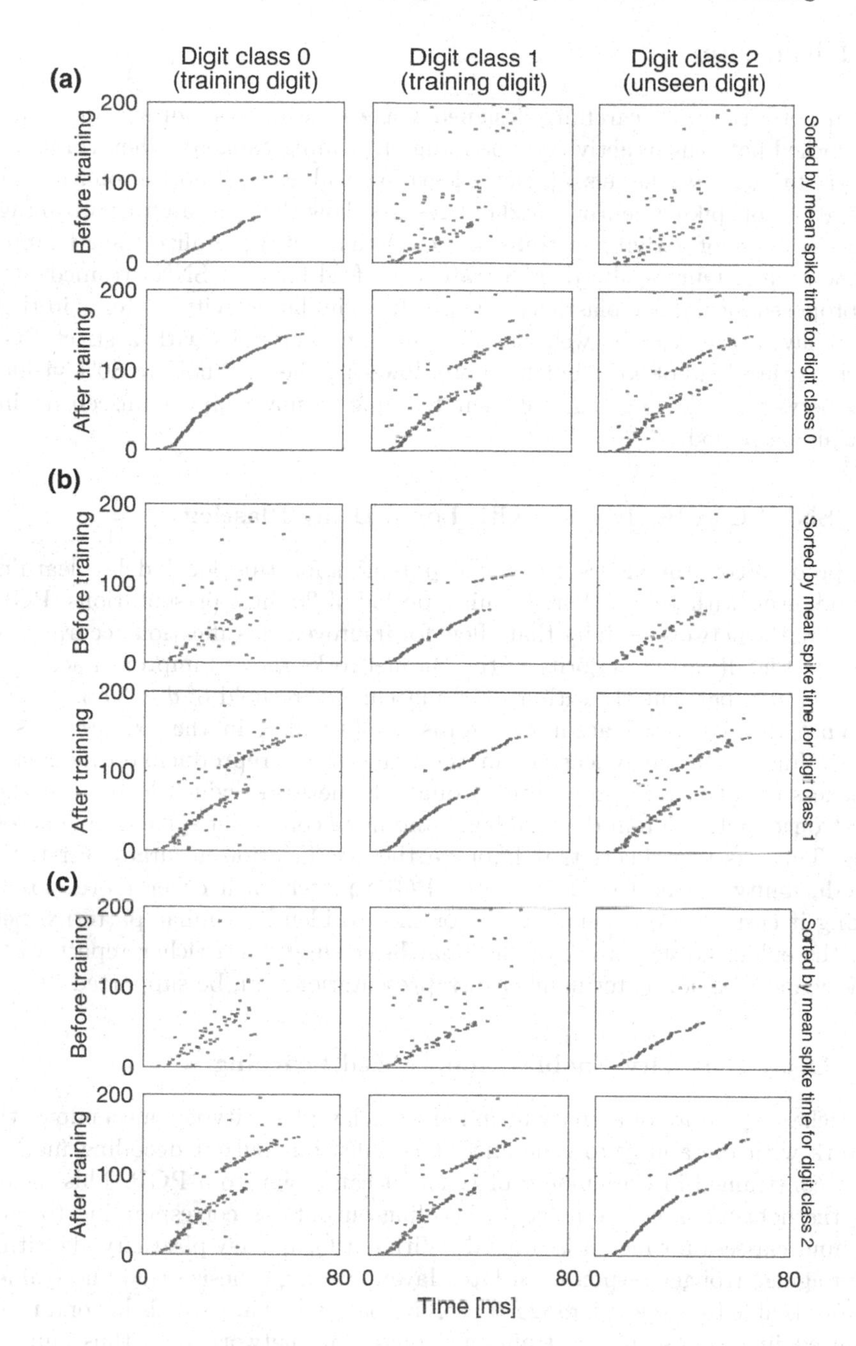


**Fig. 6.** Raster plots showing the mean spike time in layers 2 and 3 (neurons 1–100 and 101–200, respectively) before and after training for the same example network as in Fig. 5. Digit classes 0 and 1 were used for training, and 2 is an unseen third class presented only during testing. Neurons are sorted according to the mean spike time for all correct trials in digit classes (a) 0, (b) 1, and (c) 2.

# 4 Discussion

Neural networks with carefully designed spike time delays can support many time-locked patterns of activity, expanding the coding capacity when compared with traditional rate models [5]. Delay learning enables such polychronization in populations of spiking neurons, and our results show that we can take advantage of this richness of activity to train networks that can generalize their training to new inputs. Our results demonstrate that feed-forward SNNs trained with our proposed local delay plasticity rule produce similar activity patterns in their output layers that can be well classified in some networks with a strict PGP matching threshold of 90%. Furthermore, lowering the threshold to 80% yielded some networks able to generalize their training to novel inputs unseen during the training period.

## 4.1 SNNs Can Be Trained with Local Delay Plasticity

Our proof-of-concept shows the great potential for this local delay learning method; even with only a short training period of 20 digit presentations, PGPs emerge in the network activity that allow for improved classification accuracy. As shown in Fig. 4(a), the majority of the 500 networks showed improved accuracy after training, particularly with a PGP matching threshold of $\theta = 90\%$.

What this increased accuracy entails is illustrated in the example raster plots in Fig. 5. Our delay learning method encourages reproducible time-locked sequences of activity to propagate through the network, which leads to earlier and stronger activation in the final layer and more consistent spike timing across trials. This has two effects that improve the classification accuracy. First, the reproducibility means that the output PGPs match each other more closely, making it easier for our clustering algorithm to identify similar patterns. Second, the enhanced activation of the final layer supports a richer repertoire of activity, meaning a greater number of representations can be supported.

## 4.2 Delay Plasticity Enables Generalized Learning

The richer repertoire of activity attained by delay plasticity is what endows the network with the ability to generalize. Crucially, our output decoding method is not constrained in the number of labels it can assign to a PGP. This means that the network is not confined to labeling outputs as corresponding to only the input classes on which it is trained. With our local delay plasticity algorithm encouraging stronger responses in later layers, we hypothesize that the trained network is able to support a range of activity patterns that extends beyond those produced in response to the training inputs. The network may thus support representations of untrained input classes that our output clustering algorithm can recognize as distinct from those of the training classes.

### 4.3 Competition in the Output Layers to Improve Separability

Although our results are promising, there are some limitations to our current approach. Figure 4 shows that not all networks perform well after training. In most of these cases, the poor performance is largely due to non-separability of input classes, frequently accompanied by a fairly high accuracy prior to training (see Fig. 4(a) with threshold 80%). These networks are likely being over-trained and producing a homogeneous PGP that represents multiple input classes; this is further evidenced by the similarity of the representations evident even in the relatively high-performing network shown in Figs. 5 and 6.

To counteract this and improve separability, it would be beneficial to introduce a mechanism to produce stimulus-specific competition among the neurons in the population; this would make the resultant representations of each digit class sparser and avoid the similarity evident in Figs. 5 and 6. Such stimulus-specific competition could be introduced by, for example, lateral inhibition in early layers [28]. This would encourage stimulus specificity among neurons in the same layer and give preference to neurons that fire earlier in response to a given input, leading to sparser and more distinct representations of the different input classes.

### 4.4 Future Work: Mixed-Mode Learning and Neuromorphics

In future work, plastic weights and diverse neuron types can be combined with our delay learning approach to expand the computational capacity and enable mixed learning strategies. Our delay learning approach does not yield accuracies comparable with state-of-the-art weight training methods; however, training with delays in combination with conventional weight training has been shown to improve efficiency and accuracy [11]. As such, our future work will similarly combine weight and delay training as a means to evaluate how delay learning can improve conventional weight-based approaches, rather than act as a substitute for weight training.

We also expect that approach to delay learning will prove useful in the training of neuromorphic event-based hardware [15]. Although SNNs are computationally demanding to implement in conventional hardware, novel unconventional hardwares can enable a more energetically efficient implementation, and as such, compatible training algorithms will be needed for these new computational systems. Using event-based computing in this way is expected to be particularly beneficial in time-based tasks, such as forecasting, and we hope to test our delay learning method on such tasks in the future.

**Acknowledgements.** This work was partially funded by the SOCRATES project (Research Council of Norway, IKTPLUSS grant agreement 270961) and the DeepCA project (Research Council of Norway, Young Research Talent grant agreement 286558).

## References

1. Ponulak, F., Kasinski, A.: Introduction to spiking neural networks: information processing, learning and applications. Acta Neurobiol. Exp. **71**(4), 409–33 (2011)
2. VanRullen, R., Guyonneau, R., Thorpe, S.J.: Spike times make sense. Trends Neurosci. **28**(1), 1–4 (2005)
3. Grüning, A., Bohte, S.M.: Spiking neural networks: principles and challenges. In: 22nd European Symposium on Artificial Neural Networks, Computational Intelligence and Machine Learning, ESANN 2014 - Proceedings, pp. 1–10 (2014)
4. Paugam-Moisy, H., Bohte, S.: Computing with spiking neuron networks. Handb. Nat. Comput. **1–4**, 335–376 (2012)
5. Izhikevich, E.M.: Polychronization: computation with spikes. Neural Comput. **18**(2), 245–282 (2006)
6. Hebb, D.O.: The Organization of Behavior: A Neuropsychological Theory. Wiley, New York (1949)
7. Markram, H., Gerstner, W., Sjöström, P.J.: A history of spike-timing-dependent plasticity. Front. Synaptic Neurosci. **3**, 4–4 (2011)
8. Markram, H., Lübke, J., Frotscher, M., Sakmann, B.: Regulation of synaptic efficacy by coincidence of postsynaptic aps and EPSPs. Science **275**(5297), 213–215 (1997)
9. Tavanaei, A., Ghodrati, M., Kheradpisheh, S.R., Masquelier, T., Maida, A.: Deep learning in spiking neural networks. Neural Netw. **111**, 47–63 (2019)
10. Gittis, A.H., du Lac, S.: Intrinsic and synaptic plasticity in the vestibular system. Current Opinion Neurobiol. **16**(4), 385–390 (2006). sensory systems
11. Zhang, W., Linden, D.J.: The other side of the engram: experience-driven changes in neuronal intrinsic excitability. Nat. Rev. Neurosci. **4**(11), 885–900 (2003)
12. Lin, J.W., Faber, D.S.: Modulation of synaptic delay during synaptic plasticity. Trends Neurosci. **25**(9), 449–455 (2002)
13. Debanne, D.: Information processing in the axon. Nat. Rev. Neurosci. **5**(4), 304–316 (2004)
14. Taherkhani, A., Belatreche, A., Li, Y., Cosma, G., Maguire, L.P., McGinnity, T.M.: A review of learning in biologically plausible spiking neural networks. Neural Netw. **122**, 253–272 (2020)
15. Grimaldi, A., Gruel, A., Besnainou, C., Jérémie, J.N., Martinet, J., Perrinet, L.U.: Precise spiking motifs in neurobiological and neuromorphic data. Brain Sci. **13**(1), 68 (2022)
16. Schrauwen, B., van Campenhout, J.: Extending spikeprop. In: 2004 IEEE International Joint Conference on Neural Networks (IEEE Cat. No.04CH37541), vol. 1, pp. 471–475 (2004)
17. Wang, X., Lin, X., Dang, X.: A delay learning algorithm based on spike train kernels for spiking neurons. Front. Neurosci. **13**, 252–252 (2019)
18. Taherkhani, A., Belatreche, A., Yuhua, L., Maguire, L.P.: Dl-resume: a delay learning-based remote supervised method for spiking neurons. IEEE Trans Neural Netw Learn Syst **26**(12), 3137–3149 (2015)
19. Mostafa, H.: Supervised learning based on temporal coding in spiking neural networks. IEEE Trans. Neural Netw. Learn. Syst. **29**, 1–9 (2017)
20. Johnston, S.P., Prasad, G., Maguire, L.P., McGinnity, T.M.: A hybrid learning algorithm fusing STDP with GA based explicit delay learning for spiking neurons. In: 2006 3rd International IEEE Conference Intelligent Systems, pp. 632–637. IEEE (2006)

21. Kheradpisheh, S.R., Masquelier, T.: Temporal backpropagation for spiking neural networks with one spike per neuron. Int. J. Neural Syst. **30**(06), 2050027 (2020)
22. Grappolini, E., Subramoney, A.: Beyond weights: deep learning in spiking neural networks with pure synaptic-delay training. In: Proceedings of the 2023 International Conference on Neuromorphic Systems, pp. 1–4. ACM, New York, NY, USA (2023)
23. Paugam-Moisy, H., Martinez, R., Bengio, S.: Delay learning and polychronization for reservoir computing. Neurocomput. (Amsterdam) **71**(7), 1143–1158 (2008)
24. Hazan, H., Caby, S., Earl, C., Siegelmann, H., Levin, M.: Memory via temporal delays in weightless spiking neural network (2022)
25. Farner, J.J.: Activity dependent delay learning in spiking neural networks. Master's thesis, Oslo Metropolitan University (2022)
26. LeCun, Y., Cortes, C.: The mnist database of handwritten digits (2005)
27. Izhikevich, E.M.: Simple model of spiking neurons. IEEE Trans. Neural Netw. **14**(6), 1569–1572 (2003)
28. Pehlevan, C., Sengupta, A.M., Chklovskii, D.B.: Why do similarity matching objectives lead to Hebbian/anti-Hebbian networks? Neural Comput. **30**(1), 84–124 (2018)

# Improving PVC Detection in ECG Signals: A Recurrent Neural Network Approach

Fabiola De Marco(✉), Luigi Di Biasi, Alessia Auriemma Citarella, and Genoveffa Tortora

Department of Computer Science, University of Salerno, Via Giovanni Paolo II, 132, 84084 Fisciano, Italy
fdemarco@unisa.it

**Abstract.** Premature ventricular contractions (PVCs) are abnormal ventricle heartbeats that disrupt the normal QRS rhythm. Classification of PVCs is crucial for the diagnosis and management of cardiac conditions. Detecting anomalies in an electrocardiogram (ECG) is straightforward, but determining the specific number of anomaly classes related to PVCs remains challenging despite Holter monitoring. This work explores the potential of Recurrent Neural Networks (RNNs), specifically Long Short-Term Memory (LSTM) and bidirectional LSTM (BLSTM) models, to learn and reproduce the wave patterns present in ECGs without abnormalities. The working hypothesis is that these trained RNNs can accurately analyze new ECGs with normal QRS complexes by minimizing the root mean square error (RMSE) between their predictions and the actual ECG values. Applying the trained RNNs to ECGs with anomalies aims to identify intervals where the RMSE is high, indicating the presence of PVC pattern classes. Only regular ECG data in the dataset is used for training and evaluation. At the same time, both classes are included in the testing phase to assess the ability of the network to discriminate between the two ECG classes. Also, we used trained models on another dataset composed of PVCs to extract the interval with the highest RMSE. The results obtained from this study suggest the promising performance of LSTM-based approaches in PVC pattern classification.

**Keywords:** Premature Ventricular Contractions · Recurrent Neural Network · Long Short-Term Memory

## 1 Introduction

Heart and blood vessel diseases are collectively called cardiovascular diseases (CVDs) and encompass a wide range of conditions that can impact the structure

This study was carried out within the FAIR - Future Artificial Intelligence Research and received funding from the European Union Next-GenerationEU (PIANO NAZIONALE DI RIPRESA E RESILIENZA (PNRR) - MIS- SIONE 4 COMPONENTE 2, INVESTIMENTO 1.3 - D.D. 1555 11/10/2022, PE00000013).


M. Villani et al. (Eds.): WIVACE 2023, CCIS 1977, pp. 256–267, 2024.
https://doi.org/10.1007/978-3-031-57430-6_20

and function of the cardiovascular system: coronary artery disease, hypertension, stroke, heart failure, and arrhythmias account for millions of fatalities annually. Unfortunately, according to *World Health Organization* (WHO)[1], *Center for Disease Control* (CDC)[2], *American Heart Association* (AHA)[3] and *European Society of Cardiology* (ESC)[4], CVDs are rapidly increasing worldwide. CVDs are the primary cause of death throughout the world. The cardiovascular system (CVS) comprises the heart and circulatory system. It has a central role in maintaining in life the human body. The heart is the core of CVS: for the scope of this contribution, it is viewed as composed only of three components: pericardium, endocardium, and sinoatrial node (SA). The SA node can generate electrical impulses, which lead to contractions of atria and ventricles and regulate the frequency of the contraction of the heart [2]. In a healthy heart, the rate related to rest status is called sinus rhythm.

The heartbeat can be recorded and digitalized by Electrocardiogram (ECG). The ECG is a noninvasive diagnostic tool that measures the electrical activity of the heart, typically using multiple electrodes (usually 12) on different body parts. This procedure is generally called ECG or EKG. The electrical signals generated from the SA are recorded and can be displayed as waveforms on a monitor, printout [22], or the same signal can be digitized and recorded in digital format. In particular, a healthy heartbeat can be split into seven parts identified by the letters P, P-R, Q-R-S, T, ST, QT, and optionally U. Each part is characterized by a particular shape and typical value interval, as shown by Fig. 1.

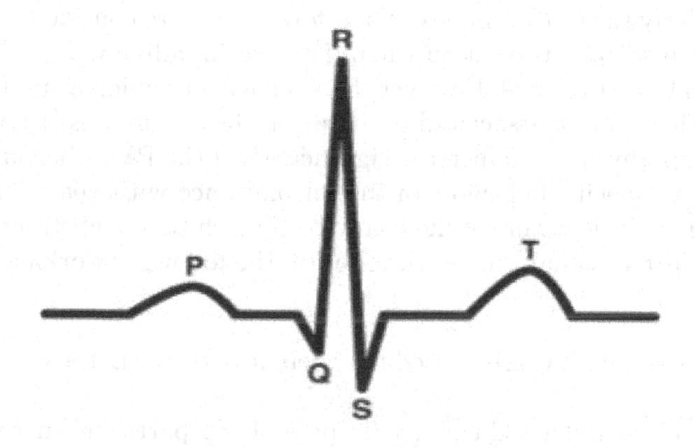


**Fig. 1.** Signal component in ECGs

The P wave portion appears small, typically rounded, and upright. It lasts on average 0.08 s (but can vary from 0.05 s to 0.12). The P-R portion (formally

[1] https://www.who.int/news-room/fact-sheets/detail/cardiovascular-diseases-(cvds).
[2] https://www.cdc.gov/heartdisease/index.htm.
[3] https://www.heart.org/en/health-topics/consumer-healthcare/what-is-cardiovascular-disease.
[4] https://www.escardio.org.

P-R interval) is a low electrical signal portion that ends in correspondence with the start of the Q-R-S portion. The P-R interval has a duration that varies between 0.16 s and 0.2 s. The Q-R-S portion comprises the Q, R, and S portion and follows the P-R interval. Together, Q-R-S waves form the QRS complex. Typically, it lasts 0.12 s. The T portion is usually a smooth, rounded wave. The ST period follows the T wave and is a horizontal segment that ends at the start of the subsequent P period. QT portion reflects the entire time for ventricular depolarization and repolarization. It is determined from the start of the QRS complex to the end of the T-wave. Finally, the U portion may be present: if present, the U period is a minor and typically faint wave that follows the T period. Its precise relevance is unknown [14]. It is important to note that in a healthy heart, electrical signals go through a specific path during a typical cardiac cycle (P-PR-QRS-T-QT-ST), starting contractions in the chambers one at a time [15]. The ECG wave related to the cardiac cycle is very similar for every healthy heart heartbeat. By analyzing the ECG, cardiologists can detect abnormal heart rhythms, identify changes in electrical activity, diagnose acute myocardial infarction, assist in analyzing heart failure, and identify abnormalities in the heart structure.

The method and results reported in this work are related to a specific CVD in the arrhythmia class, Premature Ventricular Contractions (PVCs), which originates in the ventricles and is characterized by an aberrant signal that can result in early or excess heartbeat, disrupting the normal sinus rhytm [1].

For the scope of this work, PVCs are considered to disrupt part of the cardiac cycle, particularly the QRS complex. Therefore, PVCs are considered benign and do not require medical intervention when detected in individuals without underlying structural heart disease. However, PVCs have a double nature (benign and malignant). They can be associated with various heart diseases [1]: arrhythmias and cardiomyopathy risk can increase significantly if the PVCs become recurrent or are related to specific behaviors or in concomitance with heart failure [3].

Due to the analogic nature of the heartbeat, which can be abstract as a simple time-related electric signal, this work relies on the following working hypotheses (WHs):

- different classes of PVC are related to different patterns in the electrical signal (**WH1**);
- if multiple classes of PVC exist, it is possible to partition an ECG dataset composed of PVC and non-PVC ECG acquisition in multiple clusters, each related to a possible outcome (**WH2**).

This study investigates the validity of working assumptions using the Extended Genetic Algorithms (EGA) approach proposed in [5] in conjunction with an RNN. The EGA and RNN are employed to analyze QRS complexes from the MIT-BIH Arrhythmia Database, containing non-PVC and PVC data, to learn the characteristics of standard QRS signals. Furthermore, the trained RNN is used to identify abnormal segments within PVC ECG signals. EGA is utilized to design the RNN architecture, predicting critical EEG signal portions on which these classes rely, with RMSE as the scoring metric. The main objective was to determine whether this hybrid strategy could yield a more resilient classifier, even using a relatively straightforward neural network architecture.

The paper is organized as follows: Sect. 2 reports a literature overview focusing on works that try to address similar issues; Sect. 3 define the data collection, pre-processing, and method proposed in the work. Section 4 presents the obtained results, and Sect. 5 summarizes the main findings and contributions of the paper and suggests future directions.

## 2 Related Works

Due to the double nature of the PVC and following the results reported in the literature, it is possible to discriminate between benign, dangerous, and deadly outcomes [10]. The life-threatening nature of the dangerous and fatal PVC classes make clear that early diagnosis in PVC case, particularly the correct identification of the PVC class, might be life-saving. Nowadays, an extensive evaluation is required to diagnose PVCs, including a review of the medical history of the patient, a physical exam, and an ECG to detect irregular heartbeats, which is a noninvasive way to identify different arrhythmias [16]. In [4], the reported results suggested the difficulties in correctly identifying the PVC class and the importance of allowing healthcare providers to determine the frequency, pattern, class, and severity of the PVCs. Also, the results reported in the previous work showed that detecting arrhythmias through an ECG is a problem with a simple solution. Still, the PVC classification remains an open problem due to the difficulties in obtaining labeled ECG data.

Deep Learning models, such as Convolutional Neural Networks (CNNs) or RNNs like LSTM, have emerged as powerful tools for efficiently and effectively addressing the detection problem [7]. Typically, neural network design relies on the expertise of the user, involving critical decisions like determining hidden layers, configuring neuron connections, adjusting learning rates, managing momentum, and fine-tuning weights and biases. These factors significantly impact network performance. To address this complexity, some studies explore a hybrid approach merging Genetic Algorithms (GAs) with conventional neural network training methods. Most research in the literature utilizes GAs primarily to refine neural network parameters rather than create entirely new neural network architectures [19]. In fact, GAs excel in seeking diverse solutions and adapting to find optimal configurations, making them ideal for enhancing the robustness and adaptability of CNNs. The use of genetic algorithms for building the architecture design of a neural network is becoming increasingly important. In particular, GAs optimize complex AI model architectures, reducing the manual tuning of hyperparameters, which can be time-consuming and suboptimal. GAs automate this process, freeing researchers from the need for extensive trial-and-error adjustments. The dynamic nature of GAs allows AI models to adapt to evolving data or changing clinical conditions. Adaptability is vital in healthcare, including PVC detection, where data might change over time.

In [25], GA is used to dynamically shape the architecture of a multilayer perceptron to address the challenge of fixed network structures, thereby offering an adaptable and optimal model configuration. This entails automatically determining hidden layer quantities, the number of nodes within each layer, and the synaptic connection weights. The training process unfolds in stages, comprising a searching phase and a cleanup phase, during which the ideal number of neurons in each layer is ascertained through a combination of crossover, mutation operators, and cloning techniques. This GA-driven approach specifically focuses on distinguishing premature ventricular beats from regular beats within the MIT-BIH arrhythmia database. The maximum number of nodes in each layer is constrained to three. The collective findings reveal an average recognition rate of 96.96% across all records.

The study [23] introduced a neuro-genetic approach wherein a neural classifier was combined with GA to determine the optimal interconnections between neurons, thereby enhancing the identification of PVCs within the MIT-BIH arrhythmia database. The architecture of the neural network was initially established through a trial-and-error method. Subsequently, careful attention was given to the design of genetic operators to fine-tune its structure. This approach was refined to maximize the recognition of true positives, thus reducing the occurrence of false negatives. The best-performing classifier, denoted as AG-CLS2, demonstrated impressive results with a correct classification rate of 98.86%, with a sensitivity of 99.09% and a specificity of 98.66% and for an optimal neural network structure consisting of only eight connections.

## 3 Methods

### 3.1 Dataset

This work uses a subset extracted from the MIT-BIH Arrhythmia Database dataset, which is created through cooperation between Boston, Massachusetts' Beth Israel Hospital (BIH) and the Massachusetts Institute of Technology (MIT), and it contains several forms of arrhythmias, such as PVCs. MIT-BIH is well-known and frequently utilized for research and development in arrhythmia detection and classification [17]. In particular, the subset contains 14000 samples that belong to two different classes: PVC and NON-PVC [4]. The recording came from 47 distinct patients. The average record length is approximately 169 milliseconds.

For our experimentation, we generated three subsets:

- *training set* that contains 70% of total NON-PVC samples;
- *validation set* that contains 30% of NON-PVC samples and 50% of PVC samples;
- *test set* that contains only PVC samples.

The *training set* comprises only NON-PVC samples. The *training set* and the *validation set* did not contain overlapped data.

Consequently, in our training, we exclusively use NON-PVC data, in the evaluation process, we utilize NON-PVC and PVC data in the dataset. However, during the testing phase, we deliberately used only PVC data. This inclusion evaluates the proficiency of the network in distinguishing between these two distinct classes.

### 3.2 Residual Neural Network

RNN architectures such as LSTM are used to analyze sequential data [18], thanks to the presence of a memory cell and gating mechanisms that allow them to store and forget information over lengthy sequences, modelling both short-term and long-term dependencies in sequential data. LSTM uses the input, forget, and output gates to regulate information flow. The input gate chooses which input information to store, and the forget gate chooses what to discard from the memory cell. In contrast, the output gate controls the output by selectively exposing or suppressing memory cell information [8].

Bidirectional LSTM (BLSTM) combines the properties of both LSTM and bidirectional models. BLSTM processes input sequences in forward and backward directions, capturing past and future context information. This bidirectional nature enables BLSTMs to understand temporal dependencies in the data better, making them suitable for tasks such as ECG classification [21].

### 3.3 EGA

GA are optimization methods influenced by the biological principles of evolutionary theory [11]. It permits discovering an optimal solution to an optimization problem by simulating natural Selection. Therefore, it uses a population of individuals, each with $n$ chromosomes. These individuals continue to change from generation to generation by mechanisms analogous to the natural process of evolution. Chromosomes almost always take the form of binary strings when they are stored. Each locus (a particular location on a chromosome) is composed of two alleles (various versions of genes) represented by the numbers 0 and 1 [12].

A GA aims to build a population of candidate solutions to a problem by evaluating each candidate solution using a fitness function (scoring function). Also, GA selects the best solutions to generate a new population. This process continues in an iterative way, which ultimately results in the enhancement of solutions through time. Each GA performs at least the same basic operations:

- *initialization*, which randomly populates the population of potential solutions;
- *selection*, where the GA selects the most suitable parents from the present population;
- *crossover*, where the GA recombines current population genetic material to produce new solutions;
- *mutation*, where random changes to the genetic material arise, helping prevent the GA from becoming locked in a local optimum [9] by introducing new genetic variations into the population;

- *evaluation*, where the fitness of each solution in the population compared to target Optimus; The criteria could be based on the number of generations, the best solution's fitness, or other factors [13];
- *replacement*, where GA replaces the current population with a new generation of solutions. This process is repeated until a satisfactory solution is found or for several generations.

Specifically, we establish an entity as a vector with $m$ features. Each feature of a generic entity is referred to as a "gene" of that entity, and the entire collection of genes constitutes the "genome" of that entity.

This work uses an extended version of classic GA by allowing the *merging* operation, which allows two genomes possessed by two entities of the same population to merge.

### 3.4 Map/Reduce Approach to Run EGA

Although genetic algorithms have been utilized successfully in various fields, they can be computationally costly due to the training and evaluation of many candidate solutions over successive generations [24].

To address the computation cost limitation, we used a simple map-reduced distributed system named GRIMD that relies on a Beowulf[5] approach. We annotated the Matlab script using GRIMD Explode Point definition (EP) to allow each entity (genome variation) to be evaluated on different grid nodes with different parameters. For more information regarding the explode point, it is possible to refer to [20].

To clarify the EP utilization, this work reports the EGA configuration regarding GRIMD distribution packages. Please note that each EP comprises a keyword surrounded by "at" (@) character.

The population genome can comprise one or more of the following gene types:

- 0 = sequenceInputLayer(@c = INT_RANGE(1,10)@, Normalization = "zscore")
- 1 = softmaxLayer
- 2 = regressionLayer
- 3 = fullyConnectedLayer
- 4 = dropoutLayer [0.5]
- 5 = sequenceInputLayer(14, Normalization = "zerocenter")
- 6 = sequenceInputLayer(14, Normalization = "rescale-simmetric")
- 7 = dropoutLayer [0.2]
- 8 = lstmLayer(@c@, 'OutputMode', 'last')
- 9 = bilstmLayer(@c@, 'OutputMode', 'last')
- 10 = reluLayer
- 11 = lstmLayer(@c@, 'OutputMode', 'sequence')
- 12 = bilstmLayer(@c@, 'OutputMode', 'sequence')

[5] https://en.wikipedia.org/wiki/Beowulf_cluster.

The initial size of the population was 24, while the initial genome length was 12. The EP @c = INT_RANGE(1, 10)@ forces the distribution of 10 copies of the same genome in the first evolution step, each able to perform regression using a different `sequenceLayer` with input length = @c@.

It is important to note that simple heuristics were implemented to reduce the death ratio [6]. In particular, each entity will have a `sequenceLayer` as the first gene and a `fullyConnectedLayer` and `regressionLayer` in the tail. These heuristics guarantee that each entity will be a valid network Matlab structure, avoiding startup failure.

**EGA Generated LSTM Architecture.** In this preliminary study, the EGA execution discovered two simple RNN network architectures able to reach a RMSE of 15. In particular, the GA produced four different LSTM and BLSTM network structures.

```
1  LSTM1.N = [
2      sequenceInputLayer(N)
3      lstmLayer(1500)
4      fullyConnectedLayer(N)
5      regressionLayer];
6  BLSTM1.N = [
7      sequenceInputLayer(N)
8      lstmLayer(1500)
9      bilstmLayer(1000)
10     fullyConnectedLayer(N)
11     regressionLayer];
12
```

Best performances were archived for the sequence length N with 1 and 10. LSTM1.N comprises 9M learnable and a BLSTM1.N comprises 29M learnable.

### 3.5 Anomalies Detection

To identify anomalies and locate the segments of waves that may contain PVC classes, we employed the trained network on the TEST dataset. In particular, it is possible to consider a generic PVCx as an ECG wave that undoubtedly contains a PVC pattern because an expert cardiologist annotated it.

Therefore, PVCx can be considered a sequence of samples, as described below:

$$PVCx = s1, ..., sn \tag{1}$$

Then, for each chunk of data $\{si, ..., sj\} \in$ PVCx with this conditions:

$$\begin{cases} si \geq 1 \\ sj \leq n \end{cases}$$

it is possible to use the trained networks to predict the expected value $[pi, ..., pj]$.

Due to the training dataset, the network will predict the value following the learned data based only on natural QRS.

Therefore, for each $PVCx$, it is possible to compute the expected prediction ($EPVCx$) containing the predicted waveforms that would appear if the ECG were regular:

$$(EPVCx) = \{p1, ..., pn\} \tag{2}$$

Finally, it is possible to calculate the difference between the expected value between $EPVCx$ and the real PVC waveforms contained in $PVCx$.

This information can then be used to construct a table (TABLE), where each row corresponds to an element from the TEST set (comprising all PVCs). Within each column, the RMSE is computed, representing the difference between the expected value of a typical QRS and the actual values present in the PVC signals.

## 4 Results

In this preliminary study, the application of the EGA approach identifies two straightforward RNN architectures. These architectures demonstrated the capability to achieve a Root Mean Square Error (RMSE) of 15. For this preliminary work, the sum of columns in the TABLE was used to estimate which part of the ECG wave could contain a PVC pattern.

The graph in Fig. 2 reports the sum and suggests that a pattern that could imply the presence of PVC classes being concentrated in the central segment of the ECG. This observation may have significant implications for understanding the distribution and characteristics of PVCs within the cardiac waveform. In this specific ECG signal segment, the RMSE exhibits variability, underscoring distinctions between the expected QRS complex attributes and the actual characteristics found in PVC signals.

These findings suggest that specific anomalies linked with PVCs emerge in this specific signal region. Also, these preliminary findings suggest that WH2 might be demonstrated.

Following these results, we could conclude that these arrhythmias could indicate underlying physiological or electrical disruptions in the ventricles, leading to the generation of PVCs, which have a distinct pattern in contrast to the normal rhythm observed in the ECG normal waveforms.

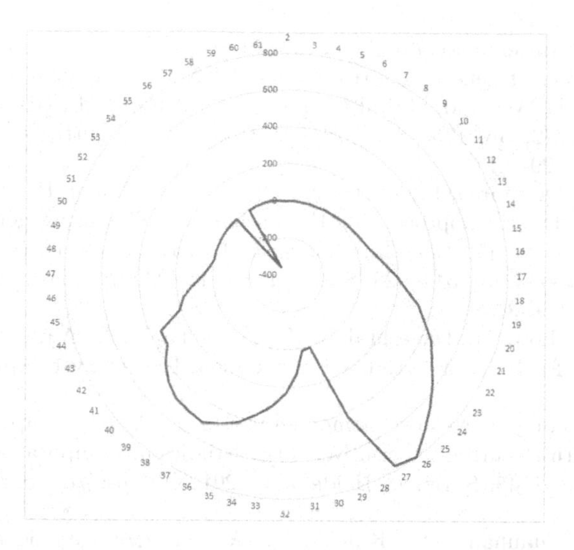


**Fig. 2.** Estimated ECG signal zones related to PVC with higher RMSE variability compared to normal QRS complexes

## 5 Conclusions

This study investigates the effectiveness of recurrent neural networks, particularly LSTM and BLSTM, in detecting PVC patterns inside an ECG waveform. The preliminary findings indicate that the PVC patterns are likely to be situated in the central segments of the ECG signals. This observation is supported by the RSME values detected in these specific portions, suggesting the significance of this region for accurate PVC pattern detection. This result could be investigated deeply, particularly by extracting these ECG portions, plotting them in graphical form, and submitting them to expert cardiologists for more accurate human and specialist validation. In the future, these results can also be used to explore Explainable Artificial Intelligence (XAI) to develop more transparent and interpretable AI models.

## References

1. Ahn, M.S.: Current concepts of premature ventricular contractions. J. Lifestyle Med. **3**(1), 26 (2013)
2. Boyett, M.R., Honjo, H., Kodama, I.: The sinoatrial node, a heterogeneous pacemaker structure. Cardiovasc. Res. **47**(4), 658–687 (2000)
3. Cojocaru, C., Penela, D., Berruezo, A., Vatasescu, R.: Mechanisms, time course and predictability of premature ventricular contractions cardiomyopathy-an update on its development and resolution. Heart Failure Rev. 1–13 (2021)
4. De Marco, F., Ferrucci, F., Risi, M., Tortora, G.: Classification of QRS complexes to detect premature ventricular contraction using machine learning techniques. PLoS One **17**(8), e0268555 (2022)

5. Di Biasi, L., Auriemma Citarella, A., De Marco, F., Risi, M., Tortora, G., Piotto, S.: Exploration of genetic algorithms and CNN for melanoma classification. In: Schneider, J.J., Weyland, M.S., Flumini, D., Füchslin, R.M. (eds.) WIVACE 2021. CCIS, vol. 1722, pp. 135–138. Springer, Cham (2021). https://doi.org/10.1007/978-3-031-23929-8_13
6. Di Biasi, L., De Marco, F., Auriemma Citarella, A., Barra, P., Piotto Piotto, S., Tortora, G.: Hybrid approach for the design of CNNs using genetic algorithms for melanoma classification. In: Rousseau, J.J., Kapralos, B. (eds.) ICPR 2022. LNCS, vol. 13643, pp. 514–528. Springer, Cham (2022). https://doi.org/10.1007/978-3-031-37660-3_36
7. Ebrahimi, Z., Loni, M., Daneshtalab, M., Gharehbaghi, A.: A review on deep learning methods for ECG arrhythmia classification. Expert Syst. Appl.: X **7**, 100033 (2020)
8. Graves, A.: Long short-term memory. In: Graves, A. (ed.) Supervised Sequence Labelling with Recurrent Neural Networks. Studies in Computational Intelligence, vol. 385, pp. 37–45. Springer, Heidelberg (2012). https://doi.org/10.1007/978-3-642-24797-2_4
9. Katoch, S., Chauhan, S.S., Kumar, V.: A review on genetic algorithm: past, present, and future. Multimed. Tools Appl. **80**, 8091–8126 (2021)
10. Klewer, J., Springer, J., Morshedzadeh, J.: Premature ventricular contractions (PVCs): a narrative review. Am. J. Med. **135**, 1300–1305 (2022)
11. Kramer, O.: Genetic algorithms. In: Kramer, O. (ed.) Genetic Algorithm Essentials. Studies in Computational Intelligence, vol. 679, pp. 11–19. Springer, Cham (2017). https://doi.org/10.1007/978-3-319-52156-5_2
12. Kumar, M., Husain, D.M., Upreti, N., Gupta, D.: Genetic algorithm: review and application. Available at SSRN 3529843 (2010)
13. Lambora, A., Gupta, K., Chopra, K.: Genetic algorithm-a literature review. In: 2019 International Conference on Machine Learning, Big Data, Cloud and Parallel Computing (COMITCon), pp. 380–384. IEEE (2019)
14. Limaye, H., Deshmukh, V.: ECG noise sources and various noise removal techniques: a survey. Int. J. Appl. Innov. Eng. Manag. **5**(2), 86–92 (2016)
15. Luisada, A.A., MacCanon, D.M.: The phases of the cardiac cycle. Am. Heart J. **83**(5), 705–711 (1972)
16. Marcus, G.M.: Evaluation and management of premature ventricular complexes. Circulation **141**(17), 1404–1418 (2020)
17. Moody, G.B., Mark, R.G.: The impact of the MIT-BIH arrhythmia database. IEEE Eng. Med. Biol. Mag. **20**(3), 45–50 (2001)
18. Ni, L., Li, Y., Wang, X., Zhang, J., Yu, J., Qi, C.: Forecasting of forex time series data based on deep learning. Procedia Comput. Sci. **147**, 647–652 (2019)
19. Nikbakht, S., Anitescu, C., Rabczuk, T.: Optimizing the neural network hyperparameters utilizing genetic algorithm. J. Zhejiang Univ.-Sci. A **22**(6), 407–426 (2021)
20. Piotto, S., Di Biasi, L., Concilio, S., Castiglione, A., Cattaneo, G.: GRIMD: distributed computing for chemists and biologists. Bioinformation **10**(1), 43 (2014)
21. Qiao, F., Li, B., Zhang, Y., Guo, H., Li, W., Zhou, S.: A fast and accurate recognition of ECG signals based on ELM-LRF and BLSTM algorithm. IEEE Access **8**, 71189–71198 (2020)
22. Sattar, Y., Chhabra, L.: Electrocardiogram. In: StatPearls [Internet]. StatPearls Publishing (2022)
23. Sekkal, M., Chikh, M.A., Settouti, N.: Evolving neural networks using a genetic algorithm for heartbeat classification. J. Med. Eng. Technol. **35**(5), 215–223 (2011)

24. Zheng, J., et al.: Optimal multi-stage arrhythmia classification approach. Sci. Rep. **10**(1), 2898 (2020)
25. Zhou, J., Li, L.: Using genetic algorithm trained perceptrons with adaptive structure for the detection of premature ventricular contraction. In: 2004 Computers in Cardiology, pp. 353–356. IEEE (2004)

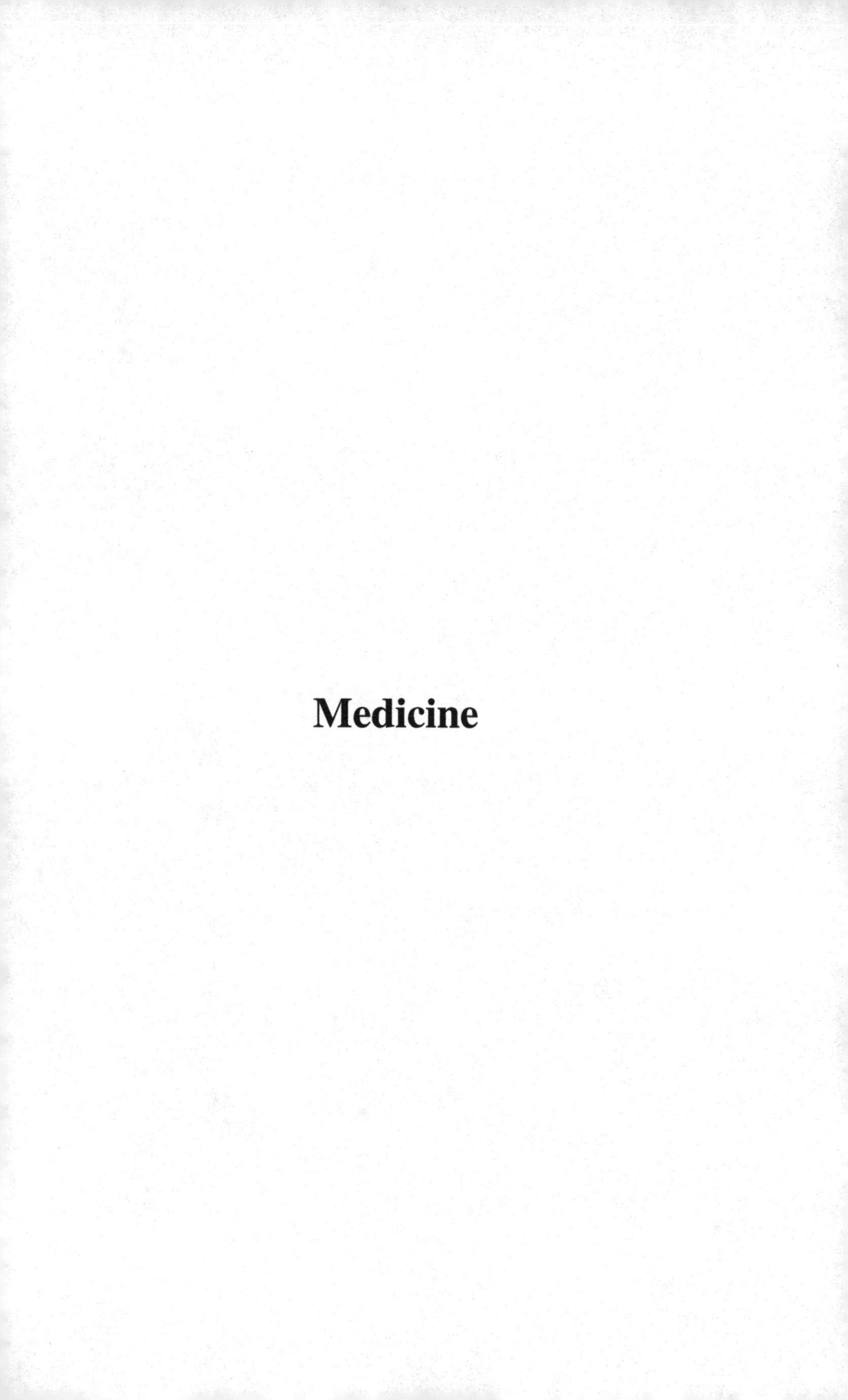

# Medicine

# Clustering Trajectories to Study Diabetic Kidney Disease

Veronica Distefano[1,2], Maria Mannone[1,3,4](✉), Irene Poli[1], and Gert Mayer[5]

[1] European Centre for Living Technology (ECLT), Ca' Foscari University of Venice, Venice, Italy
{veronica.distefano,maria.mannone,irenpoli}@unive.it
[2] Department of Economic Sciences, Università del Salento, Lecce, Italy
[3] Department of Engineering, University of Palermo, Palermo, Italy
[4] Dipartimento di Scienze Molecolari e Nanosistemi (DSMN), Ca' Foscari University of Venice, Venice, Italy
[5] Department Internal Medicine IV (Nephrology and Hypertension), Innsbruck Medical University, Innsbruck, Austria
gert.mayer@i-med.ac.at

**Abstract.** Diabetic kidney disease (DKD) is a serious complication of type-2 diabetes, defined prominently by a reduction in estimated glomerular filtration rate (eGFR), a measure of renal waste excretion capacity. However DKD patients present high heterogeneity in disease trajectory and response to treatment, making the *one-model-fits-all* protocol for estimating prognosis and expected response to therapy as proposed by guidelines obsolete. As a solution, precision or stratified medicine aims to define subgroups of patients with similar pathophysiology and response to the therapy, allowing to select the best drug combinations for each subgroup. We focus on eGFR when aiming to identify eGFR decline trends by clustering patients according to their eGFR trajectory shape-similarity.

The study involved 256 DKD patients observed annually for four years. Using the Fréchet distance, we built clusters of patients according to the similarity of their eGFR trajectories to identify distinct clusters. We formalized the trajectory-clustering approach through category theory. Characteristics of patients within different progression clusters were compared at the baseline and over time.

We identified five clusters of eGFR progression over time. We noticed a bifurcation of eGFR mean trajectories and a switch between two other mean trajectories. This particular clustering approach identified different mean eGFR trajectories. Our findings suggest the existence of distinct dynamical behaviors in the disease progression.

**Keywords:** clustering · trajectory · precision medicine · category

---

**Supplementary Information** The online version contains supplementary material available at https://doi.org/10.1007/978-3-031-57430-6_21.


M. Villani et al. (Eds.): WIVACE 2023, CCIS 1977, pp. 271–283, 2024.
https://doi.org/10.1007/978-3-031-57430-6_21

## 1 Introduction

Precision medicine [1,2] is a flourishing research area, which aims to find the best individualized treatment for patients according to their characteristics. In fact, the formula "one-model-fits-all" is unsatisfying when it comes to many diseases as far as progression and response to therapy is concerned. To find subgroups of similar patients, cluster analysis approach is a useful and informative tool, as witnessed by several studies [3–9].

Defining sub-groups of such an evolving population can help shed light on underlying common features in each sub-group, allowing physicians, if linked to pathophysiology and drug mode of action, to foster a more appropriate targeted treatment. This approach to medical research paves the way toward effective personalized or at least better stratified treatment. This approach to medical research paves the way toward effective individualized treatments. Of particular relevance is, for instance, the differentiation at the baseline, regarding different parameters. We focus on clusters of patients sharing the same disease-behavior across time, as instances of longitudinal studies. Longitudinal studies have been used also to address more general quality of life issues [10] and depression patterns across time [11], with statistical approaches such as growth mixture models.

In this article, we focus on patients with type-2 diabetes mellitus (T2DM) and its associated diabetic kidney disease (DKD) from the DC-ren dataset.[1] DKD is a serious public health problem and the main cause of end-stage renal disease (ESRD) in developed countries [26]. Longitudinal changes of renal function help inform on patients' clinical courses and if, identified by pathophysiologically relevant characteristics, help select individualized treatment according to patients' specific characteristics.

In this article, we will build clusters of patients' trajectories. This information can constitute a first step toward the development of a decision system to foster individualized strategies for DKD treatment [4,12]. We analyze trajectories of patients with respect to the dependent variable eGFR. The variation of eGFR provides an estimate of the severity of the disease and the response to treatment [8,16]. We build clusters of trajectories based on shape similarity and on eGFR range-similarity.

To group trajectories according to their shape similarity, we use the Fréchet distance, first proposed in the domain of calculus [13], and recently applied to medicine with the *kmlShape* clustering technique [14]. The Fréchet distance is evaluated upon the comparison between pairs of points following the profiles of the curves they belong to.

The approach to trajectory clustering is formalized within the framework of Category Theory [17,18]. It is an abstract branch of mathematics, initially developed to formalize the transformations between transformations, and to connect

[1] The project *Drug combinations for rewriting trajectories of renal pathologies in type II diabetes* (DC-ren), https://dc-ren.eu/, is funded by the Horizon 2020 research and innovation programme, Action RIA Research and Innovation action Call: H2020-SC1-BHC-2018-2020; Topic: SC1-BHC-02-2019.

different areas of mathematics between them. Applied category theory includes research in physics [19], chemistry [20], neuroscience [21]. A few applications also concern cluster analysis [22], with the formalization of a clustering method as a *functor*. A functor is a morphism between categories. A category is constituted by objects (points) and morphisms between them (arrows), whose composition is associative and has the identity element.

We aim to find subgroups of similar patients and build clusters of mean trajectories. We find cases of bifurcations and switch of trajectory clusters. To understand the possible pathophysiological reasons underlying patients exhibiting such a behavior, we analyze their medical and demographical variables. Coupled with drug mode of action, our results can be fed into a decision system, to find the best individualized treatments for future DKD patients. This article is the development of a first study where the Fréchet distance was applied to real data [23]. Here, we consider an extended dataset and a more refined computational approach. The novelty of our work is the use of categorical formalism for a medical real case study, and the application of a relatively-new statistical method, *kmlShape*, to a real data for a non-public dataset.

The article is organized as follows. After a review of some concepts of longitudinal cluster analysis and category theory (Sect. 2), we present a case of study with patients affected by DKD (Sect. 3), and we discuss our findings (Sect. 4).

## 2 Methodology

In this section, we present our trajectory, clustering approach using some formal tools of category theory; we then describe the *kmlShape* method, to investigate trajectory shape-similarity according to the Fréchet distance.

### 2.1 A Shape-Similarity Clustering of Longitudinal Data

Longitudinal data are measured repeatedly over time for the same individual. In this paper, we are interested in the evolution regarding the individual variation of estimated glomerular filtration rate (eGFR) in a small group of patients with type 2 diabetes and chronic kidney disease (DKD) at different stages. We used the *kmlShape* approach, that creates clusters of trajectories according to their evolution [14]. This approach is a variation of the longitudinal k-means [24] using a "shape-respecting distance" and a "shape-respecting mean."

The Fréchet distance [27] computes the shape similarity of two curves $P_1$ and $P_2$, based on the smallest of the maximum pairwise distances obtained with two respective reparametrizations, $\alpha : [0,1] \to [0,1]$ and $\beta : [0,1] \to [0,1]$, as follows:

$$F(P_1, P_2) = \inf_{\alpha,\beta \in R} \max_{t \in [0,1]} \{dist(P_1(\alpha(t)), P_2(\beta(t)))\}.$$

The approach of *kmlShape* considers the discrete version of the Fréchet distance, based on a sequence of pairs of points belonging to the two curves (represented as polylines). Since the two curves need not to have the same length, we have to "walk through them" at different speeds. The ratio between the different speeds to move along the curves is the time scale $\lambda$, discussed later.

Since we are interested in assessing the trend of the disease rather than verifying its presence, we focused on this method. The same approach has recently been followed in another application of *kmlShape* to a medical dataset [27]. In fact, *kmlShape* quantifies the differences of trend between the eGFR trajectories. In addition, the *kmlShape* method presents a highest ARI index when compared with Traj and GMM method [28].

The Fréchet distance measures the longest link between the trajectories [14]. Its computation between two trajectories does not require the same number of time-points in each trajectory.

We consider a generalization of the Fréchet mean to $n$ curves. To this aim, we implement the *kmlShape* with the *RandomAll* technique [14], with $n$ patients randomly scattered through the leaves of a binary tree.

Genolini and co-authors [14] provided a generalized definition of the Fréchet distance including a time scale $\lambda$. Indeed, in the context of real data, there can be an issue of relative scale, because the variable of interest and the time variable are measured according to different unities. The change of time scale impacts the partitioning, and thus the resulting clusters. The meaning of the scale variation is the change of "travel speed" to go through a curve. The value of $\lambda = 0.5$ is empirically determined for each research problem. We run different tests before choosing this value. More details including the precise definition of the Fréchet's mean can be found in the article by Genolini and co-authors [14].

### 2.2 Category Theory for Trajectory Clustering

Patients with similar characteristics over time can be computationally and graphically grouped together in the same cluster [15]. The comparison between processes over time can be contextualized in the framework of category theory [25]. Here, we use its diagrammatic language to describe patients' grouping according to their trajectory similarity. We also discuss the transition from a patient-based representation to a state-based representation. First, we briefly summarize the basic definitions of category theory.

A *category* is constituted by objects (points) and morphisms (arrows) between them. The composition of morphisms must be associative, and there should exist the identity morphism. A *functor* is a generalization of a function. More precisely, it is a mapping between categories (mapping objects and morphisms of a category into objects and morphisms of another category, preserving structures), and a *natural transformation* is a mapping between functors.

According to Spivak [20], category theory constitutes a powerful (i.e., precise) communication tool of ideas tool between different fields of mathematics. It can be used to compare structures and methods of different disciplines. Category theory starts being applied to several domains of science to acquire an abstract and thus general overview [20]. According to [29], category theory can also be used for medical dataset. Here, we use category theory as a bridge between clinical practice as defined by physicians, real data of patients, and information theory. We use this formalism to make more precise the comparison between each patient at different time-points, and between different patients at the same time-point. In addition, connecting the case study with the categorical framework allows one to recover all theorems and methods defined in abstract mathematics, which have the potential to make possible further applications and developments.

Let us consider a dataset composed of $n$ patients characterized by $p$ observable variables at four time points $t_0$, $t_1$, $t_2$, $t_3$. Each patient is characterized as a triplet $(\mathbf{x}_i(t_k), \mathbf{D}(t_k), y_i(t_{k+1}))$, where $i$ is the individual (the patient); $t_k$ is the time point $k = 0, 1, 2, 3$; $\mathbf{x}_i(t_k) = x_i^1(t_k), ..., x_i^p(t_k)$ is a set of values of variables, characterizing the individual; $\mathbf{D}(t_k) = D_1(t_k), D_2(t_k), D_3(t_k), D_4(t_k)$ stands for the given drug combination; $y_i(t_{k+1})$ is the value of the response variable $Y$ at $t_{k+1}$, measured after one year of treatment. The response variable is evaluated as the variation of the dependent variable, that is, the estimated glomerular filtration rate (eGFR); we thus indicate it as $E$ in the following. The trajectory over time of the $i$-th patient ($p_i$) with respect to the eGFR (E) is: $p_i^E(t_0) \rightarrow p_i^E(t_1) \rightarrow p_i^E(t_2)$. For the $i$'-th patient we have: $p_{i'}^E(t_0) \rightarrow p_{i'}^E(t_1) \rightarrow p_{i'}^E(t_2)$. We can evaluate the distance of a patient with respect of herself/himself through time, or the distance between different patients at the same time. We indicate the distance between patients $i, i'$ with respect to the variable $E$ and time $t_k$ as $d_{i,i'}^E(t_k)$, and the distance between values observed at times $t_k, t_{k'}$ of the variable $E$ for the same patient $i$ as $d_i^E(t_k, t_{k'})$, see diagram (1). In such a patient-based representation, each point is a patient at a time-point. This representation is dual to the state-based representation, which will be useful to create the *state map* (Fig. 1).

$$\begin{array}{ccc} p_i^E(t_0) & \xrightarrow{d_{ii'}^E(t_0,t_0)} & p_{i'}^E(t_0) \\ {\scriptstyle d_{ii}^E(t_0,t_1)}\downarrow & & \downarrow{\scriptstyle d_{i'i'}^E(t_0,t_1)} \\ p_i^E(t_1) & \xrightarrow{d_{ii'}^E(t_1,t_1)} & p_{i'}^E(t_1) \\ {\scriptstyle d_{ii}^E(t_1,t_2)}\downarrow & & \downarrow{\scriptstyle d_{i'i'}^E(t_1,t_2)} \\ p_i^E(t_2) & \xrightarrow{d_{ii'}^E(t_2,t_2)} & p_{i'}^E(t_2) \end{array} \tag{1}$$

In the language of categories, the construction of diagram (1), with observations and distances, can be described as an enriched *double category* with metrics in $\mathbb{R}$ [18], whose objects are the values of variable $E$, and whose morphisms are vertical and horizontal distances $d_{i,i'}^E(t_k)$, $d_i^E(t_k, t_{k'})$. The comparison between trajectories of different patients involves both of these distances.

Similar trajectories can be grouped within the same cluster of trajectories. The clustering is described as a functor [22] from the category of dataset to the category of the partitioned dataset. This concept can be applied to the trajectory-clustering, from the category of trajectories to the category of clusters of trajectories (Fig. 2). In the language of categories, the comparison between similar clustering methods corresponds to an arrow between arrows, that is, a

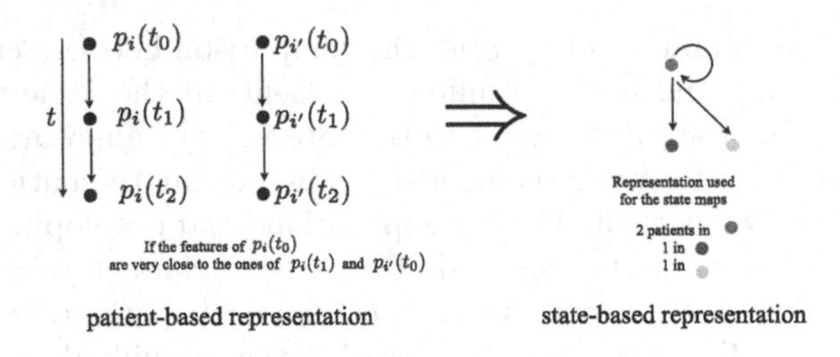


**Fig. 1.** Patient-based representation and state-based representation. The representation on the left is typical of categories. $p_i(t)$ indicates the clinical characteristics of the $i$-th patient at time $t$, and $p_{i'}(t)$ refers to the $i'$-th patient. Time flows vertically. The representation on the right neglects the detail on single-patient and time, in favor of a description of the clinical states where one or more patients can stay or return (loop arrow). The second representation can be built from the overall analysis of single-patients longitudinal data.

natural transformation. The comparison between the clusters that are obtained with slightly different methods is formalized as an arrow (morphism) in the category of clusters of trajectories. Thus, one can shift the attention from the natural transformation (comparison between clustering methods) to a morphism (comparison of clusters obtained with slightly different methods). Natural transformations (arrows between arrows) formalize the comparison between different transformations. Trajectory clustering processes, despite their differences, can be seen as processes from trajectories to clusters of trajectories, and thus we can use the language of categories to compare them.

### 2.3 Study Population

We considered 256 DKD patients observed during annual visits in a time span of four years. 48.4% were male, the mean age was 67 years. The characteristics of patients at the baseline are presented in Table 1.

The variability in eGFR decline was analyzed with cluster analysis. The eGFR is defined in the Appendix. In clinical routine the eGFR trajectory is used to judge the response to the therapeutic treatment: the *controlled disease* corresponds to an increase of eGFR or a decrease not exceeding 5% of the baseline value (the value at $t_0$), while the *uncontrolled disease* corresponds to an eGFR decrease higher than 10% of the baseline value [30].

The mean eGFR at $t_0$ ranges from 31 and 90 ml/min/1.73 m$^2$; at $t_3$ it is comprised between 19 and 120 ml/min/1.73 m$^2$, denoting an overall decrease of kidney efficiency through time. The descriptive statistics of eGFR, with mean and standard deviation at each time-point, are presented in Table 2. The mean value of eGFR decreases with time, indicating an overall worsening of the disease in the group.

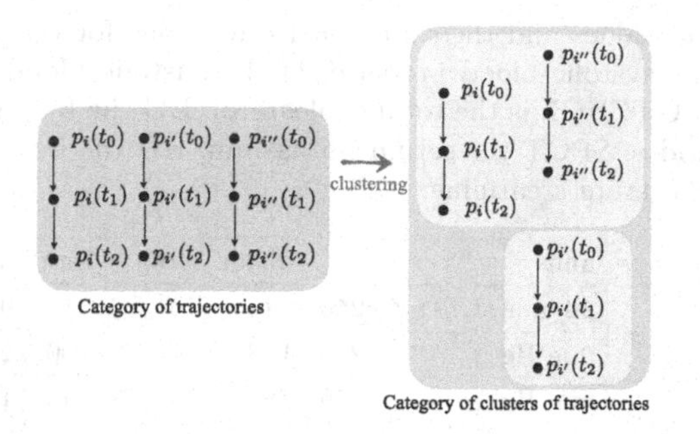


**Fig. 2.** Clustering as a functor from the category of trajectories to the category of clusters of trajectories. In category theory, a functor is a generalization of a function, mapping point and arrows from a category to another one. Here, we consider a mapping from the category of trajectories to the category of clusters of trajectories. The points are the patients at given time points, and the arrows are the comparisons of their clinical values. Trajectories are given by the comparisons of patients with themselves over time. In the second category, we group patients presenting similar trajectories inside the same cluster.

We derive the profile of patients of this population considering a set of the most relevant variables describing their characteristics. The variables are measured at the baseline (time $t_0$) and at three follow-ups ($t_1$, $t_2$, $t_3$). At $t_0$ (Table 1), the 256 patients have a mean eGFR of $64 \pm 16$. Their mean values of systolic blood pressure and diastolic blood pressure are, respectively, $138 \pm 16$ and $78 \pm 10$ mmHg, and serum triglycerides ($172 \pm 106$ mg/dl); these values are moderately high. The mean values of blood glucose ($144 \pm 46$ mg/dl) and HbA1c ($7.2 \pm 1.2\%$) are also elevated. The mean values of total cholesterol ($181 \pm 44$ mg/dl) and serum potassium ($4.5 \pm 0.5$ mmol/l) are in the normal range. The mean value of UACR is moderately elevated ($78.94 \pm 283.85$ mg/g Creatinine). The large standard deviation takes into account the great variability of UACR values across patients. In the following, we examine trajectory clusters obtained with the *kml-Shape* method.

**Table 1.** The mean values and their standard deviations for the 256 patients at the baseline. SBP is the systolic blood pressure, DBP diastolic blood pressure, SCR the serum creatinine, TOTCHOL is the total cholesterol, BG the blood glucose, STRIG is the serum triglycerides, SPOT the serum potassium, HB the hemoglobin, UACR the ratio of albumine to serum creatinine.

| | unit | N | mean | std | min | max |
|---|---|---|---|---|---|---|
| eGFR | ml/min/1.73 $m^2$ | 256 | 64 | 16 | 31 | 90 |
| SBP | mm Hg | 256 | 138 | 16 | 100 | 180 |
| DBP | mm Hg | 256 | 78 | 10 | 44 | 101 |
| BG | mg/dl | 256 | 144 | 46 | 47 | 326 |
| HbA1c | % | 256 | 7.2 | 1.2 | 4.9 | 11.8 |
| SCR | mg/dl | 256 | 1.08 | 0.30 | 0.66 | 2.12 |
| TOTCHOL | mg/dl | 256 | 181 | 44 | 85 | 363 |
| STRIG | mg/dl | 256 | 172 | 106 | 44 | 859 |
| SPOT | mmol/l | 256 | 4.5 | 0.5 | 3.2 | 6.1 |
| HB | g/dl | 256 | 13.5 | 1.5 | 9.7 | 17.6 |
| UACR | mg/g | 256 | 78.94 | 283.85 | 0.0 | 2777.14 |

**Table 2.** Values of the mean eGFR for the 256 patients at each time-point.

| time | min | max | mean | std |
|---|---|---|---|---|
| baseline ($t_0$) | 31 | 90 | 64.0 | 16.3 |
| follow-up 1 ($t_1$) | 23 | 122 | 63.3 | 18.9 |
| follow-up 2 ($t_2$) | 15 | 105 | 60.6 | 18.3 |
| follow-up 3 ($t_3$) | 19 | 120 | 58.6 | 18.2 |

## 3 Results of the Longitudinal Clustering

In this section, we present the clusters of longitudinal data obtained with the *kmlShape* method (Fig. 3). We chose 5 as the number of clusters following clinical practice to analyze heterogeneity in eGFR patients' trends, taking into account that the eGFR is computed from different variables [30]. Such a choice is also motivated by the analysis of eGFR trajectories in different follow-ups, whose mean presents a slow decline. The choice of 5 classes allowed us to highlight behaviors such as crossing and bifurcations[2] having a medical interest.

[2] Each cluster of trajectories contains the same patients. However, we noticed that there are crossings, that we called "switches", between the mean values of eGFR of specific clusters. It means that, for instance, the patients in a specific cluster had an improvement over time, while the patients belonging to a cluster of initial good values of eGFR, had later a worsening of their condition. Or, in another case, we notice that two initially-close clusters of trajectories are then moving apart (the bifurcation). This is interesting from a medical point of view, because the patients who are initially quite close, then can have a different disease behavior.

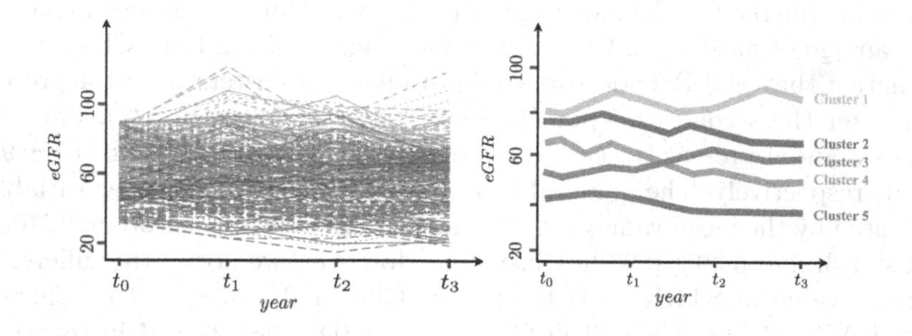


**Fig. 3.** Patients' eGFR trajectories (left), and mean eGFR trajectories obtained with *kmlShape* (right).

**Table 3.** Characteristics of patients in each of the five clusters of trajectories.

| | unit | cluster 1 ($N = 160$) | | cluster 2 ($N = 292$) | | cluster 3 ($N = 152$) | | cluster 4 ($N = 216$) | | cluster 5 ($N = 204$) | | p-value |
|---|---|---|---|---|---|---|---|---|---|---|---|---|
| | | mean | std | mean | std | mean | std | mean | std | mean | std | |
| eGFR | mg/dl | 84 | 11 | 72 | 10 | 57 | 9 | 57 | 11 | 38 | 9 | 0.000 |
| age | years | 62 | 10 | 66 | 7 | 69 | 8 | 69 | 8 | 73 | 9 | 0.000 |
| BMI | kg | 30.90 | 5.18 | 30.54 | 4.65 | 31.53 | 4.97 | 31.38 | 4.65 | 30.31 | 5.16 | 0.050 |
| BG | mg/dl | 148 | 50 | 149 | 48 | 144 | 56 | 152 | 54 | 155 | 74 | 0.363 |
| HbA1c | % | 7.4 | 1.1 | 7.2 | 1.2 | 7.3 | 1.3 | 7.4 | 1.2 | 7.3 | 1.4 | 0.182 |
| TOTCHOL | mg/dl | 176 | 41 | 184 | 44 | 186 | 49 | 178 | 53 | 172 | 39 | 0.020 |
| STRIG | mg/dl | 159 | 88 | 169 | 102 | 169 | 117 | 203 | 150 | 194 | 131 | 0.000 |
| SPOT | mmol/l | 4.5 | 0.5 | 4.4 | 0.5 | 4.4 | 0.5 | 4.5 | 0.5 | 4.9 | 0.6 | 0.000 |
| HB | g/dl | 14.3 | 1.4 | 14.0 | 1.3 | 13.5 | 1.5 | 13.2 | 1.4 | 12.7 | 1.5 | 0.000 |
| CRP | mg/l | 0.68 | 2.13 | 0.49 | 1.46 | 0.56 | 1.15 | 0.49 | 0.99 | 0.86 | 2.32 | 0.035 |
| UACR | mg/g | 34.91 | 102.07 | 43.53 | 157.58 | 54.68 | 153.43 | 88.09 | 281.23 | 122.91 | 324.18 | 0.000 |

Examining the resulting mean trajectories, we notice a bifurcation between cluster 1 and cluster 2, and a switch between cluster 3 and cluster 4. In the following, we analyze how the relevant variables describe the profile of patients, trying to understand the pattern of their dynamic behavior.

The mean values of the selected variables for patients in each cluster are shown in Table 3. They are clusters of patients, grouped according to the shape similarity of their eGFR trajectories. Patients are distributed along five different levels of eGFR, ranging from a mean value of $84 \pm 11$ in Cluster 1, and $30 \pm 9$ in Cluster 5. For the patients in each cluster, we also computed the mean values of the variables which presented a statistical significance (with the p-value test): age, body-mass index, blood glucose, HbA1c, total cholesterol, serum triglycerides, serum potassium, hemoglobin, and UACR. The information provided by these other variables can shed light on unexpected behaviors of eGFR mean trajectories.

Observing Fig. 3, we notice that the eGFR trajectories for patients in Clusters 1 and 2 start from close values of mean eGFR around 80 mg/dl, then have a different trend. Both Clusters 1 and 2 present higher eGFR values ($84 \pm 11$ and $72 \pm 10$, respectively), UACR under control ($34.91 \pm 102.07$ and $43.04 \pm 87.38$), but mean values of STRIG higher for patients in Cluster 2 ($169 \pm 102$ against

$159 \pm 88$ of Cluster 1). The mean age of patients in Cluster 2 is also higher than the mean age of patients in Cluster 1 ($66 \pm 7$ against $62 \pm 10$ of Cluster 1). We then notice that eGFR trajectories of patients in Clusters 3 and 4 present a switch after the second time-point, that is, the first follow-up ($t_1$). Patients in Clusters 3 and 4 present trajectories of eGFR slightly lower, that is, $57 \pm 9$ and $57 \pm 11$, respectively. The main difference with respect to the other variables is constituted by the mean values of CRP, that is, the C-reactive protein: $0.56 \pm 1.15$ in Cluster 3, and $0.49 \pm 0.99$ in Cluster 4. Moreover, we notice the difference of the mean value of STRIG ($169 \pm 117$ in Cluster 3, $203 \pm 150$ in Cluster 4) and of UACR ($54.68 \pm 153.43$ in Cluster 3, $88.09 \pm 281.23$ in Cluster 4). The improvement of mean eGFR across time for patients in Cluster 4 can be due to the effect of drug treatment, more effective for patients belonging to this specific subgroup. Patients in Cluster 5 present the lowest values of mean eGFR ($38 \pm 9$), and they are characterized by critical values of HB and SPOT ($12.7 \pm 1.5$ and $4.9 \pm 0.6$, respectively).

## 4 Discussion

Diabetic kidney disease is a devastating complication of type-2 diabetes mellitus that reduces quality and quantity of life of affected patients and puts an enormous burden on healthcare budget. In addition to the optimization of lifestyle, the selection of the optimum drug combination for therapy is crucial to prevent the incidence and progression of DKD. Once thought to be a uniform disease, it is now evident that there is massive inter-individual and longitudinal intra-individual heterogeneity in disease pathophysiology, clinical presentation, and response to therapy. Linking the characteristics of each patient with the features of a specific subgroup of patients can give hints about the possible effective drug combination.

This is why, starting from a DKD dataset, we built subgroups of similar patients. In particular, we noticed indeed the bifurcation and a switch between mean trajectories. From a theoretical point of view related with basic definitions of category theory, we connected the comparison between clustering methods with the comparison between their results as clusters. A clustering method can be seen as a transformation, and the comparison between clusters is a natural transformation. Here, we estimated similarities and differences of the two methods (and thus, the natural transformation between them) in terms of their effect on a given dataset.

From our analysis of the results, we emphasize that patients with similar levels of eGFR at the baseline can then present a different disease evolution. This fact can be explained with different characteristics of the other variables at each time-point. This result is found using a shape-similarity method, the *kmlShape*, using the Fréchet distance.

The considered patients were given, at each time-point, one of four possible combinations of drugs, described in the Appendix: RASI + GLP1a, RASi + SGLT2i, RASi + MCRa, and RASi only. Analyzing the eGFR mean trajectories shown in Fig. 3, and comparing them with the respective treatment received by the patients, we notice that the patients in Cluster 1 mostly received RASi only; patients in Clusters 2 and 3 were given RASi + MCRa; patients in Cluster 4 mostly received SGLT2i. On the other hand, patients in Clusters 3 and 5 did not receive GLP1a, independently by the level of eGFR.

The information achieved with trajectory clustering can be fed into a decision system, to predict the disease evolution of patient, according to their baseline clinical overview. Thus, our study can lead to a machine-learning application to help physicians deal with new cases of DKD disease. We highlighted here a connection between abstract mathematics, medical practice, and patients' real data, with a potential for further technological applications.

This research may help foster new strategies to improve DKD patients' lives.

**Funding and Data Availability.** **Funding.** This publication was supported by the European Union's Horizon 2020 research and innovation program under grant agreement Nr. 848011. Views and opinions expressed are however those of the authors only and do not necessarily reflect those of the European Union. Neither the European Union nor the granting authority can be held responsible for them. **Institutional review.** DC-ren approval number of the Ethics Committee of the Medical University Innsbruck: EK Nr. 1188/2020, 19.06.2020. **Informed consent.** The DC-ren cohort consists of patients from PROVALID and informed consent was obtained from all patients. A: Ethical approval from the Ethics Committee of the Medical University Innsbruck AN4959 322/4.5 370/5.9 (4012a); 29.01.2013 and approval of the Ethics Committee of Upper Austria, Study Nr. I-1-11; 30.12.2010. H: Approval from Semmelweis University, Regional and Institutional Committee Of Science And Research Ethics; No.12656-0/2011-EKU (421/PV11.); 17.06.2011. UK: Approval from WoSRES, NHS; Rec. Reference: 12/WS/0005 (13.01.2012). NL: Approval of the Medical Ethical Committee of the University Medical Center Groningen, ABRnr. NL35350.042.11. **Data availability.** The dataset is not publicly available.

This publication was supported by the European Union's Horizon 2020 research and innovation programme under grant agreement No 848011

## References

1. Mayer, G., Heerspink, H.J.L., Aschauer, C., Heinzel, A., Heinze, G., Kainz, A., et al.: Systems biology-derived biomarkers to predict progression of renal function decline in type 2 diabetes. Diabetes Care **40**, 391–397 (2017)
2. Park, S., Xu, H., Zhao, H.: Integrating multidimensional data for clustering analysis with applications to cancer patient data. J. Am. Stat. Assoc. **116**(533), 14–26 (2021)
3. Liu, L., Lin, L.: Subgroup analysis for heterogeneous additive partially linear models and its application to car sales data. Comput. Stat. Data Anal. **138**, 239–259 (2019)
4. Philipson, L.H.: Harnessing heterogeneity in type 2 diabetes mellitus. Nature **16**(79), 80 (2019)

5. Fuchs, S., Di Lascio, M., Durante, F.: Dissimilarity functions for rank-invariant hierarchical clustering of continuous variables. Comput. Stat. Data Anal. **159**, 107201 (2021)
6. Amiri, S., Clarke, B.S., Clarke, J.L.: Clustering categorical data via ensembling dissimilarity matrices. J. Comput. Graph Statist. **27**(1), 195–208 (2017)
7. Boucquemont, J., Loubère, L., Metzger, M., Combe, C., Stengel, B., Leffondre, K.: Identifying subgroups of renal function trajectories. Nephrol. Dial Transpl. **32**, ii185–ii193 (2017)
8. Kerschbaum, J., et al.: Intra-individual variability of eGFR trajectories in early diabetic kidney disease and lack of performance of prognostic biomarkers. Nat. Sci. Rep. **10**, 1973 (2020)
9. Karpati, T., Leventer-Roberts, M., Feldman, B., Cohen-Stavi, C., Raz, I., Balicer, R.: Patient clusters based on HbA1c trajectories: a step toward individualized medicine in type 2 diabetes. PLoS ONE **13**(11), e0207096 (2018)
10. Park, S.: Examining trajectories of early adolescents' life satisfaction in South Korea using a growth mixture model. Appl. Res. Qual. Life **17**, 149–168 (2022). https://doi.org/10.1007/s11482-020-09884-5
11. Liu, C., Wei, Y., Ling, Y., et al.: Identifying trajectories of Chinese high school students' depressive symptoms: an application of latent growth mixture modeling. Appl. Res. Qual. Life **15**, 775–789 (2020). https://doi.org/10.1007/s11482-018-9703-3
12. Perco, P., Mayer, G.: Molecular, histological, and clinical phenotyping of diabetic nephropathy: valuable complementary information? Kidney Int. **93**, 308–310 (2018)
13. Fréchet, M.: Sur quelques points du calcul fonctionnel. Rendiconti Circolo Matematico Palermo **22**, 1–72 (1906)
14. Genolini, C., Ecochard, R., Benghezal, M., Driss, T., Andrieu, S., Subtil, F.: kmlShape: an efficient method to cluster longitudinal data (time-series) according to their shapes. PLoS ONE **11**(6), e0150738 (2016)
15. Pinaire, J., Aze, J., Bringay, S., Poncelet, P., Genolini, C., Landais, P.: Hospital healthcare flows: a longitudinal clustering approach of acute coronary syndrome in women over 45years. Health Inform. J. 1–17 (2021). https://doi.org/10.1177/14604582211033020
16. Mayer, B.: Using systems biology to evaluate targets and mechanism of action of drugs for diabetes comorbidities. Diabetologia **59**, 2503–2506 (2016)
17. Mac, L.S.: Categories for the Working Mathematicians. Cambridge University Press, New York (1978)
18. Grandis, M.: Higher Category Theory. World Scientific, Singapore (2020)
19. Baez, J., Lauda, A.: A prehistory of n-categorical physics. In: Halvorson, H. (ed.) Deep Beauty: Understanding the Quantum World Through Mathematical Innovation. Cambridge University Press (2011)
20. Spivak, D.: Category Theory for the Sciences. MIT Press, Cambridge (2014)
21. Ehresmann, A., Góomez-Ramirez, E.: Conciliating neuroscience and phenomenology via Category Theory. Progr. Biophys. Mol. Biol. (PBMB) **119**, 347–359 (2015)
22. Carlsson, G., Mémoli, F.: Classifying clustering schemes. Found. Comput. Math. **13**, 221–252 (2013)
23. Mannone, M., Distefano, V., Silvestri, C., Poli, I.: Clustering longitudinal data with category theory for diabetic kidney disease. In: CLADAG 2021, Book of Abstract (2021, to appear)
24. Genolini, C., Falissard, B.: KmL: K-means for longitudinal data. Comput. Stat. **25**(2), 1–34 (2010)

25. Tran, C.S., Nicolau, D., Nayak, R., Verhoeven, P.: Modeling credit risk: a category theory perspective. J. Risk Financ. Manage. **14**(298), 1–21 (2021)
26. Alicic, R.Z., Rooney, M.T., Tuttle, K.R.: Diabetic kidney disease: challenges, progress, and possibilities. Clin. J. Am. Soc. Nephrolol. **12**, 2032–2045 (2017)
27. Pinaire, J., Aze, J., Bringay, S., Poncelet, P., Genolini, C., Landais, P.: Hospital healthcare flows: a longitudinal clustering approach of acute coronary syndrome in women over 45 years. Health Inform. J. **27**(3) (2021)
28. Verboon, P., Pat-El, R.: Clustering longitudinal data using R: A Monte Carlo study. Eur. J. Res. Methods Behav. Soc. Sci. **18**, 144–163 (2022)
29. Varoutas, P.-C., Rizand, P., Livartowski, A.: Using category theory as a basis for a heterogeneous data source search meta-engine: the Prométhée framework. In: Johnson, M., Vene, V. (eds.) AMAST 2006. LNCS, vol. 4019, pp. 381–387. Springer, Heidelberg (2006). https://doi.org/10.1007/11784180_30
30. Thöni, S., Keller, F., Denicolò, S., Buchwinkler, L., Mayer, G.: Biological variation and reference change value of the estimated glomerular filtration rate in humans: a systematic review and meta-analysis. Front. Med. (Lausanne) **6**(9), 1009358 (2022)

# Multi-classification of Alzheimer's Disease by NSGA-II Slices Optimization and Fusion Deep Learning

Ignacio Rojas-Valenzuela[1(✉)], Ignacio Rojas[1], Elvira Delgado-Marquez[2], and Olga Valenzuela[3]

[1] ETSIIT, University of Granada, Granada, Spain
irojas@ugr.es
[2] Department of Economics and Statistics, University of Leon, Leon, Spain
[3] Faculty of Science, University of Granada, Granada, Spain

**Abstract.** This paper presents a two-phase hierarchical classifier for determining the different states in Alzheimer's disease (AD). In the first phase, an evolutionary system is developed to determine the most relevant slices (in both X-axis and Y-axis) of the magnetic resonance imaging (MRI) for the construction of a classifier. To obtain the image features, the biorthogonal wavelet transform 3.3 was used at level 2. Due to the high number of coefficients, a dimensionality reduction is performed using minimum Redundancy - Maximum Relevance algorithm (mRMR) and Principal Component Analysis (PCA). An evolutionary algorithm on a high-performance computer with GPU was used to optimize the slides. Support vector machine (SVM) was used in the fitness function to estimate the features of the classifier in a computationally simple way. In the second phase, using the different solutions of the Pareto front obtained by the evolutionary algorithm, a multiple deep learning system was developed, each of the systems having as input one of the selected slices of the analyzed solution. The solution with three slices (trade-off between complexity and accuracy) was used as the solution. The obtained hierarchical deep learning system fused the information from each system and analyzed the probabilities obtained for each class. As a final result, an accuracy of 92% was obtained for the six classes. A total of 1,200 patients from the Alzheimer's disease neuroimaging initiative (ADNI) database were used, corresponding to six different classes of patients (with varying degrees of dementia).

**Keywords:** Alzheimer's Disease · Multiclass Classification · Multi-objective Genetic Algorithm · Hierarchical System · Deep Learning · Ensemble System

## 1 Introduction

Alzheimer's disease (AD) is one of the most common dementias of our century and has a major impact on elderly patients [1]. The World Health Organization


M. Villani et al. (Eds.): WIVACE 2023, CCIS 1977, pp. 284–297, 2024.
https://doi.org/10.1007/978-3-031-57430-6_22

estimates that one person in the United States of America develops Alzheimer's disease every 66 s. It is estimated that by 2050 there will be a high number of patients over the age of 65 in the U.S. (an estimated 14 million people), which is why scientific and technological advances that can help combat the disease are so important [2]. An estimated 46.8 million people worldwide are living with Alzheimer's disease or other dementias. By 2030, barring any new findings, the number of people affected is expected to rise to nearly 74.7 million [3]. By 2050, the number of people suffering from Alzheimer's disease could rise to 131.5 million. Every 3.2 s, a new case of dementia occurs somewhere in the world. Early diagnosis is crucial to enable patients suffering from Alzheimer's to live longer [4,5]. Early diagnosis (even before obvious physical symptoms of the disease appear) allows other non-pharmacological therapies, such as cognitive stimulation, to be started at an early stage [6]. Early diagnosis has been shown to improve the quality of life of patients and their families, increase or maintain their personal autonomy and preserve their cognitive abilities [7]. A great deal of research has been and is being done on the part of the scientific community, and the number of publications on Alzheimer's disease is high and growing. Alzheimer's disease has been shown to have intermediate stages before the severity of dementia becomes moderate or serious [8]. It is important to note that the development of Alzheimer's disease can be perceptible and that patients themselves occasionally suffer from the knowledge that their cognitive abilities are progressively deteriorating and worsening. However, in the early stages of the disease, when there are hardly any external symptoms, the disease goes unnoticed until the symptoms increase in an advanced stage [9]. It is therefore extremely important to have the right tools to diagnose the disease at an early stage, so that treatment and prognosis can be started as soon as possible, which will have a positive impact on the patient's health and quality of life.

In [10] an interesting contribution is presented, which takes into account that with the current progress in personalized medicine and the obtaining of omics data (genetics, proteomics, transcriptomics), together with the information provided by medical images, new opportunities to study, analyze and deal with Alzheimer's disease can be developed. In addition, the hybridization of machine learning (ML) methods is available to deal with high-dimensional data, integrate data from different sources and discover new biomarkers. [10] provides a comprehensive review of various ML methods that have been applied to the study of AD using single or multimodal platform data. An analysis of ML applications for five different topics in AD is provided: Disease classification, drug repurposing, subtyping, progression prediction, and biomarker discovery. In this context, several attempts have already been made to detect and diagnose Alzheimer's disease using artificial intelligence (machine learning) systems and brain imaging (usually MRI) [11]. An example of this approach can be found in [12], in which the authors investigated whether optimization using evolutionary algorithms (EA) can be a precision tool for diagnosing early mild cognitive impairment (EMCI) patients compared to control participants (CN). To carry out this study, using the alzheimer neuroimaging initiative (ADNI) database, a total of 54 patients in the EMCI study and 56 controls were used. Three different types of brain segmentation were analyzed, extracting volumetric features that served as input to the optimization algorithm. Thus, we obtained a classification with an accuracy of 93% for this bi-class problem.

The use of classifiers, such as SVM, has been widely used in the determination of Alzheimer's patients, versus control patients, using MRI [13] and fluorodeoxyglucose (FDG)-positron emission tomography (PET), denoted as FDG-PET, images [14]. For example, in [15], T1-weighted brain MRI for 100 AD patients and 100 normal elderly subjects were used. For these analyses, various structures consisting of the thickness of the entorhinal cortex of both hemispheres and the total grey matter volume, hippocampal volume, and amygdala volume were examined. WEKA and SVM-light were used for training the SVM, obtaining the area under the curve (AUC) values of 0.913 and 0.918 for WEKA and SVM-light respectively.

Currently, deep learning algorithms are a very fruitful and powerful alternative for the automatic analysis of medical images, and for the detection of pathologies based on thousands of previous cases and the experience accumulated by hundreds of professionals [16,17]. In fact, the training of these systems is a complex process [18], where it is necessary to either use previously trained neural structures in other problems and carry out transfer learning, or carry out learning from scratch, which implies having a very large set of images (due to the large number of synaptic connections that need to be optimized). For example, in [19], transfer learning is used to take advantage of the pre-trained models for medical image classifications, such as the VGG19 model.

In this paper, a novel methodology is proposed using multiple classifiers and machine learning tools (SVM, Multi-Objective Genetic Algorithms, mRMR, PCA, and Deep Learning) in different phases, and developing a classifier that uses the most relevant MRI slices for multiple classification of the patient sets used (cognitively normal, significant memory problems, early mild cognitive impairment, mild cognitive impairment, late mild cognitive impairment, and Alzheimer's disease). The first objective proposed in this project is to determine, for a problem with multiple classes (disease states), which are the best slices and in what matter (white matter or gray matter) for the construction of a good classifier. By using all the slices we could be able to give a more precise prediction of the status of the patient. The problem of this approach is the huge amount of data to process: to train a model we need to store the information of many images together in RAM memory which is not feasible for a regular machine. Despite this problem, it is possible to develop intelligent algorithms and extraction methods to efficiently identify the most important features, as well as single-slice models where all results are merged to create a multicriteria classifier.

Since this process is performed by an evolutionary multi-objective optimisation system (NSGA-II), it is necessary that the fitness function has a low computation time, otherwise, when performing thousands of evaluations of this function, the computation time could make the proposed methodology inaccessible.

The dilemma of choosing between neural networks, more specifically convolutional neural network (CNN) and support vector machines for biomedical image processing is solved in the proposed methodology, which SVM is using in the training phase. In the second phase, with the best slices selected, a Deep Learning Neural Network is training and tested (tested with test patients, not used in no other phase). The final Deep Learning Neural Network is an ensemble of different network with different slices of the MRI as input.

## 2 Cohort Used. ADNI Database

ADNI stands for Alzheimer's Disease NeuroImaging Initiative. It is an international project that collects information regarding AD in an open database for the study of Alzheimer's disease, in order to unite the efforts of researchers and health professionals in the research work of this neurodegenerative disorder, it also integrates a open database. It is an initiative created by the Laboratory of Neuro Imaging (LONI) of the University of Southern California, which began in 2004 and collects data from various types of medical tests of patients of different origin: images (MRI and PET) taken in different centers with different machines, clinical and genetic tests, and biological samples. ADNI joins the goal of detecting AD at the earliest possible stage, before signs of dementia. The database, included in what they call IDA (Image & Data Archive), stores tests of adult patients between 55 and 90 years old, with Alzheimer's (AD), with mild cognitive impairment (MCI) and cognitively normal (NC). It is divided into 4 subprojects that have had different durations: ADNI, ADNI GO, ADNI 2, ADNI 3. The latter is the one that is currently being carried out. For this project, 1.5T and 3T MRI images have been obtained for six different kind of patients:

The set of patients used in this paper are classified in six groups:

- CN (Cognitively Normal, class 1): corresponds to healthy individuals (control).
- SMC (Significant Memory Concern, class 2): Patients diagnosed with SMR, when performing the various tests and measures stipulated, may have a score within the normal range of cognition, whereby, the informant does not equate the expressed concern with progressive memory impairment.
- MCI (Mild Cognitive Impairment, class 3): Patients who have been classified in this category, when tested for cognitive abilities, already indicate the existence of a slight but perceptible impairment, which is reflected in the measurements obtained.
- EMCI (Early Mild Cognitive Impairment, class 4): patients classified in this stage suffer milder episodic memory impairment.
- LMCI (Late Mild Cognitive Impairment, class 5): patients diagnosed in this group already have a more advanced stage of MCI, which may be a stage prior to AD.

- AD (Alzheimer Disease, class 6): this final group are patients who are diagnosed and suffering the severe consequences of Alzheimer's disease.

In order to make a balanced system, a total of 200 MRI images have been used for each of the six classes, thus having a total of 1200 MRI images.

# 3 Methodology

## 3.1 Phase 1: Selecting the Best Slices in the X and Y Plane

Magnetic resonance imaging (MRI) of the head uses a powerful magnetic field, radio waves and a computer to produce detailed images of the brain in 3 dimensions (there are therefore 3 axes, the X, Y and Z axis in space). On each axis there are a certain number of slices, where a two-dimensional image can therefore be obtained. Every subject in the experiments run in this paper, has been spatially normalized to a bounding box of $157 \times 189 \times 136$ voxels in the X, Y, and Z directions respectively and voxels (remind that one voxel is a three dimensional pixel) reshaped so that each one measures 1 mm $\times$ 1 mm $\times$ 1 mm (this normalisation is the standard one made for ADNI images in the bibliography, using the SPM toolbox for Matlab by [20]). Brain images have also been segmented so that whole matter (also called W images), gray matter (C1 images) and white matter (C2 images) are available in different files. The effect of this segmentation is shown in Fig. 1 using "Tools for NIfTI and ANALYZE image" by Jimmy Shen [21].

There is no unanimous agreement to determine which are the most relevant slices to carry out an accurate classifier. To the best of our knowledge, there are also no contributions in the bibliography that try to determine which are the most relevant X-axis and Y-axis planes for a six-class problem like the one presented here. Therefore, we are going to use an evolutionary multi-objective algorithm such as the NSGA-II.

The X-axis slices we are going to use will be in the range [15:140]. For the Y axis, the slices of the range [15:170] will be used. This results in a total search space of 126 slices for the X-axis and 156 for the Y-axis. For each image, both the grey and white matter are examined, so that the total number of slices is 564 (282 for the grey matter and 282 for the white matter), with the evolutionary algorithm being responsible for selecting the best slices (optimising the fitness function of various defined targets).

First, it is necessary to characterize each of the images or slices extracted from the MRI for each subject. To do this, the wavelet transform will be used [22].

**Wavelet for Feature Extraction.** Wavelets can be thought of as kind of a supercharged Fourier transform. Using a Fourier transform, you can decompose a signal into a sum of sines and cosines where these sines and cosines form and orthogonal basis for the space of functions that we want to represent [23].

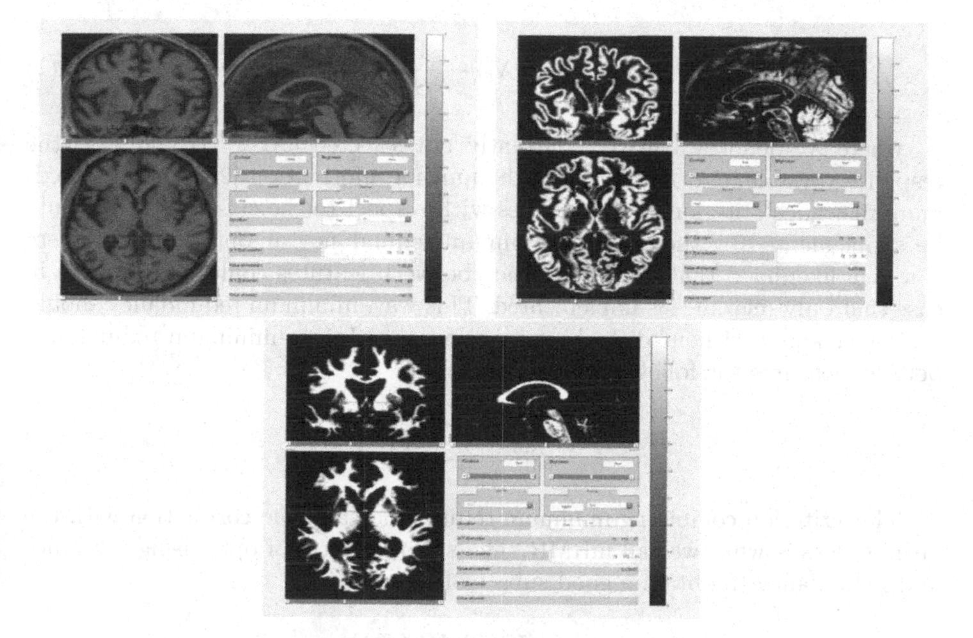

**Fig. 1.** MRI of the brain using NIfTI. Whole matter image (top left), gray matter image (top right), white matter image (bottom) (Color figure online)

This idea of sines and cosines can be generalized using wavelets to other orthogonal functions that might provide a better representation of certain types of signals [24]. Today, we can affirm that wavelets has changed the way we compress and represent signals in the digital era.

In this article, the Wavelet-2D transform has been used, with the mother function bior3.3 at level 2 [25].

As there is a total of 564 slices, and for each of the images, the Wavelet transform obtains a total of 1924 coefficients, there is therefore a large matrix of coefficients, of size 1085136. This number is enormously high for each of the 1200 MRI images corresponding to the different individuals (patients) analysed in this paper and should therefore be minimised using feature selection algorithms (both mRMR and PCA).

**Minimum Redundancy Maximum Relevance (mRMR) and PCA for Feature Reduction.** Minimum Redundancy Maximum Relevance (mRMR) is a feature selection approach that tends to select features with a high correlation with the class (output) and low correlation between themselves.

Given a selected feature set $S$ and assuming there are m features ($|S| < m$), the feature importance of $X_i$, $i \in \{1, 2, ..., m\}$ based on the mRMR criterion can be expressed as:

$$f^{mRMR}(X_i) = I(Y, X_i) - \frac{1}{|S|} \sum_{X_s \in S} I(X_s, X_i) \tag{1}$$

where $X_i$ denotes a feature currently not selected ($X_i \notin S$) and $Y$ is the response variable, and I(Y, ($X_i$)) is the mutual information [26,27]. The maximum relevance criterion of the features with respect to the class variable should be supplemented by the use of a minimum redundancy among features. There is a high likelihood that the dependency between features could be increased in case that only relevance is implemented. This way, minimum redundancy should be implemented without disturbing its relevance. To get minimum redundancy between features the following formula is used:

$$W_I(S) = \frac{1}{|S|^2} \sum_{X_i, X_j \in S} I(X_i, X_j) \tag{2}$$

This criterion combining minimum redundancy and the correlation with the output class is what we call mRMR. The simplest form of optimising relevance and redundancy to obtain a good subset of features is:

$$max\{\phi(V_I(S), W_I(S))\} \tag{3}$$

where $\phi = (V_I(S) - W_I(S))$ and:

$$V_I(S) = \frac{1}{|S|} \sum_{X_i \in S} I(Y, X_i)$$

Practically speaking, the implementation of the mRMR algorithm is high resource consuming. For the research that concerns us, this method will be applied recursively on different subsets of features instead of the full feature set directly. We will discuss its details in upcoming sections.

**SVM for the Fitness Function of NSGA-II.** Support Vector Machines [28], constitute a learning-based method for solving classification and regression problems.

It is a supervised machine learning algorithm that can be used for classification or regression problems. But it is usually used to classify. Given 2 or more labeled data classes, it acts as a discriminative classifier, formally defined by an optimal hyperplane separating all classes. The new examples that are then mapped onto that same space can be classified according to which side of the gap they are on.

Support vectors are the data points closest to the hyperplane, the points in a data set that, if removed, would alter the position of the split hyperplane. The reason for using SVM in the learning phase of the evolutionary algorithm is its fast computation for the fitness function [29].

The block diagram for this initial phase of selecting the best slides for MRI image classification, using and merging various machine learning concepts (SVM, mRMR, PCA, NSGA-II, etc.) is presented in Fig. 2. There is a binary representation for the coding of the individuals of the evolutionary algorithm, where each possible solution has a length of 564 bits and a bit value of 1 would mean that this slice is selected, and a value of 0 means that this slice is not selected. The fitness function of the evolutionary algorithm takes into account both the complexity of the system (measured by the number of selected slices) and the accuracy of the system (measured by the classification accuracy using the confusion matrix). This leads to a Pareto front.

As can be seen in Fig. 2, all MRI images come from the ADNI base, and there are three different sets: training and validation sets, which are used to optimize the selected slices, and the test set, which is used to perform the test and determine the confusion matrix of the system. In Fig. 2, during slices optimisation, feature extraction and selection is first performed for the different slices, and NSGA-II is used to optimize which of the slices are most suitable for classifying the system (using an SVM as classifier). The lower part of Fig. 2 shows the test phase. Since there are different solutions of the NSGA-II algorithm (there is a Pareto front with solutions with different number of slices and different accuracy), different solutions can be selected and their accuracy can be analysed with test images.

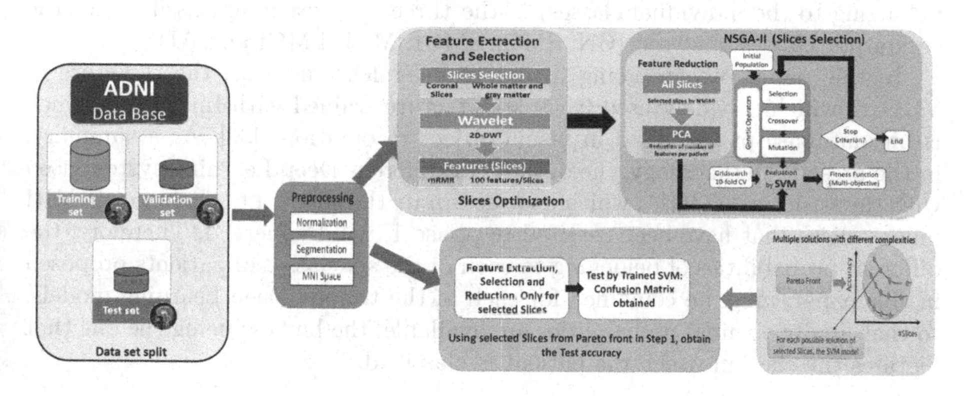


**Fig. 2.** The block diagram for this initial phase of selecting the best slides

## 3.2 Phase 2: Fusion Several Deep Learning System with Different Slices Selected

Deep Learning is defined as a structured or hierarchical automatic algorithm that emulates human learning in order to obtain certain knowledge [30,31]. It stands out because it does not require previously programmed rules, but rather the system itself is capable of "learning" by itself to carry out a task through a previous training phase. In turn, it is also characterized by being composed of intertwined artificial neural networks for information processing. It is mainly used for the automation of predictive analysis.

Deep Learning systems are a fundamental tool in artificial intelligence and are being widely used in biomedical image recognition problems [32].

In this paper, we are going to use transfer learning with the structure of VGG16, which is widely used CNN architecture trained with ImageNet (which is big database project used in visual object recognition) [33].

In the NSAGA-II algorithm used, as mentioned above, several slices were obtained that can be used to build an SVM-type classification system. To build the SVM classifier, the most relevant features must be extracted and selected (using the wavelet transform of the image). Once the best slices have been obtained, we can have a Pareto front in which there are several solution configurations (with different complexity, measured by the number of slices used) for the resolution of the system. The next step we will perform is to use the most representative slices, but now as input to a deep learning system, where it is therefore not necessary to perform the feature extraction phase. We will use the solution where there are 3 slices and therefore three images. Since each image produces a classifier, the information from the three deep learning classifiers is fused taking into account the probability given to each of the output classes. This structure is therefore a weighted union (according to the probabilities of belonging to the individual classes) of the three deep learning classifiers for the six output classes analysed (CN, SMC, EMCI, MCI, LMCI and AD).

In this paper, we are going to use transfer learning with the structure of VGG16, which is widely used CNN architecture trained with ImageNet (which is big database project used in visual object recognition) [33]. As a summary, in the present methodology, the outputs of different Deep Learning system used with the structure VGG16, which have as inputs the slices of the volumetric MRI image, slices that have been selected in phase 1, will be merged. Therefore, the different probabilities of belonging to each of the six classes of patients proposed in this paper, must be combined for each of the trained Deep Learning models, so that finally six final probabilities are available, the highest being the one that dictates the class in which the patient is classified.

## 4 Results

Because the Pareto Front offers different solutions (Fig. 3a), a configuration with few slices is used in this paper, as is the solution with three slices. In this solution, the Slice of the cut in the X plane with value 81 for gray matter is used, together with the cut in the Y plane number 134 for white matter and the cut in Y plane 146 for gray matter.

The Pareto Front Genome (which would be the slices used and their impact on the accuracy of the classifier) are presented in Fig. 3b, together with the frequency of appearance in Fig. 4

Since three slices are selected, a complex deep learning system must be built with three deep learning systems based on the VGG16 structure, but each of

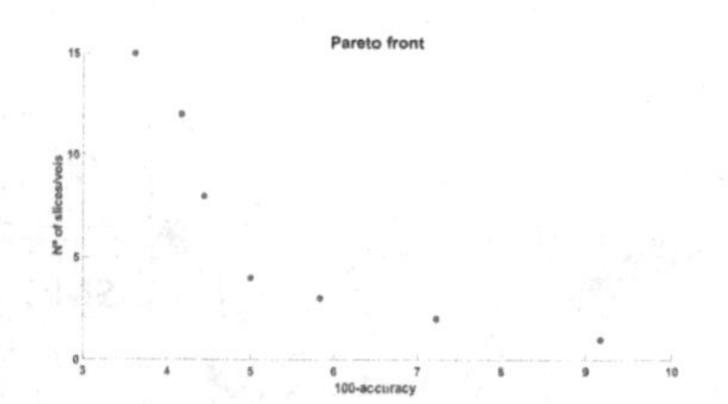


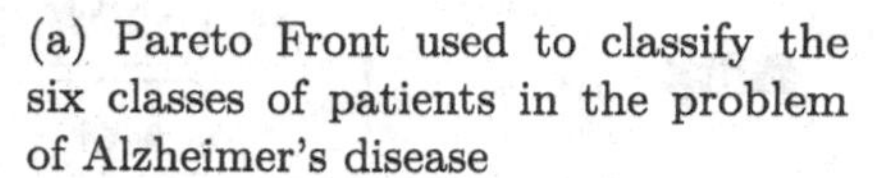
(a) Pareto Front used to classify the six classes of patients in the problem of Alzheimer's disease

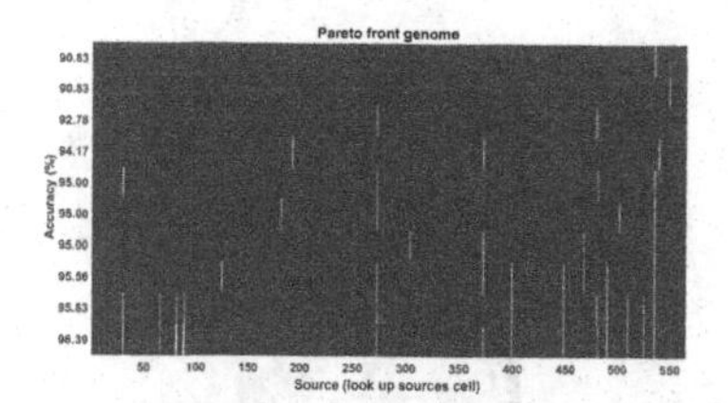


(b) Pareto Front Genome: relevance in the accuracy of the classifier depending on the slices used.

**Fig. 3.** Pareto Front used to classify the six classes and Pareto Front Genome

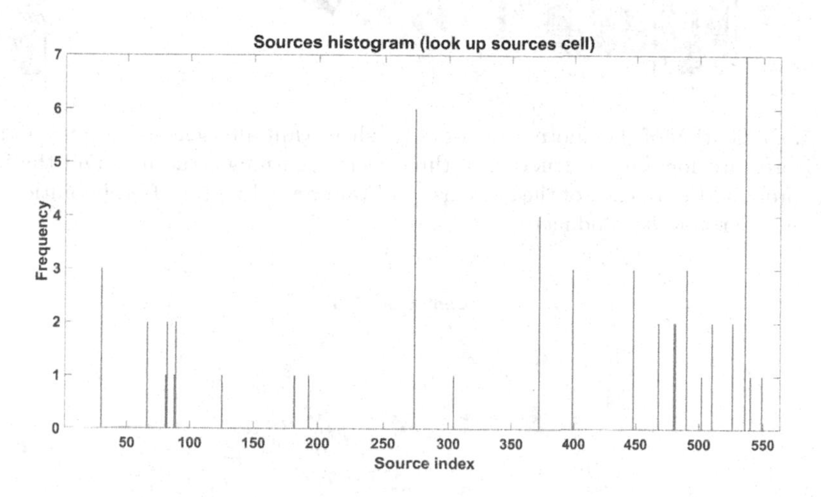


**Fig. 4.** Frequency of appearance of the slices in the solutions found by the evolutionary algorithm.

them receiving a different image of the slice and matter selected by our genetic algorithm solution (remember that it is X = 81 for gray matter, Y = 134 for white matter and finally Y = 146 for gray matter). The structure is presented in Fig. 5.

The probabilities of belonging to each of the six classes analyzed, for each of the three systems, are computed jointly, to obtain, by multiplying them, final probabilities of belonging to each of the six classes. This will be the final output of the system, the class that gets the highest probability. Using a set of test patients, not used in previous phases, the confusion table is constructed. This table is represented in Fig. 6.

As can be seen, for the hierarchical system presented in this contribution, a classification accuracy greater than 92% is obtained.

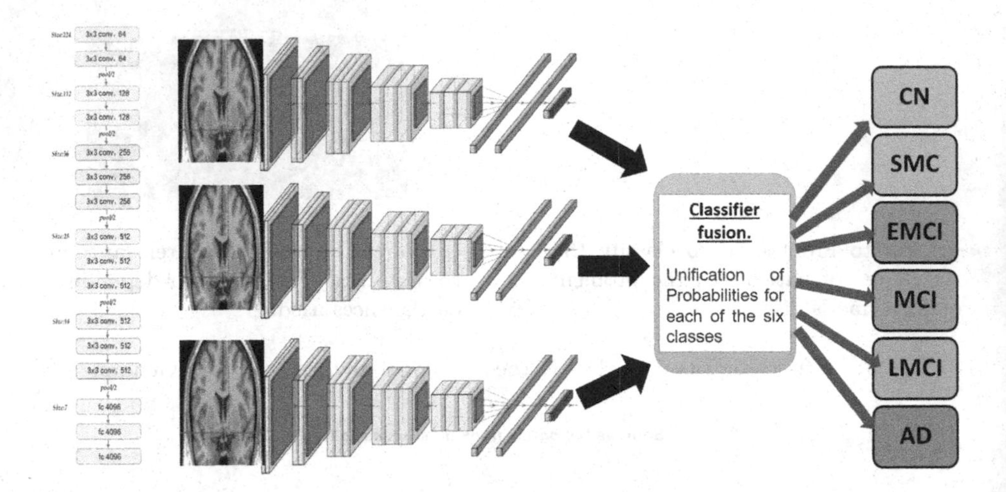


**Fig. 5.** Unification of the information of three slices (initially selected by an evolutionary algorithm), for the construction of three Deep Learning structures and the fusion of the probabilities of each of the systems to obtain a final vector of probabilities (used as merge operator the product)

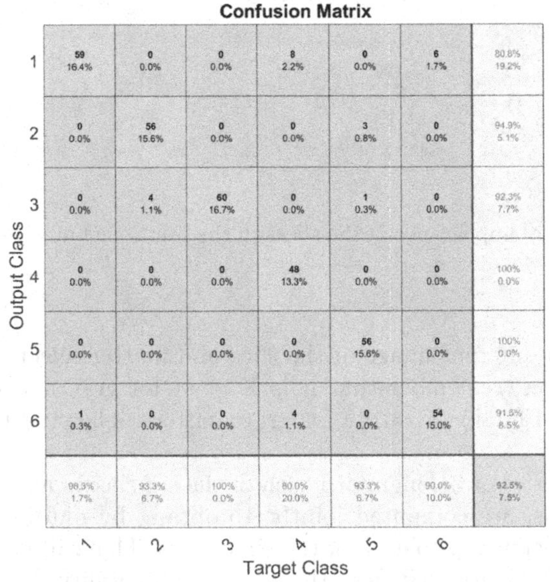

**Confusion Matrix**

| Output Class \ Target Class | 1 | 2 | 3 | 4 | 5 | 6 | |
|---|---|---|---|---|---|---|---|
| 1 | 59<br>16.4% | 0<br>0.0% | 0<br>0.0% | 8<br>2.2% | 0<br>0.0% | 6<br>1.7% | 80.8%<br>19.2% |
| 2 | 0<br>0.0% | 56<br>15.6% | 0<br>0.0% | 0<br>0.0% | 3<br>0.8% | 0<br>0.0% | 94.9%<br>5.1% |
| 3 | 0<br>0.0% | 4<br>1.1% | 60<br>16.7% | 0<br>0.0% | 1<br>0.3% | 0<br>0.0% | 92.3%<br>7.7% |
| 4 | 0<br>0.0% | 0<br>0.0% | 0<br>0.0% | 48<br>13.3% | 0<br>0.0% | 0<br>0.0% | 100%<br>0.0% |
| 5 | 0<br>0.0% | 0<br>0.0% | 0<br>0.0% | 0<br>0.0% | 56<br>15.6% | 0<br>0.0% | 100%<br>0.0% |
| 6 | 1<br>0.3% | 0<br>0.0% | 0<br>0.0% | 4<br>1.1% | 0<br>0.0% | 54<br>15.0% | 91.5%<br>8.5% |
| | 98.3%<br>1.7% | 93.3%<br>6.7% | 100%<br>0.0% | 80.0%<br>20.0% | 93.3%<br>6.7% | 90.0%<br>10.0% | 92.5%<br>7.5% |

**Fig. 6.** Confusion table for the six classes used and the accuracy values obtained during the training phase.

## 5 Conclusion

Researchers are still working on new diagnostic tools that could help doctors diagnose Alzheimer's dementia earlier in the course of the disease, when symptoms are very mild or even before they appear. For people with Alzheimer's dementia, doctors can offer drug and non-drug interventions that can ease the burden of the disease. Doctors often prescribe medications that slow the deterioration of memory and other cognitive abilities.

In this sense, the construction of intelligent and automatic classifiers, which can discern and identify the different stages and stages of Alzheimer's disease through MRI images, is of great relevance for neurological experts. In this paper, a novel contribution is presented, which analyzes the most relevant slices for classification and presents a fusion system of neural systems (based on Deep Learning) for the construction of a classifier that obtains an accuracy of 92%.

**Acknowledgements.** This work was funded by the Spanish Ministry of Sciences, Innovation and Universities under Project PID2021-128317OB-I00 and the projects from Junta de Andalucia P20-00163.

## References

1. Querfurth, H.W., LaFerla, F.M.: Alzheimer's disease. N. Engl. J. Med. **362**(4), 329–344 (2010)
2. GBD 2019 Dementia Forecasting Collaborators, "Estimation of the global prevalence of dementia in 2019 and forecasted prevalence in 2050: an analysis for the global burden of disease study 2019," Lancet Public Health (2022)
3. "Global dementia cases forecasted to triple by 2050," Alzheimer's Association International Conference®(AAIC®) (2021)
4. Leifer, B.P.: Alzheimer's disease: seeing the signs early. J. Am. Acad. Nurse Pract. **21**(11), 588–595 (2009)
5. Barczak, A.: Mossakowski Medical Research Centre, Polish Academy of Sciences, Warsaw, Poland, "The early diagnosis of alzheimer's disease", Pediatr. Med. Rodz., vol. 14(2), 157–166 (2018). Department of Neurodegenerative Disorders
6. Jonin, P.-Y., et al.: Building memories on prior knowledge: behavioral and fMRI evidence of impairment in early Alzheimer's disease. Neurobiol. Aging **110**, 1–12 (2022)
7. Selkoe, D.J.: Early network dysfunction in Alzheimer's disease. Science **365**(6453), 540–541 (2019)
8. Li, X., Coyle, D., Maguire, L., Watson, D.R., McGinnity, T.M.: Gray matter concentration and effective connectivity changes in Alzheimer's disease: a longitudinal structural MRI study. Neuroradiology **53**(10), 733–748 (2011)
9. Odusami, M., Maskeliūnas, R., Damaševičius, R., Krilavičius, T.: Analysis of features of Alzheimer's disease: detection of early stage from functional brain changes in magnetic resonance images using a finetuned ResNet18 network. Diagnostics (Basel) **11**(6), 1071 (2021)
10. Li, Z., Jiang, X., Wang, Y., Kim, Y.: Applied machine learning in Alzheimer's disease research: omics, imaging, and clinical data. Emerg. Top. Life Sci. **5**(6), 765–777 (2021)

11. Wu, H., Luo, J., Lu, X., Zeng, Y.: 3D transfer learning network for classification of Alzheimer's disease with MRI. Int. J. Mach. Learn. Cybern. **13**, 1997–2011 (2022)
12. Zamani, J., Sadr, A., Javadi, A.-H.: Diagnosis of early mild cognitive impairment using a multiobjective optimization algorithm based on T1-MRI data. Sci. Rep. **12**(1), 1020 (2022)
13. Sun, Z., Qiao, Y., Lelieveldt, B.P., Staring, M., Alzheimer's disease neuroimaging initiative.: integrating spatial-anatomical regularization and structure sparsity into SVM: improving interpretation of Alzheimer's disease classification. NeuroImage, **178**, 445–460 (2018)
14. Dukart, J., et al.: Meta-analysis based SVM classification enables accurate detection of Alzheimer's disease across different clinical centers using FDG-PET and MRI. Psychiatry Res. **212**(3), 230–236 (2013)
15. Tantiwetchayanon, K., Vichianin, Y., Ekjeen, T., Srungboonmee, K., Ngamsombat, C., Chawalparit, O.: Comparison of the WEKA and SVM-light based on support vector machine in classifying Alzheimer's disease using structural features from brain MR imaging. J. Phys: Conf. Ser. **1248**(1), 012003 (2019)
16. Romano, M.F., Kolachalama, V.B.: Deep learning for subtyping the Alzheimer's disease spectrum. Trends Mol. Med. **28**, 81–83 (2022)
17. Lei, B., et al.: Predicting clinical scores for Alzheimer's disease based on joint and deep learning. Expert Syst. Appl. **187**(115966), 115966 (2022)
18. Dai, Y., Bai, W., Tang, Z., Xu, Z., Chen, W.: Computer-aided diagnosis of Alzheimer's disease via deep learning models and radiomics method. Appl. Sci. (Basel) **11**(17), 8104 (2021)
19. Helaly, H.A., Badawy, M., Haikal, A.Y.: Deep learning approach for early detection of Alzheimer's disease. Cognit. Comput. **14**, 1–17 (2021)
20. Ashburner, J.: SPM: a history. NeuroImage **62**(2), 791–800 (2012). https://doi.org/10.1016/j.neuroimage.2011.10.025
21. Shen, J.: Tools for NIfTI and analyze image. MATLAB Central File Exchange (2023)
22. Chaplot, S., Patnaik, L., Jagannathan, N.: Classification of magnetic resonance brain images using wavelets as input to support vector machine and neural network. Biomed. Sig. Process. Control. **1**(1), 86–92 (2006). https://doi.org/10.1016/j.bspc.2006.05.002
23. Feng, J., Zhang, S.-W., Chen, L.: Identification of Alzheimer's disease based on wavelet transformation energy feature of the structural MRI image and NN classifier. Artif. Intell. Med. **108**, 101940 (2020). https://doi.org/10.1016/j.artmed.2020.101940
24. Gilles, J.: Empirical wavelet transform. IEEE Trans. Sig. Process. **61**(16), 3999–4010 (2013). https://doi.org/10.1109/tsp.2013.2265222
25. Wang, X.H., Zhao, B., Li, L.: Mapping white matter structural covariance connectivity for single subject using wavelet transform with T1-weighted anatomical brain MRI. Front. Neurosci. **16**, 1038514 (2022). https://doi.org/10.3389/fnins.2022.1038514
26. Radovic, M., Ghalwash, M., Filipovic, N., Obradovic, Z.: Minimum redundancy maximum relevance feature selection approach for temporal gene expression data. BMC Bioinform. **18**, 1–14 (2017). https://doi.org/10.1186/s12859-016-1423-9
27. Qu, K., Xu, J., Han, Z., Xu, S.: Maximum relevance minimum redundancy-based feature selection using rough mutual information in adaptive neighborhood rough sets. Appl. Intell. **53**(14), 17727–17746 (2023). https://doi.org/10.1007/s10489-022-04398-z

28. Drucker, H., Wu, D., Vapnik, V.: Support vector machines for spam categorization. IEEE Trans. Neural Netw. **10**(5), 1048–1054 (1999)
29. Devos, O., Downey, G., Duponchel, L.: Simultaneous data pre-processing and SVM classification model selection based on a parallel genetic algorithm applied to spectroscopic data of olive oils. Food Chem. **148**, 124–130 (2014)
30. LeCun, Y., Bengio, Y., Hinton, G.: Deep learning. Nature **521**(7553), 436–444 (2015)
31. He, K., Zhang, X., Ren, S., Sun, J.: Deep residual learning for image recognition. In: 2016 IEEE Conference on Computer Vision and Pattern Recognition (CVPR). IEEE (2016)
32. Litjens, G., et al.: A survey on deep learning in medical image analysis. Med. Image Anal. **42**, 60–88 (2017)
33. Dubey, A.K., Jain, V.: Automatic facial recognition using VGG16 based transfer learning model. J. Inf. Optim. Sci. **41**(7), 1589–1596 (2020)

# Exploiting the Potential of Bayesian Networks in Deriving New Insight into Diabetic Kidney Disease (DKD)

Debora Slanzi[1,2(✉)], Claudio Silvestri[1,3], Irene Poli[1], and Gert Mayer[4]

[1] European Centre for Living Technology, Dorsoduro 3911, Calle Crosera, 30123 Venice, Italy
{debora.slanzi,silvestri,irenpoli}@unive.it

[2] Department of Management, Ca' Foscari University of Venice, Cannaregio 873, 30121 Venice, Italy

[3] Department of Environmental Sciences, Informatics and Statistics, Ca' Foscari University of Venice, via Torino 155, 30172 Mestre, VE, Italy

[4] Department of Internal Medicine IV (Nephrology and Hypertension), Medical University Innsbruck, Anichstraße 35, 6020 Innsbruck, Austria
gert.mayer@i-med.ac.at

**Abstract.** Diabetic kidney disease is a serious complication of diabetes and one of the leading causes of chronic and end-stage kidney disease worldwide. The clinical course and response to therapy is complex and heterogeneous both between and over time within individuals. Therefore it is extremely important to derive even more in-depth information on what characterizes its pathophysiology and pattern of disease progression. Statistical models can help in this task by understanding the interconnections among variables clinically considered to characterize the disease. In this work we propose to use Bayesian networks, a class of probabilistic graphical models, able to identify robust relationships among a set of variables. Furthermore, Bayesian networks are able to include expert knowledge in the modeling phase to reduce the uncertainty on the phenomenon under study. We provide some evidence that the synergy between data and expert prior information is a great source of valuable help in gaining new knowledge about Diabetic Kidney Disease.

**Keywords:** Diabetic Kidney Disease · Bayesian networks · Expert knowledge · Network of relations

## 1 Introduction

Type 2 diabetes mellitus (T2DM) is a chronic metabolic disorder characterized by high levels of blood sugar (glucose) resulting from the body's resistance to insulin or its reduced secretion. The number of adults suffering from T2DM in Europe varies between countries but it is expected to increase overall from


M. Villani et al. (Eds.): WIVACE 2023, CCIS 1977, pp. 298–308, 2024.
https://doi.org/10.1007/978-3-031-57430-6_23

52.8 in 2011 to 69 million by 2045 (www.heartstats.org, accessed June 2023). About 30–40% of affected individuals develop diabetic kidney disease (DKD), a devastating complication that reduces quality as well as duration of life and imposes an enormous burden on health care budget. In developed countries DKD is the leading cause of end stage renal disease [1].

For many years kidney disease in type 2 diabetes was considered to mimic kidney disease in type 1 diabetes, a somewhat "homogenous" disorder primarily driven (at least in early stage) by genetic predisposition and quality of metabolic control. However recently it became evident that it is much more complex and multifactorial due to different comorbidities more prevalent in this elderly population (like hypertension) and deregulations in a large number of different biological pathways including metabolic, hemodynamic, and inflammatory processes have been described [2]. A consequence of this complexity is massive inter-individual and longitudinal intra-individual heterogeneity of pathophysiology on the molecular level the phenotype (i.e. clinical presentation) and the response to specific therapy is observed. Understanding these mechanisms and their interactions cross sectionally and over time is crucial for improving clinical care and developing targeted therapies and interventions to prevent or delay the onset and slow the progression of DKD. With better profiling of patients there is an increasing need of a new understanding on the framework of relationships involving some of the variables and their interactions used to judge the state of a patient with DKD and support selection of appropriate therapy.

In this work we propose a probabilistic graphical model, namely the Bayesian network, to identify the network of relationships among the selected variables of the disease pathophysiology of DKD. Ideally the results should give a consensus to the theoretical path of pathophysiology, and when combined with expert knowledge or per se, should improve the information on the actual relationships among the different considered factors. Specifically, by estimating a Bayesian network model we can contribute in

- evaluating the strength of the well-known relationships on DKD;
- proving new insights on new relationships emerged from the data on patients;
- identifying differences that could be imputed to the specific therapy.

The paper is structured as follow: in Sect. 2 we introduce the study conducted to derive the data used in the analyses and the statistical approach developed to address the proposed objectives, in particular how to include prior knowledge available from the literature and experts to produce more informative models; then in Sect. 3 we present the main results achieved in the content of DKD. Finally, in Sect. 4 we propose some concluding remarks about issues requiring further researches.

## 2 Materials and Methods

### 2.1 The PROVALID Study

The data used in this work were provided by the PROVALID study ("PROspective cohort study in patients with type 2 diabetes mellitus for VALIDation of biomarkers"), a prospective observational study that recruited over 4.000 patients with T2DM in five European countries with normal, mild or moderately reduced kidney function. Patients were followed for at last 4 years and variables holding information on clinical data, laboratory values and medication were collected on an annual basis. For a more complete description of the study and the available data we refer to [3,4]. The disease trajectories (as assessed by changes in eGFR, a measure of renal excretory capacity) were highly variable in the PROVALID participants even under stable therapy [5]. Next to drug adherence and environmental factors, heterogeneity in pathophysiology is a very likely explanation for this finding. In order to systematically approach this problem we defined two populations of patients:

- *RASi only*, population 1: a population of patients that was continuously treated with agents that block the renin angiotensin system, the current standard of care for at least 4 years.
- *Drop-in*, population 2: a selection of patients to whom other agents were added on top of RASi therapy by their clinicians in order to improve metabolic control and/or DKD (sodium glucose transporter 2 inhibitors, i.e. *SGLT2is*, glucagon like peptide 1 receptor agonists, i.e. *GLP1as*, or the mineralocorticoid receptor antagonists, i.e. *MCRAs*.

The definition of these two different populations can help in addressing the aim of identifying differences that could be attributed to the specific therapy.

Among the over one hundred variables collected within the PROVALID data, thirteen available from routine clinical care visits and considered important by physicians were selected. After a preprocessing of the data to remove incomplete cases and to adjust skewed distribution by means of log transformation if appropriated, the selected variables in the two populations are described in Table 1. We point out that the data which we analyzed are datapoints, i.e. we did not consider the longitudinal component of the data. From the Table we can highlight that differences on the mean value of some variables emerge when comparing the two populations, meaning that therapy seems to have an effect to those variables.

After the selection of the relevant variables, clinical expertise was used to construct an interaction network based on pathophysiology understanding. This network, considered a theoretical framework is presented in Fig. 1. Then, the interaction network and the suspected strength of the interactions between variables was considered as a benchmark, and compared with a purely data driven approach to determine if the latter could improve our understanding of the DKD complex interactions. However confounding of this network by changing in treatment that affects target variables with or without altering disease pathology is an obvious weakness.

### 2.2 The Bayesian Networks

To derive the network of relationships among the selected variables of the disease pathophysiology of DKD, we propose to build Bayesian networks (BNs) [6,7].

**Table 1.** Comparison of main selected variables between the two populations. Mean ± standard deviation are reported. p-value refers to a *t*-student test to evaluate the statistical difference between the two populations.

| Variable | Overall n = 1288 | RASi only n = 798 | Drop-in n = 490 | p-value |
|---|---|---|---|---|
| SBP - Systolic Blood Pressure | 135.52 ± 14.93 | 136.84 ± 15.28 | 133.37 ± 14.08 | 0.000 |
| DBP - Diastolic Blood Pressure | 77.61 ± 9.35 | 77.70 ± 9.76 | 77.46 ± 8.65 | 0.636 |
| BG - Blood Glucose | 154.00 ± 56.96 | 154.00 ± 59.30 | 153.99 ± 52.99 | 0.996 |
| HBA1C - Triglycerides $HbA_{1c}$ | 7.50 ± 1.30 | 7.43 ± 1.34 | 7.61 ± 1.24 | 0.016 |
| TOTCHOL - Total Cholesterol | 178.86 ± 47.73 | 181.83 ± 49.20 | 174.03 ± 44.88 | 0.004 |
| HDLCHOL - HDL Cholesterol | 48.29 ± 13.69 | 49.76 ± 14.45 | 45.89 ± 11.98 | 0.000 |
| STRIG - Serum triglycerides | 184.61 ± 121.89 | 178.07 ± 122.52 | 195.26 ± 120.23 | 0.014 |
| SPOT - Serum Potassium | 4.51 ± 0.51 | 4.53 ± 0.51 | 4.49 ± 0.52 | 0.204 |
| HB - Hemoglobin | 13.76 ± 1.56 | 13.46 ± 1.50 | 14.23 ± 1.54 | 0.000 |
| SALB - Serum Albumin | 4.49 ± 0.42 | 4.42 ± 0.41 | 4.60 ± 0.41 | 0.000 |
| CRP - C reactive protein (log) | −1.23 ± 1.14 | −1.22 ± 1.17 | −1.24 ± 1.10 | 0.666 |
| BMI - Body Mass Index | 31.92 ± 5.59 | 31.12 ± 5.09 | 33.23 ± 6.11 | 0.000 |
| UACR - Urinary albumin/creatinine ratio (log) | 2.53 ± 1.75 | 2.65 ± 1.69 | 2.35 ± 1.83 | 0.003 |

Bayesian networks provide a method for the representation and reasoning of uncertainty and have been widely used in the medical field [8–10]. Specifically, a BN for a set of random variables $\mathbf{X} = \{X_1, \ldots, X_p\}$ (in this case $p = 13$) is identified by

- a *network structure* $G$, a directed acyclic graph (DAG) where nodes represent the variables $\mathbf{X}$ of the system and the directed arcs between nodes represent the probability dependences between them,
- a set of *parameters*, representing conditional probability distributions $P(X_i|Pa(X_i))$ associated to each variable $X_i$, $i = 1, \ldots, p$, where $Pa(X_i)$ are the variables that correspond to the parents of $X_i$ in the DAG (i.e. the nodes with an arc pointing towards $X_i$).

The global distribution of the variables $\mathbf{X}$ is decomposed into the local distributions of the individual variables $X_i$ as

$$P(\mathbf{X}) = \prod_{n=1}^{p} P(X_i|Pa(X_i)) \quad (1)$$

The process of estimating a BN is called *learning* and typically involves two main steps: (1) the *structure learning* to identify the topological structure, i.e. which arcs are present in the graph and therefore which probabilistic relationships are supported by the data, and (2) the *parameter learning* to learn the conditional probability distributions that regulate the strength of the relationships.

There are many approaches in literature to estimate BNs from the data [11]: in this work we will focus on a *Search & Score* strategy which uses a score function in order to compare the structures of the network and then selects the

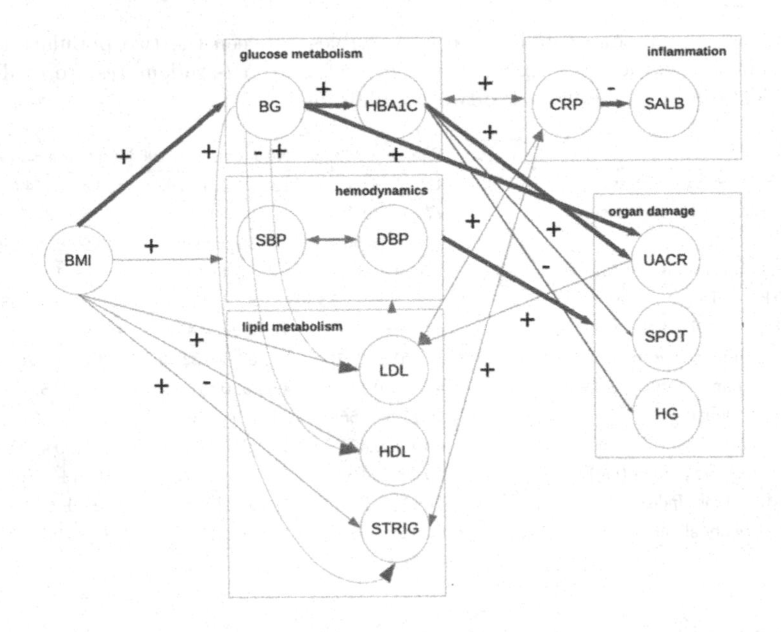


**Fig. 1.** Theoretical framework. Red arcs reflect quite known relationships between the associated variables whereas blue arcs reflect relationships which are less clear from clinical point of view. Signs indicate the direction of the associations.

structure which better fits the data. Specifically, we develop structure learning by means of hill-climbing *search* procedure and a BDe *score* [6]. Furthermore, to reduce the impact of the noise present in the data, *model averaging* learning techniques can be used to improve the reliability of structure learning [12]. The process consists in:

- perform *bootstrap resampling*, i.e. re-sample the data $k$ times using bootstrap and perform structure learning separately on each of the resulting samples, thus collecting $k$ DAGs;
- calculate *arc strength*, i.e. compute the frequency with which each arc appears in those $k$ graphs deriving an "average" consensus DAG by selecting those arcs that have a frequency above a certain threshold $t$.

In this work we fix the number of bootstrap replications to $k = 200$ and threshold to $t = 0.5$ (selection of only arcs with strength $> 0.5$). The average BN model built within this process should be less sensitive to noisy data and typically should produce more accurate predictions for new observations [8].

One more characteristic on structure learning is that BN can include prior knowledge available from the literature and the practice of the discipline to produce more informative models and to overcome the inherent noisiness and variability of data. This is possible by means of *whitelisted arcs*: they represent well-known dependencies which should be forced to be present in the graph. In this work we estimate several BNs by including and excluding prior knowledge

**Table 2.** Measures of graphical differences. *Expert* refers to BN where the structure represents the expert knowledge, *Data only* refers to BN learned using data only, *Data + Prior* refers to BN learned by using both data and expert knowledge.

| Overall population | | | | | | | |
|---|---|---|---|---|---|---|---|
| | Num. arcs | Av. MB size | Av. neighb. size | Missing priors (FN) | TP | TN | BIC |
| Expert | 32 | 6.15 | 4.92 | - | - | - | −51504.84 |
| Data only | 28 | 7.23 | 4.31 | 25 | 7 | 21 | −50990.75 |
| Data + Prior | 49 | 10 | 7.54 | 0 | 32 | 17 | −51044.46 |

representing the theoretical framework of interconnections among the selected variables in DKD. The prior information was delivered by study physicians in the form of 32 prior relationships (whitelisted arcs) derived from the pathophysiology theoretical framework in Fig. 1.

Last, BNs are derived both considering the whole dataset (Overall population) to improve the experts understanding of the pathophysiology complex interactions, and the therapy-specific populations (Rasi and Drop-in populations) to identify if any difference can be imputed to added agents.

## 3 Results

To evaluate the strength of the well-known relationships on DKD and how data can provide insights on new relationships in patients on therapy, we introduce some measure of graphical differences. In Table 2 we provide the number of arcs (Num. arcs), the average Markov Blanket size (Av. MB size), the average neighborhood size (Av. neighb. size), the number of missing priors (FN), the number of confirmed priors (TP) and the number of new arcs emerging from data (TN) with respect to the "Expert" network built with only the 32 whitelisted arcs suggested by expert clinicians. Last, a BIC measure was provided for each BN in order to compare the fit to the data. BNs in Table 2 are learned using data referred to the whole dataset (Overall population).

From the results we can see that the "Data only" BN have a less number of arcs, meaning that data provide relationships that should be considered as robust. By comparing them with the expert prior whitelisted arcs, we highlight that the 7 TP arcs detected by a purely data driven approach have a strength ranging from 1 to 0.910 meaning that the associated prior relationships are highly confirmed also from an empirical point of view (some examples are: SBP → DBP, DBP → HB and BG → HBA1C, all with associated strength equal to 1). Furthermore, 21 new emerging arcs are achieved: some of them describe prior relationships but with a reversed directions (for example, HDLCHOL → BMI with strength equals to 1 or HBA1C → BMI with strength equals to 0.975), but many others can provide new insights on the DKD pathophysiology network as, for example SALB → HB (strength = 1), SALB → UACR (strength = 1) or CPR → BMI (strength = 1).

When looking at the results of the BN learned by using prior expert information, we see that the number of emerged new relationships is 17 and most of them are the same as in the network built using only data.

To understand if therapies affect the results, the same procedure was separately developed in the Rasi only and Drop-in populations. Results are presented in Table 3. The BNs built without prior information within the Drop-in population seems to present less arcs with respect to Rasi only population. Only 3 prior relationships are confirmed in both populations (SBP $\rightarrow$ DBP, DBP $\rightarrow$ HB and BG $\rightarrow$ HBA1C, all with strength equals to 1) but what emerges is that the new relationships found in Rasi only population are mainly different compared to Drop-in population. In Fig. 2 the arcs which can be attributed to therapy are shown. Specifically, black solid lines represent relationships which are present in both Rasi only and Drop-in populations, blue dashed lines represent relationships which are present in Rasi only population but not in Drop-in population and red solid lines represent relationships which are present in Drop-in population but not in Rasi only population. When introducing prior information, despite the high number of common whitelisted arcs which can also put constraints in the search approach, there are again differences that can be attributed to the therapies as shown in Fig. 3. Most of them confirm the results obtained by a purely data-driven approach, but some new relationships also emerge. This suggest that expert prior information can guide and contribute to a better understand on the interconnection network among the variables involved in the disease.

To evaluate how expert knowledge merged with information directly extracted from the data is able to better identify the pattern of pathophysiology, we calculate the predictive accuracy of the BNs estimated from data with and without prior information in the different populations in terms of correlation between the observed and the predicted value for all the variables. This predictive accuracy is achieved by using 10-fold cross-validation [13]. 10-fold cross-validation is a model validation technique that assesses how well a statistical model accurately predict the behavior of new observations; for each variable we compute the correlation between the observed and predicted pairs and this quantity is called predictive correlation. The predictive correlations for all the variables are reported in Table 4. Both Data and Data + Prior BNs predictions for all the considered variables outperform the predictive correlations in the Expert network for all the populations, meaning that data can provide a very valuable source of additional information to better understand unknown mechanisms in the DKD. Furthermore, in differentiating by therapies we can also achieved specific directions of intervention: for example, the value of the predictive correlation of CRP is about 0.2 for Rasi only population and about 0.4 for Drop-in population meaning that the interconnections found in this last BN are able to better describe what influences the value of CRP.

**Table 3.** Measures of graphical differences among populations. *Expert* refers to BN where the structure represents the expert knowledge, *Data only* refers to BN learned using data only, *Data + Prior* refers to BN learned by using both data and expert knowledge.

| RASi only population | | | | | | | |
|---|---|---|---|---|---|---|---|
| | Num. arcs | Av. MB size | Av. neighb. size | Missing priors (FN) | TP | TN | BIC |
| Expert | 32 | 6.15 | 4.92 | - | - | - | −31963.79 |
| Data only | 23 | 5.69 | 3.54 | 27 | 5 | 18 | −31640.04 |
| Data + Prior | 44 | 8.77 | 6.77 | 0 | 32 | 12 | −31693.48 |
| **Drop-in population** | | | | | | | |
| | Num. arcs | Av. MB size | Av. neighb. size | Missing priors (FN) | TP | TN | BIC |
| Expert | 32 | 6.15 | 4.92 | - | - | - | 19505.76 |
| Data only | 19 | 4.15 | 2.92 | 26 | 6 | 13 | −19285.05 |
| Data + Prior | 44 | 9.54 | 6.77 | 0 | 32 | 12 | −19335.19 |

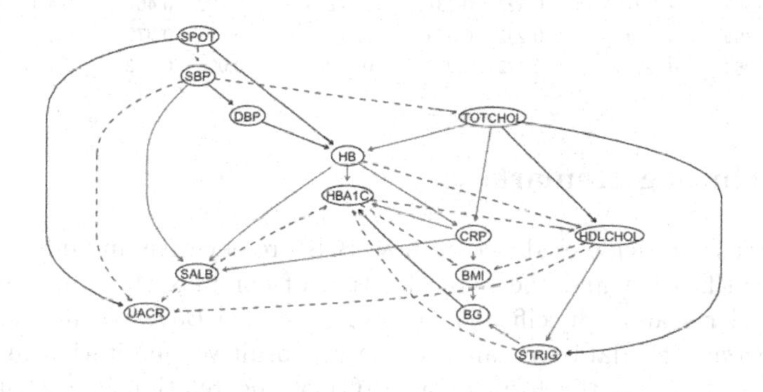


**Fig. 2.** Structural differences imputed to therapy - Data only

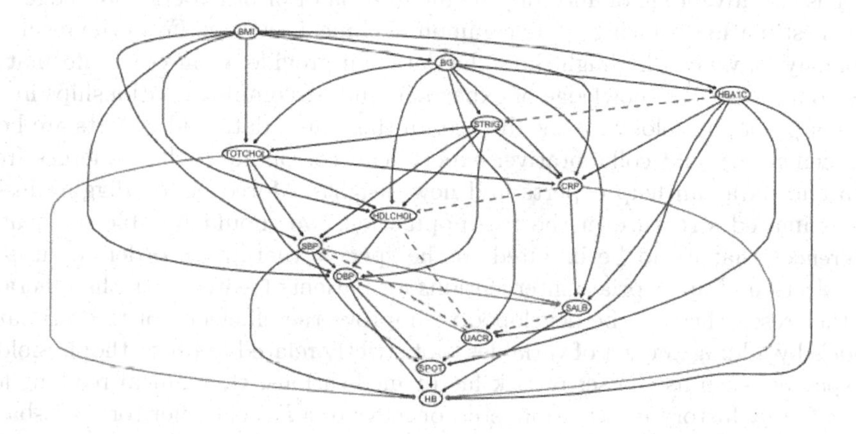


**Fig. 3.** Structural differences imputed to therapy - Data + Prior

**Table 4.** Measures of prediction performance. *Data only* refers to BNs learned using data only, *Data + Prior* refers to BNs learned by using both data and expert knowledge, and *Expert* refers to BNs where the structure represents the expert knowledge and only the parameters of the BN are estimated from data.

| | Overall population | | | Rasi population | | | Drop-in population | | |
|---|---|---|---|---|---|---|---|---|---|
| | Data only | Data + Prior | Expert | Data only | Data + Prior | Expert | Data only | Data + Prior | Expert |
| SBP | 0.424 | 0.424 | 0.385 | 0.419 | 0.416 | 0.390 | 0.417 | 0.415 | 0.337 |
| DBP | 0.416 | 0.422 | 0.416 | 0.430 | 0.426 | 0.425 | 0.369 | 0.422 | 0.381 |
| BG | 0.569 | 0.569 | 0.559 | 0.579 | 0.573 | 0.563 | 0.553 | 0.546 | 0.542 |
| HBA1C | 0.614 | 0.613 | 0.595 | 0.628 | 0.628 | 0.607 | 0.567 | 0.563 | 0.566 |
| TOTCHOL | 0.583 | 0.575 | 0.133 | 0.598 | 0.586 | 0.116 | 0.567 | 0.550 | 0.268 |
| HDLCHOL | 0.595 | 0.596 | 0.215 | 0.625 | 0.622 | 0.277 | 0.510 | 0.500 | 0.091 |
| STRIG | 0.578 | 0.574 | 0.291 | 0.577 | 0.572 | 0.294 | 0.544 | 0.555 | 0.278 |
| SPOT | 0.324 | 0.319 | 0.057 | 0.329 | 0.332 | 0.098 | 0.301 | 0.280 | -0.040 |
| HB | 0.404 | 0.396 | 0.193 | 0.368 | 0.365 | 0.213 | 0.426 | 0.421 | 0.120 |
| SALB | 0.272 | 0.280 | 0.110 | 0.208 | 0.200 | 0.091 | 0.270 | 0.311 | 0.135 |
| CRP | 0.325 | 0.321 | 0.228 | 0.231 | 0.248 | 0.179 | 0.469 | 0.480 | 0.328 |
| BMI | 0.356 | 0.354 | 0.262 | 0.359 | 0.347 | 0.321 | 0.379 | 0.378 | 0.104 |
| UACR | 0.253 | 0.248 | 0.082 | 0.282 | 0.289 | 0.098 | 0.072 | 0.146 | 0.021 |

## 4 Concluding Remarks

In this work we provide evidence on how BNs are effective and efficient models for the identification and the quantification of complex structures in medical practice and research. Specifically, by using average Bayesian network models for therapy-specific data we can provide an intuitive qualitative and quantitative description (in the form of a DAG) of the relationships that link the variables of the theoretical framework. Furthermore, this methodological strategy has the advantage of allowing the integration of prior expert knowledge into model estimation, which is quite common in clinical settings. From the results of the analysis, we can highlight how the data can provide a source of information able to increase the knowledge of experts in finding complex relationships in the path of pathophysiology for the disease. In this sense, data and experts are both complementary and collaborative: experts can corroborate what emerges from data and data can help experts find new insights. Moreover, by digging inside the estimated structure in the two populations we should be able to identify differences that could be imputed to the specific therapy in order to support the selection of appropriate interventions for patients treated with that therapy. Further researches can be developed to improve the efficiency of the estimated models by adding new set of variables (not strictly related to the pathophysiology perspective such as the set of risk factor medications, the clinical readout features, family history information, etc.) or move to a BN classifier (or a BN-based predictive model) with the main emergent relationships to derive a personalized probabilistic outcome.

**Acknowledgments.** The authors would like to acknowledge all the members of the DC-ren consortium and the ECLT for fruitful conversations and suggestions.

**Funding Information.** Funded by the European Union's Horizon 2020 research and innovation programme under grant agreement No 848011 ("DC-ren"). Views and opinions expressed are however those of the author(s) only and do not necessarily reflect those of the European Union or of the granting authority. Neither the European Union nor the granting authority can be held responsible for them.

## References

1. Perco, P., Pena, M., Heerspink, H.J.L., Mayer, G.: Multimarker panels in diabetic kidney disease: the way to improved clinical trial design and clinical practice? Kidney Int. Rep. **4**(2), 212–221 (2019)
2. Galicia-Garcia, U., et al.: Pathophysiology of type 2 diabetes mellitus. Int. J. Mol. Sci. **21**(17), 6275 (2020)
3. Eder, S., et al.: A prospective cohort study in patients with type 2 diabetes mellitus for validation of biomarkers (PROVALID) - study design and baseline characteristics. Kidney Blood Press. Res. **43**(1), 181–190 (2018)
4. Eder, S., et al.: Guidelines and clinical practice at the primary level of healthcare in patients with type 2 diabetes mellitus with and without kidney disease in five European countries. Diab. Vasc. Dis. Res. **16**(1), 47–56 (2019)
5. Kerschbaum, J., et al.: Intra-individual variability of eGFR trajectories in early diabetic kidney disease and lack of performance of prognostic biomarkers. Sci. Rep. **10**, 19743 (2020)
6. Koller, D., Friedman, N.: Probabilistic Graphical Models: Principles and Techniques. MIT Press, Cambridge (2009)
7. Scutari, M., Denis, J.B.: Bayesian Networks with Examples in R. Chapman & Hall, London (2014)
8. Scutari, M., Auconi, P., Caldarelli, G., Franchi, L.: Bayesian networks analysis of malocclusion data. Sci. Rep. **7**(1), 15236 (2017)
9. Arora, P., Boyne, D., Slater, J.J., Gupta, A., Brenner, D.R., Druzdzel, M.J.: Bayesian networks for risk prediction using real-world data: a tool for precision medicine. Value Health **22**(4), 439–445 (2019)
10. Shen, J., Liu, F., Xu, M., Fu, L., Dong, Z., Wu, J.: Decision support analysis for risk identification and control of patients affected by COVID-19 based on Bayesian networks. Expert Syst. Appl. **196**, 116547 (2022)
11. Kitson, N.K., Constantinou, A.C., Guo, Z., Liu, Y., Chobtham, K.: A survey of Bayesian network structure learning. Artif. Intell. Rev. **56**, 8721–8814 (2023). In press
12. Scutari, M., Nagarajan, R.: On identifying significant edges in graphical models of molecular networks. Artif. Intell. Med. **57**, 207–217 (2013)
13. Hastie, T., Tibshirani, R., Friedman, J.: The Elements of Statistical Learning: Data Mining, Inference, and Prediction. Springer, New York (2009). https://doi.org/10.1007/978-0-387-21606-5

# A Genetic Algorithm for Feature Selection for Alzheimer's Disease Detection Using a Deep Transfer Learning Approach

Tiziana D'Alessandro(✉), Claudio De Stefano, Francesco Fontanella, Emanuele Nardone, and Alessandra Scotto Di Freca

Department of Electrical and Information Engineering (DIEI), University of Cassino and Southern Lazio, Cassino, Italy
tiziana.dalessandro@unicas.it

**Abstract.** Alzheimer's disease (AD) is one of the most common forms of neurodegenerative impairment. It is a progressive brain disorder affecting memory, thinking, and behaviour, ultimately leading to severe impairment and loss of independence. In predicting Alzheimer's disease, it is widely recognized that handwriting is one of the first abilities affected by the onset of the disease. Most existing prediction systems focus on analyzing the dynamics of the handwriting process using online handwriting samples. However, these systems often fail to capture changes in handwritten characteristics' shape, size, and thickness, which can indicate motor control alterations caused by neurodegenerative disorders. A previous study introduced a novel approach by combining shape and dynamic information to address this limitation. Synthetic colour images were generated from online handwriting samples, where each elementary trait's colour encoded the associated dynamic information in the three RGB channels. Such a dataset was then used for classification through Deep Learning (DL). Moving from what was done, our study introduces a hybrid method, where Deep and Machine Learning (ML) techniques are used to implement a more powerful classification system to support the experts in diagnosing AD. Among the ML techniques considered, we performed two feature selections, one based on a recursive method and another on a genetic algorithm. Promising preliminary experimental results have confirmed the effectiveness of this proposed approach.

## 1 Introduction

Neurodegenerative disorders (NDs) are debilitating conditions characterized by the progressive degeneration of nerve cells. These diseases can have a profound impact on both physical movements and mental abilities, with Alzheimer's disease (AD) and Parkinson's disease (PD) being the most commonly known types.

A gradual and progressive decline in cognitive functions, including memory, thinking, judgment, and learning abilities characterizes AD. During the early


M. Villani et al. (Eds.): WIVACE 2023, CCIS 1977, pp. 309–323, 2024.
https://doi.org/10.1007/978-3-031-57430-6_24

stages of AD, individuals often exhibit impairment in episodic memory, which indicates dysfunction in the ventromedial temporal lobe [1]. As the disease progresses, there is typically a subsequent development of progressive amnesia and deterioration in other cognitive domains, indicating the involvement of more widespread neural systems.

Currently, there is no definitive cure for Alzheimer's disease, emphasizing the urgent need for effective treatments. However, early diagnosis plays a crucial role in managing the condition. Detecting Alzheimer's in its initial stages allows for timely interventions, access to available treatments, and the opportunity to participate in clinical trials. Early diagnosis enables individuals and their families to plan and implement strategies to maximize the quality of life, optimize care, and provide the necessary support. Early detection promotes a better understanding of the disease and facilitates ongoing research efforts to develop breakthrough therapies for effective management and potential future prevention.

In this context, the analysis of handwriting can be very useful since handwriting is the result of complex interactions between bio-mechanical parts (arm, wrist, hand, etc.) and brain areas devoted to the control and memorization of the elementary motor sequences used to produce handwritten traces [7]. For example, in the clinical course of AD, dysgraphia occurs both during the initial phase and in the progression of the disease [12]. It follows that handwriting alterations can be used as further evidence of the onset of AD, helping physicians make an early diagnosis, which remains a challenging task.

In many machine learning problems, there is often a large set of available features, but not all of them contribute equally to the predictive power of a model. Some features may be noisy, irrelevant, or redundant, leading to increased computational complexity and drawbacks like overfitting or underfitting. Feature selection methods address these issues by identifying the most informative features in the whole set of available features. The benefits of feature selection include improved model performance, reduced overfitting, increased interpretability, and reduced computational complexity. By selecting only the most informative features, models can better generalize unseen data and avoid the curse of dimensionality. Feature selection algorithms usually need to define an evaluation function and a search procedure. Evaluation functions can be divided into two broad classes, namely filter and wrapper. Filter methods evaluate the relevance of features based on statistical measures or scores. They evaluate a feature subset without considering the machine learning algorithm used. Common filter methods include correlation-based feature selection, mutual information, and chi-square tests. Wrapper methods, on the other hand, assess the performance of a machine learning model with different subsets of features. They involve training and evaluating the model using different combinations of features and selecting the subset that achieves the best performance. Wrapper methods can be computationally expensive but typically provide more accurate results. Given an evaluation function, the optimal subset can be obtained by exhaustively evaluating all possible solutions. However, this exhaustive search is often impracticable due to the exponential increase in the number of solutions

($2^N$, where $N$ is the number of available features). Therefore, to address this challenge, various search techniques have been developed for feature selection, including complete search, greedy search, and heuristic search. However, since these algorithms do not effectively consider complex interactions among features, they often encounter problems such as being trapped in local optima or incurring high computational costs. Evolutionary Computation (EC) techniques have been widely used in this context as they are well-known for their global search ability [2,8,9]. Furthermore, EC techniques do not need domain knowledge and do not make any assumptions about the evaluation function, such as, for example, linearity or differentiability.

Our study presents a novel hybrid method that combines the strengths of Deep Learning and Machine Learning techniques to develop a robust classification system for assisting experts in Alzheimer's disease (AD) diagnosis. In our investigation, we explored various ML techniques and specifically focused on two feature selection methods: a recursive approach and a genetic algorithm. We are pleased to report that our preliminary experimental results have demonstrated promising outcomes, providing confirmation of the effectiveness of this proposed approach. This hybrid method holds significant potential in enhancing the accuracy and reliability of AD diagnosis, empowering medical professionals with a more powerful toolset for their diagnostic endeavours.

The remainder of the paper is organized as follows: Sect. 2 provides an explanation of the data acquisition process and the methodology employed for generating the images; Sect. 3 introduces the workflow of the proposed system, outlining the individual components and providing a detailed description of each; and Sect. 4 offers a comprehensive discussion on the best performances achieved by the system. Finally, Sect. 5 concludes the paper, summarizing the findings and presenting the overall conclusions.

## 2 Data Acquisition

The data for this study were collected through the execution of a protocol [6] consisting of various types of handwriting tasks. A total of 174 individuals participated in the data acquisition phase, with 89 patients (PT) at the early stages of Alzheimer's disease (AD) and a healthy control group (HC) of 85 individuals. The recruitment process involved collaboration with the geriatrics department and Alzheimer's unit of the "Federico II" hospital in Naples. Participants were selected based on standard clinical tests, including the Mini-Mental State Examination (MMSE), the Frontal Assessment Battery (FAB), and the Montreal Cognitive Assessment (MoCA), which served as criteria for eligibility.

The following subsections describe the tasks of the dataset and the generation of images.

### 2.1 The Tasks

The protocol used for the data collection consists of 25 tasks and is thoroughly described in [6], whereas preliminary results were presented in [5]. The selection of these tasks was based on existing literature and aimed to analyze various aspects of handwriting and the potential deterioration of the skills required for their execution. It is important to note that Alzheimer's disease (AD) can compromise different abilities, such as cognitive, kinesthetic, and perceptive-motor functions [16], encompassing language comprehension, muscle control, spatial organization, and coordination. The protocol can be categorized into four distinct groups of tasks, each serving a specific objective:

- Graphic tasks: assess the patient's proficiency in writing elementary strokes, connecting points, and drawing figures of varying complexity and dimensions;
- Copy and Reverse Copy tasks: evaluate the patient's ability to replicate complex graphic gestures with semantic meaning, such as letters, words, and numbers of different lengths and spatial arrangements;
- Memory tasks: aim to test the patient's capacity to reproduce a specific graphic element, retain a word or letter in memory, or maintain a motor plan;
- Dictation: investigate how writing is influenced when working memory is required, involving phrases or numbers.

The choice of each subgroup of tasks is driven by a rationale based on the study of Alzheimer's symptoms [6]. It is well-established that the effects of the disease can vary from person to person, with some individuals experiencing more impairment in mental functions, others in motor functions, and some finding compensatory mechanisms, particularly for physical impairments. By including 25 tasks, this protocol aims to investigate whether handwriting is altered in individuals with Alzheimer's, considering the diverse range of symptoms associated with the disease. During the data processing phase, some tasks were partitioned, so 34 handwriting samples were collected for each person who executed the protocol.

### 2.2 Image Generation

The tasks mentioned in the previous subsection were performed using the Wacom Bamboo Folio tablet, which allowed the recording of spatial coordinates $(x - y)$ and pressure $(z)$ for each point at a sampling rate of 200 Hz. The acquisition tool can detect on-paper and in-air points within a maximum distance of three centimeters from the tablet. This information was stored in a *.csv* file for every task performed. We used the provided files to create artificial images by reconstructing the initial handwritten characteristics, considering only the on-paper points. This involved interpolating between successive points and incorporating kinematic data, such as pressure, velocity, and jerk, represented by the RGB channels. Further information can be found in the reference [3] and [4].

To generate these synthetic images, we followed a specific procedure. First, each trait is represented by a triplet of values $(z_i, v_i, j_i)$, corresponding to the RGB colour components. These values are associated with the points $(x_i, y_i)$ and $(x_{i+1}, y_{i+1})$ that define the trait. In detail, we computed the triplet values as follows:

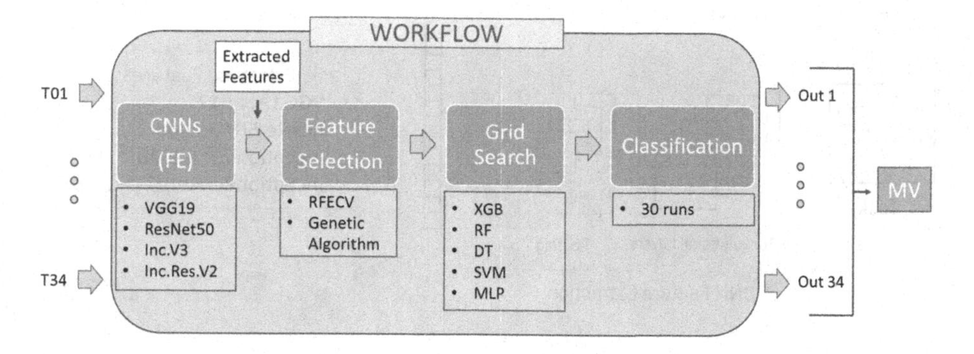


**Fig. 1.** Experimental workflow

- $z_i$ represents the pressure value at point (xi, yi) and remains constant along the trait;
- $v_i$ denotes the velocity of the trait, calculated as the ratio of the trait's length to the acquisition interval time of 5 ms (the tablet's sampling period);
- $j_i$ represents the jerk of the trait, which is defined as the second derivative of $v_i$.

To ensure consistency with the standard 0–255 colour scale, the triplet values $(z_i, v_i, j_i)$ are normalized within the range [0, 255]. This normalization is done by considering the training set's minimum and maximum values for these three quantities. All the images have been resized to the same dimensions of $299 \times 299$ pixels; this measure was set to conform to the specific format required by the Neural networks described in the following sections.

## 3 The Proposed Workflow

In this research, we proposed a hybrid approach, where DL and ML techniques are used to implement a robust and efficient support system for diagnosing AD. The following sections detail how each part of the system works and connects to the others. We suggest referring to Fig. 1, which illustrates the entire workflow. We decided to consider more options for each part of our system to compare them and find the best one according to the performance achieved.

### 3.1 Deep Feature Extraction

The first step of the workflow is the deep feature extraction, applied to the RGB images described in Sect. 2.2, for one task at a time. The term deep comes from the fact that we used different models of Convolutional Neural Networks (CNNs) to extract features from our images: VGG19 [13], ResNet50 [11], InceptionV3 [15], InceptionResNetV2 [14]. They are layered structures, different from each other in several architectural and functional aspects. The input RGB handwriting images were adjusted in size to conform to the specific format required by the

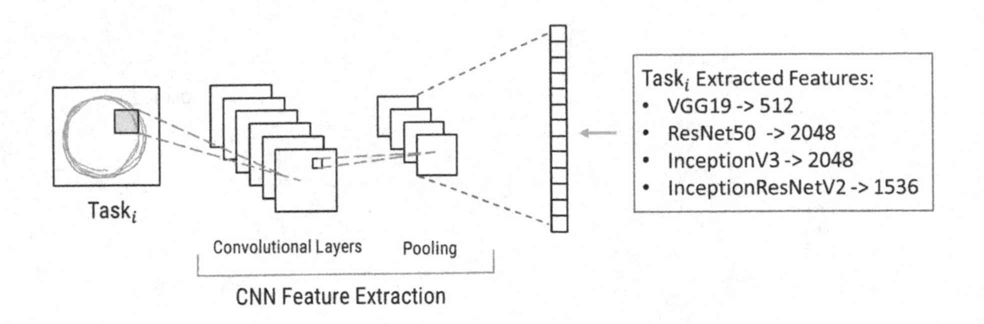


**Fig. 2.** Feature Extraction Procedure

CNN used in the analysis. Every CNN architecture comprises a feature extractor (FE) and a classifier (C). For our research, we removed the classifier from the architectures to use them as feature extractors; in this way, we obtained a feature vector, also denoted as the bottleneck, whose size depends on the CNN for every image. Figure 2 shows an example of this procedure for the $i$-th task and reports the output feature vector size for each model (right box). To obtain valuable features, we pre-trained the FE part of the CNN on ImageNet [10], following the popular technique of Transfer Learning. We adopted a 5-fold cross-validation strategy and saved the feature vectors when images were in the test set. In this manner, we extracted the features for all the images in our dataset, obtaining a new representation of our original data. In particular, initially, we had a set of 174 images for every task; this set was used to feed four different CNN models, so we obtained four sets of features for each task.

### 3.2 Feature Selection

Feature selection is a typical ML process to choose a subset of relevant features from a larger set of available features in a dataset. It aims to identify the most informative and discriminative features that contribute significantly to the predictive power of a model while discarding irrelevant or redundant features. Its goal is to improve model performance, reduce overfitting, enhance interpretability, and decrease computational complexity. Selecting a smaller set of features reduces the dimensionality of the data and eliminates noise and irrelevant information that may negatively impact the model's accuracy and efficiency. As shown in Fig. 1, we used two different ways to implement this process: the Recursive Feature Elimination (RFE) and the Genetic Algorithm.

In detail, RFE is a feature selection method that iteratively selects subsets of features and evaluates their importance by training an ML algorithm with cross-validation. We used XGBoost as an estimator. RFE offers the advantage of considering the inherent relationships between features, as it eliminates features recursively. It provides a feature importance ranking and identifies the optimal subset that generalizes well, preventing overfitting and maximizing the model's performance. It is beneficial when dealing with high-dimensional datasets as it automatically reduces the dimensionality of the data and improves the model's interpretability and efficiency.

The second method used for feature selection is the genetic algorithm (GA), which offers many advantages. First, it guarantees a comprehensive search by efficiently exploring vast search spaces. It can evaluate possible feature combinations, considering various feature interactions and dependencies. One of the main characteristics of this family of algorithms is their adaptability, as they can adapt and evolve, dynamically adjusting the population of feature subsets based on their fitness, allowing them to converge toward optimal solutions. GAs can handle non-linearity and complex relationships between features. This is an important advantage in our case as we don't know the meaning of the relations among features in our dataset since a CNN extracts them. By combining and mutating features, the algorithm can identify synergistic effects and discover valuable feature combinations. Overall, genetic algorithms for feature selection provide a powerful and flexible approach to identifying relevant features and constructing optimal feature subsets. They excel in handling complex relationships, exploring large search spaces, and automating the feature selection process, ultimately improving model performance, interpretability, and efficiency. Following is a description of how this algorithm works:

1. Initialization: The algorithm creates an initial population of potential feature subsets. Each one represents a potential solution or chromosome;
2. Evaluation: The algorithm evaluates the fitness of each chromosome in the population. Fitness is typically based on a performance metric like accuracy or error rate, which indicates how well a given subset of features performs when used by the provided estimator (a machine learning model);
3. Selection: The algorithm selects chromosomes from the population for the next generation based on their fitness. Chromosomes with higher fitness have a higher probability of being selected, mimicking the natural selection process;
4. Crossover: Selected chromosomes undergo a crossover operation. This operation involves exchanging genetic information between two parent chromosomes to create offspring chromosomes. The crossover probability parameter determines the likelihood of crossover occurring;
5. Mutation: Mutation introduces small random changes in the offspring chromosomes. This process helps explore new regions of the feature space that may lead to better solutions. The mutation probability parameter controls the likelihood of mutation occurring;
6. Elitism: The algorithm incorporates elitism by carrying over a certain number of the best-performing chromosomes from the previous generation to the next without any changes. This ensures that the best features discovered so far are preserved;
7. Repeat: Steps 2 to 6 are repeated for a fixed number of generations or until a termination criterion is met (e.g., no improvement in fitness over a certain number of generations);
8. Output: The algorithm outputs the best-performing chromosome or feature subset found during the evolutionary process.

**Table 1.** GA paramaters

| Parameter | Value |
|---|---|
| Population size | 100 |
| Max #features | 8 |
| Crossover probability | 0.5 |
| Mutation probability | 0.2 |
| #generations | 50 |
| Crossover independent probability | 0.5 |
| Mutation independent probability | 0.04 |
| Tournament size | 3 |
| Estimator | XGB |
| #folds cross-validation | 5 |
| Scoring | accuracy |

To set the values of the GA parameters, we performed a set of preliminary trials. Table 1) shows the values used in the experiments reported in the next Section. The last part of the table refers to parameters chosen to evaluate the feature set selected from the GA. The chosen estimator is XGBoost with cross-validation to estimate the fitness of each chromosome based on the accuracy evaluation metric.

The genetic algorithm iteratively explores and refines feature subsets, favouring those that yield higher performance based on the provided estimator and fitness evaluation. By the end, it identifies a subset of features that optimizes the machine learning model's performance.

### 3.3 Grid Search and Classification

To evaluate the effectiveness of the sets of extracted features, detailed in Sect. 3.1, we employed a machine learning (ML) approach by implementing five well-known classification algorithms: XGBoost (XGB), Random Forest (RF), Decision Tree (DT), Support Vector Machine (SVM), and Multi-Layer Perceptron (MLP). We performed a five-fold cross-validated grid search to identify the best combination of hyperparameters to ensure optimal performance for each classifier. Grid search is a systematic approach used in machine learning to tune the hyperparameters of a model. This technique exhaustively searches through a specified set of hyperparameter combinations to determine the optimal configuration that produces the best model performance. Table 2 shows the tested hyperparameter ranges.

Once we obtained the best hyperparameters configuration, we proceeded with the classification step with the previously cited classification algorithms. To ensure statistically significant results, we repeated the process 30 times for each set of features and each classifier. The dataset was randomly shuffled at each run and divided into a training and test set.

**Table 2.** ML classifiers and the sets of hyperparameters selected by the Grid search process

| Classifier | Hyperparameters | constraints |
|---|---|---|
| XGB | min child weight | 1, 5, 10 |
| | gamma | 0.5, 1, 1.5, 2 |
| | subsample | 0.6, 0.8, 1 |
| | colsample bytree | 0.6, 0.8, 1 |
| | max depth | 3, 5, 7 |
| | n estimators | 100, 300, 500 |
| | learning rate | 0.1, 0.01, 0.001 |
| RF | bootstrap | True, False |
| | max depth | 10, 20, 50 |
| | max features | auto, sqrt, log2 |
| | min samples leaf | 1, 2, 4 |
| | min samples split | 2, 5, 10 |
| | n estimators | 100, 200, 300 |
| DT | criterion | gini, entropy |
| | min samples split | 2, 5, 10 |
| | max depth | 2, 5, 10 |
| | min samples leaf | 1, 2, 5, 10 |
| | max leaf nodes | 2, 5, 10 |
| | max features | sqrt, log2 |
| SVM | C | 0.1, 1, 10, 100 |
| | gamma | scale, auto |
| | kernel | linear, rbf |
| MLP | hidden layer sizes | 30, 50, 100 |
| | activation | tanh, relu |
| | solver | lbfgs, adam, sgd |
| | alpha | 0.0001, 0.001, 0.01 |
| | learning rate | constant, adaptive |

### 3.4 Majority Vote

The majority vote rule (MV) is a decision-making strategy commonly used in ensemble learning and classification tasks. It focuses on the principle of aggregating predictions from multiple individual classifiers to make a final prediction, so it's particularly useful in our case, where we have the classification results from different tasks for every chosen CNN and ML classifier. In the majority vote rule, each classifier in the ensemble independently predicts the class label for a given input. The final prediction is determined by selecting the class label that receives the majority of votes from the classifiers. This rule aims to improve the

**Table 3.** Majority Vote results, achieved by combining the predictions over the 34 tasks.

| Model | XGB | | RF | | DT | | SVM | | MLP | |
|---|---|---|---|---|---|---|---|---|---|---|
| | RFE | GA | RFE | GA | RFE | GA | RFE | GA | RFE | GA |
| VGG19 | 74.59 | 85.43 | 73.83 | 82.95 | 74.09 | 76.82 | 67.98 | 74.9 | 73.71 | 76.11 |
| ResNet50 | 81.13 | 86.68 | 81.38 | 86.08 | 81.57 | 76.41 | 66.86 | 71.93 | 81.89 | 73.61 |
| Inc.V3 | 84.41 | **89.07** | 83.83 | 88.1 | 84.65 | 80.91 | 72.52 | 81.09 | 85.03 | 79.62 |
| Inc.Res.V2 | 77.11 | 83.34 | 76.78 | 81.75 | 78.06 | 76.69 | 65.91 | 73.92 | 76.8 | 75.93 |

overall accuracy and robustness of the system. It leverages the diversity of individual classifiers to collectively make more accurate predictions, benefiting from different perspectives and patterns observed by each classifier. In this research, we combined the answers obtained over the 34 tasks from each classifier on the feature extracted from every CNN, considering both the feature selection methods.

## 4 Experimental Results

The experimental part of this research involved multiple processes, and hence, only the most relevant ones will be showcased in this paper.

The final step of the workflow in Fig. 1 is the application of the majority vote combining rule. Table 3 shows the accuracy values (in percentage) obtained by combining the predictions over the 34 tasks. In particular, each row refers to the CNN model used as a feature extractor, whereas the column refers to the ML classifier tested. Every column is divided into two subcolumns, highlighting the performance differences between GA and RFE. The best majority vote performance (in bold) is an accuracy of 89.07%, reached by combining the predictions of the XGB ML classifier on the 34 tasks for features extracted through the CNN InceptionV3 and selected using the GA. The table shows that features extracted with the InceptionV3 model always performed better than the others. Also, most of the time, the GA seems to be the most valuable feature selection method, and XGB is one of the best ML classifiers.

Given the MV outcome in Table 3, we decided to show and discuss only the results related to the configuration, which allowed us to reach the best MV performance. Figure 3 illustrates two bar plots, showing the average accuracy (y-axis, in percentage) obtained by every classifier (legend) over thirty runs. For every task (x-axis), we considered features extracted with InceptionV3 from the RGB on paper images, while the feature selection method was the GA. Due to the excessive number of tasks, we split the figure into two images: (a) shows the results for tasks from 1 to 17, while (b) from 18 to 34. Looking at this Figure, it is possible to discuss many aspects. First, the best classification algorithms are XGB and RF. Only in a few cases SVM outperforms these two; instead, DT and MLP are the worst classifiers. Accuracies vary in a range that goes from 53%,

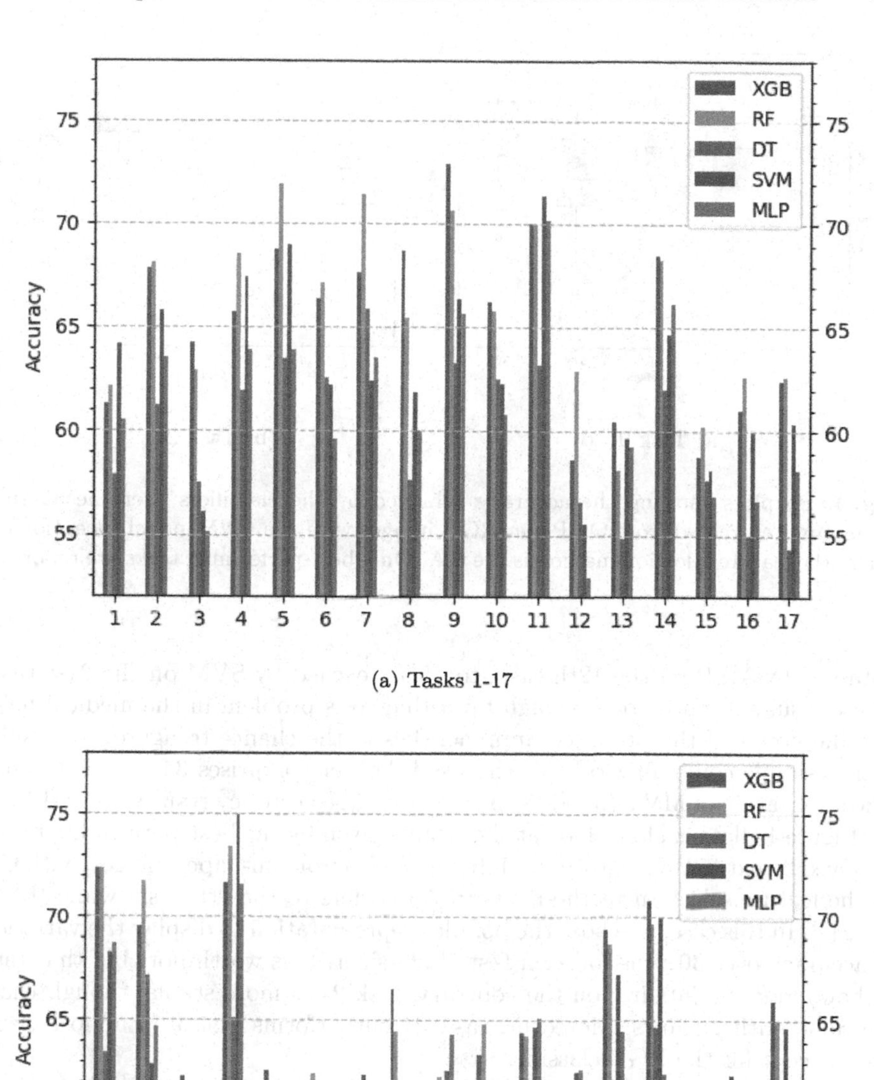


**Fig. 3.** Average accuracy achieved by the classifiers over the 30 runs, for features extracted from On Paper RGB images with the CNN model InceptionV3, where the feature selection method is the GA.

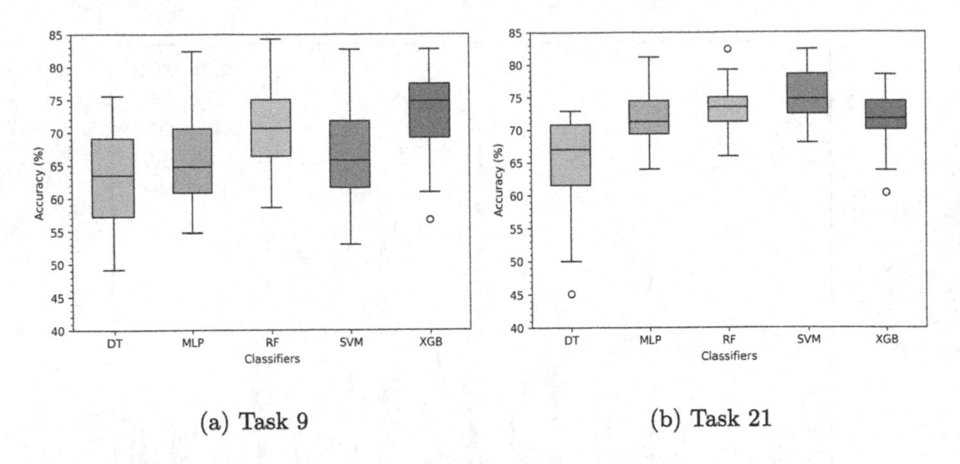


(a) Task 9 (b) Task 21

**Fig. 4.** Boxplots showing the accuracy achieved by the classifiers over the 30 runs for features extracted from On Paper RGB images with the CNN model InceptionV3, where the feature selection method is the GA. Only best-performing tasks are reported.

obtained by MLP on the 12th task, to 75%, reached by SVM on the 21st task. These results are not good enough for a diagnosis problem in the medical field, but the power of the proposed approach lies in the chance to aggregate results with a simple combining rule, as the used dataset comprises 34 different tasks. The application of MV allowed us to improve these accuracy results beyond 10%.

Figure 4 allows a closer look at the results given by the best-performing tasks, achieved by extracting features with InceptionV3 from on-paper images, with GA as the feature selection method. Figure 4(a) refers to the 9th task, while (b) to the 21st. In this case, we chose the boxplot representation to display the variation of accuracy over 30 runs for each tested classifier. It is worth noticing that task 9 shows more variability; on the contrary, task 21 is more stable, though there are some outlier values. Moreover, task 21 outperforms task 9 in almost every case, except for the XGB classifier.

Another point of view to analyze the experimental results is in Table 4, which refers to the experimental setting that allowed us to reach the best results with the Majority Vote, that is InceptionV3 as feature extractor, the Genetic Algorithm to select the features and XGB as a classifier. The Table shows a set of evaluation metrics for every task in the first column, which will allow us to better comprehend the results achieved. The metrics computed are accuracy (ACC), precision (PRE), sensitivity (SEN), specificity (SPE), F-score (F-SC), False Negative Rate (FNR), and Area Under the Curve (AUC). These values are averaged over the 30 runs and expressed in percentage, except for the AUC, which is a measure that varies from 0 to 1.0. Looking at this table, the best performance is achieved by the 9th task, where the person is asked to write the bigram 'le' four times continuously. It is a task that requires good spatial organization, motion control and coordination, and cognitive skills.

**Table 4.** Metrics evaluation for the performance of XGB classifier on InceptionV3 extracted features, selected with the GA

| Task # | ACC | PRE | SEN | SPE | F-SC | FNR | AUC |
|---|---|---|---|---|---|---|---|
| 1 | 61.29 | 63.87 | 63.79 | 59.22 | 63.36 | 36.21 | 0.66 |
| 2 | 67.82 | 70.41 | 69.66 | 66.38 | 69.41 | 30.34 | 0.74 |
| 3 | 64.24 | 66.52 | 65.98 | 63.22 | 65.59 | 34.02 | 0.71 |
| 4 | 65.72 | 68.59 | 67.00 | 65.36 | 67.04 | 33.01 | 0.72 |
| 5 | 68.73 | 72.22 | 67.21 | 71.18 | 68.94 | 32.79 | 0.75 |
| 6 | 66.36 | 67.61 | 73.16 | 60.17 | 69.39 | 26.84 | 0.72 |
| 7 | 67.64 | 70.64 | 67.85 | 67.41 | 68.63 | 32.15 | 0.73 |
| 8 | 68.71 | 70.15 | **73.81** | 63.33 | 71.34 | **26.19** | 0.74 |
| 9 | **72.88** | **74.84** | 73.29 | 72.92 | **73.62** | 26.71 | 0.79 |
| 10 | 66.23 | 69.96 | 64.06 | 69.52 | 66.39 | 35.94 | 0.71 |
| 11 | 69.99 | 73.29 | 68.41 | 72.07 | 70.21 | 31.61 | 0.76 |
| 12 | 58.66 | 62.71 | 57.56 | 60.73 | 59.15 | 42.44 | 0.63 |
| 13 | 60.46 | 62.61 | 61.34 | 60.21 | 61.31 | 38.66 | 0.64 |
| 14 | 68.44 | 70.82 | 68.51 | 69.55 | 68.92 | 31.51 | 0.75 |
| 15 | 58.74 | 61.39 | 58.96 | 59.55 | 59.34 | 41.04 | 0.62 |
| 16 | 61.05 | 63.63 | 62.73 | 60.21 | 62.45 | 37.27 | 0.66 |
| 17 | 62.46 | 65.49 | 65.04 | 61.42 | 64.02 | 34.96 | 0.67 |
| 18 | 72.31 | 74.68 | 73.03 | 72.27 | 73.27 | 26.97 | **0.82** |
| 19 | 69.51 | 69.79 | 59.67 | **78.39** | 63.41 | 40.33 | 0.77 |
| 20 | 62.25 | 64.02 | 60.48 | 64.78 | 61.35 | 39.52 | 0.68 |
| 21 | 71.64 | 73.88 | 67.34 | 76.32 | 69.97 | 32.66 | 0.79 |
| 22 | 62.59 | 64.36 | 60.54 | 65.81 | 61.45 | 39.46 | 0.66 |
| 23 | 61.78 | 62.95 | 63.11 | 61.01 | 62.33 | 36.91 | 0.61 |
| 24 | 60.23 | 61.48 | 59.38 | 61.95 | 59.58 | 40.62 | 0.65 |
| 25 | 66.48 | 69.39 | 61.65 | 72.23 | 64.16 | 38.35 | 0.75 |
| 26 | 60.33 | 60.38 | 59.68 | 61.57 | 58.94 | 40.32 | 0.61 |
| 27 | 63.34 | 65.11 | 62.36 | 65.21 | 62.84 | 37.64 | 0.68 |
| 28 | 64.42 | 66.45 | 56.03 | 73.05 | 59.99 | 43.97 | 0.72 |
| 29 | 59.03 | 61.44 | 64.33 | 54.78 | 61.89 | 35.67 | 0.63 |
| 30 | 69.39 | 72.71 | 68.55 | 71.38 | 69.66 | 31.45 | 0.77 |
| 31 | 70.78 | 73.06 | 71.84 | 70.31 | 71.69 | 28.16 | 0.74 |
| 32 | 58.27 | 61.27 | 59.93 | 57.61 | 59.82 | 40.07 | 0.63 |
| 33 | 62.33 | 65.84 | 59.46 | 65.68 | 61.81 | 40.54 | 0.66 |
| 34 | 65.95 | 68.01 | 68.55 | 64.06 | 67.55 | 31.45 | 0.72 |

## 5 Conclusions and Future Work

The objective of this study was to validate the proposed workflow and understand which variant, in terms of components, guarantees the best performance in distinguishing people affected by AD and healthy controls. The input of our system is composed of synthetic RGB on-paper images for 34 different handwriting tasks. These images are used to feed four models of CNN to extract features automatically. Once the features were identified, two feature selection methods were considered, RFE and a GA, so the selected features were forwarded to five ML algorithms whose parameters were tuned through a grid search procedure. Finally, since the workflow is iterated over the 34 tasks, combining the predictions with a majority vote rule was possible. We evaluated many configurations of our system by varying the CNN model employed as a feature extractor, the feature selection method, and the final classifier. The best performance was achieved using InceptionV3 as the CNN model, the Genetic Algorithm to select features, and the XGB classifier. Regarding feature selection, RFE is a simpler and computationally efficient approach suitable for many scenarios. It is a popular choice when the dataset has a moderate number of features and the relationship between features, and the target variable is reasonably linear. On the other hand, GA can be a more powerful but computationally expensive approach, suitable for more complex feature selection problems with larger feature spaces or non-linear relationships. As expected, the GA guaranteed a more valuable set of features that allowed our system to obtain better results with respect to RFE.

Future work will involve new experiments with new input images, which will be built by using both the in-air and on-paper movement coordinates. Such images contain more information about handwriting traits and also in-air movement, which are used in many research studies related to neurodegenerative diseases. Other improvements will see the refinement of our proposed workflow, deleting the processes that obtained the worst performance and implementing new ones.

## References

1. Armstrong, M.J., et al.: Criteria for the diagnosis of corticobasal degeneration. Neurology **80**(5), 496–503 (2013)
2. Cilia, N.D., De Stefano, C., Fontanella, F., Scotto di Freca, A.: Variable-length representation for EC-based feature selection in high-dimensional data. In: Kaufmann, P., Castillo, P.A. (eds.) EvoApplications 2019. LNCS, vol. 11454, pp. 325–340. Springer, Cham (2019). https://doi.org/10.1007/978-3-030-16692-2_22
3. Cilia, N.D., D'Alessandro, T., De Stefano, C., Fontanella, F., Molinara, M.: From online handwriting to synthetic images for Alzheimer's disease detection using a deep transfer learning approach. IEEE J. Biomed. Health Inform. **25**(12), 4243–4254 (2021)
4. Cilia, N.D., D'Alessandro, T., Stefano, C.D., Fontanella, F.: Deep transfer learning algorithms applied to synthetic drawing images as a tool for supporting Alzheimer's disease prediction. Mach. Vis. Appl. **33**(3), 49 (2022)

5. Cilia, N.D., De Stefano, C., Fontanella, F., Molinara, M., Scotto Di Freca, A.: Handwriting analysis to support Alzheimer's disease diagnosis: a preliminary study. In: Vento, M., Percannella, G. (eds.) CAIP 2019. LNCS, vol. 11679, pp. 143–151. Springer, Cham (2019). https://doi.org/10.1007/978-3-030-29891-3_13
6. Cilia, N.D., De Stefano, C., Fontanella, F., Di Freca, A.S.: An experimental protocol to support cognitive impairment diagnosis by using handwriting analysis. Procedia Comput. Sci. **141**, 466–471 (2018)
7. Cilia, N.D., et al.: Lognormal features for early diagnosis of Alzheimer's disease through handwriting analysis. In: Carmona-Duarte, C., Diaz, M., Ferrer, M.A., Morales, A. (eds.) IGS 2022. LNCS, vol. 13424, pp. 322–335. Springer, Cham (2022). https://doi.org/10.1007/978-3-031-19745-1_24
8. De Falco, I., Tarantino, E., Della Cioppa, A., Fontanella, F.: A novel grammar-based genetic programming approach to clustering. In: Proceedings of the 2005 ACM Symposium on Applied Computing, pp. 928–932 (2005)
9. De Falco, I., Tarantino, E., Della Cioppa, A., Fontanella, F.: An innovative approach to genetic programming-based clustering. In: Abraham, A., de Baets, B., Köppen, M., Nickolay, B. (eds.) Applied Soft Computing Technologies: The Challenge of Complexity. Advances in Soft Computing, vol. 34, pp. 55–64. Springer, Heidelberg (2006). https://doi.org/10.1007/3-540-31662-0_4
10. Deng, J., Dong, W., Socher, R., Li, L.J., Li, K., Fei-Fei, L.: ImageNet: a large-scale hierarchical image database. In: CVPR, pp. 248–255. IEEE Computer Society (2009)
11. He, K., Zhang, X., Ren, S., Sun, J.: Deep residual learning for image recognition. In: 2016 IEEE Conference on Computer Vision and Pattern Recognition (CVPR), pp. 770–778 (2016)
12. Impedovo, D., Pirlo, G.: Dynamic handwriting analysis for the assessment of neurodegenerative diseases: a pattern recognition perspective. IEEE Rev. Biomed. Eng. 1–13 (2018)
13. Simonyan, K., Zisserman, A.: Very deep convolutional networks for large-scale image recognition. CoRR abs/1409.1556 (2015)
14. Szegedy, C., Ioffe, S., Vanhoucke, V.: Inception-v4, inception-ResNet and the impact of residual connections on learning. In: AAAI (2016)
15. Szegedy, C., Vanhoucke, V., Ioffe, S., Shlens, J., Wojna, Z.: Rethinking the inception architecture for computer vision. In: 2016 IEEE Conference on Computer Vision and Pattern Recognition (CVPR), pp. 2818–2826 (2016)
16. Tseng, M.H., Cermak, S.A.: The influence of ergonomic factors and perceptual-motor abilities on handwriting performance. Am. J. Occup. Ther. **47**(10), 919–926 (1993)

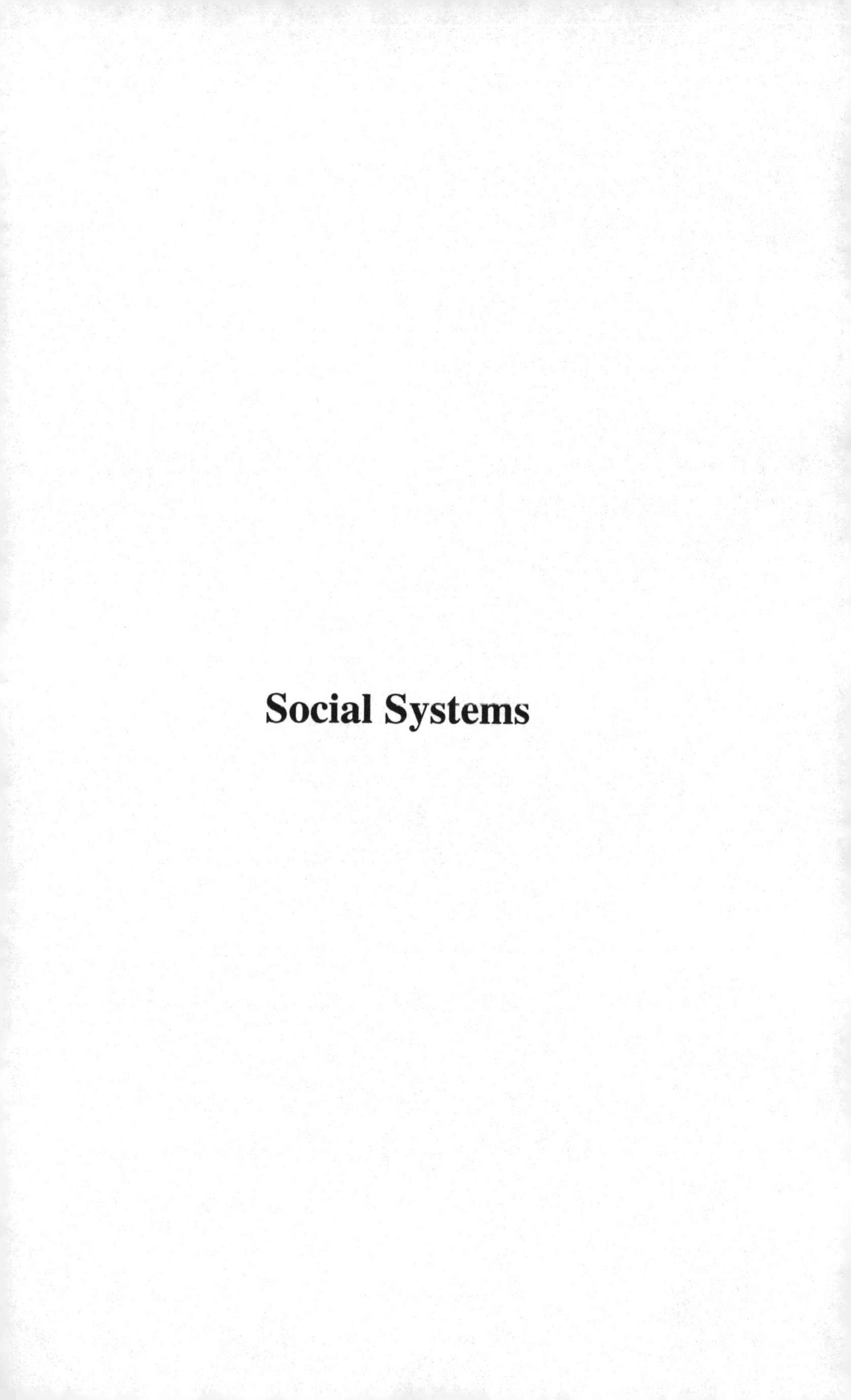

# Social Systems

# Learning Whether to be Informed in an Agent-Based Evolutionary Market Model

Paolo Pellizzari(✉)

Department of Economics, Ca' Foscari University, Venice, Italy
paolop@unive.it

**Abstract.** Can traders in a financial market learn whether to be informed and which information to use in their demand for risky assets? We describe in this paper an agent-based model where heterogeneous traders seek short-term profits and differ in their choices to use or discard some signals. In the model, a vector of fresh news/signals is available at every period and some (but not all) the signals affect the stochastic payoff of the stock.

Under an evolutionary dynamics favouring higher myopic returns we find that, in equilibrium, traders mostly end up in either discarding *all* signals or being (perfectly) informed using *all* the relevant signals (paying the related costs). Moreover, the rate of use of information strongly depends on the "complexity" of the market: an excessively large abundance of signals to be screened or a high volatility of the market, result in large shares of passive agents who overestimate the market's risk; conversely, low market complexity is associated with a more intense use of information and aggressiveness of informed traders.

**Keywords:** Evolutionary models · Agent-based models · Information in financial markets

## 1 Introduction

Many investors acquire information on their investments and try to make some sense of the markets' situations and prospects. We refer, in what follows, especially to "fundamental" information regarding what is typically believed in economic textbooks to be relevant to explain to some extent the movements of equity prices such as, say, interest rates, inflation, GDP growth in developed end emerging markets, geopolitical events, international imbalances and breaking news on firms or events of (potential) broad impact[1].

As an example, on April 29th 2023 the most common Italian financial newspaper, "Il Sole 24 ore" printed that:

[1] We discard "technical" information, mostly derived from time-series and historical data. Many traders may use such "information" but the model has little to say in this respect as no past observation is used, see [4] for an evergreen examination of technical trading.


M. Villani et al. (Eds.): WIVACE 2023, CCIS 1977, pp. 327–338, 2024.
https://doi.org/10.1007/978-3-031-57430-6_25

- Wall Street "bets" on (forthcoming) interest rates cuts of 300 basis points (first page);
- Taipei has denounced the intrusion of 38 Chinese military airplanes in its space. Chinese authorities pleaded that they were "monitoring" one US military fighter flying in the area (first page);
- Jerome Powell, chairman of the FED, declared "We will not reduce the interest rates in 2023" (page 7);

The previous news (or signals) have the capability to provide valuable investment insights and, yet, it is hard to pick the most significant or decide whether to use them all. It is even more difficult to unambiguously interpret the news. There are basically opposite statements on the trend of US interest rates: an hypothetical personification of Wall Street expects a drop of 3%, but Powell stated this is not going to happen. Well, at least in 2023! In principle, exploiting the lack of a clear timeline, both news could be correct as rates' cuts may come in 2024. Indeed, there is a good joke stating that wise forecasters should never provide a number *and* a date... Geopolitical strained relations involving China, Taipei and USA are hinted at, with no clear implications on asset prices. Whether and how to use such information is an interesting, as well as far from trivial, issue.

Many analogous examples can be drawn, virtually any day, from other financial newspapers or websites, official and informal reports by public and private institutions, central banks' statements, and various intelligence from advisory firms and respected professionals or gurus. We present in this model a stylised depiction of investors who are similarly flooded with information and have little guidance on how to use such body of insights. They try to use and interpret a stream of signals in order to decipher how to behave, most of them attempt to select relevant information, weight it properly and discard irrelevant news, are aware there is no easy recipes and are willing to imitate strategies or practices put in place by others.

We assume that traders are boundedly rational and learn to change their investment strategy by imitating other agents who had better (i.e., higher) returns. In other words, they copy the pattern of use of information made by more successful peers. A strategy is a vector of bits (bit-string) where 1 means that the news is used and 0 means that it is discarded. We include mutation allowing a small fraction of agents to occasionally flip one of their bits at random. In our simple setup, given their strategies, all agents have to decide in every period is how much to buy of a risky asset (while the rest of their endowments will be put in a safe bond paying a constant interest rate). In this sense, there is a one-to-one correspondence between a strategy (a bit-string listing signals to be used or discarded) and a demand function for the risky asset (as the demand depends on the used information/signals).[2] Broadly speaking, demands

[2] Admittedly, the agents in our model learn in a very basic way, as they have no memory or expectations and update their behaviour based on a single random match. A discussion of more sophisticated reinforcement learning approaches (with an extended bibliography) is in [3], where a form of collective intelligence is built to maximise returns. In contrast, we assume agents are selfish and myopic.

are based on perceived mean and variance of returns (the demand of equity is directly/inversely proportional to the mean/variance). Once all agents submit their demands, a noisy clearing price can be computed in this market and all transactions will occur at this local-in-time equilibrium price, realising gains and losses. Actual returns are also used to assess the quality of the strategies and fuel learning through a very simple mechanism: couples of agents will be matched, they compare the realised returns and the worst performer copies the strategy (or demand function, if you wish) of the best performer, beginning to use it in the following period when everyone's endowment is replenished[3].

Having defined a population of agents, a game that is repeatedly played by traders and a process to revise old strategy (or learn better ones), our agent-based model can be interpreted as a canonical evolutionary model. Such models were first introduced in biology, where genotypes are inherited and not chosen by individuals, [10], but were increasingly applied to social and economic environments where strategies are selected consciously in such a way that the ones with greater payoff tend to prevail, see the classic [8], or [6] where recent applications are surveyed. We aim at identifying the set of strategies that will thrive in the long-run, simulating the market for many periods and examining the final shares resulting from this evolutionary dynamics.

In brief, our evolutionary model robustly show two main results: first, most traders end up either in being passive (i.e., discard all signals) or being (wholly) informed (i.e., acquire all the relevant signals); second, information usage depends on the "complexity" of the market, as measured by its volatility or by the quantity of the information that traders have to screen and process. Overall, the combination of the above outcomes explains why only some of the relevant information is used by the agents in the market, with the informed traders holding notably riskier positions than passive ones. Several of these findings, driven by short-term evolutionary pressure and inability to deal with the overwhelming complexity of the market, appear to have a realistic flavour that is somewhat difficult to get in standard rational expectation equilibrium models where, for instance, it is difficult to justify why relevant information should be discarded.

The paper is organised as follows. The next section describes the model of the market. An example is used to illustrate the flow of decisions/actions, from strategies/demands to prices/profits and evolution through learning, that are executed in every period. In Sect. 3, the main results are presented and some conclusive reflections are given in Sect. 4.

---

[3] It is useful to add to the description of what our agents do, a list of things they do not do: they do not explicitly maximise any utility function, they do not have memory, they do not search in a set of possible alternative strategies or, if they do so, they may need several periods in which they compare the outcomes with a single strategy, they do not try to anticipate the equilibrium price based on the shares of strategies in the population, they do not save or accumulate wealth strategically. In a nutshell, they keep a strategy till they stumble on concrete evidence that someone else makes higher returns and occasionally flips some bits.

## 2 The Model

Consider $M$ agents in a market with two assets, a risk-less bond with unit cost and payoff $R > 1$ and a risky stock that for a price $p_t^*$, to be endogenously determined based on the demands of the traders, will pay a random payoff $\tilde{D}_t$. Agents are endowed with a constant periodic endowment $w_{0t} \equiv w_0$, care about return and variance of their portfolio, and have to decide how many units of the stock to buy or sell in any period (what is not spent in stocks will be invested in the bond).

A stream of $N$ news $\theta_t = (\theta_1, \theta_2, ..., \theta_N)_t \in \mathbf{R}^N$ is available to traders in any period $t = 1, ..., T$. We assume that each of the $N$ signals is identically and independently distributed as $N(0, v_\theta)$. The careful reader should notice the lapse between *texts*, such as the ones listed in the introduction, and a vector $\theta_t$ of *numeric values*. For simplicity, we just suppose that some judgemental or mechanic procedure translates news (sentences, comments...) into a vector of estimates.

As it will be made clear below, only $S \leq N$ individual signals $\theta_j$ will truly affect $\tilde{D}_t$. With no loss of generality, we will assume in this treatment that the first signals $\theta_1, ..., \theta_S$ are relevant. This simplifies the exposition but is unknown to agents who must decide by trial-and-error whether to use signals at all and which ones to use. To keep track of this *learning process*, each traders has a strategy $b_{it} = (b_1, b_2, ..., b_N)_{it}$, where each bit $b_j, j = 1, ..., N \in \{0, 1\}$ denotes if the $j$-th signal is used: a value of 1/0 means the signal is used/discarded. Equipped with the strategy $b_{it}$ at time $t$, the demand schedule of the $i$-th agent is

$$x_{it}(p) = \frac{d + b_{it}\theta_t' - pR}{a v_{it}}, \tag{1}$$

where $d > R$, $v_{it}$ is an individual assessment of the variance of $D_t$, and $a$ can be thought as a risk-aversion coefficient. For tractability, we assume that the deterministic component $d$ of the payoff is exogenous and known to agents and $a$ is constant across them. Given the price $p$, the demand (1) is, essentially, a ratio of expected excess return of one unit of stock (in excess of $pR$ that could be gained with the bond) over perceived risk. Hence, the expected $\tilde{D}_t^{(i)}$ for agent $i$ is given by

$$d + b_{it}\theta_t' = d + \sum_{j=1}^{N} b_{ijt}\theta_{jt},$$

and depends on which bits are switched-on in the strategy $b_{it}$. The individual demand function $x_{it}(p)$ is then readily obtained.

A unique transaction price $p_t^*$ is determined matching the demand functions of all the agents and solving

$$\sum_{i=1}^{M} x_{it}(p) = 0. \tag{2}$$

Equation (2) is linear in $p$ and the solution $p_t^*$ can be numerically computed to allow the agents to determine the realised purchases/sales of the stock, $x_{it}^* = x_{it}(p_t^*|b_{it}, v_{it})$, where we stress that the quantity (plainly) depends also on the individual strategy $b_{it}$ and on $v_{it}$.

Let the realised stock's payoff be determined "by nature" as

$$\tilde{D}_t = d + b^* \theta_t' + \epsilon_t, \tag{3}$$

where $\epsilon_t \sim N(0, v_\epsilon)$ and, as said before, $b^*$ has the first $S$ bits set to one, $(\underbrace{1, ..., 1}_{S \text{ bits}}, \underbrace{0, ..., 0}_{N-S \text{ bits}})$, so that only $S$ bits out of $N$ affects the payoff.

The profits of agents can now be computed and used to evolve a new population of strategies, or demand functions, moving from $\mathcal{P}_t = \{b_{1t}, ..., b_{Mt}\}$ to $\mathcal{P}_{t+1}$ that differs from the old one because some agents are offered the chance to imitate and mutate their strategy. In detail, denote the profit of $i$-th agent at $t$ as:

$$w_{it} = x_{it}^* D_t + (w_0 - x_{it}^* p_t^*) R - cost \cdot b_{it} \mathbf{1}',$$

where *cost* is the cost of acquiring or processing one signal, $\mathbf{1}$ is the vector with $N$ ones, $(1, ..., 1)$, and $b_{it}\mathbf{1}'$ is the number of used bits. The amount $w_{it}$ is consumed or spent elsewhere (and, therefore, agents start afresh, in terms of wealth, in the next period). Evolution and competition in the market occurs matching $h$ couples of agents, comparing the profits and changing the strategic profile $(b_i, v_i)$. When, say, traders $r$ and $s$ are matched:

$$\text{If } w_{rt} \geq w_{st} \text{ then } \begin{cases} b_{s,t+1} = b_{rt}, \\ v_{s,t+1} = v_{rt} \tilde{U}(1-q, 1+q), \end{cases} \tag{4}$$

where $q > 0$ is a small number and $U(a, b)$ is a uniform random variable in $]a, b[$. Formula (4) describes how, if agent $r$ outperformed agent $s$, the latter copies the strategy of the former and replaces his $v_s$ with a random multiplicatively shocked $v_r$. Observe that each pair $(r, s)$ of agents is randomly formed and, hence, there is no deliberate attempt to imitate or cherry-pick successful traders. Moreover, the straight comparison of revenues in (4) is entirely justified by the constant risk-aversion parameters. In each period, we also allow a single mutation, flipping a random bit of a random trader's strategy.

Among the $2^N$ strategies that can be evolved, two turn out to be prominent in the following: we will refer to agents with $b_i = b^*$ as *informed*, in that they come to know and use in their demand all the relevant bits in Eqs. (1, 3); we call *passive* the agents with $b_i = (0, 0, ..., 0)$ as they do not use any signal and resort to a very simple constant expected value for $D$, namely $d$.

The following example will clarify the mechanics: at (the beginning of) a given time $t$ (omitted in the sequel), with $N = 5, cost = 0.01, d = 1.1, R = 1.01, a = 2$ and $b^* = (1, 1, 1, 0, 0)$, the first and second agents have $v_1 = 0.025, b_1 = (1, 1, 1, 0, 0), v_2 = 0.055$ and $b_2 = (0, 0, 0, 0, 0)$. The first agent is informed, employing all relevant bits and paying a total cost of 0.03 per period, whereas the second agent does not use any information (and has null cost). If the vector of signals is $\theta_t = (0.03, -0.05, 0.07, 0.01, -0.05)$ and, based on the demand schedule of all agents, $p_t^* = 1.11$, accordingly the realised demands are

$$x_1^* = \frac{d + b_1\theta_t' - p_t^* R}{av_1} = \frac{1.1 + 0.05 - 1.11 \cdot 1.01}{2 \cdot 0.02} = 0.578,$$
$$x_2^* = \frac{d + b_2\theta_t' - p_t^* R}{av_2} = \frac{1.1 + 0 - 1.11 \cdot 1.01}{2 \cdot 0.055} = -0.192,$$

meaning that, at the equilibrium price 1.11 prevailing at time $t$, the informed trader buys 0.578 units of the stock and the passive one sells 0.192 units. Now, $\epsilon_t$ is drawn and payoffs can be computed: let the random value be, for instance, $\epsilon_t = -0.1$ so that

$$D_t = d + b^*\theta_t' + \epsilon_t = 1.1 + 0.05 - 0.1 = 1.05.$$

Observe that the payoff is smaller than the price and, as a consequence, net buyers/sellers will experience a loss/gain. Indeed,

$$\begin{aligned} w_{1t} &= x_{1t}D_t + (w_0 - x_{1t}p_t^*)R - 3 \cdot cost \\ &= 0.578 \cdot 1.05 + (1 - 0.578 \cdot 1.11) \cdot 1.01 - 0.03 = 0.939, \\ w_{2t} &= x_{2t}D_t + (w_0 - x_{2t}p_t^*)R - 0 \cdot cost \\ &= -0.192 \cdot 1.05 + (1 + 0.192 \cdot 1.11) \cdot 1.01 = 1.024. \end{aligned}$$

Hence, due to the (relatively large) negative $\epsilon$ and to other "unlucky" events, the first agent happens to record a loss and the second agent a gain. We stress that this outcome holds at time $t$, due to the values taken by the random variables involved in this period (i.e., $\theta_t, \epsilon_t$) and to the shares of different strategies in the population that ultimately contribute to determine *the current* $p_t^* = 1.11$. Other realisations would obviously have produced different $w_{1t}$ and $w_{2t}$ for the two agents in our example. Assume now that in the learning phase the previous two agents are randomly matched: the first trader (painfully) realises that the second trader outperformed him by about 8% in period $t$ and, therefore, he imitates the other strategy and variance, so that $b_{1,t+1} = (0, 0, 0, 0, 0)$ and his variance will move to $v_{1,t+1} = v_{2t}\tilde{U}$, for a random $\tilde{U}$. The second trader does not change in any way his strategy/parameters and is ready to begin period $t + 1$.

In a standard application of evolutionary game theory, we are interested in looking at the features of the stationary population $\mathcal{P}_t$, for $t \to \infty$.

## 3 Results

We run 100 simulations for each $N = 5, 10, 15$, using $T = 10000$ periods and setting the other parameters as in Table 1.

Both $N$ and $v_\epsilon$ can, to some extent, quantify the complexity of informational extraction in a market: a large $N$ corresponds to situations where agents are

**Table 1.** Values and description of the parameters.

| Param | Value | Description | Param | Value | Description |
|---|---|---|---|---|---|
| $v_\epsilon$ | $\{\frac{2}{100}, \frac{3}{100}, \frac{4}{100}\}$ | Variance of $\epsilon$ | $v_\theta$ | 0.01 | Variance of $\theta$ |
| $R$ | 1.01 | Bond yield | $d$ | 1.10 | Stock yield |
| $a$ | 2 | Risk aversion | $q$ | 0.1 | Variance adjustment |
| $S$ | $\{1, ..., 5\}$ | # of bits | M | 1000 | # of agents |
| $w_{0i} \equiv w_0$ | 1 | Endowment | cost | 0.01 | Cost of information |
| $b_i$ | Initialised with random bits, then subject to learning and mutation | | | | |
| $v_i$ | Initialised at $v_\epsilon$, then subject to random shocks | | | | |

exposed to many signals and, especially for small values of $S$, this means that relatively few relevant signals must be carefully picked (out of the possible $N$). Besides being a direct measure of the volatility of the payoff, $v_\epsilon$ affects the signal-to-noise ratio that is proxied by $\frac{v_\theta}{v_\epsilon}$ or, in other words, *ceteris paribus*, signals are expected to be more valuable when $v_\epsilon$ is smaller and, in such sense, $v_\epsilon$ can be thought as an "adjusted price" of information.

**Table 2.** Number of agents using the most common strategies (All figures are averages over 100 simulations).

| | $N=5$ | $N=10$ | $N=15$ |
|---|---|---|---|
| Top 1 strat | 387.20 | 375.53 | 373.19 |
| Top 2 strat | 574.74 | 563.17 | 577.09 |
| Top 5 strat | 765.86 | 694.56 | 691.06 |
| Inf + Pas | 549.64 | 554.37 | 561.34 |

Generally speaking, the strategies surviving at equilibrium (i.e., in $\mathcal{P}_T$ that proxies $\mathcal{P}_\infty$)) are only a tiny fraction of the possible ones and this is a robust finding holding in all parametrizations. Table 2 shows the number of agents using the most common, the two and the five most common strategies. For instance, when $N = 5$, the two most frequent strategies are used on average by 575 agents (out of 1000). In other words, a fraction of $2/32 \approx 6\%$ of the strategies account for 57.5% of agents in equilibrium. Usually, the most used 5 strategies are taken by about 70% of agents or more. The concentration into very few strategies is striking if one thinks that there are $2^{10} = 1024$ and $2^{15} = 32768$ strategies when $N = 10$ and 15.

The last row of Table 2 shows the average number of agents who selected either the informed ($b_i = b^*$) or the passive strategy ($b_i = \mathbf{0}$). A comparison of the second and the 'Inf + Pas' row reveals that, essentially, the two most used strategies are precisely the informed and the passive one. Hence, evolution drives most of the agents to pick exactly one between these two strategies, despite the availability of tens (or hundreds or thousands) of alternatives.

Not only agents concentrate on 2 (or very few) strategies but use less information than may be naively expected. From the seminal work in [2] we know that enough informed traders should in principle allow the others to deduce or "smell" what is needed even with no direct access to the information itself. However, the extent to which this happens is probably surprising.

**Table 3.** Use of information: shares of relevant bits set and overall share (All figures are averages over 100 simulations.

| | | $N = 5$ | | $N = 10$ | | $N = 15$ | |
|---|---|---|---|---|---|---|---|
| | | Overall | Relevant | Overall | Relevant | Overall | Relevant |
| | 0.02 | 0.241 | 0.225 | 0.251 | 0.216 | 0.256 | 0.214 |
| $v_\epsilon$ | 0.03 | 0.165 | 0.149 | 0.187 | 0.156 | 0.189 | 0.152 |
| | 0.04 | 0.154 | 0.141 | 0.166 | 0.141 | 0.165 | 0.133 |

Table 3 shows the fraction of bits set to 1 by the whole population of traders and the fraction of relevant bits[4], as a function of $N$ and $v_\epsilon$. The number of used bits does not depend much on $N$ but it is quite sensitive to the variance (price of information): regardless of $N$, the fraction of 1-bits drops, say, from one-quarter to about 16% as $v_\epsilon$ varies from 0.02 to 0.04.

Put differently, the model points to a low use of information that is discarded by many traders in our noisy setup where strategies compete on short-term profits. Table 4, showing the fraction of informed and passive agents, provides additional details.

**Table 4.** Percentage of informed and passive traders. All figures are averages over 100 simulations.

| | | $N = 5$ | | $N = 10$ | | $N = 15$ | |
|---|---|---|---|---|---|---|---|
| | | Informed | Passive | Informed | Passive | Informed | Passive |
| | 0.02 | 0.440 | 0.067 | 0.402 | 0.101 | 0.369 | 0.111 |
| $v_\epsilon$ | 0.03 | 0.346 | 0.232 | 0.299 | 0.272 | 0.291 | 0.291 |
| | 0.04 | 0.253 | 0.325 | 0.221 | 0.391 | 0.245 | 0.380 |

First, scanning the table horizontally, it can be seen that the fraction of informed traders decreases with increasing complexity, as measured by $N$ (for any level of $v_\epsilon$). The effect is more pronounced when $v_\epsilon$ is low or medium.

[4] The "overall" number of set bits is (# of set bit)$/NM$ and the number of "relevant" bits is (# of set bits$/SM$).

Second, the fraction of informed agents sharply decreases with $v_\epsilon$. For instance, when $N = 10$, doubling $v_\epsilon$ roughly halves the fraction of informed traders (from 40.2% to 22.1%). This hints at the fact that (the same) information is less useful when embedded in noisier markets or, if you wish, when it is more expensive in relative terms.

Third, a similar portrait surfaces looking at the number of passive traders, the ones who decide not to use *any* signal at equilibrium. Their number is inversely proportional to that of informed traders and, hence, in general there are more such agents for large $N$ and $v_\epsilon$. The intuition is that passive agents are better equipped to survive in more volatile markets, flooded with plenty of information. Simply put, in such "difficult" environments, discarding all signals and avoiding any cost is often the most commonly evolved strategy (picked in equilibrium by nearly 40% of agents in some cases). Interestingly, there are several values of the parameters (depicting somewhat realistic markets) where the passive traders outnumber the informed one, a finding that goes against the conventional wisdom that using (good) information should be better than discarding it.

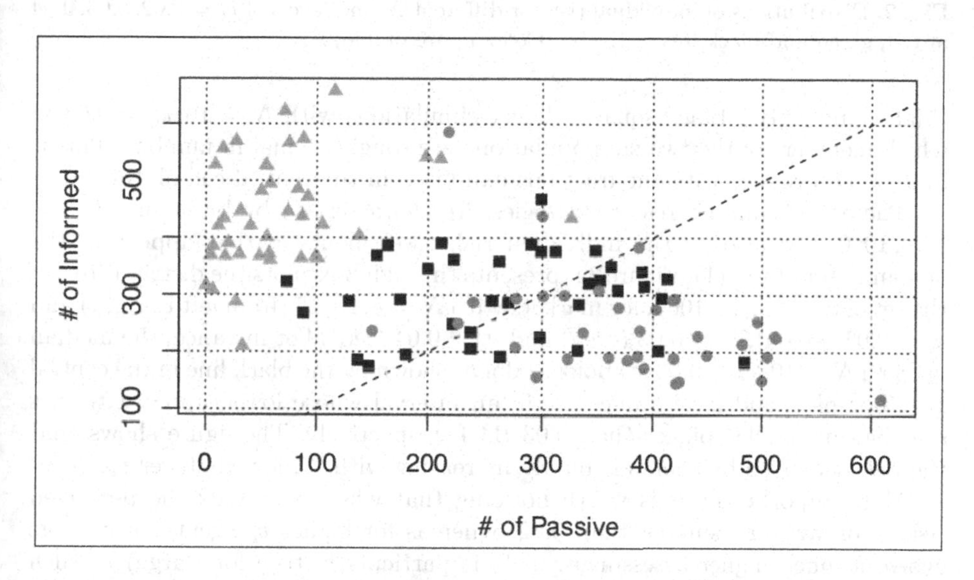


**Fig. 1.** Number of passive (on the $x$-axis) and informed traders (on the $y$-axis) in three different markets: low informational complexity on the top-left corner, for $N = 5, v_\epsilon = 0.02$, and high informational complexity moving down to the bottom-right corner, where $N = 15, v_\epsilon = 0.04$. The dashed diagonal line is where the two numbers are the same.

Figure 1 depicts the number of passive and informed agents (out of 1000) and visually reinforces the previous claims: on the top left corner there is one green triangle for each simulation, with $N = 5, v_\epsilon = 0.02$ (low complexity), and few passive agents are outnumbered by many informed ones. At the other extreme, red circles show that the situation reverses when $N = 15, v_\epsilon = 0.04$

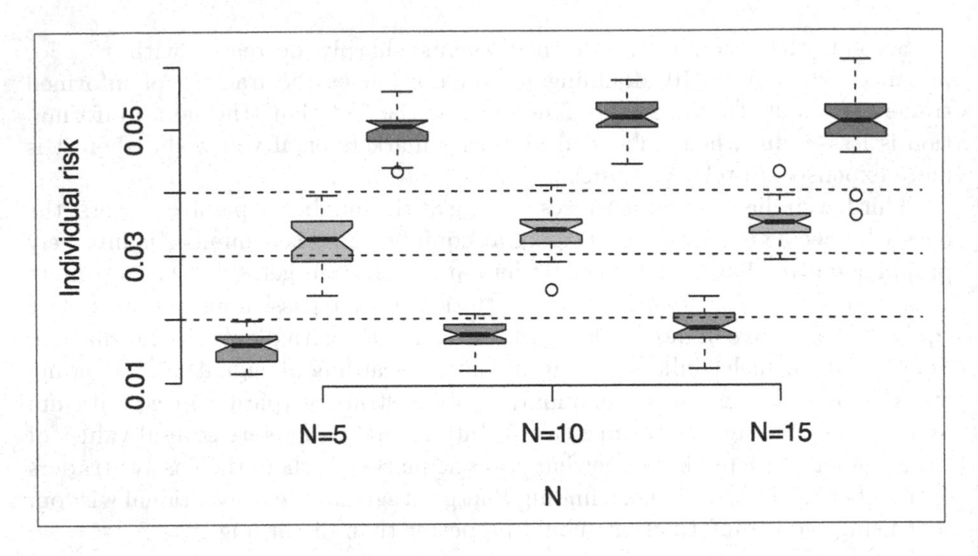


**Fig. 2.** Distributions of individual risk for different $N$ and levels of $v_\epsilon = 0.02, 0.03, 0.04$ in red, green and black, respectively. (Color figure online)

(high complexity). Black squares depict simulations with $N = 10, v_\epsilon = 0.03$ in which members of the two subpopulations are roughly equal in number. This is in good accordance with the fractions exhibited in the central cell of Table 4.

The equilibrium shares of strategies are characterised by bit-strings $b_i, i = 1, ..., 1000$, but also by the individual risk assessments $v_i$ that appear in the denominator of Eq. (1). Figure 2 represents through box-plots the distributions of the set of $v_i, i = 1, ..., 1000$, for markets with $N = 5, 10, 15$ (from left to right) and $v_\epsilon = 0.02$ (green), $v_\epsilon = 0.03$ (grey) and $v_\epsilon = 0.04$ (red). For instance, the median $v_i$ when $N = 10, v_\epsilon = 0.03$ is about 0.035, as shown by the black line in the central grey box-plot, and most values are in an interval whose lower/upper extremes are slightly smaller/bigger than 0.03/0.04, respectively. The figure shows that the risk perceived by agents is mildly increasing with $N$, for whatever $v_\epsilon$.

More importantly, it is worth noticing that when $v_\epsilon = 0.02$ the perceived risk is, on average, smaller than 0.02, whereas for higher $v_\epsilon$ agents on average evolve a much higher assessment. This is particularly true for (large) $v_\epsilon$ such as 0.04, as the red box-plots show substantially large medians about 0.05. In other words, when the volatility of the markets is low, agents learn to slightly underestimate the noise level in the market; conversely, when the volatility is higher, and especially for $v_\epsilon = 0.04$, they learn to overestimate the riskiness of the stock. As a consequence, because the $v_i$ are in the denominator of the demand function, they take larger equity positions (than would perhaps be expected) when the volatility of the market is low, and reduce the risky component of their investments when the volatility is high. This behaviour, aggressive as well as cautious in different cases, appears to curb the probability of financial extinction of agents' current strategy in our setup where sustained evolutionary competition is present.

As seen before, the number of passive traders is relatively large in markets with high $v_\epsilon$ and, consistently with this fact, passive traders evolve higher $v_i$. Hence, they not only discard information in building their portfolio but demand less units of the risky stock (other things being fixed) in an attempt to take into account the residual risk of not being informed. Symmetrically, informed agents use all the relevant information *and* boost their demand through lower $v_i$s[5].

# 4 Conclusion

The model described in this paper depicts a market where only $S$ pieces of news out of $N$ affects the stochastic payoff of a risky stock. News/information are available at a cost in any period $t$ and boundedly rational agents must figure out (or learn through evolution) which signal to use in forming their demand for the risky asset. Agents' demands (phenotypes) are driven by strategies (genotypes) prescribing the signal to use/discard and by an adjustable assessment of risk. Updates occur through pairwise comparisons (or tournaments) aiming at favouring the strategies yielding the highest payoffs. A small rate of mutation ensures that adequate diversity is preserved in learning.

We have examined which strategy prevail in the long run and their shares in the population $\mathcal{P}_t, t \to \infty$ of heterogeneous agents. This can be thought as a canonical evolutionary model where fitter strategies are determined by sharp pairwise comparison of profits, and tend to grow along time.

We found that most traders evolve (or learn) either a passive strategy discarding all signals or a fully informed strategy, where all relevant signals are incorporated in the estimation of the future payoffs (sustaining the costs). Whether the passive or informed strategy takes the lion's share mostly depends on the "complexity" of the market: if the exogenous volatility or the number of news $N$ are large, most traders will be passive and use no information whatsoever; if, instead, the volatility is low and the informational landscape has manageable size and costs, then a majority of agents will develop informed strategies. This polarity between full usage and full disregard of information fits well with models in the spirit of [9], where it is shown that returns are U-shaped in terms of information and, hence, being entirely uninformed (passive in our setup) or fully informed is more profitable than being half-way on either side.

Overall, the model is demonstrating that it may impossible for boundedly rational agents to exploit all the information or disentangle relevant from irrelevant news in volatile market setups or when too much information is provided and must be screened. The model shows that fact-based learning by trial-and-error and imitation (plus mutation), does not allow full exploitation of relevant information. It is left to future research to investigate whether the same results

[5] Another interpretation leads to overconfidence on the part of informed traders: in principle, once all relevant component of $\theta_t$ are used, there is no intrinsic noise other than $\epsilon$ and super-rational agents should set $v_i = v_\epsilon$. The model shows that there is evolutionary pressure to adjust downwards the individual risk assessment, or, that it pays off for the informed to be overconfident at equilibrium.

hold for other learning schemes (or using other ways to spread fitter strategies). For example, the probability to adopt a better strategy in a pair may be proportional to the returns, instead of being 100%. While our switching rule may appear dummy, agents have a nice way to rationalise their behaviour as, in equilibrium, all strategies have the same median returns and, say, passive agents have a 50% chance to over-perform informed trader, i.e. they would fare first one time out of two in a race.

A byproduct of the model possibly hints at a novel way to explain why many investors appear to hold an excessively small share of equity[6], see [7] for an explanation based on biological evolution. In our model, for many values of the parameters, the passive traders, who demand little equity due to their large $v_i$'s, are the majority share and this leads in aggregate to a limited (relative) share of risky holdings with respect to risk-less investments, in line with the historical simulation in [1].

**Acknowledgements.** We thank the audiences at WEHIA 2022 in Catania, City University London and AAU Klagenfurt for their comments and remarks. Luca Gerotto and Marco Tolotti provided many insightful comments and ideas on related previous work.

## References

1. Benartzi, S., Thaler, R.: Myopic Loss aversion and the equity premium puzzle. Q. J. Econ. **110**(1), 73–92 (1995)
2. Grossman, S.J., Stiglitz, J.E.: On the impossibility of informationally efficient markets. Am. Econ. Rev. **70**(3), 393–408 (1980)
3. Huang, Y., Zhou, C., Cui, K., Lu, X.: A multi-agent reinforcement learning framework for optimizing financial trading strategies based on timesnet. Expert Syst. Appl. **237**, 121502 (2024)
4. Lo, A.W., Mamaysky, H., Wang, J.: Foundations of technical analysis: computational algorithms, statistical inference, and empirical implementation. J. Finan. **55**(4), 1705–1765 (2000)
5. Mehra, R., Prescott, E.C.: The equity premium: a puzzle. J. Monet. Econ. **15**(2), 145–161 (1985)
6. Newton, J.: Evolutionary game theory: a renaissance. Games **9**(2), 31 (2018). https://doi.org/10.3390/g9020031 ,https://www.mdpi.com/2073-4336/9/2/31
7. Robson, A.J., Orr, H.A.: Evolved attitudes to risk and the demand for equity. Proc. Natl. Acad. Sci. **118**(26), e2015569118 (2021)
8. Sandholm, W.H.: Population Games And Evolutionary Dynamics. MIT Press, Cambridge (2010). Economic learning and social evolution
9. Schredelseker, K.: Pascal's wager and information. J. Forecast. **33**(6), 455–470 (2014)
10. Smith, J.M.: Evolution and the Theory of Games. Cambridge University Press, Cambridge (1982). https://doi.org/10.1017/CBO9780511806292

[6] The reason why this happens is not entirely clear and this phenomenon is known as the "equity premium puzzle", see [5].

# Heterogeneous Mean-Field Analysis of Best-of-n Decision Making in Networks with Zealots

Thierry Njougouo[1(✉)], Timoteo Carletti[2], Andreagiovanni Reina[3], and Elio Tuci[1]

[1] Faculty of Computer Science and Namur Institute for Complex Systems, naXys, University of Namur, Rue Grandgagnage 21, 5000 Namur, Belgium
thierry-sainclair.njougouo@unamur.be, thierrynjougouo@ymail.com
[2] Department of Mathematics and Namur Institute for Complex Systems, naXys, University of Namur, Rue Grafé 2, 5000 Namur, Belgium
[3] Institute for Interdisciplinary Studies on Artificial Intelligence, IRIDIA Université Libre de Bruxelles, Brussels, Belgium

**Abstract.** Humans and animals often choose between options with different qualities. When the decisions are not determined by one or a few individuals leading a group, a collective can achieve a consensus through repeated interactions among the individuals. Collective decision-making is widely studied in the context of opinion dynamics, showing that individual mechanisms of option selection and the underlying social network affect the outcome. Mathematical techniques, such as the heterogeneous mean-field (HMF) theory, have been developed to systematically analyse the collective behaviour of interconnected agents. Based on the HMF theory, we propose a mathematical model that looks at the combined effects of multiple elements bearing upon the collective decision dynamics, such as the individuals' cognitive load, the difference in the quality of the options, the network topology, and the location of the zealots in the network. The results of this study show that, in scale-free networks, when individuals employ specific opinion selection mechanisms, characterised by a low cognitive load, the zealots have the ability to steer the consensus towards the option with the lowest quality or to group indecision. This result is reversed when the interaction network is sparsely connected and quite homogeneous – that is, most nodes have few neighbours – and cognitively simple individuals make accurate collective decisions, mostly unaffected by zealots voting for the option with the lowest quality.

**Keywords:** Opinion dynamics · Best-of-n Problem · Zealots · Heterogeneous mean-field

## 1 Introduction

Human beings are every day faced with the problem of choosing among different options. Limited information or noisy conditions can make such decisions even


M. Villani et al. (Eds.): WIVACE 2023, CCIS 1977, pp. 339–351, 2024.
https://doi.org/10.1007/978-3-031-57430-6_26

more difficult; a possible way to overcome the issue is to exploit social interaction. Collective decision-making (CDM) is hence characterised by the fact that once the decision is made it is no longer attributable to any individual of the group. Achieving a consensus is the result of multiple interactions in which individuals choose an option according to some opinion formation mechanisms that can be relatively simple. For example, a general agreement can be reached using social feedback, by which consensus emerges among individuals that select an option by copying the preferences of one or more group mates.

CDM is not exclusive to humans but is also observed in other social species [6]. For example, groups of baboons collectively decide in which direction to move [27]; flocks of birds collectively decide their motion direction [3]; and swarms of bees collectively decide where to build a new nest site [21,25]. Investigating CDM is important for understanding the behaviour of many biological systems, and for enabling autonomy in artificial systems such as robots [20]. For example, swarms of robots are programmed with collective decision-making algorithms to cooperatively perform a variety of tasks [13,24,28,30]. Therefore we can conclude that different scientific disciplines are interested in investigating CDM and unveiling the elements that influence and contribute to determining the outcome of various decision-making processes.

CDM problems have been studied with different methods such as experimental methods [6,7], computational modelling and simulation methods [12], and social network analysis [4,18]. These studies have focused on different issues such as: i) the effect of different individual opinion selection mechanisms, each of which is associated with different cognitive costs (e.g., linked to memory, perception, attention) [10,22]; ii) the effect of the homogeneity/heterogeneity in the group behaviour (i.e. individuals have equal/different behaviours) [8,23]); iii) the effects of different topologies of the interaction network between the individuals [14,26]; iv) the effects of the cost/benefit trade-off associated to the selection of each option (e.g., the quality of the chosen option and the time spent selecting it) [17,19].

The objective of this study is to develop a mathematical model to analyse the combined effects of multiple factors (i.e., the cognitive load, the option qualities, the network topology, and the location in the network of zealots voting for the inferior option) bearing upon the opinion dynamics. More precisely, we model an asymmetrical binary collective decision-making process in which both options have equal costs, but one option has better quality than the other. Moreover, we model the exchange of information among agents as happening on a finite-size network composed of $N$ nodes and $L$ undirected edges, i.e., each node represents an agent and an edge the interaction existing among two agents. We also consider that certain individuals use conformism rules through which they agree with the opinion of their peers (which we call susceptible agents), and the rest never change their opinion and are normally called zealots [8,23] or stubborn agents [17]. In our study, we only consider zealots with an opinion in favour of the inferior option, with the lower quality. Finally, we study different behaviours of the susceptible agents with respect to their cognitive load, that in our model translates into different pooling errors when an agent processes the opinions of her neighbours. The cognitive load is considered relatively low when an individual simply copies the preference of a randomly selected neighbour among the agents within her first connections, this behaviour corresponds to the voter model [26]. The cognitive load progressively increases for social feedback mechanisms in which each agent has to sample a progressively higher number of peers within her network of connections in order to select an option, e.g., to apply the local majority rule [11].

The original contribution of this study is to illustrate how the interactions between i) the agents' cognitive load, ii) the interaction network topology, and iii) the location of zealots in the network, influence the decision-making process, i.e., consensus, or not, for the opinion with the best quality. Given the asymmetry in quality (i.e., one option is better than the other and therefore is shared more often [29]), we study under which conditions, the zealots (who only share opinions for the inferior option) manage to counterbalance the difference in qualities and drive the population toward a consensus on the lowest quality option. Our study shows that when the susceptible agents follow a simple behaviour with relatively high pooling errors, the zealots voting for the inferior option lead the population into either an indecision state or a consensus for the inferior option. However, our results also show that this result can be reversed when connectivity and heterogeneity of the interaction (social) network reduce (i.e., the network becomes more homogeneous with most nodes with few neighbours).

## 2 Method and Methodology

The aim of this section is to introduce the basic rules upon which the agents possibly update their opinion and then to build a mathematical model based on the heterogeneous mean-field assumption to unravel the role of some main model parameters, namely the fraction of zealots present in the population, their location in the network and the network topology.

### 2.1 Model Description

Let us thus consider a group of $N$ agents interacting in an undirected scale-free network [1,16], where the probability for an agent to have $k$ neighbours is given by $p_k \sim 1/k^\gamma$, with $\gamma > 2$. Let us recall that the closer $\gamma$ is to 2 the more heterogeneous the degree distribution is, indeed nodes with a very large degree can be present because $\langle k^2 \rangle$ is unbounded; on the other hand, if $\gamma \gg 3$ very high degree nodes are very rare and the degree spread is well described by finite variance of the degree distribution. Assume also the network to be connected, to avoid to consider the trivial case of a population split into several non-communicating groups, and simple, namely among every couple of agents there is at most one communication channel. The network topology is thus encoded by the $N \times N$ adjacency matrix, $\mathbf{M}$, whose entries satisfy $M_{ij} = M_{ji} = 1$ if and only if agents $i$ and $j$ can exchange opinion, and 0 otherwise.

We classify agents, i.e., the nodes of the network into susceptible and zealots, the former change their opinion over time in response to social interactions while the latter are inflexible and never change their initial choice. In this work, we consider the scenario of the best-of-2 problem where each node holds an opinion that can take one of two different values, $A$ or $B$, modelling the choice between two beliefs on a particular issue or topic. We also associate to each opinion the corresponding *quality*, i.e., $Q_A > 0$ and $Q_B > 0$. The quality defines the strength or the probability that the option is communicated to the neighbours [8,23,29]. Without lack of generality, for the rest of the work, we assume $Q_A = 1$, $Q_B \leq Q_A$ and hence $Q = Q_B/Q_A \leq 1$. We only consider the scenario where zealots hold an opinion in favour of the opinion with the lower quality (i.e., opinion $B$). In fact, it is less interesting to introduce zealots voting for the option with the highest quality (option $A$) because the group already votes more frequently for options with better quality and its is more frequently selected by the group. Here, we study the ability of the group to select the best option despite the presence of zealots voting for the inferior option.

To specify how a susceptible individual updates her belief based on the weighted opinions of her neighbours, we consider the model (1) from [22] which we display in Fig. 1 for some representative values of the parameter $\alpha \in [0, 1.5]$.

$$P_\alpha(x) = \begin{cases} \frac{1}{2} - \frac{1}{2}(1-2x)^\alpha & \text{if } 0 \leq x \leq \frac{1}{2} \\ \frac{1}{2} + \frac{1}{2}(2x-1)^\alpha & \text{if } \frac{1}{2} < x \leq 1. \end{cases} \tag{1}$$

Let us observe that $\alpha$ is negatively correlated to the *cognitive load*: as $\alpha$ increases the agents makes more pooling error. More precisely, for $\alpha = 0$, the function $P_0$ models agents that make no errors and change their opinion based on the weighted average of all their neighbours, i.e., agents adopt a majority rule. This requires a larger cognitive load than when $\alpha > 0$ in which case agents make errors. In the case of $\alpha = 1$, the function $P_1$ models agents that update their opinion by copying the one of a randomly selected neighbour, namely this behaviour is the (weighted) voter model. Our model generalises thus two prominent models of opinion dynamics, the (weighted) voter model [26,29] and the (weighted) majority model [2,11,12]. For generic values of $\alpha > 0$ and $\alpha \neq 1$, the proposed model allows us to explore behaviours with intermediate levels of cognitive cost and pooling error.

The system evolves asynchronously: each time step an agent $i$ is randomly selected with a uniform probability and makes a social interaction. If the agent is a zealot nothing happens; otherwise if the selected agent is susceptible, she updates her opinion as a function of the weighted fractions of local opinions

$$n_{i,A}^{\#} = \frac{Q_A n_{i,A}}{Q_A n_{i,A} + Q_B n_{i,B}} \quad \text{and} \quad n_{i,B}^{\#} = \frac{Q_B n_{i,B}}{Q_A n_{i,A} + Q_B n_{i,B}}, \tag{2}$$

which are based on the number of $i$'s neighbours $n_{i,A}$ and $n_{i,B}$, with opinion $A$ and $B$, respectively, and the options qualities $Q_A$ and $Q_B$. Note that we trivially have $n_{i,A}^{\#} + n_{i,B}^{\#} = 1$, $\forall i$.

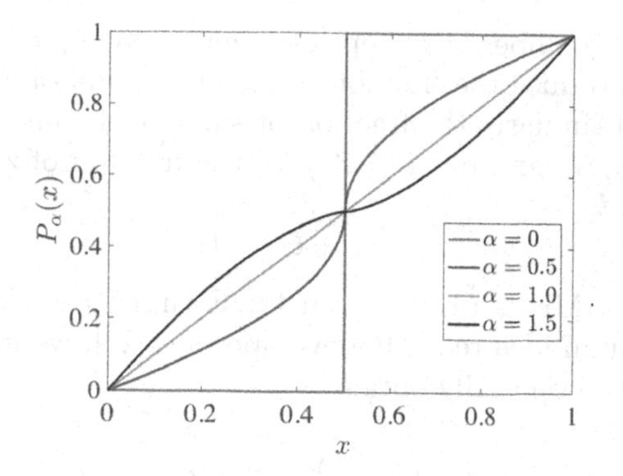


**Fig. 1.** The function $P_\alpha(x)$ for several values of $\alpha$.

Let $k_i$ be the degree of node $i$, namely the number of agent $i$'s neighbours, thus $n_{i,A} + n_{i,B} = k_i$. Then by recalling $Q = Q_B/Q_A$ we can rewrite Eq. (2) as

$$n^{\#}_{i,A} = \frac{n_{i,A}/k_i}{(1-Q)n_{i,A}/k_i + Q} \quad \text{and} \quad n^{\#}_{i,B} = 1 - n^{\#}_{i,A}\,. \tag{3}$$

Assume the selected $i$-th agent holds opinion $A$ (resp. opinion $B$), then with probability $P_\alpha(n^{\#}_{i,B})$ (resp. $P_\alpha(n^{\#}_{i,A})$), she can change her opinion to $B$ (resp. $A$). Let us also observe that because of the functional form of (1) and because $n^{\#}_{i,A} + n^{\#}_{i,B} = 1$, we can conclude that $P_\alpha(n^{\#}_{i,A}) + P_\alpha(n^{\#}_{i,B}) = 1$. The process continues by iteratively selecting one agent at a time and by updating its opinion; eventually the system reaches a stationary state.

### 2.2 A Mathematical Model with Option's Quality and Zealots

The objective of this subsection is to propose a simple mathematical model defined by an ordinary differential equation (ODE) allowing us to study the evolution of group opinion, but also to unravel the role of the involved parameters, the cognitive load, the ratio of the opinion qualities $Q = Q_B/Q_A$, the fraction of zealots, and the network topology $\gamma$.

To make some analytical progress we rely on the heterogeneous mean-field assumption (HMF) [5,15], namely we hypothesise that nodes with the same degree are dynamically equivalent and their evolution can be described by using the degree conditional probability $p(k'|k)$, namely the probability that a node with degree $k$ is connected to another node of degree $k'$. Therefore, nodes are grouped into degree classes, more precisely we define $A_k$ (resp. $B_k$), as the number of nodes with degree $k$ and opinion $A$ (resp. opinion $B$). To distinguish between susceptible agents with opinion $B$ and zealots, let us introduce $Z_k$ to denote the number of zealots with opinion $B$ and degree $k$. Therefore, letting $N_k$ to denote the total number of nodes with degree $k$, we have:

$$A_k + Z_k + S_k = N_k, \tag{4}$$

where $S_k$ denotes the number of susceptible agents having opinion $B$ and degree $k$. Eventually we introduce the fraction of agents having opinion $A$ and degree $k$, $a_k = A_k/N_k$, and similarly the fraction of susceptible having opinion $B$ with degree $k$ by $b_k = S_k/N_k$ and by $\zeta_k = Z_k/N_k$ the fraction of zealots with degree $k$. Therefore, for all $k$,

$$a_k + b_k + \zeta_k = 1. \tag{5}$$

The goal of the HMF is to derive an ODE ruling the evolution of $a_k$ and $b_k$. By starting from an idea recently developed in [22], we improve it with the addition of zealots to eventually obtain

$$\frac{d\langle a\rangle}{dt} = -\langle a\rangle + \sum_k q_k(1-\zeta_{k+1}) \sum_{\ell=0}^{k+1} \binom{k+1}{\ell} \langle a\rangle^{k+1-\ell} (1-\langle a\rangle)^{\ell} P_\alpha\left(\frac{k+1-\ell}{k+1-\ell+\ell Q}\right), \tag{6}$$

where we defined $\langle a\rangle = \sum_k q_k a_{k+1}$, being $q_k$ the probability for a node to have an excess degree $k$, namely

$$q_k = \frac{(k+1)\, p_{k+1}}{\langle k\rangle} \quad \forall k \geq 0,$$

with $\langle k\rangle = \sum_k k p_k$ the average node degree. Equation (6) contains the relevant parameters of the model, the zealots ($\zeta_k$), the model of opinion dynamics ($P_\alpha$), the opinion quality ($Q$) and the network topology ($q_k$). The aim of the next section is to determine the equilibria of such equation and determine their stability, hence the system fate. Before to do this let us observe that knowing $\langle a\rangle(t)$ from (6) one can obtain the evolution of $a_k$ for all $k$ by using the following equation [22]:

$$\frac{da_k}{dt} = -a_k + (1-\zeta_k) \sum_{l=0}^{k-1} \binom{k}{l} \langle a\rangle^{k-l} (1-\langle a\rangle)^{l} P_\alpha\left(\frac{k-l}{k-l+lQ}\right).$$

### 2.3 Equilibria and Stability of the Analytical Model

The equilibria of the system are obtained by setting to zero the right hand side of (6). Let us thus define the function

$$f_\alpha(x) := -x + \sum_k q_k(1-\zeta_{k+1}) \sum_{\ell=0}^{k+1} \binom{k+1}{\ell} x^{k+1-\ell} (1-x)^{\ell} P_\alpha\left(\frac{k+1-\ell}{k+1-\ell+\ell Q}\right), \tag{7}$$

hence by denoting $\langle a^*\rangle$ a system equilibrium, we have by definition

$$f_\alpha(\langle a^*\rangle) = 0.$$

A direct inspection of (7) allows to prove that $f_\alpha(0) = 0$, hence $\langle a^*\rangle = 0$, i.e., absence of agents with opinion $A$, is an equilibrium of the system. On the other hand, $f_\alpha(1) = -\sum_k q_k\zeta_{k+1} \neq 0$, hence the presence of zealots (with opinion $B$) prevents the system from converging to a population where only agents $A$ will exist. Finally the existence of nontrivial solution $0 < \langle a^* \rangle < 1$ to the equation $f(\langle a^* \rangle) = 0$ will determine a coexistence of opinions $A$ and $B$ in the network.

The stability of the above-mentioned equilibria can be determined by looking at the derivative of the function $f_\alpha$ evaluated on the same equilibria. Such analysis will be presented in the following section where we also discuss the impact of the main model parameters.

## 3 Results

In this section, we present the results obtained for the analytical model described in Sect. 2.2. As already mentioned, we focus on the impact of the parameter $\alpha$, the network topology, hereby summarised into the exponent $\gamma$ of the power law, and the social influence of the zealots. More precisely, regarding the zealot analysis, we are interested in both their relative abundance and their position in the network, namely if they sit onto high-degree (hubs) or small-degree (leaves) nodes. To place zealots in hubs, we set $\zeta_k = 1$ for all $k \geq k_M$, for some sufficiently large $k_M$, this accounts to add into the model an average number of zealots equal to $Z_{tot} = \sum_{k \geq k_M} N_k \sim \sum_{k \geq k_M} N c_\gamma / k^\gamma$, where $c_\gamma$ is a normalisation constant such that $\sum_k p_k = 1$ and $N$ is the total number of nodes in the network. When we assume zealots to lie on leaves nodes and to fair compare this condition with the previous one, we consider the same number of zealots, that we set into the hubs by assuming $\zeta_{k_{min}} = Z_{tot}/N_{k_{min}}$, where $k_{min}$ is a small enough degree; more precisely:

$$\zeta_{k_{min}} = \frac{Z_{tot}}{N_{k_{min}}} \sim \frac{Z_{tot}}{N p_{k_{min}}} = k^\gamma_{min} \sum_{k \geq k_M} \frac{1}{k^\gamma} \sim \frac{k_M}{\gamma - 1} \left( \frac{k_{min}}{k_M} \right)^\gamma .$$

Note that the above strategy does not imply adding an infinite number of zealots, indeed in any network realisation, e.g., by using the configuration model, there is a finite number of nodes with a degree larger than $k_M$ and thus $Z_{tot}$ is also a finite quantity. These finite-size effects can be studied in future research.

Figure 2 summarises our main results. We fix the values of $Q = Q_B/Q_A = 0.9$, the power law exponent $\gamma$, and the zealot location in the network, and we numerically compute the zeros of the function $f_\alpha$ for values of $\alpha \in [0, 2]$ to obtain the equilibria of the system. Once the equilibria have been found, we evaluate the derivative of $f_\alpha$ and we determine its sign, if $f'_\alpha(\langle a^* \rangle) > 0$ then the equilibrium $\langle a^* \rangle$ is unstable and marked with a red points in Fig. 2, on the other hand if $f'_\alpha(\langle a^* \rangle) < 0$ then the equilibrium $\langle a^* \rangle$ is stable and we represent it in green. The three top panels refer to the strategy consisting of setting the zealots in the leaves (here $k_{min} = 1$), and the three bottom panels refer to the opposite strategy with the zealots in the hubs, $k_M = 100$. Moving from left to right we increase $\gamma$, passing from $\gamma = 2.5$ (left panels a) and d)), $\gamma = 3.0$ (middle panels b) and e)) and $\gamma = 3.5$ (right panels c) and f)).

Several conclusions can be drawn from those results. For large enough $\alpha$, the system always sets into a state where opinions $A$ and $B$ coexist, the closer to

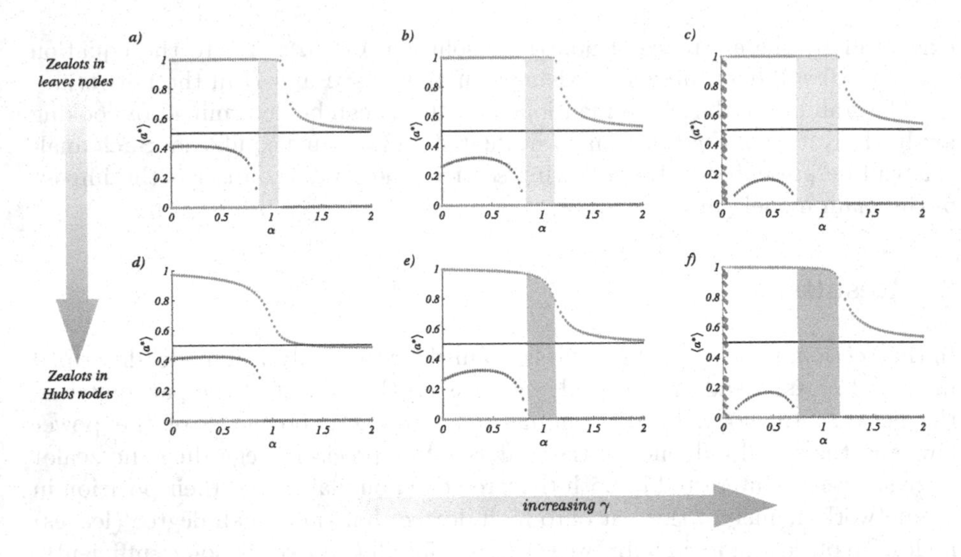


**Fig. 2.** Bifurcation diagrams of the HMF model. We report the equilibria $\langle a^* \rangle$ of Eq. (6) as a function of $\alpha$; stable equilibria, i.e., associated to $f'_\alpha(\langle a^* \rangle) < 0$, are coloured in green while unstable ones, i.e., associated to $f'_\alpha(\langle a^* \rangle) > 0$, in drawn red. The top panels correspond to zealots set into leaves nodes while the bottom panels to the strategy of placing the zealots into the hubs. Panels a) and d) correspond to $\gamma = 2.5$, panels b) and e) $\gamma = 3.0$, panels c) and f) to $\gamma = 3.5$. The remaining parameters have been fixed to $k_{min} = 1$, $k_M = 100$ and $Q = 0.9$.

0.5 the larger $\alpha$; this behaviour is independent of where the zealots are placed in the network or the network topology, i.e., $\gamma$. Hence a too-large pooling error $\alpha$ by the agents (which corresponds to a very small cognitive load) prevents the group from choosing the opinion with the highest quality.

For intermediate values of the pooling error, e.g., close to $\alpha = 1$, the impact that zealots have on the opinion dynamics depends on the interaction network topology. For scale-free networks with strong degree heterogeneity, e.g., $\gamma = 2.5$, the location where zealots are placed has a strong impact on the system fate. Putting the zealots into the leaves does not prevent the groups from selecting the best option (Fig. 2a), instead when zealots sit in the hubs, the susceptible agents are unable to make consensus decisions and remain polarised between the two options (in Fig. 2d the system for $\alpha \approx 1$ converges to the stable equilibrium $\langle a^* \rangle \sim 0.5$). The situation changes when the network heterogeneity decreases (i.e. for higher $\gamma$). Here, regardless of the location of the zealots in the network (leaves or hubs), the stable equilibrium is $\langle a^* \rangle \sim 1$, representing a consensus decision for the best option. This can be further appreciated by comparing the grey rectangles in the top and bottom panels of Fig. 2, which have the same horizontal size. The cause of this effect – to be investigated in future research – can be due to the rare presence of hubs in networks with large $\gamma$.

We also observe a strong impact of $\gamma$ on the system outcome for small pooling error $\alpha$, however having zealots in different locations (leaves or hubs) does not impact the system's equilibria. Indeed for $\alpha \approx 0$ we can observe that the unstable equilibrium branch (red dots) is below 0.5, this means that an initial population with few $A$ agents, e.g., 40% of $A$ and 60% of $B$, is capable to converge to a consensus toward $A$, and this result holds true despite the presence of zealots and their placement. This effect amplifies with increasing $\gamma$, e.g., for $\gamma = 3.0$ the faction of $A$ can be as small as ~30%, and for large enough $\gamma$, i.e., scale-free networks with a relatively homogeneous degree distribution, any initial arbitrarily small fraction of agents with opinion $A$ will be able to prevail and spread in the whole population (see the tiny dashed rectangles in panels c and f associated to $\gamma = 3.5$, where the equilibrium $\langle a^* \rangle = 0$ is unstable and thus the system converges to the only remaining possible equilibrium $\langle a^* \rangle \sim 1$).

To obtain a more global view of the complex interplay of the parameters, we studied the equilibrium $\langle a^* \rangle$ as a function of $\alpha$ and $\gamma$ for a fixed value of $Q = 0.9$ (see Fig. 3). Moreover in each considered case we studied the impact of the strategy of placing the zealots on the leaves nodes (top panels) or on the hubs (bottom panels). In the two panels on the left, we colour-code the equilibrium reached by the system (yellow high values of $\langle a^* \rangle$ close to 1 and blue $\langle a^* \rangle$ close to 0), starting from an initial population with half agents holding opinion $A$ and half opinion $B$. One can observe a striking difference between the top panel Fig. 3a), where zealots sit into leaves, and the bottom panel Fig. 3b), where zealots sit into hubs. In the first case, the equilibrium $\langle a^* \rangle$ is almost independent of $\gamma$ and the system exhibits two main behaviours: for $\alpha \lesssim 1$ the whole group converges to a consensus to $A$, while for $\alpha \gtrsim 1$ the population is deadlocked at indecision with two similar-sized groups of agents with opinion $A$ and $B$ that coexist. In the second case, when zealots are placed into the hubs (Fig. 3c): the population converges to a consensus for the opinion with the lower quality when $\alpha \sim 1$ and $\gamma \lesssim \gamma_*$, where $\gamma_* \sim 2.33$. To better visualise this qualitative difference in the dynamics, we report on the right panels the bifurcation diagram with the three equilibria $\langle a^* \rangle$ as a function of the cognitive load for $\gamma = 2.2$, which is lower than the critical $\gamma_* \sim 2.33$. In the top panel, Fig. 3b), with zealots set into the leaves, the population converges to an almost consensus (large majority) for option $A$ for $\alpha \lesssim 1.1$. On the other hand, in the bottom panel, Fig. 3d), with zealots set into hubs, for $\alpha \sim 1$ the group chooses the opinion with the lower quality. These results show that a population of agents using the (weighted) voter model as decision-making behaviour can be driven to adopt the opinion with the lower quality by zealots placed into hubs of a sufficiently heterogeneous scale-free network, i.e., $\gamma < \gamma_*$.

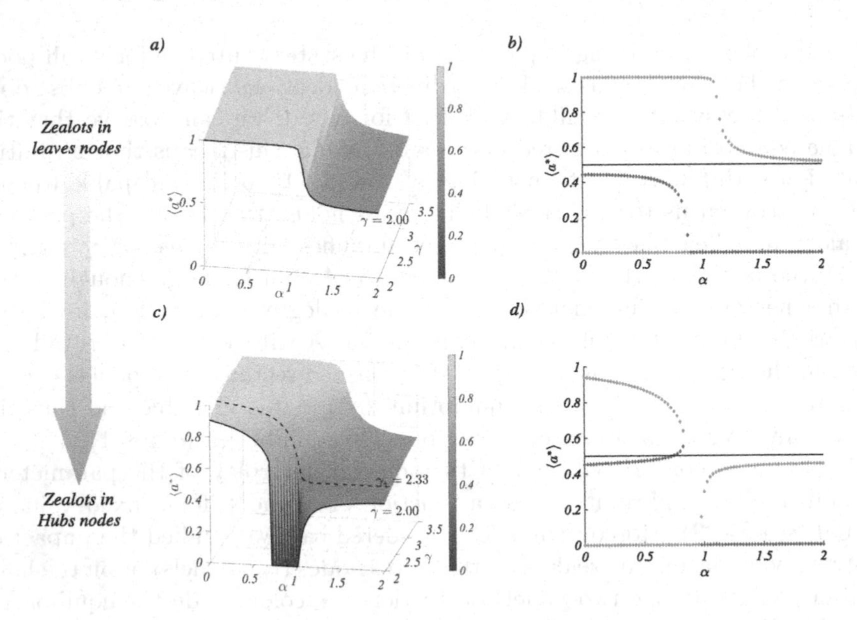


**Fig. 3.** Bifurcation diagrams of the HMF. We report the equilibrium $\langle a^* \rangle$ given by (6) as a function of $(\alpha, \gamma)$ for a fixed value of $Q = 0.9$ (left panels), and the same equilibrium where we also fix $\gamma = 2.2 < \gamma_*$ (right panels). Top panels correspond to zealots set into leaves nodes while the bottom panels to the strategy of placing the zealots into the hubs. The remaining parameters have been fixed to $k_{min} = 1$ and $k_M = 100$.

## 4 Conclusion

In this paper, we presented the results of a study focused on a best-of-n collective decision-making problem, with $n = 2$ options of different quality, and a heterogeneous population comprising a majority of the agents that have a conformist behaviour and change their opinion based on the social feedback and a minority of agents – referred to as zealots – who never change their opinion. The interactions among the agents happen over a social network whose nodes are the agents and the edges are the possible interaction links. We analyse the opinion dynamics for populations of agents with voter-like behaviours. We consider a continuum of behaviours characterised by the pooling error $\alpha$ that agents make when processing social information; making more errors reduces the agent's cognitive load. Our model extends [22] and generalises through a single function a number of known voter-like models, such as the (weighted) voter model [26,29] and the majority model [9,11,12].

We build our mathematical model of the collective decision process using the heterogeneous mean-field (HMF) theory. The determination of the system equilibria and their stability allowed us to study the combined effect of the model parameters, characterised by the cognitive load, the opinion quality, the network topology, and the location of zealots in the network. In particular, we have studied populations of agents with a given cognitive load (pooling error $\alpha$) that interact on a scale-free network. In our analysis, we varied both the network topology and the location (leaves or hubs) where zealots—all holding the lowest quality opinion—are placed. The results have shown that the combined effect of these factors generated an articulated landscape characterised by different outcomes of the collective decision process. In case agents employ a high level of cognitive load, the collective decision follows the one of the majority (i.e., democratic decisions) with a bias for the best alternative that grows as the network degree distribution becomes more homogeneous (i.e., high $\gamma$). When the cognitive load is minimal, the group is unable to make a decision due to high pooling errors. The most interesting outcomes are obtained when the parameter $\alpha \sim 1$, corresponding to populations of agents employing decision-making mechanisms requiring a medium level of cognitive load, similar to the (weighted) voter model. In this case, zealots placed in the hubs, i.e., nodes with a large degree, are able to drive the entire population away from the consensus for the best quality option and lock it into indecision or to consensus for the inferior option. When zealots are placed in the leaves (i.e., nodes with a small degree) rather than in hubs, the population shows different decision dynamics and zealots are not able to interfere with the selection of the best option. This phenomenon is amplified by the parameter $\gamma$, i.e., the one ruling the heterogeneity of the scale-free network, in terms of node degree distribution. Large enough $\gamma$ allows the better quality opinion to spread in the whole population even if initially a relatively small minority of agents has such opinion (Fig. 2). In the future, we aim to generalise these results by exploring a larger range of parameters and finding unifying patterns. In particular, we intend to study, in addition to the location in the network, how the quantity of the zealots and the option qualities influence the decision-making dynamics and how all these parameters interplay with each other in determining the opinion dynamics of the population.

**Acknowledgements.** T.N. thanks the University of Namur for the financial support. A.R. acknowledges support from the Belgian F.R.S.-FNRS, of which he is a Chargé de Recherches.

## References

1. Barabási, A.L., Albert, R.: Emergence of scaling in random networks. Science **286**(5439), 509–512 (1999)
2. Canciani, F., Talamali, M.S., Marshall, J.A.R., Reina, A.: Keep calm and vote on: swarm resiliency in collective decision making. In: Workshop Resilient Robot Teams of 2019 IEEE ICRA (2019)
3. Cavagna, A., et al.: Natural swarms in 3.99 dimensions. Nat. Phys. **19**(7), 1043–1049 (2023)
4. Centola, D.: How Behavior Spreads: The Science of Complex Contagions, vol. 3. Princeton University Press, Princeton (2018)
5. Colizza, V., Pastor-Satorras, R., Vespignani, A.: Reaction-diffusion processes and metapopulation models in heterogeneous networks. Nat. Phys. **3**(4), 276–282 (2007)

6. Conradt, L., List, C.: Group decisions in humans and animals: a survey. Phil. Trans. of The Royal Soc. B **364**(1518), 719–742 (2009)
7. Couzin, I.D., et al.: Uninformed individuals promote democratic consensus in animal groups. Science **334**(6062), 1578–1580 (2011)
8. De Masi, G., Prasetyo, J., Tuci, E., Ferrante, E.: Zealots attack and the revenge of the commons: quality vs quantity in the best-of-*n*. In: Dorigo, M., et al. (eds.) ANTS 2020. LNCS, vol. 12421, pp. 256–268. Springer, Cham (2020). https://doi.org/10.1007/978-3-030-60376-2_20
9. Galam, S.: Majority rule, hierarchical structures, and democratic totalitarianism: a statistical approach. J. Math. Psychol. **30**(4), 426–434 (1986)
10. Hillemann, F., Cole, E.F., Sheldon, B.C., Farine, D.R.: Information use in foraging flocks of songbirds: no evidence for social transmission of patch quality. Anim. Behav. **165**, 35–41 (2020)
11. Krapivsky, P.L., Redner, S.: Dynamics of majority rule in two-state interacting spin systems. Phys. Rev. Lett. **90**(23), 238701 (2003)
12. Lambiotte, R.: Majority rule on heterogeneous networks. J. Phys. A: Math. Theor. **41**(22), 224021 (2008)
13. Masi, G.D., Prasetyo, J., Zakir, R., Mankovskii, N., Ferrante, E., Tuci, E.: Robot swarm democracy: the importance of informed individuals against zealots. Swarm Intell. **15**(4), 315–338 (2021)
14. Momennejad, I.: Collective minds: social network topology shapes collective cognition. Phil. Trans. of the Royal Soc. B **377**(1843), 20200315 (2022)
15. Pastor-Satorras, R., Vespignani, A.: Epidemic spreading in scale-free networks. Phys. Rev. Lett. **86**(14), 3200 (2001)
16. Pastor-Satorras, R., Vespignani, A.: Evolution and Structure of the Internet: A Statistical Physics Approach. Cambridge University Press, Cambridge (2004)
17. Prasetyo, J., De Masi, G., Ranjan, P., Ferrante, E.: The best-of-*n* problem with dynamic site qualities: achieving adaptability with stubborn individuals. In: Dorigo, M., Birattari, M., Blum, C., Christensen, A.L., Reina, A., Trianni, V. (eds.) ANTS 2018. LNCS, vol. 11172, pp. 239–251. Springer, Cham (2018). https://doi.org/10.1007/978-3-030-00533-7_19
18. Redner, S.: Reality-inspired voter models: a mini-review. C R Phys. **20**(4), 275–292 (2019)
19. Reina, A., Bose, T., Srivastava, V., Marshall, J.A.R.: Asynchrony rescues statistically-optimal group decisions from information cascades through emergent leaders. R. Soc. Open Sci. **10**, 230175 (2023)
20. Reina, A., Ferrante, E., Valentini, G.: Collective decision-making in living and artificial systems: editorial. Swarm Intell. **15**(1–2), 1–6 (2021)
21. Reina, A., Marshall, J.A., Trianni, V., Bose, T.: Model of the best-of-n nest-site selection process in honeybees. Phys. Rev. E **95**(5), 052411 (2017)
22. Reina, A., Njougouo, T., Tuci, E., Timoteo, C.: Studying speed-accuracy tradeoffs in best-of-n collective decision-making through heterogeneous mean-field modelling. arXiv:2310.13694 (2023)
23. Reina, A., Zakir, R., De Masi, G., Ferrante, E.: Cross-inhibition leads to group consensus despite the presence of strongly opinionated minorities and asocial behaviour. Commun. Phys. **6**, 236 (2023)
24. Schmickl, T., Möslinger, C., Crailsheim, K.: Collective perception in a robot swarm. In: Şahin, E., Spears, W.M., Winfield, A.F.T. (eds.) SR 2006. LNCS, vol. 4433, pp. 144–157. Springer, Heidelberg (2007). https://doi.org/10.1007/978-3-540-71541-2_10

25. Seeley, T.D., Visscher, P.K.: Group decision making in nest-site selection by honey bees. Apidologie **35**(2), 101–116 (2004)
26. Sood, V., Antal, T., Redner, S.: Voter models on heterogeneous networks. Phys. Rev. E **77**(4), 041121 (2008)
27. Strandburg-Peshkin, A., Farine, D.R., Couzin, I.D., Crofoot, M.C.: Shared decision-making drives collective movement in wild baboons. Science **348**(6241), 1358–1361 (2015)
28. Valentini, G., Ferrante, E., Dorigo, M.: The best-of-n problem in robot swarms: formalization, state of the art, and novel perspectives. Front. Robot. AI **4**, 9 (2017)
29. Valentini, G., Hamann, H., Dorigo, M.: Self-organized collective decision making: the weighted voter model. In: Proceedings of AAMAS 2014, pp. 45–52 (2014)
30. Zakir, R., Dorigo, M., Reina, A.: Robot swarms break decision deadlocks in collective perception through cross-inhibition. In: Dorigo, M., et al. (eds.) ANTS 2022. LNCS, vol. 13491, pp. 209–221. Springer, Cham (2022). https://doi.org/10.1007/978-3-031-20176-9_17

# Does 'Community Detection' Find Real Emerging Meso-structures? A Statistical Test Based on Complex Networks Methods

Riccardo Righi(✉)

Centro Analisi Politiche Pubbliche (CAPP), Università degli Studi di Modena e Reggio Emilia, Modena, Italy
riccardo.righi@unimore.it

**Abstract.** Multiple systems that can be represented in network terms usually present areas where the nodes are densely connected among themselves. Community detection analysis is precisely pointing at revealing these areas, thus providing a partition of the network under investigation. Usually, the results of such analysis are discussed in descriptive terms, either with the aid of some statistics, or by listing and discussing the nodes that belong to the different communities. Thus, a statistical evaluation of the detected community structure still missing in literature. In this work, we design a series of tests to assess if the community detection results are compatible with random processes of tie formation, or if what emerges from such analysis cannot be ascribable to randomness. As community detection naturally points at uncovering the presence of meso-structures within a system, what needed is a statistical tool to test if these are just areas that have randomly formed or if, behind what is detected, there is a real emerging phenomenon. In order to provide an example, we run the tests on the network of UK Faculty and discuss the results.

**Keywords:** data science · networks · community detection · within-community connections · configuration model · Kolmogorov-Smirnov test

## 1 Introduction

Community detection is one of the most frequently used techniques for analyzing social networks and complex networks [1,3,4,11]. It makes it possible to split networks into internally densely connected sub-parts (i.e., the 'communities'), while connections between the detected sub-parts are more scarce. Depending on the nature of the network, this technique can be used to achieve different goals. In the study of social systems, community detection is implemented to unveil groups of agents that are socially tied and bounded (e.g., groups of friends within a class). In other types of systems, typically more associated with complexity, such as economic, biological and physical networks, community detection makes it


M. Villani et al. (Eds.): WIVACE 2023, CCIS 1977, pp. 352–365, 2024.
https://doi.org/10.1007/978-3-031-57430-6_27

possible to identify sub-parts that, even if not necessarily detached from the rest, can reveal some natural divisions of the whole system. In any case, independently from the nature of the network that is considered, community detection allows grouping nodes based on their closeness and the density of the ties that are connecting them. Therefore, it can be said that community detection naturally points at investigating the meso-level structures that can occur in a system.

Several algorithms are available to perform such analysis [3]. Typically the evaluation of the output is based either on a punctual investigation of the meaningfulness of the detected communities (in terms of the agents that are involved), or on considering of some statistics describing the cohesiveness of the communities (e.g., the density of the communities[1]). However, instead of immediately trying to characterize and describe the detected groups, the first questions to be explored should rather be: *do the detected communities really exist? Do they reveal some emergent and non-random cohesiveness? Or these just groups have to be considered as nothing more than the output of an algorithm that in whatever circumstances will always provide a complete network partition? Are the detected communities real meso-level structures characterizing the system under investigation, or are just denser areas randomly happening?*

Community detection algorithms are exhaustive. So, for any type of network, they will always provide a complete partition of it. However, in the investigation of real systems, we should first ask ourselves if the structures detected (at the micro, macro, or meso level, like in this case) are something more than the consequences of some mere random process. If they are not, then they become interesting to be investigated: their formation and emergence are determined by some underlying principles yet to be discovered. Therefore, we ask ourselves if '*communities are real*' in the sense of investigating if their presence and structure are somehow related to something non-random that has occurred in the process of formation of the network. To address this point, we outline a methodology that, apart from community detection analysis, is based on generating random networks, on non-parametric goodness-of-fit test and, eventually, on cluster analysis. The aim is to statistically investigate if the detected communities appear as emerging meso-structures of the system under investigation or if they have just to be considered as random denser areas with no further meaning in real terms. In other words, we want to assess whether the unveiled communities reveal some significant property in terms of structure and cohesiveness, or if the obtained network partition is just the result of some random process of tie formation.

## 2 The Methodology

As a starting point, we need a network $G = \{N, E\}$, where $N$ is the set of nodes present in the network, and $E = \{\omega_{i,j} : G_{i,j} > 0\}$ is the set of existing weighted connections between the nodes of the network, being $i, j$ two nodes of network

[1] Number of ties detected among the nodes belonging to the community, divided by the number of all possible ties that can exist among those nodes.

$G$ that are connected (if $G_{i,j} > 0$), and with $\omega_{i,j}$ indicating the weight of the connection between them. We also indicate with $\Omega_G$ the sum of all $\omega_{i,j} \in G$.[2]

### 2.1 Focus on *Within-Community* Connections

At first stage, we perform a community detection of $G$, so to get its structure of communities, i.e., information on which nodes belong to which community. A discussion about assessing community detection algorithm to consider is not part of this work, as our objective is not to evaluate community detection algorithms. The only crucial element for the proposed methodology is to select one algorithm and keep using it from the beginning of the process until its end, as multiple community detections must to be performed.[3]

Once the community detection is run, we are then able to split the set of connections $E$ into $E^W$ and $E^B$, where $E^W$ are all the connections between nodes of the same community (namely, *within-community* connections), and $E^B$ are all the connections between nodes of different communities (namely, *between-community* connections).[4] This is done by simply checking if the two nodes involved in the connection belong (or not) to the same community. By focusing on $E^W$, we can compute the statistic $\eta^W(G)$, as follows:

$$\eta^W(G) = \frac{\sum \omega_{i,j}}{\Omega_G} \qquad \forall\ i \in c_\lambda \wedge j \in c_\lambda \tag{1}$$

where $c_\lambda$ indicates any of the communities that are detected in $G$. In qualitative terms, $\eta^W(G)$ describes the sum of weights of the *within-community* connections, i.e., $E^W$, in fractional terms with respect to the sum of all the connections belonging to the network $G$.[5]

### 2.2 Clarifications on $\eta^W(G)$

First, it has to be clarified that with this work we are not debating on the optimality of statistic $\eta^W(G)$ for testing communities. Indeed, the statistic $\eta^W(G)$ is just one possible statistic to be considered. Nonetheless, it has been formalized in this work—where we propose it as an initial starting point for testing community structure' adscription to randomness—for its straightforwardness and its simplicity. As network communities are (by definition) characterized by their dense internal connectivity, one of the first things to be checked is exactly the

[2] If the network is unweighted, it just means that $\omega_{i,j} = \{1\}$, i.e., in absence of weights any existing connection is considered to have value 1.

[3] There are several contributions explaining, discussing and comparing the performances of community detection algorithms, e.g., [3], but our work does not directly point at proposing a new method to evaluate them.

[4] $E^W \cup E^B = E \quad \text{and} \quad E^W \cap E^B = \emptyset$.

[5] By definition, $0 \leq \eta^W(G) \leq 1$.

observed level of *within-community* connections. Thus, from this point of view $\eta^W(G)$ is very focused, as it expresses the percentage of the sum of the weights of the *within-community* connections over the sum of the weights of all the connections belonging to $G$. At the same time, $\eta^W(G)$ is also very simple as it does not require much information and computational effort to be calculated.

Second, we anticipate that in the rest of the work we will not study or propose any reference value for $\eta^W(G)$, as the scope of this work is not to assess when $\eta^W(G)$ has to be considered large or small. Indeed, for the way network communities are defined—and consequently investigated and detected—, the majority of network connections will always be *within-community* connections. Then $\eta^W(G)$ is always expected to be relatively large.[6] Also, it seems complicated to think of some specific theoretical threshold values for $\eta^W(G)$, as its magnitude may depend on various characteristics of the system under analysis. Rather, this work aims to develop a methodology to assess whether the observed value of $\eta^W(G)$ is somehow resulting from a random process, or not. We focus on investigating whether the observed level of communities' internal cohesiveness has to be considered as expected, or if the level of $\eta^W(G)$ is significantly different from what it should be likely to be. To do so, we propose comparing the statistic $\eta^W$ for the observed network $G$, with the same statistic but computed in many null networks serving as terms of comparison.

### 2.3 Generation of Null Networks as Term of Comparison

In order to get a term of comparison for $\eta^W(G)$, we generate a set, namely $\boldsymbol{G^*}$, made of a large number of random networks (any of which we generically indicate as $G^*$) by means of the 'configuration model' by Molloy and Reed [7]. This is a method that, based on the random re-wiring of the connections observed in the original network $G$, allows the generation of a null network having (i) the same nodes of $G$, (ii) the same density of $G$, i.e., the same percentage of number of edges over number of possible edges, and (ii) the same degree sequence of $G$, i.e., each node maintains the same number of connections as it has in the original network $G$. In addition, if the network is weighted, but if and only if the connections' weight values $\in \mathbb{N}^+$, the 'configuration model' makes it possible to preserve the same strength sequence, i.e., any node preserves the same strength as in the original network, instead of preserving the same degree

[6] Community detection algorithms identify areas of the networks, i.e., groups of nodes, within any of which the connections are many, so making that part of the network dense. On the opposite, connections between nodes of different communities are scarcer.

sequence.[7] For any $G^*$, we calculate $\eta^W(G^*)$ repeating the same process developed to compute $\eta^W(G)$. Therefore, for any $G^*$:

- a new and independent (from what is done in $G$) community detection is performed,
- *within-community* connections are then separated from *between-community* connections,
- $\eta^W(G^*)$ is calculated.

After generating many networks $G^*$, we are able to compute multiple $\eta^W(G^*)$, i.e., one for each random network, allowing us to have a distribution of values. As we calculate this distribution, we can finally test whether $\eta^W(G)$ fits it well. Since the objective of the methodology is to provide statistical evidences to discuss whether the community structure detected in $G$ is ascribable to randomness or not, we want to compare it with other community structures emerging from networks that: (i) on the one hand, maintain some characteristics of the original one (i.e., nodes and degree sequence, which also implies same density and same degree distribution), and (ii) on the other hand, have random connective structure (edges are established randomly according to previous condition).

It is important to highlight that in any of the generated null networks, a new community detection analysis is run. Since what we want to test precisely the community structure of the original network $G$, we need to compare it with other community structures. In the proposed methodology, we do not constrain in any sense the community detection analyses that are performed in the networks of $\boldsymbol{G}^*$. The point is exactly to compute the same exact statistic in networks that, being similar but different from the original one, will reveal a community structure that is also supposed to have different characteristics from the one

[7] If the network is weighted with connections' weight values $\in \mathbb{N}^+$, it is only necessary to repeat any connection for a number of times equal to its weight, and to assign to all resulting connection a weight equal to 1. The network will be no longer a simple network, i.e., one and only one connection between any couple of node, but this is irrelevant. We can randomize the peers involved in the connections (included those repeated because of a weight larger than 1), and still the network will result having the same strength sequence, i.e., any node will have the same strength as in the original network. Clearly, since connections are repeated based on their weights and then randomized, this does not make it possible to preserve the same degree sequence as in the original network.

If the original network's weights $\notin \mathbb{N}^+$, then connections can be directly randomized (i.e., no repetition of them) and their weight can just be sampled (or probabilistically inputed) based on the weights of the original network. The resulting random network will preserve the degree sequence and the sum of connections' weight, but not the strength sequence.

Finally, in case connections' weight $\notin \mathbb{N}^+$ and the network is bipartite, at least the strength sequence can be preserve for one type of node, but not for the other: connections' weights can be randomly sampled from those displayed by the node in the original network. However, when this is done for one type of node, then it cannot be done also for the other type. Nonetheless, in this case the degree sequence is preserved for all nodes.

observed in $G$. Then, the scope of the next and last step will be to check whether these differences have to be considered statistically significant. In any way, the only constraints that our methodology imposes are related to the definition of the level of 'randomness' that we want to include in the null systems of $\boldsymbol{G}^*$. These constraints, which in our work are essentially related to the degree sequence of the nodes, may be further specified, thus leading to the generation of null systems that can be even more similar to the original network $G$, and hence to stronger statistical tests.

Finally, it is also important to remark that the selection of the 'configuration model' by Molloy and Reed [7] as method for generating null systems answers to two needs. First, to have random networks that can have more similarities with the original network than just the edge density. This would be the case of random networks generated with the Erdős-Rényi model [2,9], a benchmark in the field. However, this model most relevant limitation is that it is not able to replicate the degree sequence, i.e., distribution of number of connections per node. This is a feature that cannot be considered as secondary, especially in a context of generation of null systems to be used as a term of comparison for the study of real system properties. For this reason, the 'configuration model' has received attention [8]. Second, from a computational point of view, the configuration model is not based on probabilities but only on the re-wiring of the connections observed. So, in other terms, the generation of a null network with this method is relatively simple, as it requires to randomly sample combinations of two nodes at a time (without replacement), from a set made of the nodes of $G$ (those with at least one connection[8]), and with any of these nodes repeated as many times as number of connections it has in $G$. To conclude, alternative ways to generate null networks may be considered as more pertinent depending on the type of system under analysis. Still, at the same time the 'configuration model' includes enough sophistication to be pertinently considered as a fair starting point for the methodology.

## 2.4 Test of *Within-Community* Connections

Finally, in the last stage of the methodology, we implement a goodness-of-fit (GoF) test between $\eta^W(G)$ and the distribution of $\eta^W(\boldsymbol{G}^*)$. Since we have no prior hypothesis regarding the shape and the parameters of this distribution, we consider the Kolmogorov-Smirnov (KS) two-sided GoF test [5,6].[9] This is a non-parametric test that allows us to determine whether the value observed in

[8] A node can be part of a network even if not displaying any connections in it.

[9] The test is based on the computation of the empirical cumulative distribution function (eCDF) of the values $\eta^W(\boldsymbol{G}^*)$, and it calculates the percentile in which $\eta^W(G)$ falls with respect of the mentioned eCDF.

the original network, i.e., $\eta^W(G)$, is compatible with $\eta^W(\boldsymbol{G}^*)$. Hence, we can finally determine whether $\eta^W(G)$ is ascribable to randomness.[10]

Suppose the test reveals that the *within-community* connections observed in $G$ are compatible with what observable in the null systems $\boldsymbol{G}^*$. In that case, should surmise that the communities of $G$ arise as a consequence of the basics assumptions considered to build the random networks, i.e., same density and same degree sequence. This would mean that the agglomeration and the cohesiveness that is observed in certain parts of $G$ (i.e., where the communities are detected) could have occurred in other parts of it. Thus, no emerging phenomenon is really taking place at the level of the network structure, and we would have to conclude that what is observed in terms of community structure simply is the result of a random process. On the opposite, if the test is negative, i.e., $\eta^W(G)$ is not compatible with the distribution of $\eta^W(\boldsymbol{G}^*)$, it means that the cohesiveness detected within the communities of $G$ is significantly larger (or lower) than what would have been if ties were formed randomly. In this case, it would be confirmed that some emergent phenomenon is really happening.

### 2.5 A Further Development: Controlling Random Tie Formation Based on Nodes' Heterogeneity

In network theory, there are several concepts, distinct but similar, all pointing to the fact that nodes are typically more (or less) prone to develop connections with other nodes depending on some category/type to which they belong. For instance, network assortativity and homophily, are theoretical aspects describing different ways in which the heterogeneity of nodes can have a role in the network formation process [10]. The point is that this can also have an effect in the process of emergence of communities. Therefore, in order to control for nodes' heterogeneity, we propose a further development of the designed methodology.

A stronger test can be developed by considering nodes' categorization. This categorization can already be present in the data (e.g., a categorical variable, like gender or ethnic group), or it can be created by implementing a cluster analysis.[11] In both cases, what matters is that we have to classify all nodes $N$ in $K$ non-overlapping sub-sets.[12] We label any of these sub-sets with the letter $k$, followed by a subscript indicating its specific identity. In other words, information is needed on how to split $N$ in $\{k_1, k_2, k_3, ..., k_N\}$, where any $k$ includes at least one node of $N$ and with all nodes $N$ belonging to a specific $k$. We use this information to divide $G$ in all its possible $G_{k_\alpha,k_\beta}$ sub-graphs, where

[10] It is important to highlight that the described test allows us also to assess if $\eta^W(G)$, in case it is statistically significant, has a value which is larger (or smaller) than expected. This would reveal the existence of a non-random community structure characterized by *within-community* connections that are enhanced (or inhibited), respectively.

[11] It is not scope of this work to discuss how to implement a cluster analysis and which kind of variables to consider.

[12] $K \in \mathbb{N}^+ \wedge K < N$.

the combination $(k_\alpha, k_\beta)$ indicates the categories of the nodes to be considered at a time.[13]

Then, we proceed by implementing the 'configuration model' method for any sub-graph $G_{k_\alpha,k_\beta}$, exactly as described above for the generation of an entire null network $G^*$. We indicate with $G^\dagger_{k_\alpha,k_\beta}$ the random sub-graphs that are generated by this part of the process. Once all the combinations of $(k_\alpha, k_\beta)$ are taken into account, then all the newly generated $G^\dagger_{k_\alpha,k_\beta}$ are put back together in a single new random network $G^\dagger$, which presents the following characteristics with respect to the original network $G$: (i) same density, (ii) same degree sequence, (iii) same 'preferential matching' in terms of the considered nodes' categorization. Indeed, the most important point is the last. We listed 'preferential matching' among the elements that are preserved in the new null systems, because in any $G^\dagger$, any node not only maintains the same number of connections as in $G$, but it also maintains the same number of connections by type of peers. For instance, if node $i$ has three connections with some nodes of type $k_\alpha$, and two connections with some nodes of type $k_\beta$, then in any $G^\dagger$ node $i$ will always have three connections with some nodes of type $k_\alpha$ and two connections with some nodes of type $k_\beta$. Clearly, which specific nodes, among those belonging to type $k_\alpha$ and those belonging to type $k_\beta$, will be connected to $i$ in $G^\dagger$ will be randomly selected. Then, the methodology continues as previously described: a large number of $G^\dagger$ is generated and the value of $\eta^W(G)$ is tested against the final distribution of $\eta^W(\boldsymbol{G}^\dagger)$ with the non-parametric Kolmogorov-Smirnov GoF test.

Since with this option each node always preserves the same centrality degree, as well as the same number of connections by type of nodes with which it is connected, this has some implications for the analysis of the statistical significance of $\eta^W$. Given that the null networks are generated with a lower degree of randomness, the test is now stronger. As the new null systems $G^\dagger$ are more similar to $G$ than $G^*$ null systems are similar to $G$, the test on communities based on $\eta^W(\boldsymbol{G}^\dagger)$ is stronger. More specifically, if this last test results negative, i.e., the $\eta^W(G)$ is not compatible with the distribution of the $\eta^W(\boldsymbol{G}^\dagger)$, it will reveal the presence of non-random meso-structures that cannot be attributed to any connectivity propensities based on nodes' typology.[14] The randomization, but with fixed amount of connections based on the categories of nodes involved, allows us to exclude any effect related to specific propensities based on nodes' heterogeneity from the elements that are significantly characterizing the process of generation of the communities.

Finally, it is relevant to remark that this last part of the procedure also opens the room for the computation and the statistical evaluation of $\eta^W(G, k_\alpha, k_\beta)$, i.e., the same statistic $\eta^W$ but calculated only for what concerning the connections that are involving one node of type $k_\alpha$ and one node of type $k_\beta$. Formally,

[13] For instance, if $K = 3$, then $G$ is going to be split in all the possible sub-graphs based on nodes' categorization, i.e., $G_{k_1,k_1}$, $G_{k_2,k_2}$, $G_{k_3,k_3}$,$G_{k_1,k_2}$, $G_{k_1,k_3}$, and $G_{k_2,k_3}$.

[14] It is important to observe that with the first test proposed, i.e., $\eta^W(G)$ vs. distribution of $\eta^W(\boldsymbol{G}^*)$, this element could not be excluded as a constitutive element for the communities.

$$\eta^W(G, k_\alpha, k_\beta) = \frac{\sum \omega_{i,j}}{\Omega_{G,k_\alpha,k_\beta}} \qquad \forall\ i \in k_\alpha\ \wedge j \in k_\beta\ \wedge\ i \in c_\lambda\ \wedge\ j \in c_\lambda \tag{2}$$

where $\Omega_{G,k_\alpha,k_\beta}$ indicates the sum of the weights of those connections of $G$ that involve one node belonging to category $k_\alpha$ and one node belonging to category $k_\beta$. This makes it possible to explore the presence of significant community cohesiveness exclusively for what concerning the nodes belonging to specific categories of nodes, i.e., $k_\alpha, k_\beta$, and not for all of them, as it was previously proposed in a purely systemic perspective.

And clearly, not necessarily $\eta^W(G, k_\alpha, k_\beta)$ has to be computed with $k_\alpha \neq k_\beta$. Indeed, the test is suitable also to focus on one and only one specific type of nodes, e.g., those belonging to $k_\alpha$, and to investigate *within-community* connections exclusively among nodes of that category. Formally,

$$\eta^W(G, k_\alpha) = \frac{\sum \omega_{i,j}}{\Omega_{G,k_\alpha}} \qquad \forall i, j \in k_\alpha\ \wedge\ i \in c_\lambda\ \wedge\ j \in c_\lambda \tag{3}$$

where $\Omega_{G,k_\alpha}$ indicates the sum of the weights of those connections of $G$ that are involving two nodes both belonging to category $k_\alpha$. This opens the possibility to test if nodes of a specific category are prone to significantly develop communities among them.

## 3 Analysis of UK Faculty Network

We implement the described tests on the UK Faculty network (data included in the R Package *igraph*), which we indicate as $G$. It is a network consisting of 81 vertices (members of a Faculty Department in the UK) and 817 directed and weighted connections ($\Omega_G$ equals to 3,730) describing friendship relationships. The nodes are also characterized for the fact that any of them belongs to a specific School (there are four of them) of the Faculty Department. We implement the *walktrap.community* algorithm (considering connections' weight), which is based on random walks and which is included in the R Package *igraph*. This allows us to detect 6 communities, as represented in Fig. 1, where *within-community* connections are highlighted with a bolder color.

First of all, 5,000 $G^*$ null systems are computed. Since $G$ is a weighted network and the weights $\in \mathbb{N}^+$, we constrain the generation of null systems in order to preserve the strength sequence, i.e., any node preserves the same strength as in the original network $G$.[15] The result of the Kolmogorov-Smirnov GoF test between $\eta^W(G)$ and the empirical cumulative distribution function (eCDF) of the values $\eta^W(G^*)$ is negative, with a p-value equal to 0.072 (significance

[15] What needed is to repeat any connection for a number of times equal to its weight, and then to assign to all resulting connections a weight equal to 1, and then connections can be randomized according to the 'configuration model'. By doing so, each node maintains its strength, even if its degrees may vary.

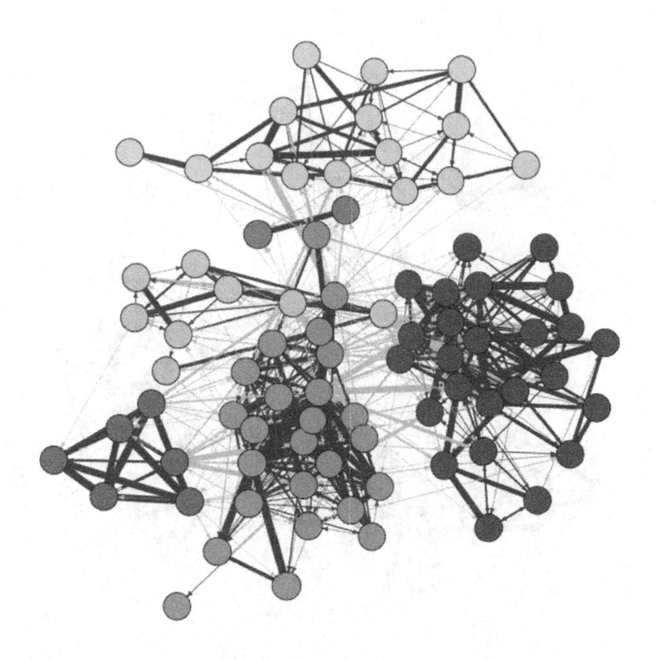

**Fig. 1.** The UK Faculty network. Color of the nodes represents the communities detected with the *walktrap.community* algorithm (considering connections' weight). Connections' width represents friendship intensity. Connections' color is darker for *within-community* connections.

smaller than 10%), and with positive sign ($\eta^W(G)$ equals 0.850, while the average of $\eta^W(\boldsymbol{G}^*)$ equals 0.512). This means that the level of *within-community* connections that is observed is significantly larger from what we should expect assuming the hypothesis that the network ties have formed randomly under the only condition of keeping as fixed the strength of each node. In other words, as expected in most of cases, the community structure that is observed is not ascribable to randomness.

Second, we consider information about the School of the Faculty Department to which the members belong (there are four Schools, #1 with 33 members, #2 with 27, #3 with 19, and #4 with 2 members). This is a categorical variable (attribute "Group" in the data) which is represented by nodes' colors in Fig. 2. 5,000 $G^{\dagger}$ null systems are computed based on the subnetworks determined by the Schools of the two members involved in the friendship connection. Since the network is directed, all the possible permutations of the four Schools (made of two elements and with allowed repetitions) are considered to identify the

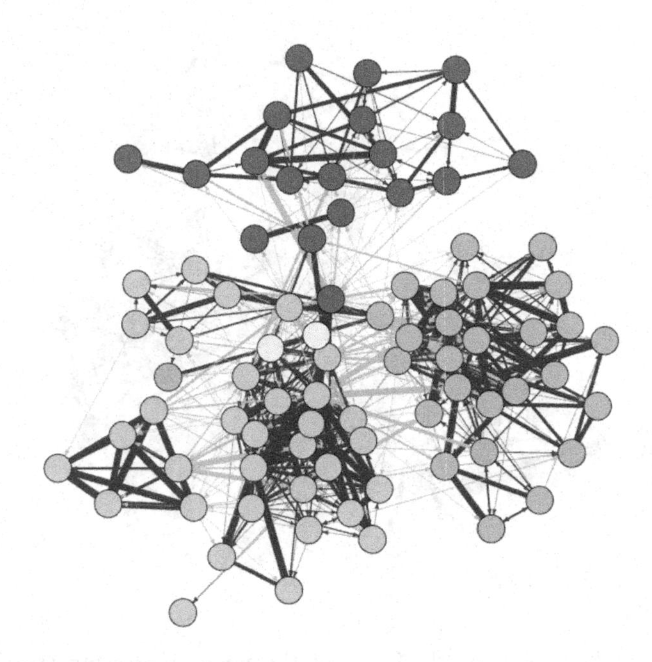

**Fig. 2.** The UK Faculty network. Color of the nodes represents the School to which members belong (School #1 in yellow, #2 in orange, #3 in green, and #4 in white). Connections' width represents friendship intensity. Connections' color is darker for *within-community* connections (from Fig. 1).

subnetworks of $G$ and to compute any $G^\dagger$.[16] The Kolmogorov-Smirnov GoF test between $\eta^W(G)$ and the empirical cumulative distribution function (eCDF) of the values $\eta^W(\boldsymbol{G}^\dagger)$ is run and the result is negative. P-value is equal to 0.028 (significance smaller than 5%) and the sign is negative ($\eta^W(G)$ equals 0.850, while the average of $\eta^W(\boldsymbol{G}^\dagger)$ equals 0.9141). This means that the level of *within-community* connections that is observed is significantly smaller from what we should expect assuming the hypothesis that the network ties have formed randomly under the only condition that any node preserves the same number of connections by School as in $G$. In other words, the observed community structure in not random with respect to the number of connections by School. Inter-

[16] In other terms, $G$ is split in the 16 subnetworks determined by the permutations (with repetitions) of the four Schools. These are $G_{1,1}$, $G_{1,2}$, $G_{1,3}$, $G_{1,4}$, $G_{2,1}$, $G_{2,2}$, $G_{2,3}$, $G_{2,4}$, $G_{3,1}$, $G_{3,2}$, $G_{3,3}$, $G_{3,4}$, $G_{4,1}$, $G_{4,2}$, $G_{4,3}$, and $G_{4,4}$, where the two subscripts indicate the School of the member from whom the friendship starts, and the School of the member that is considered as a friend, respectively. For the creation of a single $G^\dagger$, then each of these subnetworks is randomized and then all of them are put back together so to have a null system in which any node preserves the same strength as in $G$, as well as the same strength by School as in $G$.

**Table 1.** Results for the test of $\eta^W(G, k_\alpha, k_\beta)$ vs. the distribution of $\eta^W(\mathbf{G}^\dagger, k_\alpha, k_\beta)$. In each row, a different permutation of the School is considered. For the level of significance, in the seventh column, we report (***) when p-value $< 0.01$, (**) when p-value $< 0.05$, and (*) when p-value $< 0.1$. Last column indicates whether the observed value of $\eta^W(G, k_\alpha, k_\beta)$ is larger (+) or smaller (–) than expected, or as expected (=).

| School from | School to | Conn. Type | $\eta^W(G, k_\alpha, k_\beta)$ | Average of $\eta^W(\mathbf{G}^\dagger, k_\alpha, k_\beta)$ | P-value | Signif. | Sign |
|---|---|---|---|---|---|---|---|
| 1 | 1 | $1 \to 1$ | 0.897 | 0.999 | 0.016 | ** | – |
| 1 | 2 | $1 \to 2$ | 0.177 | 0.000 | 0.000 | *** | + |
| 1 | 3 | $1 \to 3$ | 0.652 | 0.683 | 1.000 | | – |
| 1 | 4 | $1 \to 4$ | 0.931 | 1.000 | 1.000 | | - |
| 2 | 1 | $2 \to 1$ | 0.205 | 0.000 | 0.000 | *** | + |
| 2 | 2 | $2 \to 2$ | 0.929 | 1.000 | 1.000 | | – |
| 2 | 3 | $2 \to 3$ | 0.000 | 0.000 | 1.000 | | = |
| 2 | 4 | $2 \to 4$ | 0.000 | 0.000 | 1.000 | | = |
| 3 | 1 | $3 \to 1$ | 0.666 | 0.874 | 0.380 | | – |
| 3 | 2 | $3 \to 2$ | 0.000 | 0.000 | 1.000 | | = |
| 3 | 3 | $3 \to 3$ | 0.839 | 0.867 | 0.619 | | – |
| 3 | 4 | $3 \to 4$ | 1.000 | 0.891 | 1.000 | | + |
| 4 | 1 | $4 \to 1$ | 1.000 | 1.000 | 1.000 | | = |
| 4 | 2 | $4 \to 2$ | 0.000 | 0.000 | 1.000 | | = |
| 4 | 3 | $4 \to 3$ | 1.000 | 0.810 | 1.000 | | + |
| 4 | 4 | $4 \to 4$ | 1.000 | 1.000 | 1.000 | | = |

estingly, if nodes had formed random friendships only paying attention to keep a certain number of connections by School (as in $G$), we would have got a stronger community structure than the one we have. This is probably because, as it is possible to observe by visually comparing Fig. 1 and Fig. 2, the Schools present a considerable correspondence with the communities, and in some cases a single School basically includes all the nodes of two (or more) communities: it is like Schools provide a higher-level partition than communities. Therefore, the level of *within-community* connections results smaller than expected because nodes have not formed friendships exclusively according to certain propensities to interact with people based on their School of affiliation.

Finally, we run the test for all possible permutations (of size equal to two) of the Schools, as described in Table 1. This allows us to investigate if the members of the different Schools have significantly developed *within-community* connections with members of their same School and with members of other Schools. This test makes it possible to understand which Schools are associated with community-behaviors, and with whom. From these tests, Schools #1 and #2 are revealed to show significant high levels of *within-community* connections between them in both directions, i.e., members of School #1 considering friends members of School #2 (connection type "$1 \to 2$"), as well as members of School #2 considering friends members of School #1 (connection type "$2 \to 1$"). Therefore, we can say that these two Schools show some special relationship between them, as their members significantly get involved in the same network communities. Interestingly, while this behavior has no significant effect on the level of *within-community* connections of members of School #2 with members of the same School, we can find that members of School #1 develop significantly less *within-community* connections among themselves (connection type "1 → 1"). Qualitatively speaking, it looks like members of School #1 tend to structure communities with members of School #2, but to do so they pay a cost in terms of the cohesiveness of the communities they are able to form among themselves.

## 4 Conclusions

In the present work we formalize a series of statistical tests to investigate whether community detection results are compatible with hypotheses of random tie formation. The tests are based on the computation of null systems using the 'configuration model', which are used as term of comparison to statistically evaluate what observed in the original network. We run the tests on a publicly available dataset, i.e., UK Faculty network, in order to show a possible implementation and the kind of results they can provide.

To conclude, some considerations on the presented work are recalled. First, the selection of the algorithm used to perform the community detection is not relevant to the general development of this methodology. What relevant is to use the same algorithm for all community detections that are implemented during the process. Second, it is not relevant if the network is weighted or un-weighted. We have developed the formalization of the test for a weighted network since this can be intended as a more detailed specification of a unweighted network. Third, it is not relevant how many communities are detected and neither which is the specific allocation of nodes to them. Since any connection is basically a dyad of nodes $i$ and $j$, what in the end exclusively matters for the computation of $\eta$ is whether $i$ and $j$ are detected to belong to the same community or not. Finally, for any system considered (the original one, i.e., $G$, as well as the generated null systems) a separate community detection is run.

## References

1. Bedi, P., Sharma, C.: Community detection in social networks. Wiley Interdisc. Rev.: Data Min. Knowl. Discov. **6**(3), 115–135 (2016)
2. Erdős, P., Rényi, A., et al.: On the evolution of random graphs. Publ. Math. Inst. Hung. Acad. Sci. **5**(1), 17–60 (1960)
3. Fortunato, S., Hric, D.: Community detection in networks: a user guide. Phys. Rep. **659**, 1–44 (2016)
4. Girvan, M., Newman, M.E.: Community structure in social and biological networks. Proc. Natl. Acad. Sci. **99**(12), 7821–7826 (2002)
5. Karson, M.: Handbook of Methods of Applied Statistics (1968)
6. Kolmogorov, A.N.: Sulla determinazione empirica di una legge didistribuzione. Giorn. Dell'inst. Ital. Degli Att. **4**, 89–91 (1933)

7. Molloy, M., Reed, B.: A critical point for random graphs with a given degree sequence. Random Struct. Algorithms **6**(2–3), 161–180 (1995)
8. Newman, M.: 369 the configuration model. In: Networks. Oxford University Press (2018). https://doi.org/10.1093/oso/9780198805090.003.0012
9. Newman, M.: Random graphs. In: Networks. Oxford University Press (2018). https://doi.org/10.1093/oso/9780198805090.003.0011
10. Newman, M.E.: Mixing patterns in networks. Phys. Rev. E **67**(2), 026126 (2003)
11. Porter, M.A., Onnela, J.P., Mucha, P.J., et al.: Communities in networks. Not. AMS **56**(9), 1082–1097 (2009)

# Self-loops in Social Networks: Behavior of Eigenvector Centrality

J. J. Merelo[1(✉)] and M. Cristina Molinari[2]

[1] Department of Computer Engineering, Automatics and Robotics and CITIC University of Granada, Granada, Spain
jmerelo@ugr.es

[2] Dipartimento di Economia, Ca' Foscari University of Venice, Venice, Italy
cmolinar@unive.it

**Abstract.** Centrality measures are an essential tool in understanding complex networks, since they give researcher insights on the role the different nodes/actors play in them. Among them, eigenvector centrality is a principled approach to these measures, using a mathematical operation on the connection matrix. This connection matrix includes connections from an actor to itself (the diagonal); however, as it is the case with most centrality measures, this fact is seldom used in social studies to compute the standing or influence of one node over others. In this paper we will analyze the difference in EV centrality with or without these self connections or self-loops and how the change depends on the actual number of these self-loops or the weight of these self-connections. Finally, we will characterize in which cases, if any, it is effective to drop self-loops and what kind of information it will give us on the nature and dynamics of the network.

**Keywords:** Complex networks · Social networks · Graph Theory · Eigenvector centrality

## 1 Introduction

Eigenvector centrality [1] measures how an actor in a network or graph influences other actors by computing the eigenvalues of the adjacency or connectivity matrix, that has as components the weight of the connection of every agent to all the others, including itself. It is a centrality measure because, along with other such measures (such as betweenness or degree centrality [17]), it is a micro-level assessment of the power or influence of a node within a network through the analysis of its connections. It has been used extensively in social network analysis [5,12,14]; but also in other fields, such as biology [9], economics [15], or even in the analysis of the spread of opinions in internet forums during the COVID-19 pandemic [11].

In its widespread use for the computation of the value for a single node it is not very different from other centrality measures; however, unlike them,


M. Villani et al. (Eds.): WIVACE 2023, CCIS 1977, pp. 366–380, 2024.
https://doi.org/10.1007/978-3-031-57430-6_28

it considers the whole network [2] taking into account all weights (including negative ones); and, more importantly for the purposes of this paper, all weights include the connections of a node to itself, or self-loops.

As a matter of fact, many complex networks, including social networks, include links from one actor/agent to itself; we can find them in commuting networks [8], where self-loops would indicate trips that start and end in the same city, county or state; in family networks [8], with self-loops indicating marriage between members of the same (extended) family; opinion networks [11], where they would indicate interaction among members of the same group; sport teams transfer networks [7], indicating transfers between teams in the same country; when freight traffic networks [10] are analyzed at a regional level, self-loops would represent shipments that start and end in different parts of the region; commercial networks [15] where self-loops indicate a contract between members of the same family, or even biological networks [9], where self-loops could indicate auto-trophic (members of a species feeding on other members) or auto-catalytic (reactive agents catalyzing reactions where they participate).

The main issue and the one we are trying to address in this paper, is that when centrality analysis is performed on these networks, most centrality measures cannot work with self-loops; thus, in most cases, they are dropped [11,14,15]. Even when a specific measure can include self-loops, like the aforementioned eigenvector centrality that works on the connection matrix, self-loops are usually dropped mainly to work on an uniform set of data, that is, not use two different connection matrices with diagonal values (for EV centrality) and other without (for other measures).

In the cases where self-loops do not have a clear interpretation (or at least a common interpretation with the rest of the connections) there are good reasons to work that way: betweenness centrality [6], for instance, models how one node is needed to transmit information to other parts of the network; how often it transmits information to itself cannot easily be accommodated within this framework, and it can be argued that it could be simply ignored when computing the geodesics from which this measure is computed. That is not the case for EV centrality, which does in fact use self-loops to compute the eigenvalues that are presented as the EV centrality measure. And these intra-links can, however, give us better insights on the dynamics of a social network, and should probably be taken into account. So far, however, there has been little analysis on what is the actual effect of considering these intra-links for computing the eigenvector centrality [2] in social networks. This paper will try, using well-chosen examples, what is that effect and how it could help to better understand social systems.

The main research question that we ask, then, is if self-loops should be included, when available, in the computation of eigenvector centrality. As an accessory question, we try to investigate what would be the effect of doing so in EV centrality measurements, and how it affects the value and the ranking of nodes in the corresponding network.

The rest of the paper is organized as follows: next we will be describing the state of the art, to proceed to describe the datasets and the experiments performed in Sect. 3. We will then discuss the results and conclude with our conclusions and future lines of work.

## 2 A Brief Literature Survey

The main motivation behind this paper is to shed a bit of light on the use of self-loops in social network analysis, mainly because it has been used so rarely in the past. Some papers acknowledge its importance in the propagation of information in social networks [22], with self-loops representing simply re-posts of some content previously created by the same person; other papers, however [11] dealing with the same subject, explicitly do not use them, thus missing a good amount of the dynamics created by these self loops. In other opinion formation papers [19] self-loops represent the amount of attention a person pays to its own opinion as opposed to others. In general, even theoretical models of the spread of information in social networks [3] include self-loops.

In general we can say that while self-loops are sometimes acknowledged and used in social network *models*, they are dropped when making centrality analysis of complex or social networks. This why we have made it the main focus of this paper.

## 3 Experiments

We will be using two datasets to perform the experiments. The first one is the Venetian matrimonial dataset [15], a social network of the noble families of the Republic of Venice from the XIV to the XIX century[1]; the second is a dataset of freight traffic among the states of the United States of America extracted from the government data portal [20]. These will be examined in turn in the next two subsections.

### 3.1 Analyzing the Venetian Matrimonial Dataset

This dataset was published by Puga and Treffler as support for their paper [15]. It consists of marriages celebrated in the Republic of Venice (and successor polities during the late XVIII and XIX century) where the groom is a noble[2], registered with the *Avvogaria del Comune* of the Republic. Families (called *casate* in the original Venetian and Italian) were the political and social unit in the republic

[1] Some marriages are not dated, but we can assume they took place in the same range of years.

[2] Since the dataset includes some marriages that happened after the fall of the Republic in 1796, the concept of "noble" in this case corresponds to families that were considered noble *during the existence of the Republic*; during French and Austrian control, as well as during the brief period of the Republic of San Marco, such nobility titles no longer had any value; however, since they were included in the original dataset there was no good reason to eliminate them.

[16,18], which explains the interest this dataset has for insights into the history of the republic. Nodes in this dataset are, thus, families with nobility titles in the Republic of Venice; we eliminated from this dataset those marriages where brides did not belong to any patrician family.[3].

This dataset has been chosen because it includes a good number of intra-family marriages, as well as because it has been studied extensively by Puga and Treffler in their paper, providing a basis for the comparison of results.

**Table 1.** Intra-family marriages per century

| Century | Intra-family marriages | Total marriages | Percentage |
|---|---|---|---|
| XIV | 5 | 51 | 9.80 |
| XV | 82 | 2740 | 2.99 |
| XVI | 118 | 4265 | 2.77 |
| XVII | 104 | 2770 | 3.75 |
| XVIII | 31 | 1732 | 1.79 |
| XIX | 3 | 108 | 2.78 |
| NA | 42 | 561 | 7.49 |

**Table 2.** Ranking of top ten families according to number of intra-family marriages (left).

| Family | #Intra-family marriages |
|---|---|
| Contarini | 65 |
| Morosini | 23 |
| Corner | 19 |
| Martinengo | 16 |
| Querini | 15 |
| Balbi | 12 |
| Donato | 11 |
| Malipiero | 10 |
| Zorzi | 10 |
| Zancaruol | 9 |

The absolute and relative number of intra-family marriages in each century is shown in Table 1, including marriages whose date is unknown. These marriages

[3] There were many restrictions to this kind of marriage, but they occurred with regularity, at least until the so-called "Second Serrata" [4], during the XV century; in this case, however, we eliminate them because they are irrelevant to the main point of the paper, not having any influence in the EV centrality of a specific node.

are not evenly distributed per family; the top 10 families according to number of intra-family marriages is shown in Table 2. This is an example where the number of self-loops is not excessive, although it is certainly significant, hovering around a few percentage points per century for an amount of 385 intra-family marriages (self-loops) over 12227 total.

**Table 3.** Ranking of top ten families according to EV centrality values, with (right) or without (left) self-loops.

| Family | EV | Family | EV + self-loops |
|---|---|---|---|
| Contarini | 1.0000000 | Contarini | 1.0000000 |
| Morosini | 0.8175570 | Morosini | 0.6827832 |
| Corner | 0.6425724 | Corner | 0.5131463 |
| Querini | 0.5045249 | Querini | 0.3852909 |
| Priuli | 0.4286878 | Priuli | 0.3139764 |
| Dolfin | 0.3771834 | Giustinian | 0.2806501 |
| Giustinian | 0.3719008 | Dolfin | 0.2775836 |
| Michiel | 0.3692852 | Michiel | 0.2743379 |
| Zorzi | 0.3629267 | Zorzi | 0.2723025 |
| Loredan | 0.3593036 | Pisani | 0.2649323 |

This dataset has been transformed into two different graphs; eliminating self-loops in one and leaving them in the other. In both cases we use undirected edges joining the two families of the partners in every marriage; the edge is weighted with the number of marriages between the families in the nodes; eigenvector centrality has then been computed for the two resulting graphs. A ranking of the top families according to their EV centrality is shown in Table 3.

Looking at Table 2 together with the left hand side of Table 3, we can see that there seems to be some correlation between the number of intra-family marriages and the EV centrality, even when we do not include it in the computation. Six out of ten families are the same, and the first five: Contarini, Morosini, Corner, Querini and Priuli, also appear in the same order. This might indicate either a common cause for both rankings (size of the family, for instance) or a cause-effect, or even a combination of the two: a family gets bigger since it is wealthy, and is wealthy due to its social capital; high EV centrality implies a lot of influence, and this begets wealth, that literally increases the fitness of the family making it big enough that intra-family marriages become possible and even common. This discussion, however, is beside the main point of this paper, although it should be noted that, in a way, self-loops are *factored in* in this specific case since intra-family marriages tend to appear more frequently in families with high EV centrality.

Including self-loops in the computation we see that there are small, but significant, variations: The Dolfins and Giustinians change their order in the ranking, plus the Loredan family is dropped and substituted by the Pisani family[4]. It can also be seen that the difference between the first and second family in the ranking has doubled, and that, in general, the value of the normalized EV centrality has also decreased; since the EV values are normalized, this simply indicates that the difference between Contarini and the other families has increased, a fact in which, of course, self-loops have had a decisive influence.

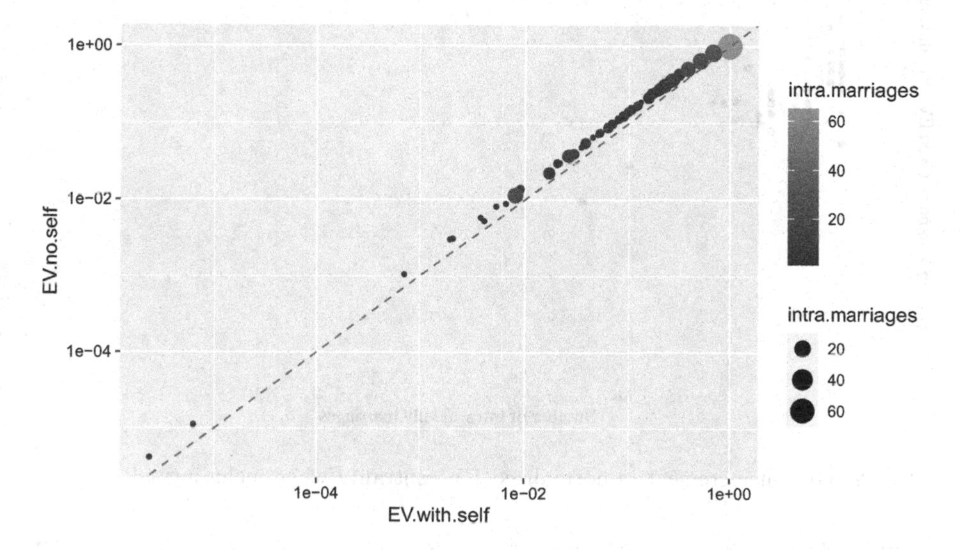


**Fig. 1.** EV centrality considering self-loops (x axis) or not (y axis). The size of the dot and color signal the number of intra-family marriages.

To better highlight the impact of self-loops, in Fig. 1 we have analyzed the differences between the EV values with ($x$ axis) or without self-loops ($y$ axis) for every node; it plots the values with self-loops ($x$ axis) vs. those without ($y$ axis) in a log scale. The plot shows that all values are over the $x = y$ line. The size and color of the dot representing every family is related to the number of intra-marriages; big, lighter blue dots will have the bigger number of self-loops, small, darker ones the lowest.

All dots are placed over the $x = y$ line, indicating that normalized EV values are lower without self-loops. Besides, looking at the colors and sizes, we can see that the smaller the size (number of marriages), the bigger the increase in value; that is, more intra-family marriages make the inclusion of self-loops decrease EV centrality *less*. However, this is due in this case to the fact that what is actually increased is the difference between the family with the higher EV (the Contarinis) and the rest, so we can look at this result from the other side: the

[4] The Pisani family is certainly more "central" than the Loredan, at least looking at the number of nobles in important offices; [15] mentions them as one of the family with the greatest amount of shares in shipping contracts.

presence of a family with a high value in the diagonal of the connection matrix (i.e. self-loops with a high weight) increases its EV centrality much more than that of the rest of the nodes/families.

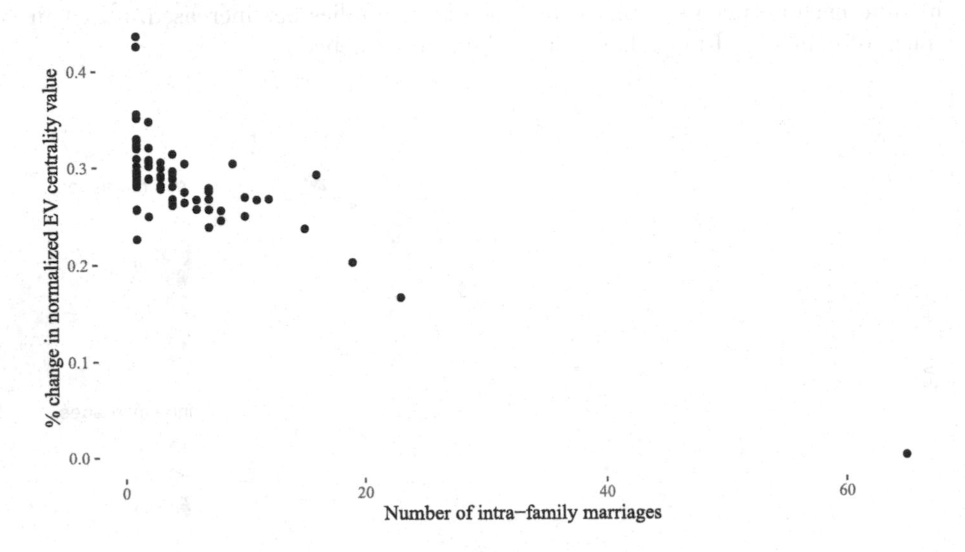


**Fig. 2.** Percent increment in normalized EV centrality when including self loops.

To investigate how the number of intra-family marriages impact on the EV centrality, in Fig. 2, we plot the number of intra-family weddings and the change in EV centrality. As this chart reveals, as expected, the value of intra-family marriages/self-loops in the computation of EV centrality certainly gives us better insights on the dynamics of the social network.

Finally, Fig. 3 shows the relationship between the family ranking with ($y$ axis) and without self-loops ($x$ axis). The red line is the $x = y$ line. The plot shows how changes in rank have a wider span in the mid-ranks, far away from the beginning and from the end. The more central families (close to (1,1)) barely move a position up or down; however, beyond rank 150, there are more changes and they have a bigger impact, with some families moving down several positions, and others (fewer) moving up a few positions. In general, including self-loops makes has a bigger impact in the mid-ranks, an interesting fact that proves the importance of taking then into account when family networks are researched.

The inclusion of self-loops, however, could have other positive and quantitative impacts in social and historical study. We will again refer to [15] as the baseline study; they show (in their Figure VIII) how EV centrality across a century is a good predictor for the same measure in the next century, showing the stability and resilience of the patrician social network in the Republic of Venice. They mention that there is a "good correlation" in this case. We have re-rendered the data for this figure in our Fig. 4 (left panel), showing also as a red line a linear fit to the data.

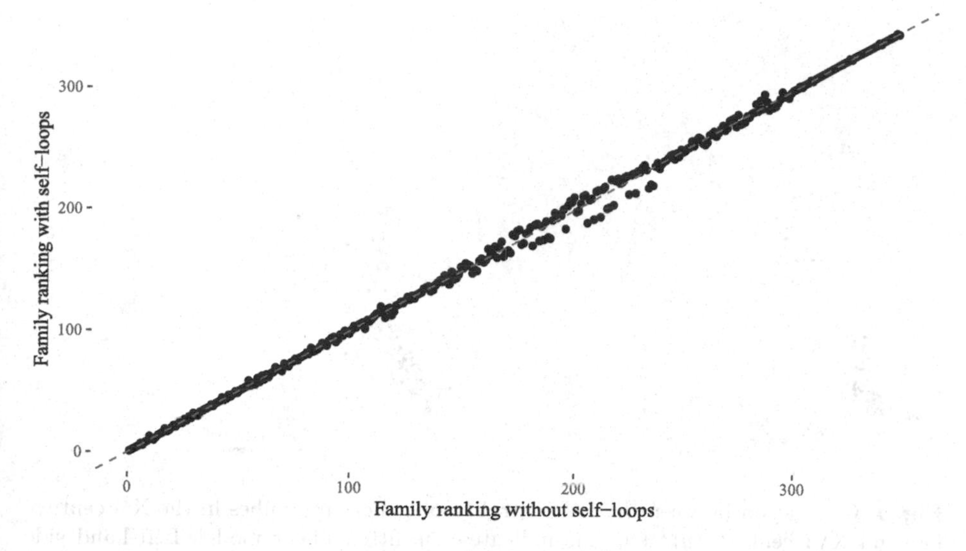


**Fig. 3.** EV centrality rank reached by families with ($y$ axis) or without ($x$ axis) self loops.

**Table 4.** Summary of linear fit of EV centrality for families in the XIV and XV century.

| | RSE | R2 | adjR2 | Fstat | intercept.t | slope.t |
|---|---|---|---|---|---|---|
| With self-loops | 0.056 | 0.825 | 0.823 | 733.504 | 7.306 | 27.083 |
| Without self-loops | 0.065 | 0.834 | 0.833 | 785.478 | 6.632 | 28.026 |

The influence of using self-loops (shown in the right-hand side panel) is clear, with points representing families with the highest EV seeming *closer* to the fit; it should be noted that these are the families that have a high-number of intra-family marriages. To quantify numerically the difference in Table 4 we show a summary of the coefficients of the regression model for the two data sets. The RSE column shows the difference of residuals between the model and the data; the top row is smaller, showing a better fit for the EV centrality values if self-loops are taken into account. This improvement is due mainly, looking at the two right-most columns, to the improvement in the fit of the intercept with 0 (the t value is better); the effect of this can be observed in Fig. 4, which shows how the red line that represents the model seems a bit more centered than the one on the left. The values of R-squared and its adjusted value, as well as the F-value, are slightly better for the model without self-loops; however, they are very similar and very high in both cases, so this difference is not considered significant.

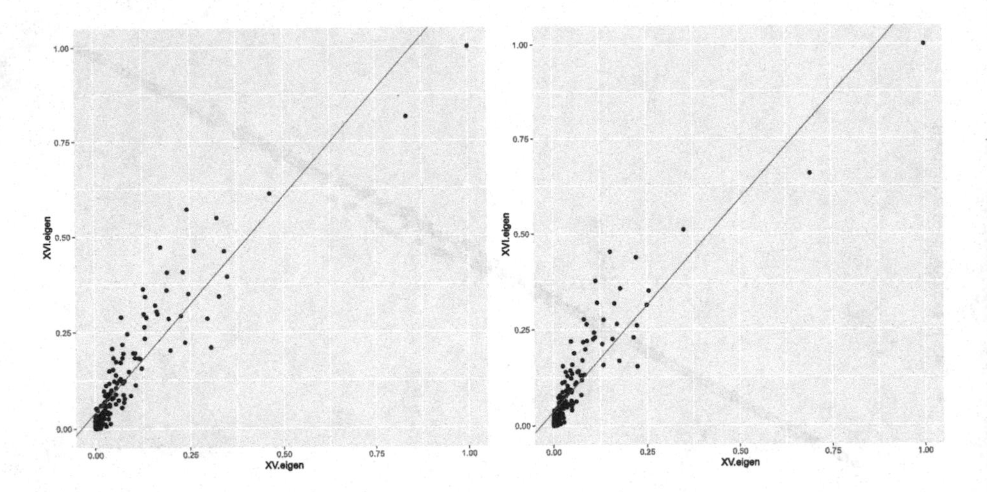


**Fig. 4.** Correlation between the EV centrality of Venetian families in the XV century ($x$) and XVI century ($y$). The line indicates the fitted linear model. Left-hand side chart does not include self-loops, right-hand side does.

At any rate, and to the extent that EV centrality in a century is a predictor for EV centrality in the next century, including self-loops in its computation makes it a better fit, even if the two models represent the data very well in both cases.

### 3.2 Freight Traffic Network

The freight traffic network is a directed network that includes as nodes different US states, and as edges the amount of freight traffic between them. The data is available from the US Bureau of Transportation Statistics on its page "Freight Flow by State"[5]. The data is available for the years 2017–2021.

We will be doing some additional processing on this data. All types of traffic (import, export, domestic) will be added into a single traffic flow; this quantity will be used as a weight in the transportation network. Also, we will use the data for the year 2021 only; the resulting processed data set is available from the GitHub repository for this paper https://github.com/JJ/redes-venecia. The resulting network is shown in Fig. 5. This network is totally different from the one analyzed in the previous subsection: all states have internal traffic, so all of them have self-loops, as the Figure shows. It is a directed network, since traffic between two states can be asymmetric. The network is very dense, with almost all states connected with all others. Finally, the network structure is quite different, with big differences between the most central states and the rest. Using self loops or not is bound to influence the vision we have of the centrality of the states.

[5] https://www.bts.gov/browse-statistical-products-and-data/state-transportation-statistics/freight-flows-state.

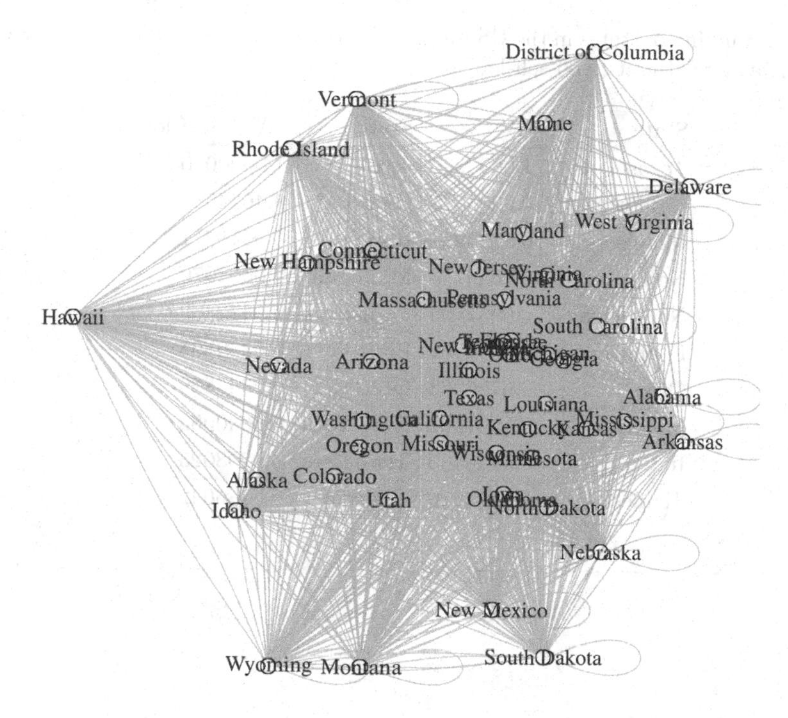


**Fig. 5.** US freight traffic network. Weight value is not represented in this graph.

In Table 5 We show the differences in the first positions in the ranking according to EV centrality. As was the case for the Venice matrimonial network, the top 3 positions of the top ten states do not change. All others (except for, curiously, Ohio) do, however, as does the quantitative difference between the first and the rest of the positions; using self-loops highlights the differences in EV centrality, and thus influence, between the first and the rest of the states, as well as the differences among themselves; that is, the measured difference in influence as measured by EV centrality is much bigger when self-loops are considered.

The stark change in EV centrality values is shown in Fig. 6, where we plot the EV centrality values with and without self-loops for all states (and the District of Columbia. Except for the two top values (Texas and California), when self-loops are considered the value of EV centrality drops below (roughly) 0.05, with changes whose value is obviously decreasing with the initial value (without self-loops).

Figure 7 shows the relationship between the decrease in (normalized) EV centrality when self-loops are considered and the value of intra-state traffic. In general, the greater intra-state traffic, the bigger the difference in EV centrality. This occurs in a scale that is totally different to that shown in Fig. 2, although it is remarkable to note that changes go in the same direction, that is, a decrease in

**Table 5.** Ranking of states in the US freight network according to EV centrality values, with (right) or without (left) self-loops.

| State | EV | State | EV + self-loops |
|---|---|---|---|
| Texas | 1.0000000 | Texas | 1.0000000 |
| California | 0.9908871 | California | 0.4624587 |
| Illinois | 0.8105954 | Illinois | 0.0518211 |
| New York | 0.7649173 | Michigan | 0.0508070 |
| New Jersey | 0.6937505 | Louisiana | 0.0472489 |
| Pennsylvania | 0.6931582 | Florida | 0.0350895 |
| Ohio | 0.6796268 | Ohio | 0.0347809 |
| Michigan | 0.6607562 | New York | 0.0340062 |
| Indiana | 0.5740545 | Tennessee | 0.0298025 |
| Tennessee | 0.5117970 | Georgia | 0.0286042 |

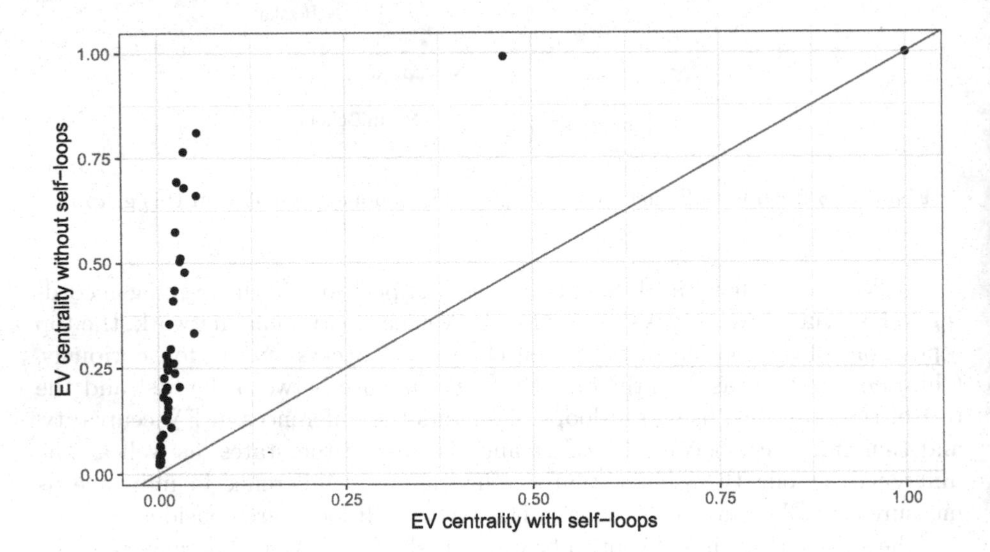


**Fig. 6.** Relationship between EV centrality with or without self-loops for the US freight network.

EV centrality, and also in general a higher self-connection weight will imply less change. This clearly indicates a correlation between the two measures, EV centrality and intra-connection weight, but it is not clear if it is a causal relationship or not, and discussing it is beyond the point of the paper.

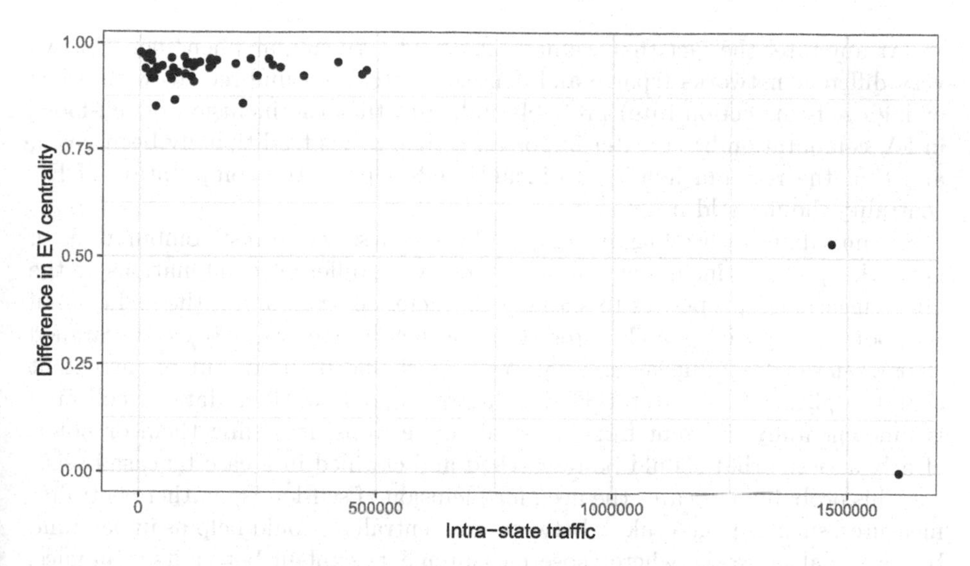


**Fig. 7.** Change in EV centrality vs. value of intra-state traffic.

## 4 Conclusions

The main research question we asked in the introduction was whether it was worth the while to take into account self-loops whenever possible in the computation of EV centrality. Using real-world networks, we have shown that, in general, the answer is yes. In the case of the Venetian matrimonial network it helps EV become a better predictor of the social status of the families involved, and in the case of the US freight network highlights the differences between the position in the network of states, boosting the difference between Texas, California, Illinois and the rest of the states, and boosting Michigan and Louisiana over New York and New Jersey given the importance of internal freight in these two states, that is, the weight of self-connections (diagonal values in the connection matrix). In both cases, the differences are not only quantitative, but also qualitative, as the ranking of the most important nodes changes when self-loops are included.

The direction and quantity of those changes is very similar in both networks examined. Complex networks reach a state that is the consequence of many different internal and external processes, and in general, the same processes that make a node in the network reach a high status will produce a high number of internal connections, that is, self-loops with a high internal weight. This makes the inclusion of self-loops *increase* the differences in status between high-ranked members, but also produce changes in ranking that are more pronounced in the case of mid-ranking members, as observed in the case of the Venetian marital network.

At any rate, the fact that we have observed very similar phenomena in two very different networks (sparse and dense, directed and undirected, with a low or high self-connection rate) probably indicates that the inclusion of self-loops in EV computation has greater importance than it has traditionally been given, and thus the recommendation to include self-loops in the computation of EV centrality should hold in general.

A immediate future line of work would of course try to test combinations of networks, possibly including synthetic ones, with different combinations of the above mentioned properties to actually characterize when and if the inclusion of self-loops is essential, merely interesting, or simply irrelevant. It can be argued that whenever they appear in a dataset they should be used, but to the extent that the physical interpretation of self-loops and how their data is collected is fundamentally different from external connections, including them or not is clearly an issue that should be researched and clarified in a case by case basis.

This probably indicates the need for inclusion of self-loops in other centrality measures, such as page-rank or betweenness centrality, would help us understand better social networks where those measurements explain better its dynamics; however, this will need a modification of the algorithms used to compute them, which is left as a future line of work. In order to be comprehensive, too, it would be interesting to see the relationship between the slope of the model shown in Fig. 1 and the relationship between the number of self-loops for all actors involved. Finally, investigating other social networks with the same characteristics would help us generalize these results and thus recommend to *never* eliminate self-loops in the analysis of social networks where they exist.

The most important line of work, however, is to check against some external measurement (number and importance of positions reached by a family, for instance, in line with [21]) which version of the EV centrality would be a better match, thus proving that self-loops/diagonal values in the connection matrix go beyond mere mathematical artifacts to have a precise and grounded interpretation; once that is proved, it would pave the way to an extension of the employment of self-loops in some way in other centrality values such as betweenness centrality.

Finally, we should note that this paper and the data used in it have been developed following the principles of Agile science [13] and its development can be observed in the repository https://github.com/JJ/redes-venecia together with the data used in it, making it fully reproducible using the same software. It is of course free software, and released under the terms of the GNU General Public License v3.0.

**Acknowledgements.** This work is supported by the Ministerio español de Economía y Competitividad (Spanish Ministry of Competitivity and Economy) under project PID2020-115570GB-C22 (DemocratAI::UGR).

## References

1. Bonacich, P.: Factoring and weighting approaches to status scores and clique identification. J. Math. Sociol. **2**(1), 113–120 (1972). https://doi.org/10.1080/0022250X.1972.9989806
2. Bonacich, P.: Some unique properties of eigenvector centrality. Soc. Netw. **29**(4), 555–564 (2007)
3. Chierichetti, F., Lattanzi, S., Panconesi, A.: Rumor spreading in social networks. Theoret. Comput. Sci. **412**(24), 2602–2610 (2011)
4. Chojnacki, S.: Social identity in renaissance Venice: the second Serrata. Renaissance Stud. **8**(4), 341–358 (1994). https://www.jstor.org/stable/24411934
5. Cotta, C., Mora, A.M., Merelo-Molina, C., Guervós, J.J.M.: FIFA world cup 2010: a network analysis of the champion team play. CoRR abs/1108.0261 (2011)
6. Freeman, L.C.: A set of measures of centrality based on betweenness. Sociometry 35–41 (1977)
7. GürsakaL, N., Sevilmiş, A., Aksan, A., Yılmaz, F.: Comparison of country-based high and low market valued transfer networks. J. Curr. Res. Soc. Sci. **10**(2), 417–430 (2020)
8. He, M., Glasser, J., Pritchard, N., Bhamidi, S., Kaza, N.: Demarcating geographic regions using community detection in commuting networks with significant self-loops. PLoS ONE **15**(4), e0230941 (2020)
9. Horstmeyer, L., Pham, T.M., Korbel, J., Thurner, S.: Predicting collapse of adaptive networked systems without knowing the network. Sci. Rep. **10**(1), 1223 (2020)
10. Iyengar, D., Rao, S., Goldsby, T.J.: The power and centrality of the transportation and warehousing sector within the US economy: a longitudinal exploration using social network analysis. Transp. J. **51**(4), 373–398 (2012)
11. Kuikka, V.: Opinion formation on social networks - the effects of recurrent and circular influence. Computation **11**(5) (2023). https://doi.org/10.3390/computation11050103, https://www.mdpi.com/2079-3197/11/5/103
12. Merelo, J., Cotta, C.: Who is the best connected EC researcher? Centrality analysis of the complex network of authors in evolutionary computation. Arxiv preprint arXiv:0708.2021 (2007)
13. Merelo-Guervós, J.J.: Agile (data) science: a (draft) manifesto. CoRR abs/2104.12545 (2021). https://arxiv.org/abs/2104.12545
14. Merelo-Guervós, J.J.: What is a good doge? Analyzing the patrician social network of the republic of Venice (2022). https://doi.org/10.48550/ARXIV.2209.07334, https://arxiv.org/abs/2209.07334
15. Puga, D., Trefler, D.: International trade and institutional change: medieval Venice's response to globalization. Q. J. Econ. **129**(2), 753–821 (2014). https://doi.org/10.1093/qje/qju006
16. Pullan, B.: 'Three orders of inhabitants': social hierarchies in the Republic of Venice. In: Denton, J. (ed.) Orders And Hierarchies In Late Medieval And Renaissance Europe. PFM, pp. 147–168. Macmillan Education UK, London (1999). https://doi.org/10.1007/978-1-349-27580-9_10
17. Rodrigues, F.A.: Network centrality: an introduction. In: Macau, E.E.N. (ed.) A Mathematical Modeling Approach from Nonlinear Dynamics to Complex Systems. NSC, vol. 22, pp. 177–196. Springer, Cham (2019). https://doi.org/10.1007/978-3-319-78512-7_10
18. Romano, D.: The Limits of Kinship: Family Politics, Vendetta, and the State in Fifteenth-Century Venice. Firenze University Press (2014)

19. Salehi-Abari, A., Boutilier, C.: Empathetic social choice on social networks. In: AAMAS, pp. 693–700 (2014)
20. Southworth, F.: Freight flow modeling in the United States. Appl. Spat. Anal. Policy **11**, 669–691 (2018)
21. Telek, Á.: Marrying the right one - evidence on social network effects in politics from the Venetian Republic (2017). https://editorialexpress.com/cgi-bin/conference/download.cgi?db_name=SAEe2017&paper_id=520
22. Zhao, Z., Zhu, X., Han, T., Chang, C.: The fake news propagate with more self-loop in online social networks. In: 2021 IEEE 2nd International Conference on Information Technology, Big Data and Artificial Intelligence (ICIBA), vol. 2, pp. 839–843 (2021). https://doi.org/10.1109/ICIBA52610.2021.9688320

# Evolutionary Story Sifting over the Log of a Social Simulation

Pablo Gervás(✉) and Gonzalo Méndez

Facultad de Informática, Universidad Complutense de Madrid, Madrid, Spain
{pgervas,gmendez}@ucm.es
http://nil.fdi.ucm.es

**Abstract.** The process of selecting subsets out of a sequence of events on the grounds that told together they constitute an interesting narrative–known as *story sifting*–has become a topic of interest due to its applicability in video games that automatically develop large scale simulations of story worlds. Existing approaches to story sifting operate by matching subsequences of the available events onto patterns of plot considered to be of interest. The present paper proposes a two stage approach that combines a process of matching small strings of events connected by common sense relations–such as asking someone on a date and having them accept, or developing an attachment to someone who has given us a present–and an evolutionary search procedure that explores combinations of this type of paired events into longer sequences that constitute small plot lines about romantic entanglement. This procedure is run over the log of a multi-agent system that simulates a set of characters that develop a set of affective affinities between them as a result of social interactions of a romantic nature.

**Keywords:** narrative generation · multi-agent simulation · story sifting · evolutionary procedure

## 1 Introduction

As the average person goes through their day, they witness a thousand small events and take part in a further set of events. At any point during that time, if called upon, they can very easily isolate a very concise subset of those events as worth telling, and they can build the resulting selection into a story that seems cohesive, makes sense, and appears to have a connecting thread that makes it interesting. This is what any of us does when asked about our day, or what we have been doing, or what is happening back where we came from. In computational terms, these tasks are far from trivial.

This paper has been partially funded by the projects CANTOR: Automated Composition of Personal Narratives as an aid for Occupational Therapy based on Reminescence, Grant. No. PID2019-108927RB-I00 (Spanish Ministry of Science and Innovation), the ADARVE (Análisis de Datos de Realidad Virtual para Emergencias Radiológicas) Project funded by the Spanish Consejo de Seguridad Nuclear (CSN).


M. Villani et al. (Eds.): WIVACE 2023, CCIS 1977, pp. 381–393, 2024.
https://doi.org/10.1007/978-3-031-57430-6_29

In practical terms, these tasks involve a number of cognitive operations: (1) identify relations between the events that have happened (usually of causality or intention), (2) identify chains of relations between the events, (3) consider which events in a given chain would be relevant to mention in an interesting story, and (4) consider which chains of events, included together in a given story, give rise to an interesting plot.

The specific challenge of identifying what may constitute an interesting plot ultimately relies on the ability to model the reactions of a reader when reading the corresponding story. The present paper brings together a number of prior achievements in the computational modeling for plot and the dynamics of reading the discourse for a story with an evolutionary algorithm that operates over the set of events from a log for a simulation about social interactions between a set of characters. These social interactions basically involve one character proposing an activity to another character, who decides whether to accept the invitation or not.

## 2 Previous Work

For the present paper, we will review previous work on social simulations as possible sources from which to extract narratives, story sifting solutions to select story-worthy events from a set of facts, and narrative composition for telling stories about facts that are already known rather than invented for the purpose. Finally, efforts to model computationally the reaction of a reader to a given narrative are reviewed in search of solutions that can provide means for quantifying the relative merit of different candidate narratives.

### 2.1 Social Simulations as Sources for Narrative Renderings

A multi-agent system (MAS) consists of a set of software entities, the agents, that are autonomous–they can make their own decisions—and interact with their environment and among themselves in terms of cooperation, coordination, negotiation or competition.

The work described in [12] presents a multi-agent planner system that is capable of generating stories taking into account plot coherence and character believability. A similar approach is used in the Sabre system [15] where a centralized planner is used to generate the story. Comme il Faut (CiF) [7] is a knowledge-based system that models the complex interplay between social norms, character desires, cultural background, and social interactions. Based on this information it can be used to support simulations of agents engaging in social settings.

Charade [10] is a multi-agent simulation designed to express relations and interactions between characters in a storyworld based on the existing affinities between the characters, and to model the evolution of these affinities through a given period of time. Based on the model of affinities between them, agents propose activities to other agents, who accept or reject them. The result is a log of interactions and evolutions of affinity levels. An example of a fragment of a log for the Charade system is shown in Table 1. These logs are subsequently used to generate episodes within a narrative [1].

**Table 1.** An example of a fragment of the log generated by the Charade multi-agent simulation system.

```
Megan PROPOSE friend_have_lunch Meredith
Lester PROPOSE friend_chat Robert
Suzette PROPOSE friend_chat Silvy
Betty PROPOSE friend_weekend_out Clark
Meredith PROPOSE mate_watch_tv Lester
Clark REJECT-PROPOSAL friend_weekend_out Betty
Lester REJECT-PROPOSAL mate_watch_tv Meredith
Meredith ACCEPT-PROPOSAL friend_have_lunch Megan
(...)
```

### 2.2 Story Sifting

Early work on computational creativity generated literary texts by selecting a subset of lines from an extensive source file [11]. A refinement on this technique that mines sequences of events corresponding to interesting stories from the logs of agent-based simulations has become a line of research known as *story sifting*. James Ryan's PhD thesis [13] outlines how, rather than automatically inventing stories, narrative may emerge from the activity of characters set in motion in a simulated story world, and defines the task of curating such narratives out of simulation logs as story sifting. The Felt story sifting and simulation engine [6] introduced the concept of *story sifting patterns*, which are descriptions of sequences of events that exhibit high potential to be part of interesting narratives. This line of research lead to the development of Winnow [5], a domain-specific language for specifying story sifting patterns that can be run on ongoing simulations to identify event sequences with narrative potential.

### 2.3 Evolutionary Solutions for Exploring Search Spaces of Plot

Evolutionary solutions rely on fitness functions that measure the quality of the final output with no consideration of how particular individuals might be constructed. This makes them particularly suitable to explore a search space of narratives based on existing work analysing narrative as a product. Examples of the use of evolutionary solutions to validate particular narratives in relation to other candidate drafts are: [14] which relies on an evolutionary approach to identify optimal candidates from an initial population built using knowledge-based heuristics, [4] which explores a search space of combination of plot templates using metrics for story consistency from a semantic point of view, or [8] which identifies combinations of partially ordered graphs of events associated with particular entities to maximise story coherence and story interest.

Of particular relevance for the approach in this paper is the work of [3] which applies an evolutionary search process to identify optimal combinations of plot-relevant units of abstraction–called *axes of interest*–into narratives. The axes of interest have free variables that need to be instantiated with particular characters. The procedure involved separate processes for selecting a set of axes of interest with unbounded variables for the roles of the characters, establishing a relative ordering between them, and creating instantiations of the unbound variables with characters for the story. The instantiations of these variables are chosen to ensure meaningful connections across events in the story, both in terms of causal relations between events and certain characters being involved in related events. An adaptation of this process is used in this paper to sift stories from simulation logs.

## 3 Evolutionary Story Sifting from the Log of a Simulation

The present paper operates on the log of events generated by the Charade multi-agent simulation system [10] as described in Sect. 2.1. The set of events in the log is read into a conceptual representation to allow further processing. For the current prototype, the following types of events are considered: `PROPOSE`, `ACCEPT-PROPOSAL` and `REJECT-PROPOSAL`.

### 3.1 Capturing Plot Relevant Connections in the Representation Format

The present paper bases its approach to story sifting in the identification of plot relevant connections between events in the log. To achieve this, a plot is represented in terms of *plot atoms*, which are abstract descriptions of an event (such as a character proposing an activity to another) that specify how the roles specific to the plot atom (proposer, proposee) are related to the set of characters in a given story. The causal connections between plot atoms in a story is captured in our representation by the concept of *axes of interest* [2]. The axes of interest being considered for the present prototype are shown in Table 2.

Axes of interest are used to parse the sequence of events in a log of the Charade system into subsets of events grouped together by virtue of being instantiations of the plot atoms in a given AoI, appearing in the log in the correct relative order and with the characters involved matching the constraints of the AoI in terms of roles. Such subsets of events are referred to as *plot projections*. Each plot projection shows the plot atoms in the axis of interest that it instantiates, the set of assignments to the roles corresponding to the AoI, and the position in the input discourse in which the corresponding plot atoms appear. An example of the parse of a Charade log into a set of plot projections is shown in Table 3.

**Table 2.** The two basic Axes of Interest employed by the system. Each axis of interest defines the connection between a proposal event and either an acceptance or a rejection of it, in terms of the co-instantiation of the variables involved (shown in bold in the table).

| Axis of Interest | Activities | Participating characters |
|---|---|---|
| ProposalAccepted | ProposeActivity | proposer = **x**, proposee = **y** |
| | ActivityAccepted | proposer = **x**, proposee = **y** |
| ProposalRejected | ProposeActivity | **x**, proposee = **y** |
| | ActivityRejected | **x**, proposee = **y** |

**Table 3.** A example of a fragment of the parse of a Charade log into a set of plot projections.

| Plot Element | Arguments | Position |
|---|---|---|
| ProposeActivity | [Betty, Clark, friend_weekend_out] | 4 |
| ProposedActivityRejected | [Betty, Clark, friend_weekend_out] | 6 |
| (...) | | |
| ProposeActivity | [Suzette, Silvy, friend_chat] | 3 |
| ProposedActivityAccepted | [Suzette, Silvy, friend_chat] | 11 |
| (...) | | |

### 3.2 Evolutionary Content Selection Based on Plot Projections

By selecting only events which are part of some plot projection we ensure that the set of characters appearing is fairly coherent and the relative order in which things happen makes some sense. To ensure full satisfaction we need to consider a further filter that guarantees full connectivity between all the events present, and which establishes some constraint on the relative order in which they appear. So we need means to identify when a selection of plot projections makes sense as a story. The present paper proposes the use of an evolutionary solution for this task.

Our procedure for identifying interesting sub-sequences of events from a log for the Charade system is defined as an adaptation of the solution by [3] described in Sect. 2.3. Because our present tasks operates from an input–the log of the Charade system–that already established the set of events, their relative order, and the characters that take part in them, most of the procedure needs to be defined anew. However, we will retain the mechanisms for establishing the relative quality of a given story draft.

Since logs may run for long periods, the search space of all possible sub-sequences of events that can be extracted from a log is very large. To focus on output narratives of an acceptable length, the system takes as input an upper bound on the number of events that can be selected out of a log to build a narrative.

The adapted evolutionary procedure is initialised with a genetic representation in the form of a numerical vector that encodes the choice of which of the plot projections obtained from the parsing of the log to include in the story draft under construction. All the other characteristics of the story come already decided in the log, namely: relative order of events and assignment of characters to play the roles in each one of them.

A population of individual drafts is created by random assignment of values to this genetic representation. Values of 1 indicate the plot projection in the given position is to be included in the story (gene activated), value of 0 indicates a plot projection to be ommitted (gene deactivated). To focus on output narratives of an acceptable length, the system constructs the initial population with a combination of possible lengths. The upper bound on story size is set to 16. The number of genes activated in the genetic representation vector must take this value into account. Because all the axes of interest currently employed connect a fixed number of two events, the number of activated genes is set to half the desired story size. The set of story sizes explored is currently set to 6, 8, 10 and 12, which correspond to drafts with 3, 4, 5 and 6 projections. With each projection involving two characters, three projections is the minimal set that allows for interactions of at least two different types between two characters or interactions between more than two characters. The upper bound of 12 is set empirically to avoid too large drafts.

Mutation operators are defined so that they deactivate some gene in the representation for every gene that they activate.

Cross-over operators are defined to select at random a point in the genetic vector, divide the gene vector for two different drafts at that point, and combine each initial half with the final half of the other draft.

Whereas the size for the story drafts is respected during mutation, cross-over operators do sometimes result in drafts of different size, either shorter or longer. For this reason, some restriction on draft size is required as part of the selection process.

The population is evolved over a desired number of generations. The fitness function applied is described in the following section.

### 3.3 Metrics on Story Draft Quality

The fitness function for the evolutionary procedure is based on metrics of three types:

- metrics that restrict draft size to manageable proportions
- metrics that measure story cohesion
- metrics the measure character variety

Metrics to control draft length assign score of 100 to drafts under 16 events in length, and 0 otherwise.

The approach in this paper assumes that story drafts will have a higher quality if constraints of a certain type hold between the projections selected to appear in the draft. These constraints specify the relative position in which the events in the selected plot projections appear in the overall sequence of the story draft with respect to other events in which the same characters participate. To provide quantitative measure of these aspects we rely on the measures of correct sequencing of events, and acceptable occurrence of characters sharing roles across AoIs proposed in [3]. For completeness, a brief summary is included here.

A given combination of AoIs, such as for instance Betty repeatedly proposing activities to Clark in spite of his continued rejections, acquires value if the reader can infer some kind of connections between the two AoIs involved by virtue of the relative sequencing and the specific instantiation of the characters. In the example shown, the interest of the particular combination in such a story draft arises from the perseverance exhibited by Betty.

A set of metrics is defined to capture this type of constraints for each combination of AoIs for which relevant connections can be established. The metrics we are considering now are driven by constraints of the type presented in Table 4. In each case, an informative label has been added in the first column to identify the feature that justifies the interest. The full set of constraints includes further types such as: `RefugeSomewhereElse`, `ChangeOfTarget`, `HappyStreak`, `ReturnInvite` or `CatchOnTheRebound`.

**Table 4.** Example of constraints. For each entry, the first row describes role-sharing constraints and the second row describes sequencing constraints.

| | | |
|---|---|---|
| Perseverance | ProposalRejected | ProposalRejected |
| | proposer = proposer | proposee = proposee |
| SuccessAfterFailure | ProposalRejected | ProposalAccepted |
| | proposer = proposer | proposee = proposee |
| HappyStreak | ProposalAccepted | ProposalAccepted |
| | proposer = proposer | proposee = proposee |

Finally, a metric has been introduced that scores 100 to drafts that have between 2 and 4 characters, 50 for drafts of 2 characters or characters above 4 but with less characters than the number of events in the draft, and 0 otherwise. This compensates the tendency of the more demanding metrics on story cohesion to force the drafts in the population towards stories with only two characters in them, and focuses the results on stories about specific characters.

A number of combinations of these metrics have been tested to use as fitness functions. They are discussed in the following section.

## 4 Discussion

The discussion includes a quantitative comparison with a number of baselines for the fitness metrics and a qualitative discussion of relations with prior approaches.

### 4.1 Comparative Evaluation

Any solutions provided by the system must be tested against some objective function. Such an objective function should require that the draft tell a story about a particular set of characters (the same set of characters recur throughout the draft) and that all events in the draft be tied in to some other event in the draft by some link of relevance.

These requirements allow us to establish some baselines for our fitness function. We consider the following alternatives for the fitness functions:

- percentage of projections in the draft that share at least one character with some other
- percentage of projections in the draft that are connected by consistency constraints
- ratio of size of largest subset of projections completely connected together by consistency constraints to overall number of projections
- ratio of size of largest subset of projections completely connected together by consistency constraints to overall number of projections weighted with percentage of applicable constraints that are actually satisfied

These metrics provide a progressive scoring, so that drafts where the sequencing constraints are not met are scored relative to how far they need to be modified for the constraints to be met. This allows mutations that modify the sequence in the right direction to be scored progressively higher, allowing evolution to converge towards optimal solutions.

Table 5 shows results for different versions of the system, each configured to run with a different fitness function. For each configuration the averages for a set of 6 runs are presented. The features shown include: maximum score found in population, number of generations required to converge to that top score, minimum score found in population, number of generations required to converge to that top score, length of draft in number of events, number of distinct characters that appear in the draft, average of the number of times that characters are mentioned.

The configuration of the evolutionary process was kept the same across the runs: initial population of 20 individuals, evolution over 20 generations, selection by accumulated fitness of a population of 20 for the next generation, and the mutation and cross-over operators described in Sect. 3.2.

**Table 5.** Results for different metrics.

| Metric | Max | conv | Min | conv | Length | Chars | Slots/char |
|---|---|---|---|---|---|---|---|
| every aoi shares variable | 100.0 | 2.0 | 100.0 | 10.7 | 10.0 | 4.0 | 5.0 |
| every aoi in constraint | 100.0 | 7.0 | 100.0 | 17.0 | 10.0 | 3.8 | 5.2 |
| all aois connected by constraints | 100.0 | 5.5 | 100.0 | 15.3 | 10.0 | 3.7 | 5.5 |
| all aois connected by constraints (weighted) | 100.0 | 5.0 | 100.0 | 15.8 | 10.0 | 3.3 | 5.7 |

Several observations can be made from these results. Although all the fitness functions produce populations that reach the top score, as the complexity of the metrics increases, the number of generations required for the scores to converge increases. The point of convergence for the minimum scores provides an indication of the relative difficulty involved in reaching the top score for all individuals in the population. This is relevant because it guarantees a complete population of quality drafts rather than a single one. The selection procedure guarantees that all drafts in a population be genetically different, so all individuals with the same score are valid alternative solutions. There is clearly a preferred size of 10 events per draft that apparently maximises all of these fitness functions. This is understandable because it allows two pairs of characters, with two projections connected together for each pair, and at least one additional projection to connect one character from one pair to a character of the other. For larger drafts it is probably much more difficult to find instances of projections in the simulation that satisfy a sufficient number of constraints. The number of characters progressively decreases, as the events in the drafts tend to focus on a smaller set of characters that are the protagonists of the story. The average number of appearances in the draft per character also increases for the more demanding metrics.

In terms of efficiency, the system has been tested with various configurations of the evolutionary parameters to identify the choice leading to best performance. Table 6 shows best average scores over six runs with different values for population size and number of generations, while all other parameters were kept unchanged. These results show that the algorithm is sensitive to increase both in the size of the population and the number of generations. They also show that slightly higher gains in performance are obtained by the increase in the size of the population. As the increase in number of generations carries a significant overhead in increased execution times, we settle for a configuration of population size of 20 and 30 generations.

Table 7 shows an example of system output expressed in terms of the internal representation format, together with the corresponding story rendered as text automatically by the system using basic templates for each of the actions involved. The evolutionary solution was run for 30 generations, with a population size of 20 individuals. The final score for this draft is 100/100.

**Table 6.** Best maximum and minimum scores (over 100) and execution times (in milliseconds) averaged over 6 runs on the same log. Rows show number of generations and columns show population size.

| | 10 | 20 | 50 |
|---|---|---|---|
| 10 | 95.0 | 100.0 | 98.3 |
| | 76.7 | 84.3 | 74.7 |
| | 357.3 | 742.7 | 1049.7 |
| 20 | 100.0 | 100.0 | 100.0 |
| | 95.0 | 96.7 | 100.0 |
| | 743.8 | 1174.3 | 2660.8 |
| 30 | 100.0 | 100.0 | |
| | 98.3 | 100.0 | |
| | 984.2 | 2170.8 | |

**Table 7.** Example of story draft obtained by story sifting from a Charade log followed by the automated template-based rendering of the story

| | |
|---|---|
| PR-0-ProposeActivity-100 | friend_play_tennis/Mary/Silvy |
| PR-0-ProposedActivityRejected-101 | friend_play_tennis/Mary/Silvy |
| PA-0-ProposeActivity-113 | friend_go_out/John/Mary |
| PA-0-ProposedActivityAccepted-117 | friend_go_out/John/Mary |
| PA-1-ProposeActivity-532 | friend_play_tennis/Mary/Meredith |
| PA-1-ProposedActivityAccepted-533 | friend_play_tennis/Mary/Meredith |
| PA-2-ProposeActivity-596 | friend_serious_talk/Mary/Meredith |
| PA-2-ProposedActivityAccepted-597 | friend_serious_talk/Mary/Meredith |
| PR-1-ProposeActivity-691 | friend_day_out/Mary/Silvy |
| PR-1-ProposedActivityRejected-692 | friend_day_out/Mary/Silvy |
| PR-2-ProposeActivity-752 | friend_serious_talk/John/Mary |
| PR-2-ProposedActivityRejected-753 | friend_serious_talk/John/Mary |

Silvy proposes to Mary to play tennis as friends. Mary rejects Silvy's invitation to play tennis as friends. Mary proposes to John to go out as friends. John accepts Mary's invitation to go out as friends. Meredith proposes to Mary to play tennis as friends. Mary accepts Meredith's invitation to play tennis as friends. Meredith proposes to Mary to serious talk as friends. Mary accepts Meredith's invitation to serious talk as friends. Silvy proposes to Mary to day out as friends. Mary rejects Silvy's invitation to day out as friends. Mary proposes to John to serious talk as friends. John rejects Mary's invitation to serious talk as friends.

This example has been selected on the basis of the perseverance of Silvy in the face of Mary's rejections, the decision of Mary to take refuge from Silvy in John, the appearance of Meredith to support Mary, and John's final change of mind later in rejecting Mary's proposals.

In general terms, it is important to point out that the quality of the stories that can be sifted out of a story log is constrained by the interest of the events in the log used as input. A possible way of taking this into account would be to develop an additional set of metrics to measure the interest of the events in the whole log, and to consider that as a baseline in the sense that stories sifted from the log cannot add interest other than by intelligent selection.

### 4.2 Relationship with Prior Work

The original work on the Charade simulation system [10] considered outputs of the simulation in terms of threads for specific characters. That is, a pair of related characters would be chosen, and an extract of all the events from the log in which the two characters were involved would be considered. This provided a span of events too long to be considered a real story. The original work explored the possible interest of these spans in terms of the evolution of the affinity value for relation between the chosen characters.

The Felt story sifting and simulation engine [6] used story sifting patterns that allowed it to represent combinations of events such as a repetition of betrayals, and parse a log to identify sequences of actions that matched them. The patterns used by Felt are expressed in a linear logic programming language that allows them to represent more complex combinations than those captured in the constraints used here. These patterns allow the user to perform very detailed searches. The solution proposed in this paper, combines a simpler set of constraints with the power of evolutionary approaches to search.

Now that Large Language Models (LLMs) are starting to be used to perform tasks that were previously carried out using other techniques, we have also tried to shed some light on their performance for story sifting [9]. We have tested ChatGPT using different prompts in order for it to process the original log used in this contribution and we found several limitations: it cannot use log files as input and has a restricted capacity for input data during conversations; it tends to forget instructions quickly, making it challenging to work with large amounts of data; it also overlooks sequential inputs, focusing more on the latest batch of information; when summarizing a set of events, ChatGPT's response is limited to a concise summary without the ability to select specific subsets based on narrative qualities; ChatGPT tends to introduce its own content, which can make it difficult to confine the output to the provided input data; ChatGPT's criteria for story sifting operations are unclear, generic, and challenging to influence for obtaining results aligned with different criteria or specific domains. As a result, our impression is that there is still room for improvement in relation to LLM-based story sifting, and consequently there is still need to keep on using and researching on other techniques for this task.

## 5 Conclusions

Story sifting has become an area of interest for its potential to automatically develop narratives emerging from large-scale simulations of story worlds. In this paper, we have proposed a two-stage approach that combines matching small strings of events connected by common sense relations with an evolutionary search procedure to explore combinations of paired events into longer sequences that form small plot lines about romantic entanglement. By running this procedure over a multi-agent system that simulates characters and their social interactions of a romantic nature, the resulting set of narrative plots show both coherence and an interesting chaining of events. The order of events is maintained as it was generated in the original simulation. Refinements on the mutation operators to allow for drafts of different sizes will be considered as further work. To improve the set of proposed metrics, potential extensions to cover additional features of the Charade system will also be explored. The consideration of affinities between characters as a relevant aspect to the perceived quality of stories will be explicitly examined. Furthermore, the process outlined in this paper for generating stories can be combined with a method for producing multi-plotline stories, as described in [4].

## References


1. Concepción, E., Gervás, P., Méndez, G.: Ines: a reconstruction of the Charade storytelling system using the Afanasyev framework. In: Proceedings of the Ninth International Conference on Computational Creativity, Salamanca, Spain, pp. 48–55 (2018)
2. Gervás, P.: Generating a search space of acceptable narrative plots. In: 10th International Conference on Computational Creativity (ICCC 2019). UNC Charlotte, North Carolina, USA (2019)
3. Gervás, P.: Evolutionary stitching of plot units with character threads. In: De Stefano, C., Fontanella, F., Vanneschi, L. (eds.) WIVACE 2022. CCIS, vol. 1780, pp. 254–265. Springer, Cham (2022). https://doi.org/10.1007/978-3-031-31183-3_21
4. Gervás, P., Concepción, E., Méndez, G.: Evolutionary construction of stories that combine several plot lines. In: Martins, T., Rodríguez-Fernández, N., Rebelo, S.M. (eds.) EvoMUSART 2022. LNCS, vol. 13221, pp. 68–83. Springer, Cham (2022). https://doi.org/10.1007/978-3-031-03789-4_5
5. Kreminski, M., Dickinson, M., Mateas, M.: Winnow: a domain-specific language for incremental story sifting. In: Proceedings of the AAAI Conference on Artificial Intelligence and Interactive Digital Entertainment, vol. 17, no. 1, pp. 156–163 (2021)
6. Kreminski, M., Dickinson, M., Wardrip-Fruin, N.: Felt: a simple story sifter. In: Cardona-Rivera, R.E., Sullivan, A., Young, R.M. (eds.) ICIDS 2019. LNCS, vol. 11869, pp. 267–281. Springer, Cham (2019). https://doi.org/10.1007/978-3-030-33894-7_27
7. McCoy, J., Treanor, M., Samuel, B., Reed, A.A., Mateas, M., Wardrip-Fruin, N.: Social story worlds with Comme il Faut. IEEE Trans. Comput. Intell. AI Games **6**, 97–112 (2014)
8. McIntyre, N., Lapata, M.: Plot induction and evolutionary search for story generation. In: Proceedings of the 48th Annual Meeting of the Association for Computational Linguistics, pp. 1562–1572. Association for Computational Linguistics, Uppsala (2010)

9. Méndez, G., Gervás, P.: Using ChatGPT for story sifting in narrative generation. In: 14th International Conference on Computational Creativity, Waterloo, Ontario, Canada (2023)
10. Méndez, G., Gervás, P., León, C.: On the use of character affinities for story plot generation. In: Kunifuji, S., Papadopoulos, G.A., Skulimowski, A.M.J., Kacprzyk, J. (eds.) Knowledge, Information and Creativity Support Systems. AISC, vol. 416, pp. 211–225. Springer, Cham (2016). https://doi.org/10.1007/978-3-319-27478-2_15
11. Montfort, N., Fedorova, N.: Small-scale systems and computational creativity. In: International Conference on Computational Creativity, vol. 82 (2012)
12. Riedl, M.O., Young, R.M.: An intent-driven planner for multi-agent story generation. In: International Joint Conference on Autonomous Agents and Multiagent Systems, vol. 2, pp. 186–193. IEEE Computer Society (2004)
13. Ryan, J.: Curating simulated storyworlds. Ph.D. thesis, University of California Santa Cruz (2018)
14. de Silva Garza, A.G., y Pérez, R.P.: Towards evolutionary story generation. In: Colton, S., Ventura, D., Lavrac, N., Cook, M. (eds.) Proceedings of the Fifth International Conference on Computational Creativity, ICCC 2014, Ljubljana, Slovenia, 10–13 June 2014, pp. 332–335. Computationalcreativity.net (2014)
15. Ware, S.G., Siler, C.: The sabre narrative planner: multi-agent coordination with intentions and beliefs. In: Proceedings of the 20th International Conference on Autonomous Agents and MultiAgent Systems, AAMAS 2021, pp. 1698–1700. International Foundation for Autonomous Agents and Multiagent Systems, Richland (2021)

# Evolution of Attraction for Cooperation

Elpida Tzafestas(✉)

Laboratory of Cognitive Science, Department of History and Philosophy of Science, National and Kapodistrian University of Athens, University Campus, 15771 Ano Ilisia, Greece
etzafestas@phs.uoa.gr

**Abstract.** We are investigating an attraction mechanism that models interpersonal relations such that an individual acts differently, and presumably more cooperatively, with those to whom it is attracted than with others. This is a generic mechanism and the types of relations envisaged can range from regular physical attraction to attachment of toddlers to caregivers. We have shown in the past that the introduction of this attraction mechanism in an iterated prisoner's dilemma game yields higher average agent scores in tournaments within uniform or mixed populations. In the present work, we show that this mechanism has an evolutionary advantage and that it can evolve in various ways (probability of attraction, network of attraction). These results show that cooperation and social stability can be enhanced by psychological mechanisms that are external to the game setting but interfere with it and that these mechanisms may be further selected and reinforced by evolution and catalyze cooperation. We further show that evolution of attraction in the context of cooperation is possible in fairly small networks of homogeneous agents with noise but not in networks of heterogeneous agents without noise. This remark has interesting implications for the study of the origins of human cognition.

**Keywords:** Cooperation · Evolution of cooperation · Tit for tat · Attraction

## 1 Introduction

From our everyday experience, we know that interpersonal relations influence our behavior as well as our understanding of the world around us. Psychology has been traditionally interested in the subject of 'attraction', where (dyadic) attraction indicates affect, as well as its opposite 'repulsion' and their effect on attitude, for example whether behavioral or personality similarity is a cause or an effect of attraction (starting from the pioneering work of Byrne, for example [6, 7, 17, 25]). Psychology is also interested in the personality roots of attraction (for example [16]), in the relation of attraction with social identity [12] and in other assorted issues. We use attraction in a broader sense that may also include neonatal or toddler attachment [5, 18], friendship [13], habituation [8], interpersonal commitment [4] and other such phenomena. All these phenomena have in common that the effect of attraction is beneficial to the social interaction and to the participating agents. Any mechanism of attraction, whether in the narrow or in the broad


M. Villani et al. (Eds.): WIVACE 2023, CCIS 1977, pp. 394–408, 2024.
https://doi.org/10.1007/978-3-031-57430-6_30

sense, will by definition be outside the realm of rational decision making, and it will instead constitute a reactive component capable of responding fast and at low cognitive cost to conditions of the social environment. This can be a very handy tool for human behavior in hostile, harsh or stressful natural environments.

We use this idea and study its application to the classic cooperation problem modeled as an iterated prisoner's dilemma (IPD, see next section) and its implications. We want to show that if the mechanism is beneficial, then it can be selected by evolution. In that case we start from the hypothesis that this mechanism can evolve in small groups of mostly kin. This would make sense for the human evolution history of the distant past and it could also provide an indication that attraction can be developed and reinforced as a tool for young humans within a more modern family-and-friends environment.

The organization of the paper is as follows: in Sects. 2 and 3 we describe the basic cooperation problem and we introduce the attraction mechanism in presence of noise. We also give indicative reference results in Sect. 4. In Sects. 5 to 7 we describe evolutionary experiments with the attraction mechanism to show that it is favored by evolution. In Sect. 8 we perform evolutionary experiments in more general societies but without noise to simulate non-kin or adult environments and study the differences. We finally sum up and discuss our findings in the last section.

## 2 General Cooperation Modeling

A major research theme in both theoretical biology and social science is the emergence and domination of cooperative behavior between selfish agents. The cooperation problem states that each agent has a strong personal incentive to defect, while the joint best behavior would be to cooperate. This problem is traditionally modeled as a special two-party game, the Iterated Prisoner's Dilemma (IPD).

At each cycle of a long interaction process, the agents play the Prisoner's Dilemma. Each of the two may either cooperate (C) or defect (D) and is assigned a score defined as follows.

| Agent | Opponent | Score |
|---|---|---|
| C | C | 3 (= Reward) |
| C | D | 0 (= Sucker) |
| D | C | 5 (= Temptation) |
| D | D | 1 (= Punishment) |

The first notable behavior for the IPD designed and studied by Axelrod [1, 2] is the Tit For Tat behavior (TFT, in short):

*Start by cooperating,*

*From there on return the opponent's previous move*

This behavior has achieved the highest scores in early tournaments and has been found to be fairly stable in ecological settings. TFT demonstrates three important properties, shared by most high scoring behaviors in IPD experiments.

- *It is good (it starts by cooperating)*
- *It is retaliating (it returns the opponent's defection)*
- *It is generous (it forgets the past if the defecting opponent cooperates again).*

Further strategies include stochastic ones [19], the Pavlov or Win-Stay-Lose-Shift strategy [20] that cooperates when it has played the same move as its opponent etc. In the literature we may also find studies in an evolutionary perspective [10], theoretical or applied biological studies [3, 9] and studies of modified IPD versions [21].

We adopt the noisy version of IPD in which there is a nonzero probability that an agent's action will be switched to the opposite, i.e. from *COOPERATE* to *DEFECT* or vice versa. It has been shown that retaliating strategies such as TFT can score quite badly in the presence of noise, despite their superiority in the non-noisy domain [14, 15]. This happens because even accidental defections may lead to a persistent series of mutual defections by both players, thus breaking cooperation. The usual approach is to introduce some degree of explicit generosity to account for opponent's misbehaviors or to attempt opponent modeling.

Our approach is based on the observation that an independent psychological or social factor can allow agents in a society to cooperate fairly well despite noise and without explicit opponent modeling or other intricate reasoning behavior. In our case, attraction plays this role.

We are using the benchmark iterated prisoner's dilemma (IPD) in its noisy version as a study vehicle with a stronger bias toward defection, where we feel it could make sense to introduce such an independent external attraction factor. More specifically, we believe that biological evolution or, equivalently, social experience would spontaneously exploit any such factor that would induce better agent scores. This is particularly true for noisy environments where agent scores may degrade abruptly, and especially when interactions are lengthier.

## 3 Attraction Modeling

We propose an attraction mechanism that relies on our everyday experience that people tend to be good and cooperative with other people that attract them and tend to be "regular" with the rest and reason about them. This translates in our model as:

*If (attracted by the opponent) then play ALLC (always cooperate) with a probability P.*

*In all other cases play as usual (for example, TFT)*

The mechanism thus is defined as directed generosity toward selected others, the ones that attract the individual in question. We should note that noise can also affect the outcome of this behavior as well. The agents are interconnected via a "web of attraction" where each agent carries a Boolean value showing whether it is connected to (attracted by) a number of others. The normal "reasoned" behavior of an agent, when unaffected by attraction, is usually one of ALLC, ALLD (always defect), TFT and Adaptive TFT [22], but we have also experimented occasionally with others. In previous work [23, 24], we performed tournament or spatial experiments with populations of agents playing a noisy IPD. We experimented with both uniform or mixed populations, whose agents have the same or diverse normal strategies. The reason we use mostly ALLC, ALLD, TFT and Adaptive TFT is that we want to make sure we explore the limits of our attraction mechanism by studying its effect on the extreme behaviors (ALLC and ALLD that act without feedback) as well as on the most intelligent/high-scoring ones (TFT that retaliates immediately and Adaptive TFT that tries to make sense of a situation). Our results have shown that the attraction mechanism yields higher average agent scores in tournaments within uniform or mixed populations than if it were not present. Benefits are higher for higher attraction factors, bigger populations or populations of "irrational" agents (that do not retaliate or reason, such as ALLD). We have also studied the impact of the attraction pairing type: reciprocal and exact (one to one) or statistical (and not necessarily reciprocal). We have found that statistical (random) not necessarily reciprocal pairing is the best, because even in the absence of reciprocity a rational agent can cooperate consistently with a cooperating attracted, even if irrational or occasionally misbehaving, opponent. In Sects. 4 to 7 we use TFT agents in a noisy environment, whereas in Sect. 8 we address general strategies (exemplified by ALLC, ALLD and TFT) in perfect non-noisy environments.

Before proceeding to describe the experimental setup and the results obtained, we should stress the fact that the attraction mechanism described is "irrational" in that it does not depend on any real feedback of the agent. Our results then suggest that the coupling of reasoning mechanisms with reactive ones (such as attraction, be it physical, emotional, social or other) may be advantageous to social behavior and this is in line with current trends in cognitive and social science.

## 4 Basic Reference Experiments

The parameters of the attraction mechanism are:

| | |
|---|---|
| M | Number of attracted agents |
| K | Number of agents that an agent is attracted to |
| P | Probability of spontaneous cooperation if attracted |

We have run experiments to confirm that the size of the social attraction network (M attracted agents of degree K, with K not necessarily uniform) and the value of the probability P (again not necessarily uniform for all agents) have an impact on the scope of cooperation: bigger network sizes and larger P values are expected to lead to higher cooperation rates as evidenced by the corresponding scores. For all experiments presented in this and the following sections, we give results that are averages of 50 runs. In all cases we give averages and we have checked the standard deviations to make sure that the differences are significant. One example is given in Table 3. Moreover, all experiments are performed for three population sizes, 20, 50 and 100, but for presentation conciseness we present these for size 20 only. In Sect. 8, however, we give comparative results for the three sizes to assess homogeneous versus heterogeneous populations.

In the reference experiments shown in Tables 1 and 2, we give average and maximum scores in societies of various network sizes and various values of P. Our results confirm our expectations.

**Table 1.** Population size N = 20, all TFT, noisy IPD with noise = 0.1, random pairing, P = 1. Average and maximum agent scores for several values of M and K. The score of an agent is the sum of scores obtained against all agents including itself in IPD games lasting 200 rounds each. All results are averages of 50 runs.

| M | K | Avg. (max) score |
|---|---|---|
| No attraction | | 9145.8 (9396.16) |
| 5 | 2 | 9250.74 (9594.12) |
| 10 | 3 | 9446.04 (9903.46) |
| 10 | 5 | 9659.86 (10198.4) |
| No noise, No attraction | | 12000 |

**Table 2.** Population size N = 20, all TFT, noisy IPD with noise = 0.1, random pairing, M = 10, K = 5. Average and maximum agent scores for several values of P. All results are averages of 50 runs.

| P | Avg. (max) score |
|---|---|
| No attraction | 9145.8 (9396.16) |
| 0.1 | 9289.42 (9563.62) |
| 0.3 | 9379.12 (9693.22) |
| 0.5 | 9548.44 (9952.7) |
| 0.7 | 9596.98 (10057.86) |
| 0.9 | 9648.42 (10136.34) |
| 1.0 | 9655.5 (10179.62) |
| No noise, No attraction | 12000 |

These results suggest that an initial small attraction network and an initial small P value may be reinforced by evolution, spread in the population and grow larger with time.

## 5 Experiment 1: Network Evolution

In our first evolutionary experiment, we start with a small network (5 out of 20 agents connected to two others each) and allow the society to evolve the network parameters. Our evolutionary algorithm is very simple and is as follows:

The half lowest ranking agents (in terms of score) adopt the network values of one of the half highest ranking agents (randomly chosen): the Boolean value of using attraction and the K value. Then with 10% probability these values are mutated (the Boolean value is reversed and K is increased or decreased by a value from 1 to 5). The half highest ranking agents mutate with 10% probability. This is a rather fast algorithm that can boost a beneficial feature – attraction, in our case. The goal is not to be realistic but to be illustrative and unravel the evolutionary tendencies. On the one hand, a mechanism or feature, found under these conditions to spread and become associated with higher fitness, is one that has good chances to evolve in realistic conditions. On the other hand, a mechanism that gives no fitness improvement within this favorable setup is close to impossible to spread and is thus candidate for extinction.

**Table 3.** Population size N = 20, all TFT, noisy IPD with noise = 0.1, random pairing, P = 1. Evolution of network (M,K) for 100 generations. Average and maximum agent K, M and scores every 10 generations. All results are averages of 50 runs.[1]

| t | M | Avg. (max) K | Avg. (max) score |
|---|---|---|---|
| No attraction | | | 9145.8 (9396.16) |
| 1 | 5 | 2 | 9461.694 (9909.74) |
| 5 | 14.86 | 3.058 (5.92) | 9622.145 (10090.88) |
| 10 | 16.76 | 5.66 (9.18) | 10048.459 (10581.88) |
| 20 | 17.26 | 10.705 (14.6) | 10713.117 (11190.52) |
| 30 | 16.24 | 14.299 (17.56) | 11001.352 (11392.6) |
| 40 | 15.64 | 16.877 (19.3) | 11153.94 (11447.64) |
| 50 | 15.26 | 17.52 (19.64) | 11167.35 (11454.2) |
| 60 | 15.04 | 17.9 (19.88) | 11176.523 (11474.12) |
| 70 | 15.0 | 18.057 (19.9) | 11176.599 (11473.26) |
| 80 | 15.0 | 17.766 (19.92) | 11177.588 (11463.74) |
| 90 | 15.34 | 18.165 (19.88) | 11226.472 (11501.92) |
| 100 | 15.2 | 18.404 (19.92) | 11224.075 (11502.64) |
| No noise, No attraction | | | 12000 |

The results given in Table 3 show that there is significant improvement in cooperation scores that are obtained with higher values for M and K, i.e., with larger networks, despite the expected fluctuations that are due to the presence of noise in the IPD game implementation.

[1] Note that the standard deviations for M, avg. K and avg. Score at generation 100 are 2.66, 1.183 and 210.672, respectively. The averages at generation 100 are therefore many standard deviations away from the averages at generation 0, therefore the results are significant in the long term, despite the very common occasional fluctuations inherent in all evolutionary systems.

For example, in one sample run the number of attraction connections for the 11 connected agents in the 100$^{th}$ generation are { 15 18 19 19 20 20 20 20 20 20 20}, which means that half the agents are extremely and even fully attracted by others and these allow cooperation to be reciprocated and persist in the population.

## 6 Experiment 2: P Evolution

In our second evolutionary experiment, we start with a fair-sized network (10 out of 20 agents connected to five others each, that is a little smaller network than the ones that evolve in the previous section) and a small uniform P (0.1). Again we allow the society to evolve its P. Our evolutionary algorithm is analogous to the previous one and is as follows:

The half lowest ranking agents (in terms of score) adopt the P value of one of the half highest ranking agents (randomly chosen). Then with 10% probability P is mutated to a new value from 0.1 to 1. The half highest ranking agents mutate with 10% probability.

**Table 4.** Population size N = 20, all TFT, noisy IPD with noise = 0.1, random pairing, M = 10, K = 5. Evolution of P for 100 generations. Average and maximum agent P and scores every 10 generations. All results are averages of 50 runs.

| t | Avg. (max) P | Avg. (max) score |
|---|---|---|
| No attraction | | 9145.8 (9396.16) |
| 1 | 0.1 | 9287.824 (9546.58) |
| 5 | 0.328 (0.798) | 9411.715 (9773.4) |
| 10 | 0.581 (0.889) | 9536.594 (9979.6) |
| 20 | 0.71 (0.918) | 9595.705 (10030.54) |
| 30 | 0.704 (0.919) | 9605.668 (10081.46) |
| 40 | 0.699 (0.911) | 9590.619 (10052.16) |
| 50 | 0.692 (0.914) | 9585.341 (10054.66) |
| 60 | 0.697 (0.91) | 9581.958 (10029.76) |
| 70 | 0.68 (0.917) | 9580.692 (10040.9) |
| 80 | 0.699 (0.906) | 9583.98 (10073.88) |
| 90 | 0.689 (0.914) | 9580.329 (10011.82) |
| 100 | 0.7 (0.901) | 9596.035 (10036.86) |
| No noise, No attraction | | 12000 |

The results given in Table 4 show that, as with the network size, there is significant improvement in cooperation scores that are obtained with higher values for P, despite the expected fluctuations that are due to the presence of noise in the IPD game implementation.

For example, in one sample run the exact values of P differ for each of the 10 connected agents in the 100$^{th}$ generation: {0.884 0.953 0.953 0.953 0.953 0.953 0.953 0.953 0.953 0.953}. Again, all connected agents are extremely generous (all but one have inherited the same high-functioning P value of 0.953 and the last one has also a high P value of 0.884), which allows cooperation to be reciprocated and persist in the population.

## 7 Experiment 3: Simultaneous Network and P Evolution

In our third evolutionary experiment, we start both with a small network (5 out of 20 agents connected to two others each) and a small uniform P (0.1). We allow the society to evolve both its the network parameters and the agents' P. Our evolutionary algorithm is a combination of the two previous ones:

The half lowest ranking agents adopt the network values and the P value of one of the half highest ranking agents (randomly chosen). Then with 10% probability these values are mutated. The half highest ranking agents only mutate with 10% probability.

**Table 5.** Population size N = 20, all TFT, noisy IPD with noise = 0.1, random pairing, initially M = 5, K = 2, P = 0.1. Evolution of network (M,K) and P for 100 generations. Average and maximum agent M, K, P and scores every 10 generations. All results are averages of 50 runs.

| t | M | Avg. K (Max. K) | Avg. P (Max. P) | Avg. (max) score |
|---|---|---|---|---|
| No attraction | | | | 9145.8 (9396.16) |
| 1 | 5 | 2 | 0.1 | 9237.501 (9516.02) |
| 5 | 14.32 | 2.746 (5.18) | 0.289 (0.775) | 9332.547 (9696.56) |
| 10 | 15.94 | 4.953 (8.84) | 0.55 (0.88) | 9755.001 (10215.06) |
| 20 | 16.82 | 10.524 (14.32) | 0.681 (0.897) | 10465.772 (10900.72) |
| 30 | 17.0 | 14.398 (17.64) | 0.638 (0.867) | 10798.493 (11150.22) |
| 40 | 17.18 | 16.865 (18.66) | 0.555 (0.821) | 10912.584 (11139.28) |
| 50 | 16.54 | 18.001 (19.64) | 0.532 (0.815) | 10924.817 (11137.96) |
| 60 | 16.92 | 18.594 (19.86) | 0.489 (0.845) | 10934.755 (11138.98) |
| 70 | 17.48 | 18.974 (19.92) | 0.474 (0.794) | 10977.707 (11152.98) |
| 80 | 16.72 | 19.105 (20.0) | 0.462 (0.782) | 10926.982 (11108.38) |
| 90 | 16.72 | 19.031 (19.98) | 0.492 (0.764) | 10971.013 (11162.6) |
| 100 | 16.78 | 19.194 (20.0) | 0.487 (0.752) | 10970.073 (11159.2) |
| No noise, No attraction | | | | 12000 |

The results given in Table 5 show that, as with network size only, there is significant improvement in cooperation scores that are obtained with larger networks and higher (but not overly high) values for P, despite the expected fluctuations that are due to the presence of noise in the IPD game implementation.

For example, in a sample evolutionary run, the number of attraction connections and the values of P for the 17 connected agents in the 100th generation are:

{16 16 17 17 17 18 18 18 18 18 20 20 20 20 20 20 20}.

{0.324 0.471 0.591 0.591 0.82 0.324 0.324 0.471 0.471 0.471 0.471 0.311 0.311 0.471 0.471 0.471 0.471}.

We observe a persistent agent pattern, that is apparently the result of evolutionary selection. In our example, almost all agents (17 of 20) are highly connected (16 to 20 connections) but moderately generous (with one exception of a single 0.82 all Ps are from 0.311 to 0.591), which allows cooperation to be reciprocated and persist in the population. It is noteworthy that when the evolution can act on both network size and P, different kinds of networks evolve, namely almost fully connected networks with moderate P values and this allows the total cooperation score to be closer to the level of the score of evolutionary experiment 1, which evolved slightly fewer numbers of slightly less connected agents but with full attraction. The solution found by evolution when all M, K and P coevolve simultaneously is better and more robust for individual agents because none may be exploited consistently (as is the case of an almost fully cooperative agent attracted to almost everyone else) and hence abrupt societal changes may be absorbed without score collapse.

We have also performed experiments with only M and P evolving (and K fixed across generations) as well as with K and P evolving (and M fixed across generations). The results are given in the following Tables 6 and 7 and confirm our former conclusion, namely that attraction evolves and gives fitness improvements. They also show that the connectivity K is a more crucial parameter than either of M and P, because when this is not allowed to evolve (Table 6), the population achieves lower scores.

**Table 6.** Population size N = 20, all TFT, noisy IPD with noise = 0.1, random pairing, K = 5. Evolution of M and P for 100 generations. Average and maximum agent M and P and scores every 10 generations. All results are averages of 50 runs.

| t | M | Avg. (max) P | Avg. (max) score |
|---|---|---|---|
| No attraction | | | 9145.8 (9396.16) |
| 1 | 10 | 0.1 | 9375.089 (9658.06) |
| 5 | 16.42 | 0.34 (0.827) | 9605.445 (10029.78) |
| 10 | 16.84 | 0.598 (0.853) | 9799.91 (10257.66) |
| 20 | 16.06 | 0.67 (0.896) | 9821.412 (10284.86) |
| 30 | 16.24 | 0.674 (0.905) | 9825.571 (10293.98) |
| 40 | 16.9 | 0.677 (0.9) | 9864.398 (10365.24) |
| 50 | 16.76 | 0.682 (0.91) | 9853.775 (10333.18) |
| 60 | 17.04 | 0.693 (0.918) | 9874.357 (10365.34) |
| 70 | 16.5 | 0.699 (0.909) | 9842.819 (10339.74) |

*(continued)*

**Table 6.** (*continued*)

| t | M | Avg. (max) P | Avg. (max) score |
|---|---|---|---|
| 80 | 16.6 | 0.691 (0.923) | 9849.722 (10315.92) |
| 90 | 16.76 | 0.69 (0.906) | 9851.968 (10310.9) |
| 100 | 16.78 | 0.705 (0.912) | 9859.323 (10341.28) |
| No noise, No attraction | | | 12000 |

**Table 7.** Population size N = 20, all TFT, noisy IPD with noise = 0.1, random pairing, M = 10. Evolution of P for 100 generations. Average and maximum agent K, P and scores every 10 generations. All results are averages of 50 runs.

| t | K | Avg. (max) P | Avg. (max) score |
|---|---|---|---|
| No attraction | | | 9145.8 (9396.16) |
| 1 | 5 | 0.1 | 9279.639 (9536.94) |
| 5 | 5.426 (8.14) | 0.332 (0.782) | 9441.607 (9852.0) |
| 10 | 6.764 (10.78) | 0.559 (0.853) | 9664.798 (10182.6) |
| 20 | 11.304 (14.74) | 0.652 (0.883) | 10054.296 (10573.72) |
| 30 | 14.132 (17.66) | 0.669 (0.927) | 10270.617 (10780.88) |
| 40 | 16.182 (19.02) | 0.666 (0.931) | 10402.984 (10860.08) |
| 50 | 17.356 (19.6) | 0.647 (0.915) | 10457.033 (10901.78) |
| 60 | 18.112 (19.86) | 0.656 (0.937) | 10521.699 (10940.48) |
| 70 | 18.19 (19.96) | 0.646 (0.931) | 10502.413 (10929.04) |
| 80 | 18.32 (20.0) | 0.637 (0.929) | 10512.312 (10926.36) |
| 90 | 18.342 (20.0) | 0.614 (0.929) | 10488.661 (10910.0) |
| 100 | 18.322 (20.0) | 0.618 (0.93) | 10489.852 (10910.42) |
| No noise, No attraction | | | 12000 |

We note that in a social developmental, rather than an evolutionary context, where subsequent generations correspond to actually the same generation in different points in life time as the population develops, it might make sense for some people to be M-P developing, or K-P developing or P-only and so on. This aligns with different personalities that may be present in the population. For example, a shy individual may not develop new social trust connections (thus maintain a fixed K) but may enhance significantly its current connections (thus develop its P).

## 8 Experiment 4: General Strategies and Sizes

In our last series of evolutionary experiments, we compare the three above settings (network MK evolution, P evolution, simultaneous network MK and P evolution) for TFT agents in presence of noise or for mixes of general strategies in various population sizes (N = 20,50,100). For each case we compare start performance (at generation-1) with final performance (at generation-100) for uniform TFT population with noise and mixed population without noise (Tables 8, 9 and 10). The consistent finding across population sizes and evolution types is that attraction gives significant performance advantage in the case of a uniform TFT population but gives minor to no performance advantage for mixed populations. Also as seen before, P evolution alone is of little benefit. Attraction evolves easily in the case of a TFT population with noise, as shown by the final high values for M, K and P in generation-100. But attraction hardly evolves in the case of mixed populations without noise, as shown by the final values for M, K and P in generation-100 that are very low compared to their TFT counterparts and sometimes lower than their initial values (for example, K ends up lower than at start for network evolution if N = 50 or N = 100). Given that the evolutionary algorithm boosts attraction, these results show that attraction can evolve for small populations of TFT agents, which is typical of prehistoric settlements or extended families, but not for high populations of arbitrary strategies, which is typical of more complex, advanced societies.

**Table 8.** Comparative results for population sizes N = 20, 50, 100. (Left) All TFT, noisy IPD with noise = 0.1, random pairing. (Right) Mixes of ALLC-ALLD-TFT (distributions given), regular IPD without noise, random pairing. MK evolution. All results are averages of 50 runs.

| t | M | Avg. (max) K | Avg. (max) score | M | Avg. (max) K | Avg. (max) score |
|---|---|---|---|---|---|---|
| **N = 20**, P = 1, No attraction | | | 9145.8 (9396.16) | **N = 20 (5-5-10)**, P = 1, No attraction | | 9265.0 (9990) |
| 1 | 5 | 2 | 9461.694 (9909.74) | 5 | 2 | 9441.182 (10436.44) |
| 100 | 15.2 | 18.404 (19.92) | 11224.075 (11502.64) | 7.52 | 2.98 (6.56) | 9499.688 (10383.08) |
| **N = 50**, P = 1, No attraction | | | 22849.96 (23344.62) | **N = 50 (10-10-30)**, P = 1, No attraction | | 24036.0 (25980) |
| 1 | 15 | 5 | 23654.38 (24467.92) | 15 | 5 | 24278.716 (26575.08) |

*(continued)*

**Table 8.** (*continued*)

| t | M | Avg. (max) K | Avg. (max) score | M | Avg. (max) K | Avg. (max) score |
|---|---|---|---|---|---|---|
| 100 | 38.26 | 43.904 (49.3) | 27992.72 (28548.64) | 12.46 | 3.811 (11.58) | 24197.12 (26398.08) |
| **N = 100**, P = 1, No attraction | | | 45705.44 (46443.06) | **N = 100 (15-15-70)**, P = 1, No attraction | | 50313.0 (53970) |
| 1 | 20 | 5 | 46651.674 (47792.66) | 20 | 5 | 50538.419 (54609.48) |
| 100 | 87.78 | 63.438 (72.4) | 54908.35 (56217.6) | 26.62 | 3.018 (12.08) | 50438.813 (54468.48) |

**Table 9.** Comparative results for population sizes N = 20, 50, 100. (Left) All TFT, noisy IPD with noise = 0.1, random pairing. (Right) Mixes of ALLC-ALLD-TFT (distributions given), regular IPD without noise, random pairing. P evolution. All results are averages of 50 runs.

| t | Avg. (max) P | Avg. (max) score | Avg. (max) P | Avg. (max) score |
|---|---|---|---|---|
| **N = 20**, M = 10, K = 5, No attraction | | 9145.8 (9396.16) | **N = 20 (5-5-10)** M = 10 K = 5, No attraction | 9265.0 (9990) |
| 1 | 0.1 | 9287.824 (9546.58) | 0.1 | 9336.519 (10096.62) |
| 100 | 0.7 (0.901) | 9596.035 (10036.86) | 0.498 (0.845) | 9592.444 (10497.04) |
| **N = 50**, M = 20, K = 5 No attraction | | 22849.96 (23344.62) | **N = 50 (10-10-30)** M = 20, K = 5, No attraction | 24036.0 (25980) |
| 1 | 0.1 | 22977.054 (23502.98) | 0.1 | 24084.524 (26092.44) |
| 100 | 0.698 (0.952) | 23218.486 (23926.38) | 0.477 (0.913) | 24246.936 (26508.82) |
| **N = 100**, M = 40, K = 20 No attraction | | 45705.44 (46443.06) | **N = 100 (15-15-70)**, M = 40, K = 20, No attraction | 50313.0 (53970) |
| 1 | 0.1 | 46179.187 (47097.14) | 0.1 | 50468.44 (54180.84) |
| 100 | 0.672 (0.984) | 47163.324 (48691.68) | 0.438 (0.952) | 50932.75 (54830.76) |

**Table 10.** Comparative results for population sizes N = 20, 50, 100. (Left) All TFT, noisy IPD with noise = 0.1, random pairing. (Right) Mixes of ALLC-ALLD-TFT (distributions given), regular IPD without noise, random pairing. MK and P evolution. All results are averages of 50 runs.

| t | M | Avg. (max) K | Avg. (max) P | Avg. (max) score | M | Avg. (max) K | Avg. (max) P | Avg. (max) score |
|---|---|---|---|---|---|---|---|---|
| **N = 20,** No attraction | | | | 9145.8 (9396.16) | **N = 20 (5-5-10)** No attraction | | | 9265.0 (9990) |
| 1 | 5 | 2 | 0.1 | 9237.501 (9516.02) | 5 | 2 | 0.1 | 9289.587 (10054.66) |
| 100 | 16.78 | 19.194 (20) | 0.487 (0.752) | 10970.073 11159.2) | 8.98 | 4.161 (9.02) | 0.514 (0.814) | 9479.14 (10296.96) |
| **N = 50,** No attraction | | | | 22849.96 (23344.62) | **N = 50 (10-10-30)** No attraction | | | 24036.0 (25980) |
| 1 | 15 | 5 | 0.1 | 23108.602 (23613.52) | 15 | 5 | 0.1 | 24065.026 (26066.3) |
| 100 | 43.14 | 47.046 (49.8) | 0.471 (0.891) | 27426.004 (27782.26) | 12.32 | 3.039 (9.6) | 0.529 (0.887) | 24096.846 (26255.54) |
| **N = 100,** No attraction | | | | 45779.39 (46466) | **N = 100 (15-15-70)** No attraction | | | 50313.0 (53970) |
| 1 | 20 | 5 | 0.1 | 45970.783 (46782.32) | 20 | 5 | 0.1 | 50340.196 (54075.22) |
| 100 | 88.66 | 62.492 (71.62) | 0.714 (0.977) | 53959.469 (55229.7) | 27 | 3.101 (12.48) | 0.541 (0.956) | 50393.435 (54360.84) |

## 9 Discussion

We have presented a simple attraction mechanism that influences social behavior and is applied to the Iterated Prisoner's Dilemma model of cooperative behavior. We show that this mechanism has an evolutionary advantage and we demonstrate how it can be selected in societies of agents playing a noisy IPD. Each agent is connected in an attraction network with other agents. In this setting, evolution favors bigger attraction networks and bigger attraction influence. Finally, the mechanism appears to be evolvable for societies of TFT agents with noise, which represent small groups of kin, or for another reason inter-committed individuals, in harsh environments, that is in presence of noise. But attraction appears very difficult and even impossible to evolve in groups of general not necessarily cooperative behaviors, which is more akin to bigger, more complex societies.

Further issues may be studied in the future. Firstly, although the advantage is clear, it is not obvious where attraction originates from. It is possible that it evolves independently for some other reason, and it is subsequently exploited and "recruited" during a cooperative/conflictual exchange. Secondly, attraction may be a serious candidate as a facilitator of further social evolution, such as for example for evolution of pure generosity and relevant norms. Another question concerns the relation of attraction (and other such reactive mechanisms) to reasoning and rationality. One idea is that everyday reciprocal reasoning could emerge from the interaction of attraction and social imitation. Thus simple attraction in smaller societies may be a step in the pathway to large-scale ultra-sociality [11] of human societies. Overall, the potential and repercussions of spontaneous, not reasoned, behaviors such as the ones emerging by attraction are significant for understanding the evolution of human cognition.

## References


1. Axelrod, R., Hamilton, W.D.: The evolution of cooperation. Science **211**, 1390–1396 (1981)
2. Axelrod, R.: The evolution of cooperation. Basic Books (1984)
3. Axelrod, R., Dion, D.: The further evolution of cooperation. Science **242**, 1385–1390 (1988)
4. Back, I., Flache, A.: The adaptive rationality of interpersonal commitment. Ration. Soc. **20**(1), 65–83 (2008)
5. Bowlby, J. (1975). Attachment (Attachment and loss, Volume I), Basic Books
6. Byrne,: Interpersonal attraction and attitude similarity. J. Abnorm. Soc. Psychol. **62**(3), 713–715 (1961)
7. Byrne, D.: Attitudes and attraction. Adv. Exp. Soc. Psychol. **4**, 35–89 (1969)
8. Davies, A.P., Watson, R.A., Mills, R., Buckley, C.L., Noble, J.: "If you can't be with the one you love, love the one you are with": how individual habituation of agent interactions improves global utility. Artif. Life **17**, 167–181 (2011)
9. Feldman, M.W., Thomas, E.A.C.: Behavior-dependent contexts for repeated plays of the prisoner's dilemma II: dynamical aspects of the evolution of cooperation. J. Theor. Biol. **128**, 297–315 (1987)
10. Fogel, D.: Evolving behaviors in the iterated prisoner's dilemma. Evol. Comput. **1**, 77–97 (1993)
11. Gowdy, J., Krall, L.: The ultrasocial origin of the anthropocene. Ecol. Econ. **95**, 137–147 (2013)
12. Hogg, M.A: Group cohesiveness: a critical review and some new directions. Eur. Rev. Soc. Psychol. **4**(1):85–111 (1993)
13. Hruschka, D.J., Henrich, J.: Friendship, cliquishness and the emergence of cooperation. J. Theor. Biol. **239**, 1–15 (2016)
14. Kraines, D., Kraines, V: Evolution of learning among Pavlov strategies in a competitive environment with noise. J. Conflict Resolut. **39**(3), 439–466 (1995)
15. Molander, P.: The optimal level of generosity in a selfish, uncertain environment. J. Conflict Resolut. **31**(4):692–724 (1987)
16. Montoya, R.M., Horton, R.S.: On the importance of cognitive evaluation as a determinant of interpersonal attraction. J. Pers. Soc. Psychol. **86**(5), 696–712 (2004)
17. Montoya, R.M., Horton, R.S.: Understanding the attraction process. Soc. Pers. Psychol. Compass **14**, e12526 (2020)
18. Mooney, C.G.: Theories of Attachment: An Introduction to Bowlby, Ainsworth, Gerber, Brazelton, Kennell, and Klaus. Redleaf Press (2009)

19. Nowak, M.A., Sigmund, K.: Tit for tat in heterogeneous populations. Nature **355**, 250–253 (1992)
20. Nowak, M.A., Sigmund, K.: A strategy of win-stay, lose-shift that outperforms tit-for-tat in the prisoner's dilemma game. Nature **364**, 56–58 (1993)
21. Stanley, E.A., Ashlock, D., Tesfatsion, L.: Iterated prisoner's dilemma with choice and refusal of partners, Artificial Life III, Addison-Wesley (1994)
22. Tzafestas, E.: Toward adaptive cooperative behaviour. In: Proceedings of the Simulation of Adaptive Behavior Conference, Paris (2000)
23. Tzafestas, E.: Attraction and cooperation in noisy environments. In: Information Sciences 2007, Proceedings of the 10th Joint Conference, Salt Lake City, Utah, USA, 18–24 July, pp. 411–417 (2007)
24. Tzafestas, E.: Attraction and cooperation in space. In: IEEE Congress on Evolutionary Computation, pp. 3698–3705 (2007)
25. Wetzel, C.G., Insko, C.A.: The similarity-attraction relationship: Is there an ideal one? J. Exp. Soc. Psychol. **18**, 253–276 (1982)

# Author Index





M. Villani et al. (Eds.): WIVACE 2023, CCIS 1977, pp. 409–410, 2024.
https://doi.org/10.1007/978-3-031-57430-6

Printed in the United States
by Baker & Taylor Publisher Services

Printed in the United States
by Baker & Taylor Publisher Services